The Encyclopedia of Global Warming Science and Technology

The Encyclopedia of Global Warming Science and Technology

Volume 2: I–Z

Bruce E. Johansen

GREENWOOD PRESS
An Imprint of ABC-CLIO, LLC

A B C  C L I O

Santa Barbara, California • Denver, Colorado • Oxford, England

Library of Congress Cataloging-in-Publication Data

Johansen, Bruce E. (Bruce Elliott), 1950–
   The encyclopedia of global warming science and technology / Bruce E. Johansen.
      v. cm.
   Includes bibliographical references and index.
   Contents: v. 1. A–H—v. 2. I–Z.
ISBN 978-0-313-37702-0 (hard copy : alk. paper) — ISBN 978-0-313-37703-7 (ebook)
1. Global warming—Encyclopedias. 2. Climatic changes—Encyclopedias.
I. Title.
   QC981.8.G56J638   2009
   577.27′603—dc22   2009005295

13   12   11   10   9   1   2   3   4   5

This book is also available on the World Wide Web as an eBook.
Visit www.abc-clio.com for details.

ABC-CLIO, LLC
130 Cremona Drive, P.O. Box 1911
Santa Barbara, California 93116-1911

This book is printed on acid-free paper ∞

Manufactured in the United States of America

Contents

List of Entries

Guide to Related Topics

Climate and Weather

Extremes of Heat and Cold around the World
Heat-Island Effect, Urban
Heat Waves
Jet Streams
Medieval Warm Period, Debate over Temperatures
Pliocene Paleoclimate
Solar Influences on Climate
Temperatures, Cold Spells
Temperatures, Recent Warmth
Temperatures and Carbon-Dioxide Levels
Thunderstorms and Tornadoes, Frequency and
  Severity of
"Urban Heat-Island Effect" and Contrarians
Wintertime Warming and Greenhouse-Gas Emissions
Worldwide Climate Linkages

Human Health

Asthma
Dengue Fever
Diseases and Climate Change
Hay Fever
Human Health, World Survey
Kidney Stones
Malaria in a Warmer World
West Nile Virus and Warming

Human Sources of Greenhouse Gases

Agriculture and Warming
Air Conditioning and Atmospheric Chemistry
Air Travel
Automobiles and Greenhouse-Gas Emissions
Coal and Climatic Consequences
Ecotourism
Fossil Fuels

Greenhouse-Gas Emissions, Worldwide
Hydrofluorocarbons (HFCs)
Jet Contrails, Role in Climate Change
Oceans, Carbon-Dioxide Levels
Tar Sands
War, Carbon Footprint

Hydrological Cycle

Deforestation
Desertification
Drought
Drought and Deluge: Anecdotal Observations
Drought and Deluge: Scientific Issues
Drought in Western North America
Hadley Cells
Hydrological Cycle
Jet Streams
Tropical Zones, Expansion of
Water Supplies in Western North America
Wildfires

Ice Melt

Glacial (Ice Age) Cycle, Prospective End of
Ice Melt, World Survey

Ice Melt—Antarctic

Antarctica and Climate Change
Antarctica and Debate over Inland Cooling
Antarctica and Speed of Ice Melt
Antarctic Oscillation
Antarctic Paleoclimatic Precedents
Antarctic Peninsula and Ice Shelf Collapse
Antarctic Peninsula and Warming

The Encyclopedia of Global Warming Science and Technology

I

Ice Cores, Record of Climate and Greenhouse-Gas Levels

A 3.19-kilometer column of ice drilled from Antarctica by European scientists during 2004 provided researchers with the oldest and most detailed record of climate change obtained to date, stretching back more than 740,000 years. The ice core indicated that today's greenhouse-gas concentrations in the atmosphere are by far the highest for at least that long, by at least 30 percent. Ice cores are valuable records of the Earth's past climate because they record variations in temperatures as well as concentrations of gases such as carbon dioxide and methane that contribute to the greenhouse effect. The new core, drilled in Eastern Antarctica, doubled the available paleoclimatic record when it was analyzed during 2005.

Furthermore, the rate at which humankind's burning of fossil fuels is changing the composition of the atmosphere is extraordinary by natural standards. "The rate of increase [in greenhouse gases] is more than 100 times faster than any rate we can detect from the ice cores we have seen so far," said Thomas Stocker, a member of the European Project for Ice Coring in Antarctica (EPICA), a consortium of laboratories and Antarctic logistics operators from 10 nations (Henderson 2004b, 4).

The new core confirms evidence from ocean sediments that the Earth has endured several ice ages during the last 740,000 years, each separated by a warmer interglacial. While ice ages typically last about 100,000 years, the interglacial periods usually are much shorter, averaging (very roughly) 10,000 years each (Henderson 2004b, 4). EPICA's findings indicate an "extremely strong" 100,000-year cycle for ice ages during at least the last 500,000 years, with the present interglacial most closely resembling another about 430,000 years ago (Wolff et al. 2004, 623).

The proportion of carbon dioxide in the atmosphere follows a well-defined cycle, rising to about 280 parts per million (ppm) during interglacial periods, declining to between 180 and 200 ppm during ice ages. In 2008, the carbon-dioxide level was 385 ppm.

The interglacial 430,000 years ago lasted about 28,000 years, longer than most. Why has the present interglacial been so long? It partially may be a matter of "how much, where, and during what season the Sun's energy reaches the planet," according to one measurement used by the EPICA scientists (White 2004, 1610). The shape of the Earth's orbit around the sun (which determines Earth's distance from the sun during different seasons) and its interaction with the seasonal cycle seem to fit the waxing and waning of ice ages noted in ice cores.

Because the present interglacial is about 12,000 years old, the EPICA scientists put little credibility in global warming contrarians' assertions that burning fossil fuels may forestall a new ice age any time soon. "Given the similarities between this earlier warm period and today, our results may imply that without human intervention, a climate similar to the present one would extend well into the future" (Wolff et al. 2004, 623). Given rising levels of greenhouse gases, however, the team finds any assumption that climate will remain stable "highly unlikely" (Wolff at al. 2004, 627).

The fact that human activity has raised the atmosphere's carbon-dioxide level to about 30 percent above the natural range for the last 740,000 years or more provokes scientists to wonder just how much the Earth's temperature will rise in coming years, especially if carbon-dioxide levels continue to rise. "In everything we have got up to now, temperature and greenhouse gases are absolutely in step with each other," Wolff said. "I don't see any particular reason this shouldn't continue into the future. It is worrying" (Henderson 2004b, 4).

FURTHER READING

Henderson, Mark. "Hot News from 740,000 Years Ago Tells Us to Get Ready for Catastrophic Climate Change." *London Times*, June 10, 2004b, 4.

"NASA Study Predicts More Severe Storms with Global Warming." NASA Earth Observatory, August 30, 2007. http://www.nasa.gov/centers/goddard/news/topstory/2007/moist_convection.html.

White, James C. "Do I Hear a Million?" *Science* 304 (June 11, 2004):1609–1610.

Wolff, Eric W., Laurent Augustin, Carlo Barbante, Piers R.F. Barnes, Jean Marc Barnola, Matthias Bigler, Emiliano Castellano, Olivier Cattani, Jerome Chappellaz, Dorthe Dahl-Jensen, Barbara Delmonte, Gabrielle Dreyfus, Gael Durand, Sonia Falourd, Hubertus Fischer, Jacqueline Fluckiger, Margareta E. Hansson, Philippe Huybrechts, Gerard Jugie, Sigfus J. Johnsen, Jean Jouzel, Patrik Kaufmann, Josef Kipfstuhl, Fabrice Lambert, Vladimir Y. Lipenkov, Genevieve C. Littot, Antonio Longinelli, Reginald Lorrain, Valter Maggi, Valerie Masson-Delmotte, Heinz Miller, Robert Mulvaney, Johannes Oerlemans, Hans Oerter, Giuseppe Orombelli, Frederic Parrenin, David A. Peel, Jean-Robert Petit, Dominique Raynaud, Catherine Ritz, Urs Ruth, Jakob Schwander, Urs Siegenthaler, Roland Souchez, Bernard Stauffer, Jorgen Peder Steffensen, Barbara Stenni, Thomas F. Stocker, Ignazio E. Tabacco, Roberto Udisti, Roderik S.W. van de Wal, Michiel van den Broeke, Jerome Weiss, Frank Wilhelms, Jan-Gunnar Winther, and Mario Zucchelli. "Eight Glacial Cycles from an Antarctic Ice Core." *Nature* 429 (June 10, 2004): 623–627.

Ice Melt, World Survey

The most widespread indication that the Earth is steadily warming has been the steady erosion of ice in the Arctic, Antarctic, and on mountain glaciers. Although a few exceptions do exist, the worldwide erosion of ice leaves little doubt that that the Earth has experienced steady warming for at least a century.

The causes of this warming are still open to debate: is it human-caused, a product of changes in atmospheric circulation, or both? The following global survey indicates that ice melt is accelerating. The melting of ice has profound implications not only for Arctic, Antarctic, and mountain ecosystems, but also for hundreds of millions of people living at lower elevations who depend on glacier melt for water and electricity generation. Billions more people around the Earth who live on or near continental coasts and

islands have (and will) feel the effects of global ice melt through gradually rising sea levels.

Human-induced climate change may be ending the cycle of glaciation that has been the norm on Earth since the end of the age of the dinosaurs about 65 million years ago. Viewed on a timescale of the last 540 million years, however, glacial cycles have been relatively rare events. For three-quarters of the last 540 million years, the Earth's ice caps have been negligible or nonexistent (Huber, MacLeod, and Wing 2000, xi).

Rivers and Lakes Freeze Later, Thaw Earlier

Today, many lakes and rivers in the Northern Hemisphere usually freeze about a week later and thaw out 10 days sooner than a century and a half ago, according to John J. Magnuson and colleagues. In some areas (such as the harbor of Toronto, Ontario) heat contributed by urbanization, as well as general warming, may have added a month or more to the ice-free season. The warming trend appears to have begun at least half a century before the significant buildup of greenhouse gases caused by the burning of fossil fuels, leading to speculation that a long-term climate cycle is accelerating changes induced by infrared forcing.

An international team of scientists analyzed written accounts, some of them centuries old, including newspaper reports, fur traders' records, ships' navigation logs, and descriptions of religious events. These results illustrate "a very clear record of the response of aquatic systems to [global] warming," said Magnuson (Suplee 2000, A-2). The authors of this study estimate that the changes in freezing and thawing dates that they found required an average temperature increase of about 3°F during the last 150 years (Magnuson et al. 2000). "We know that the regions of Eurasia and North America, where most of these [lake and river sites] are, has indeed warmed according to thermometers at a rate of about double that for the globe—both recently and for the past century," said Kevin Trenberth, head of climate analysis at the National Center for Atmospheric Research, who did not participate in the research (Suplee 2000, A-2).

Mountain Glaciers in Rapid Retreat

Mountain glaciers are in rapid retreat around the Earth, with very few exceptions. The Intergovernmental Panel on Climate Change's (IPCC) models project that between one-third and one-half of existing mountain glacial mass could disappear over the next hundred years. Sometime during the present century, the last glacier may melt in Glacier National Park.

Climbers are being plucked from the Matterhorn in the Swiss Alps as thawing mountainsides crumble under them. During the summer of 2003, Mont Blanc, Europe's tallest mountain, was closed to hikers and climbers because its deteriorating snow and ice was too unstable to allow safe passage. The mountain was crumbling as ice that once held it together melted during a record-warm summer in Europe. In the Swiss Alps, scientists have estimated that by 2025 glaciers will have lost 90 percent of the volume they contained a century earlier. Roger Payne, a director of the Swiss-based International Mountaineering and Climbing Federation, said global warming was emerging as one of the biggest threats to mountain areas. "The evidence of climate change was all around us, from huge scars gouged in the landscapes by sudden glacial floods to the lakes swollen by melting glaciers," said Payne (Williams 2002, 2).

About 1,500 Swiss glaciers have lost some 10 cubic kilometers of ice in nine years. Several glaciers are losing a meter of thickness per year on average, according to reports from the Swiss Federal Institute of Technology. "The trend is negative, but what we see is that the trend is also steepening" because of lengthening melting seasons, said Matthias Huss from the Zurich university's Laboratory of Hydraulics, Hydrology, and Glaciology (Amos 2009). The two studies were presented at the American Geophysical Union's 2008 meeting.

The U.S. Geological Survey has collected a digital library that describes the state of more than 67,000 glaciers around the world. Using historical photographs, images from space satellites, precision laser measurements, and other tools, the archive tells a story of glacial retreat around the world (Toner 2002b, 4-A). The rate of ice loss doubled from 1988 to 2002, said Mark Meier, a geologist at the University of Colorado at Boulder. "Some glaciers around the world are now smaller than they have been in the last several thousand years" (Toner 2002b, 4-A). Glaciologist Mark Dyurgerov of the University of Colorado's Institute of Arctic and Alpine Research estimated that mountain glaciers around the world by 2002 were losing about 25 cubic miles of ice each year due to climate

change (Erickson 2002, 6-A). Worldwide, the "equilibrium-line altitude" of the average glacier—the point below which it loses mass on an annual basis due to melting—rose an average of about 200 meters between 1960 and 1998 (Lynas 2004a, 233).

Generally, the only glaciers gaining mass have been in wet maritime areas of the world, such as parts of Norway and Sweden, where melting has been offset by increased snowfall, another facet of climate change. Alaska's Hubbard Glacier "is advancing so swiftly that it threatens to seal off the entrance to Russell Fiord near Yukatat and turn the fiord into an ice-locked lake. Like a handful of other Alaska glaciers, the Hubbard is fed by a high-altitude snowfield that has not yet been affected by warmer temperatures" (Toner 2002b, 4-A). "The Hubbard is definitely an exception," said the Geological Survey's Bruce Molnia, who has been tracking 1,500 Alaska glaciers. "Every mountain group and island we have investigated is seeing significant glacier retreat, thinning or stagnation, especially at lower elevations. Ninety-nine percent of the named glaciers in Alaska are retreating" (Toner 2002b, 4-A).

Eighty-five percent of the glaciers in Spain's Pyrenees melted during the twentieth century, according to Greenpeace, which reported:

> The surface of the glaciers of the Pyrenees on the Spanish side went from 1,779 hectares (4,394 acres) in 1894 to 290 acres in 2000.... That infers a loss of 85 percent of the surface of the glaciers in the last century, with the process accelerating in the last 20 years. ("More Than" 2004)

Glaciers in that area are expected to vanish by the year 2070. Melting glaciers are revealing a large number of previously buried historical artifacts. For example, a 450-year-old bison skull was found in a melting snow bank in the Colorado Rockies. Human cadavers, airplanes, dead birds, caribou carcasses, mining equipment, and prehistoric weapons have been uncovered (Erickson 2002, 6-A).

The snow and ice crown of Mount Kilimanjaro in equatorial Africa, made famous by Ernest Hemmingway a century ago, may vanish by the mid-twenty-first century. Kilimanjaro will no longer live up to its name, which in Swahili means "mountain that glitters." Mount Kenya's ice fields have lost three-quarters of their entire extent during the twentieth century (Lynas 2004c, 12). By the end of the twenty-first century, Glacier National Park in Montana may lose the last of its last permanent glaciers; its name will be a daily reminder of what humankind has done to the Earth's climate. The original 150 glaciers within Glacier National Park had been reduced to 37 by 2002, and most of these were small remnants of once-mighty ice masses. Glaciers in the New Zealand's Southern Alps lost 25 percent of their surface area during the last century (McFarling 2002b, A-4).

Scientists reported during October 2003 that the Patagonian ice fields of Chile and Argentina have been thinning so swiftly that this 6,500-square-mile region of South America now exhibits a pace of glacial retreat that is among the most rapid on Earth. During the period 1995–2000, rate of volume loss for 63 glaciers in the area doubled, compared with the 1968–2000 average (Rignot, Rivera, and Casassa 2003, 434). Early in 2004, Greenpeace International released results of an aerial survey confirming the rapid recession of the Patagonia glaciers, which was estimated at 42 cubic kilometers a year, an amount that could fill a large sports stadium 10,000 times. Greenpeace compared photographs taken during January 2004 with images from 1928, showing that "the glaciers had significantly thinned and retreated several kilometers" (Hodge 2004, 8).

At the December 2007 annual meeting of the American Geophysical Union, Ohio State University professor Lonnie Thompson said that the Quelccaya Glacier in the Peruvian Andes retreated on average 20 feet per year before the 1990s. For the past 15 years, he said, it had retreated an average of nearly 200 feet per year. "Glaciers, and especially the high-elevation tropical glaciers, are a real canary in the coal mine," said Thompson. "They're telling us that major climatic changes are occurring" (Kaufman 2008a).

"These losses are not just regrettable but actually threaten the health and well-being of us all. Mountains are the water towers of the world, the sources of many rivers. We must act to conserve them for the benefit of mountain people, for the benefit of humankind," said Klaus Toepfer, head of the United Nations Environment Programme (Vidal 2002d, 7). Glaciologist Lonnie G. Mosley-Thompson, a senior research scientist at the Byrd Polar Research Center of Ohio State University, said that the rate of glacial retreat is most pronounced in equatorial areas, such as along the spine of the Andes, where hydrological consequences may be most significant.

Glacial Ice Melt Portends Water Shortages

Late in 2003, the World Wildlife Fund (WWF) said that millions of people around the world will face severe water shortages as glaciers around the world melt, unless governments take urgent action to deal with global warming. "Increasing global temperatures in the coming century will cause continued widespread melting of glaciers, which contain 70 percent of the world's fresh water reserves.... An overall rise of temperature of four degrees Celsius before the end of the century would eliminate almost all of them," warned the WWF (Billions 2003). Ecuador, Peru, and Bolivia, where major cities rely on glaciers as their main source of water during dry seasons, would be worst affected, the WWF forecast. Areas of the Himalayas face grave danger of flooding, said the WWF, as it noted that glacier-fed rivers in the region supply water to one-third of the world's population, mainly in India and China.

Chinese glaciologist Yan Tandong estimated that during the past four decades China's glaciers have shrunk an average of 7 percent per year, an amount equivalent to all the water in the Yellow River. Yan previously told local media that as many as 64 percent of China's glaciers (mostly in Tibet) may be gone by 2050 if current trends continue. The human cost could be immense, because 300 million Chinese live in the country's arid west and depend on water from the glaciers for their survival ("Global Warming Makes" 2004).

Temperatures are rising quickly at higher elevations. For example, David L. Naftz and colleagues from the U.S. Geological Survey reconstructed trends in air temperature from alpine areas of the Wind River range in Wyoming and found, from ice-core samples, that temperatures had increased about 3.5°C from 1960 to the early 1990s. Furthermore, temperatures had increased about 5°C from the end of the Little Ice Age about 1850 to the early 1990s. Naftz and colleagues wrote in the *Journal of Geophysical Research* that their readings were "in agreement with [temperature] increases observed at selected high-altitude and high-latitude sites in other parts of the world" (Naftz et al. 2002).

In New Zealand, as in many other mountainous areas, the banks of South Island rivers are filled with water from melting glaciers. National Institute for Water and Atmospheric Research consultant glaciologist Trevor Chinn said that many of the South Island's largest glaciers have been retreating rapidly (Robson 2003, 13). As they have melted, the glaciers have been adding millions of gallons of water to local rivers. "Current flows are higher than what you would expect from rainfall. What we are seeing is borrowing water from glacier storage for the Waitaki and Clutha rivers," Chinn said (Robson 2003, 13). Scenarios for a 0.5° to 3°C rise in average global temperature would reduce New Zealand's glacial mass by 25 to 50 percent, according to one estimate. The glaciers have lost 20 percent of their area in the past 100 years and many larger ones are collapsing into lakes (Robson 2003, 13).

Ice Melt Dynamics Near the Poles

The Arctic Ice Cap has been thinning rapidly, and scientists have been speculating over how many years will pass before a summer during which all of it melts. The thinning of ice in the Arctic affects life all along the food chain. When the ice breaks up earlier than usual, polar bears have less time to build fat reserves and must rely on these reduced reserves for a longer period of time before ice forms again in the fall. Calculations indicate that a mean air temperature increase of 1°C could advance the date of ice breakup nearly a week in the western Hudson Bay (Mathews-Amos and Berntson 1999). Hunger among adult bears will be reflected in their offspring, who will be born smaller, and more prone to die at an early age.

In the Bering Sea and Hudson Bay, evidence of stress in polar-bear populations is mounting as sea ice retreats (Stirling and Derocher 1993; Stirling and Lunn 1997). Some native people in the Arctic have reported difficulty in hunting marine mammals such as walrus in recent years as sea ice has diminished.

Before 2005, increasing precipitation provoked by warming was adding to ice caps inland in Greenland and on the high plains of Antarctica. Since then, however, the world's entire range of ice deposits have sustained net loss, a trend that will continue, and accelerate, as temperatures rise.

Andrew Shepard of the University of Edinburgh and Duncan Wingham of University College London conducted some of the research that described this change. In 2007, they wrote in *Science*:

After a century of polar exploration, the past decade of satellite measurements has painted an altogether new picture of how Earth's ice sheets are changing. As global temperatures have risen, so have rates of snowfall, ice melting, and glacier flow. Although the balance between these opposing processes has varied considerably on a regional scale, data show that Antarctica and Greenland are each losing mass overall. Our best estimate of their combined imbalance is about 125 gigatons per year of ice, enough to raise sea level by 0.35 millimeters per year. This is only a modest contribution to the present rate of sea-level rise of 3.0 millimeters per year. However, much of the loss from Antarctica and Greenland is the result of the flow of ice to the ocean from ice streams and glaciers, which has accelerated over the past decade. In both continents, there are suspected triggers for the accelerated ice discharge surface and ocean warming, respectively and, over the course of the 21st century, these processes could rapidly counteract the snowfall gains predicted by present coupled climate models. (Shepard and Wingham 2007, 1529)

Their conclusions were supported by Mark Serreze and colleagues:

Linear trends in arctic sea-ice extent over the period 1979 to 2006 are negative in every month. This ice loss is best viewed as a combination of strong natural variability in the coupled ice-ocean-atmosphere system and a growing radiative forcing associated with rising concentrations of atmospheric greenhouse gases, the latter supported by evidence of qualitative consistency between observed trends and those simulated by climate models over the same period. Although the large scatter between individual model simulations leads to much uncertainty as to when a seasonally ice-free Arctic Ocean might be realized, this transition to a new arctic state may be rapid once the ice thins to a more vulnerable state. Loss of the ice cover is expected to affect the Arctic's freshwater system and surface energy budget and could be manifested in middle latitudes as altered patterns of atmospheric circulation and precipitation. (Serreze, Holland, and Stroeve 2007, 1533)

Portending the future, other observers added, at about the same time: "Satellite data show that ice sheets can change much faster than commonly appreciated, with potentially worrying implications for their stability" (Truffer and Fahnestock 2007, 1508).

Changes in Greenland are probably provoked by rising global temperatures, but scientists were unsure why glaciers of interior were losing ice to the surrounding ocean (Kaufman 2007b). "In Greenland we know there is melting associated with the ice loss, but in Antarctica we don't really know why it's happening," said Wingham. "With so much of the world's ice captured in Antarctica, just the fact that we don't know why this is happening is a cause of some concern." Antarctica's loss of ice has not been caused by melting to date, "but rather by the pushing of ice streams into the ocean by several glaciers in the west of the continent, has picked up speed in recent years." Wingham told Kaufman that because researchers did not have good measures of the depth of the Antarctic Ice Shelf until about 10 years ago, scientists do not know whether this is a natural variation or a result of human activity (Kaufman 2007b).

Some parts of Antarctica have been gaining ice depth through snowfall at the same time that temperatures rise more rapidly at the tip of the Antarctic Peninsula than anywhere else in the world (excepting, perhaps parts of Alaska and Siberia). Glaciers largely on the West Antarctic Ice Sheet are speeding up as they approach the ocean. Wingham and Shepherd said satellite radar readings show that, overall, each year the ice loss from Greenland and Antarctica accounts for about 10 percent of global sea-level rise, about a tenth of an inch per year.

Even with some ice accumulation in Eastern Antarctica, by 2007, that continent was losing an estimated 25 billion metric tons a year, despite the growth of the ice sheet in East Antarctica. While causes of increasing glacial movement in Antarctica are not well known, Greenland's glaciers are moving more quickly as melting ice has allowed them to slide more easily over rock and dirt below (Kaufman 2007b).

Rapid Changes in the Arctic

Water temperatures in parts of the Arctic Ocean increased an astonishing 1.6°C during the decade of the 1990s alone (Hodges 2000, 35). If the Arctic continues to warm at anywhere near this rate, by the year 2020 the Northwest Passage may open through the thawing Arctic Ocean year-round for the first time in human history. The passage was sought by early century explorers of North America before construction of the Panama Canal allowed shorter oceanic transport of goods between the coasts of North America. Canadian military, police, and customs officials already have begun to plan ways to use the passage, as they examine how to manage the new waterway.

Colin Woodard described the retreat of ice sheets in Iceland during the last century:

From the gravelly, newborn shores of this frigid lagoon, Iceland's Vatnajokull ice cap is breathtaking. The vast dome of snow and ice descends from angry clouds to smother jagged 3,000-foot-tall mountains. Then it spills out from the peaks in a steep outlet glacier 9.3 miles wide and 12.4 miles long—an insignificant appendage of Europe's largest ice cap despite its impressive size. A hundred years ago, however, there was no lagoon here. The shoreline was under one hundred feet of glacial ice. The outlet glacier, known as Breidamerkurjokull, extended to within 250 yards of the ocean, having crushed medieval farms and fields in its path during the preceding centuries, a time now referred to as the Little Ice Age. Today, Breidamerkurjokull's massive snout ends about two miles from the ocean. In its hasty retreat, the glacier has left the rapidly expanding lagoon, which is filled with icebergs calved from its front. The lagoon is 350 feet deep and has nearly doubled its size during the past decade. Every year, it grows larger, threatening to wash out Iceland's main highway. (Woodard 2000, A-1)

Ice Melt, Speed in the Arctic

In colloquial English, change is said to be "glacial" when one means slow, nearly unchanging. This connotation stems from a long-held belief that climate has changed slowly, especially during cold (glacial) periods. Contemporary research, however, indicates that climate changes during glacial periods often have occurred abruptly. Work by Ganopolski and Rhamstorf (2001) and Hall and Stouffer (2001) indicate that changes in ocean circulation may have been instrumental in these abrupt changes. Temperatures in Greenland are said to have changed by up to 10°C within a matter of decades (Paillard 2001, 147). "The models are not nearly as sensitive as the real world," said Richard B. Alley, an expert on Greenland's climate history at Pennsylvania State University. "That's the kind of thing that makes you nervous" (Revkin 2004b).

Late in the summer of 2003, the Ward Hunt Ice Shelf, the largest such formation in the Arctic and quite literally the "mother" of all icebergs in Canada's extreme north, split in half and started breaking up. Lying along the north coast of Ellesmere Island, the 443-square-kilometer ice shelf had supported the Arctic's largest "epishelf" lake—an ice-entombed body of fresh water that provided scientists a unique natural "cyrohabitat" for some of the Earth's rarest microorganisms, including algae and plankton from both fresh water and saltwater. The lake was once hailed as a possible model for explaining how extraterrestrial life might have evolved on Europa, one of Jupiter's frozen moons. The 43-meter-deep reservoir of fresh water drained away through the new crack.

Anecdotal observations of the Inuit are borne out by scientific studies documenting the rapid nature of glacial change in the Arctic. Eric Rignot and Robert H. Thomas, for example, writing in *Science*, described changes in the mass balance of polar ice sheets:

> Perhaps the most important finding of the last 20 years has been the rapidity with which substantial changes can occur on polar ice sheets. As measurements become more precise and more widespread, it is becoming increasingly apparent that change on relatively short time scales is commonplace: Stoppage of huge glaciers, acceleration of others, appreciable thickening, and far more rapid thinning of large sectors of ice sheet, rapid breakup of large areas of ice shelf, and vigorous bottom melting near grounding lines. (Rignot and Thomas 2002, 1505)

From their study of Greenland glaciers, H. Jay Zwally and colleagues concluded that warming temperatures accelerate the movement of glaciers, adding to the speed with which they melt. Water pools on top of glaciers during the summer, which is conducted to the glaciers' bases, speeding overall movement. In their words, "The indicated coupling between surface melting and ice-sheet flow provides a mechanism for rapid, large-scale dynamic responses of ice sheets to climate warming" (Zwally et al. 2002, 218). The Greenland Ice Sheet is losing mass at a rate sufficient to raise sea level by at least 0.13 millimeters per year because of rapid near-coastal thinning (Rignot and Thomas 2002, 1505).

Laser altimetry measurements published in March 1999 showed rapid thinning of the eastern portion of the Greenland ice sheet, particularly in the coastal regions, at rates exceeding a meter a year in some regions. The IPCC has concluded that a 3°C warming over Greenland could make melting of the ice cap irreversible, leading to a seven-meter rise in global sea level (Leggett 2001, 324).

Accelerating Decline of Arctic Sea Ice

Arctic sea ice cover shrank more dramatically between 2000 and 2007 than at any time since detailed records have been kept. A report produced by 250 scientists under the auspices of the Arctic Council found that Arctic sea ice was half as thick in 2003 as it had been 30 years earlier. If present rates of melting continue, there may be no summer ice in the Arctic by 2070, according to the study. Pal Prestrud, vice-chairman of the steering committee for the report, said, "Climate change is not just about the future; it is happening now. The Arctic is warming at twice the global rate" (Harvey 2004, 1). Since that report, ice loss has accelerated, most notably during 2007 and 2008, giving rise to forecasts of an ice-free Arctic Ocean in summer between 2020 and 2040.

The dramatic loss of ice in the Arctic, coupled with other new work showing the advance of trees and shrubs across once-barren Arctic tundra lands, "presents a compelling case that something is changing very rapidly over a wide area," said Larry Hinzman, an expert in Arctic change at the University of Alaska (McFarling 2002c). Weather in the Arctic was unusually warm and stormy during 2001, which broke up ice and melted it more readily (Serreze et al. 2003). "It's the kind of change we'd expect to see," said James Morrison of the University of Washington.

On Alaska's tundra, where air temperatures have been rising at an average of 0.5°C per decade for at least 30 years, shrubs have been observed spreading across land that heretofore has not supported them. A team reporting in *BioScience* during 2005 described processes by which the spread of shrubs tends to reinforce itself: "Increasing shrub abundance leads to deeper snow [shrubs hold more snow than tundra], which promotes higher soil temperatures, greater microbial activity, and more plant-available nitrogen," favoring greater shrub growth in subsequent years (Strum et al. 2005, 17). "With climate models predicting continued warming," Strum and colleagues wrote, "large areas of tundra could become converted to shrubland" (Strum et al. 2005, 17).

Projections into the Future

Ice melt observed to date probably accounts for only a fraction of the impact to come, as the world's oceans and atmosphere factor increased levels of "greenhouse forcing" into their dynamics. Jonathan Overpeck, a paleoclimate specialist with the National Oceanic and Atmospheric Administration (NOAA) in Boulder, Colorado, speculates that human activity may be short-circuiting the glaciation cycle that has dominated the Earth's climate for millions of years.

> If we left Mother Nature to do what she wanted to do, we would going back into another ice age in the next 10,000 years. Now, because of what humans are doing, it's unlikely that we'll be going back into another ice age. Instead, glaciers around the world are receding. The Earth is warming up, and it will likely continue to warm up. (Sudetic 1999, 102)

The melting of Arctic and Antarctic ice will do more than inconvenience coastal urban dwellers. Melting is already destroying an ecosystem built around sea ice. A report by the WWF and the Marine Conservation Biology Institute sketches the vital role of sea ice in polar ecosystems:

> Sea ice is fundamental to polar ecosystems: it provides a platform for many marine mammals and penguins to hunt, escape predators, and breed.... Its edges and undersides provide vital surfaces for the growth of algae that forms the base of the polar food web. In areas with seasonal ice cover, spring blooms of phytoplankton occur at ice edges as the ice cover melts, boosting productivity early in the season. But sea ice is diminishing in both the Arctic and the Antarctic. As this area diminishes, so does the food available to each higher level on the web, from zooplankton to seabirds. Higher temperatures predicted under climate change will further diminish ice cover, with open water occurring in areas previously covered by ice, thereby diminishing the very basis of the polar food web. (Mathews-Amos and Berntson 1999)

Reduced ice cover changes the Earth's albedo (reflectivity), which could become a factor in global warming, causing the Earth to absorb more solar energy, as ice and snow (which reflect 75 percent or more of incoming sunlight) is replaced by bare soil, which reflects 10 to 25 percent. Ice and snow, in some cases, may be replaced by liquid seawater, which reflects 10 to 70 percent of incoming sunlight, depending on the sun's angle. In both hemispheres, glacier discharge to the sea has increased markedly in recent years as warm water from intermediate depths is melting the floating ends of glaciers from below, accelerating them.

An analysis of ice-age cycles over the last million years, published in *Nature*, concluded that the climatic swings are the gyrations of a system poised to settle into a permanent colder state—with expanded ice sheets at both poles. Thomas J. Crowley of the University of Edinburgh and William T. Hyde of the University of Toronto believe that ice-age cycles during the past million years have been trending colder and that, without humankind's combustion of fossil fuels, the Earth would now be positioned at the edge of a new ice age. Earth is, they believe, in the final stages of a 50-million-year transition from a persistently warm climate to one with greatly expanded polar ice sheets. Crowley said that their argument does not obviate the need to curb greenhouse-gas emissions, but shows how powerful humankind's influence on climate has become (Revkin 2008j).

FURTHER READING

Amos, Jonathan. "Swiss Glaciers 'in Full Retreat.'" *BBC News*, accessed January 7, 2009. http://news.bbc.co.uk/2/hi/science/nature/7770472.stm.

"Billions of People May Suffer Severe Water Shortages as Glaciers Melt: World Wildlife Fund." *Agence France Presse*, November 27, 2003, n.p. (LEXIS)

Erickson, Jim. "Glaciers Doff Their Ice Caps, and as Frozen Fields Melt, Anthropological Riches Are Revealed." *Rocky Mountain News*, August 22, 2002, 6-A.

Ganopolski, Andrey, and Stefan Rahmstorf. "Rapid Changes of Glacial Climate Simulated in a Coupled Climate Model." *Nature* 409 (January 11, 2001): 153–158.

"Global Warming Makes China's Glaciers Shrink by Equivalent of Yellow River." *Agence France Presse*, August 23, 2004. (LEXIS)

Hall, Alex, and Ronald J. Stouffer. "An Abrupt Climate Event in a Coupled Ocean-Atmosphere Simulation without External Forcing." *Nature* 409 (January 11, 2001):171–174.

Harvey, Fiona. "Arctic May Have No Ice in Summer by 2070, Warns Climate Change Report." *London Financial Times*, November 2, 2004, 1.

Hodge, Amanda. "Patagonia's Big Melt 'Sign of Global Warming.'" *The Australian* (Sydney), February 12, 2004, 8.

Hodges, Glenn. "The New Cold War: Stalking Arctic Climate Change by Submarine." *National Geographic*, March 2000, 30–41.

Huber, Brian T., Kenneth G. MacLeod, and Scott L. Wing. *Warm Climates in Earth History*. Cambridge: Cambridge University Press, 2000.

Kaufman, Marc. "Antarctic Glaciers' Sloughing of Ice Has Scientists at a Loss." *Washington Post*, March 16, 2007b, A-2. http://www.washingtonpost.com/wp-dyn/content/article/2007/03/15/AR2007031501063_pf.html.

Kaufman, Marc. "Escalating Ice Loss Found in Antarctica: Sheets Melting in an Area Once Thought to Be Unaffected by Global Warming." *Washington Post*, January 14, 2008a. http://www.washingtonpost.com/wp-dyn/content/article/2008/01/13/AR2008011302753_pf.html.

Leggett, Jeremy. *The Carbon War: Global Warming and the End of the Oil Era*. New York: Routledge, 2001.

Lynas, Mark. *High Tide: The Truth about Our Climate Crisis*. New York: Picador/St. Martins, 2004a.

Lynas, Mark. "Vanishing Worlds: A Family Snap[shot] of a Peruvian Glacier Sent Mark Lynas on a Journey of Discovery—with the Ravages of Global Warming, Would It Still Exist 20 Years Later?" *London Guardian*, March 31, 2004c, 12.

Magnuson, John J., Dale M. Robertson, Barbara J. Benson, Randolph H. Wynne, David M. Livingstone, Tadashi Arai, Raymond A. Assel, Roger G. Barry, Virginia Card, Esko Kuusisto, Nick G. Granin, Terry D. Prowse, Kenton M. Stewart, and Valery S. Vuglinski. "Historical Trends in Lake and River Ice Cover in the Northern Hemisphere." *Science* 289 (September 8, 2000):1743–1746.

Mathews-Amos, Amy, and Ewann A. Berntson. "Turning up the Heat: How Global Warming Threatens Life in the Sea." World Wildlife Fund and Marine Conservation Biology Institute, 1999. http://www.worldwildlife.org/news/pubs/wwf_ocean.htm.

McFarling, Usha Lee. "Glacial Melting Takes Human Toll: Avalanche in Russia and Other Disasters Show That Global Warming Is Beginning to Affect Areas Much Closer to Home." *Los Angeles Times*, September 25, 2002b, A-4.

McFarling, Usha Lee. "Shrinking Ice Cap Worries Scientists." *Los Angeles Times* in *Edmonton Journal*, December 8, 2002c. http://www.canada.com/regina/story.asp?id={54910725-535A-4B0E-9A7E-FD7176D9C392}.

"More than 80 Percent of Spain's Pyrenean Glaciers Melted Last Century." Agence France Presse. September 29, 2004. (LEXIS)

Naftz, David L., David D. Susong, Paul F. Schuster, L. DeWayne Cecil, Michael D. Dettinger, Robert L. Michel, and Carol Kendall. "Ice Core Evidence of Rapid Air Temperature Increases since 1960 in the Alpine Areas of the Wind River Range, Wyoming, United States." *Journal of Geophysical Research* 107 (2002):4171. http://www.agu.org/pubs/crossref/2002/2001JD000621.shtml.

Otto-Bliesner, Bette L., Shawn J. Marshall, Jonathan T. Overpeck, Gifford H. Miller, Aixue Hu, and CAPE Last Interglacial Project Members. "Simulating Arctic Climate Warmth and Icefield Retreat in the Last Interglaciation." *Science* 311 (March 24, 2006):1751–1753.

Overpeck, Jonathan T., Bette L. Otto-Bliesner, Gifford H. Miller, Daniel R. Muhs, Richard B. Alley, and Jeffrey T. Kiehl. "Paleoclimatic Evidence for Future Ice-Sheet Instability and Rapid Sea-Level Rise." *Science* 311 (March 24, 2006):1747–1750.

Paillard, Didler. "Glacial Hiccups." *Nature* 409 (January 11, 2001):147–148.

Revkin, Andrew C. "An Icy Riddle as Big as Greenland." *New York Times*, June 8, 2004b. http://www.nytimes.com/2004/06/08/science/earth/08gree.html.

Revkin, Andrew C. "Will the Next Ice Age Be Permanent?." Dot.Earth, *New York Times*, November 12, 2008j. http://dotearth.blogs.nytimes.com/2008/11/12/will-next-ice-age-be-permanent/.

Rignot, Eric, and Robert H. Thomas. "Mass Balance of Polar Ice Sheets." *Science* 297 (August 30, 2002):1502–1506.

Rignot, Eric, Andres Rivera, and Gino Casassa. "Contribution of the Patagonia Icefields of South America to Sea Level Rise." *Science* 302 (October 17, 2003):434–437.

Robson, Seth. "Glaciers Melting." *Christchurch Press* (New Zealand), February 26, 2003, 13.

Serreze, Mark C., Marika M. Holland, and Julienne Stroeve. "Perspectives on the Arctic's Shrinking Sea-Ice Cover." *Science* 315 (March 16, 2007):1533–1536.

Serreze, M. C., J. A. Maslanik, T. A. Scambos, F. Fetterer, J. Stroeve, K. Knowles, C. Fowler, S. Drobot, R. G. Barry, and T. M. Haran. "A Record Minimum Arctic Sea Ice Extent and Area in 2002." *Geophysical Research Letters* 30 (2003), doi: 10.1029/2002GL016406.

Shepard, Andrew, and Duncan Wingham. "Recent Sea-Level Contributions of the Antarctic and Greenland Ice Sheets." *Science* 315 (March 16, 2007):1529–1532.

Stirling, I., and A. E. Derocher. "Possible Impacts of Climate Warming on Polar Bears." *Arctic* 46, no. 3 (1993):240–245.

Stirling, I., and N. J. Lunn. "Environmental Fluctuation in Arctic Marine Ecosystems as Reflected by Variability in Reproduction of Polar Bears and Ringed Seals." In *Ecology of Arctic Environments*, ed. S. J. Woodlin and M. Marquiss, 167–181. Oxford: Blackwell Science Ltd., 1997.

Strum, Matthew, Josh Schimel, Gary Michaelson, Jeffery M. Welker, Steven F. Oberbauer, Glen E. Liston, Jace Fahnestock, and Vladimir E. Romanovsky. "Winter Biological Processes Could Help Convert Arctic Tundra to Shrubland." *BioScience* 55, no. 1 (January 2005):17–26.

Sudetic, Chuck. "As the World Burns." *Rolling Stone*, September 2, 1999, 97–106, 129.

Suplee, Curt. "For 500 Million, a Sleeper on Greenland's Ice Sheet." *Washington Post*, July 10, 2000, A-9.

Toner, Mike. "Meltdown in Montana: Scientists Fear Park's Glaciers May Disappear within 30 Years." *Atlanta Journal-Constitution*, June 30, 2002b, 4-A.

Truffer, Martin, and Mark Fahnestock. "Rethinking Ice Sheet Time Scales." *Science* 315 (March 16, 2007):1508–1510.

Vidal, John. "Mountain Cultures in Grave Danger Says U.N.: Agriculture, Climate and Warfare Pose Dire Threat to Highland Regions around the World." *London Guardian*, October 24, 2002d, 7.

Williams, Frances. "Everest Hit by Effects of Global Warming." *London Financial Times*, June 6, 2002, 2.

Woodard, Colin. "Slowly, but Surely, Iceland Is Losing its Ice: Global Warming Is Prime Suspect in Meltdown." *San Francisco Chronicle*, August 21, 2000, A-1.

Zwally, H. Jay, Waleed Abdalati, Tom Herring, Kristine Larson, Jack Saba, and Konrad Steffen. "Surface Melt-induced Acceleration of Greenland Ice-Sheet Flow." *Science* 297 (July 12, 2002):218–222.

Ice Melt Velocity: A Slow-Motion Disaster for Antarctica?

Evidence from Antarctica suggests that melting ice may flow into the sea much more easily than earlier believed, perhaps leading to an accelerating rise in sea levels. A study published March 7, 2003, in the journal *Science* suggested that seas might rise as much as several meters during the next several centuries, given projected global warming based on business-as-usual usage of fossil fuels. The study called that prospect "a slow-motion disaster," the cost of which—in lost shorelines, salt inundation of water supplies, and damaged ecosystems—"would be borne by many future generations" (Revkin 2003, A-8; de Angelis and Skvarca 2003, 1560).

This analysis focused on the disintegration of ice shelves at the edges of the Antarctic Peninsula following decades of warming temperatures. The loss of the coastal shelves caused a drastic speedup in the seaward flow of inland glaciers. The peninsula, which stretches north toward South America, has warmed an average of 4.5°F over the last 60 years, so much so that ponds of melted water now form during summer months atop the flat ice shelves (Revkin 2003, A-8).

Two Argentine researchers described aerial surveys they conducted during 2001 and 2002, which indicated that the collapse of the Larsen A Ice Shelf during 1995 led to a sudden surge in the seaward flow of five of the six glaciers—as if a doorstop had been removed or a dam breached. Geological evidence indicated no signs of similar ice breakups along the peninsula in many thousands of years, the researchers and other experts said. Indeed, the recent disintegration of ice shelves along both coasts of the peninsula occurred after thousands of years of

relative stability, according to Pedro Skvarca, an author of the study and the director of glaciology at the Antarctic Institute of Argentina (Revkin 2003, A-8).

"We are witnessing a very significant warning sign of climate warming," Skvarca said (Revkin 2003, A-8). "This discovery calls for a reconsideration of former hypotheses about the stabilizing role of ice shelves (de Angelis and Skvarca 2003, 1560).

> It should be emphasized that the grounded ice on the northeastern Antarctic Peninsula is rapidly retreating and therefore substantially contributing to the global rise in sea level. The risk increases when the possible surging response of the Kektoria-Green-Evans and Crane glaciers is considered; these glaciers formerly nourished the section of the Larsen B Ice Shelf that disintegrated in early 2002. (de Angelis and Skvarca 2003, 1562)

Andrew Revkin wrote in the *New York Times*:

> The sliding could be abetted not only by the loss of the ice-shelf blockade, they said, but also by another unpredicted result of warming noted by other scientists in Antarctica and in Greenland: the rapid percolation of water from summertime ponds high on the ice sheets down through cracks to the base. There the water acts as a lubricant, facilitating the slide of glacial ice over the earth below. (Revkin 2003, A-8)

Many of Antarctica's ice shelves may behave in ways similar to those that were studied. This similarity gives the findings great significance, said Ted Scambos of the National Snow and Ice Data Center. The probability and timing of such an outcome remains unknown, but the new work has shed some light on the question, and it has increased some scientists' level of urgency (Revkin 2003, A-8).

Ice Sheet Thinning in Antarctica

Speculation regarding whether retreat of west Antarctic glaciers could accelerate ice flow from the continent's interior has engaged glaciologists for a quarter century. The stakes of this debate are enormous for the many hundreds of millions of people who live in urban areas on and near the world's coasts. Disintegration of the West Antarctic Ice Sheet could raise world sea levels by roughly six meters, or 20 feet.

Researchers from University College London and the British Antarctic Survey reported in

Science that the Pine Island Glacier, the largest on the West Antarctic Ice Sheet, and an important influence on the movement of the entire West Antarctic Ice Sheet, has lost 32 cubic kilometers of ice over a 5,000-square-kilometer area since 1992. The glacier was losing between one and two meters of thickness per year during the 1990s, according to this study. If thinning continued at such a rate, the entire glacier could disappear into the ocean within a few hundred years.

"It is possible," wrote Andrew Shepherd and colleagues, "that a retreat of the Pine Island Glacier may accelerate ice discharge from the West Antarctic Ice Sheet interior.... The grounded PIG thinned by up to 1.6 meters per year between 1992 and 1999, affecting 150 kilometers of the inland glacier" (Shepherd et al. 2001, 862). "If sufficiently prolonged, the present thinning could affect the flow of what is now slow-moving ice in the interior, increasing the volume of rapidly drained ice," they explained (Shepherd et al. 2001, 864). During November 2001, scientists discovered a 15-mile crack in the Pine Island Glacier, which split off a 150-square-mile iceberg (Toner 2002a, A-1). This break indicated that the glacier may be sliding toward the ocean more quickly than previously thought.

Andrew Shepherd, who led the study, said, "We have shown for the first time that such a retreat is indeed occurring. It is of paramount importance to determine whether the thinning is accelerating. Our present the theoretical understanding is not sufficient to predict firmly the future evolution of the Pine Island glacier" (Radford 2001a, 9; Shepard et al. 2001). This study added weight to the argument that small changes at the coast of the continent provoked by global warming could be transmitted swiftly inland, leading to an acceleration of glacial melting and, ultimately, significant worldwide sea-level rise.

Eric Rignot and Stanley S. Jacobs described the process in *Science*:

As continental ice from Antarctica reaches the grounding line and begins to float, its underside melts into the ocean. Results obtained from satellite radar interferometry reveal that bottom melt rates experienced by large outlet glaciers near their grounding lines are far higher than generally assumed. The melting rate is positively correlated with thermal forcing, increasing by 1 meter per year for each 0.1°C rise in ocean temperature. Where deep water has direct access to grounding lines, glaciers and ice shelves are vulnerable to ongoing increases in ocean temperature. (Rignot and Jacobs 2002, 2020)

Shepard and colleagues have explored the dynamics relating to the disintegration of the Larsen Ice Shelves on the Antarctic Peninsula. While air temperatures have been rising rapidly in the area, they have not risen quickly enough to wholly explain the rapid collapse of the ice shelves. The researchers believe that melting is being accelerated not only by rising air temperature above the ice shelves, but also by warming water temperatures below (Shepherd et al. 2003, 856). "If so," wrote Jocelyn Kaiser in *Science*, "[t]he rest of the Larsen [ice shelf] is doomed, and other Antarctic ice shelves could be more endangered than has been thought" (Kaiser 2003, 759). This one-two punch of warming from atmosphere and ocean during summertime causes pools of melted water to force their way through the ice shelves, causing crevasses that lead to collapse. While some glaciologists caution that evidence of ocean warming is scanty, others believe that the theories of Shepherd and colleagues explain why some ice masses are collapsing so quickly (Kaiser 2003, 759).

Are West Antarctic Glaciers Thickening?

Some studies seem to contradict others. Is the West Antarctic Ice Sheet increasing its velocity toward the ocean, thinning along the way, or is the opposite occurring? Some of each may be happening at the same time, in different parts of the ice cap. A study by authors Ian R. Joughin, an engineer at the Jet Propulsion Laboratory of the National Aeronautics and Space Administration in Pasadena, California, and Slawek Tulaczyk, a professor of earth sciences at the University of California–Santa Cruz, argued that the flow of these great rivers of ice is slowing, and as a result, they are growing thicker (Joughin and Tulaczyk 2002, 476–480). According to Andrew Revkin, writing in the *New York Times*, "The change means that this part of western Antarctica is likely to serve as a frozen bank for water instead of a source, slightly countering an overall trend toward rising seas" (Revkin 2002a, A-17).

This change appears to be related to the slow warming that has been going on since the end of the last ice age, 12,000 years ago, and not to the accelerated warming trend in the last five decades that many scientists have ascribed in part to human activities, according Joughin and Tulaczyk (Revkin 2002a, A-17).

Joughin and Tulaczyk built a case that parts of the ice sheet are growing thicker:

> We have used ice-flow velocity measurements from synthetic aperture radar to reassess the mass balance of the Ross Ice Streams, West Antarctica. We find strong evidence for ice-sheet growth (+26.8 gigatons per year) in contrast to earlier estimates (−20.9 gt/yr).... The overall positive mass balance may signal an end to the Holocene retreat of these ice streams. (Joughin and Tulaczyk 2002, 476)

While there is "ample evidence for a large retreat of the West Antarctic Ice Sheet over the last several thousand years.... If the current positive imbalance is not merely part of decadal or century-scale fluctuations, it represents a reversal of the long-term Holocene retreat" (Joughin and Tulaczyk 2002, 479).

Joughin and Tulaczyk acknowledged that other parts of the West Antarctic Ice Sheet are behaving in the opposite way, speeding up and sending more ice toward the sea, where great icebergs split from the broad ice shelf and eventually melt. "This teaches us more about the system," said glaciologist Richard B. Alley. "But it shouldn't affect what a coastal property owner thinks one way or the other" (Revkin 2002a, A-17).

In their study, Joughin and Tulaczyk said that the slowing and eventual thickening of the ice might result from thinning during thousands of years since the last ice age. As the rivers of ice lose mass, extreme cold at the surface may more easily migrate through the ice and freeze any water acting as a lubricant deep beneath, causing the ice to stick to the earth and grind to a halt (Revkin 2002a, A-17). "In the long run, this could produce a stop-and-start rhythm," said Robert A. Bindschadler, a glaciologist at the NASA Goddard Space Flight Center in Greenbelt, Maryland. Once the ice streams start to thicken, the chill will not reach the bottom as easily, and the geothermal heat from below might melt some ice, allowing the glacier to start sliding toward the sea again, Bindschadler said (Revkin 2002a, A-17). Whatever is happening in the streams feeding the Ross Ice Shelf is not happening in warmer areas farther north, where the Pine Island and Thwaites Glaciers are calving more icebergs than ever, said Rignot (Revkin 2002a, A-17).

Alley cautioned that the instrumental record is too short to trace long-term trends from this study: "It is important to remember how short the instrumental record is and how poorly characterized the natural variability. Sedimentary records indicate that ice streams have paused or even re-advanced during the retreat since the last ice age" (Alley 2002b, 452). Factors other than global warming (such as basal melting stemming from heat leaking from the Earth's interior) also may influence the advance or retreat of ice sheets over short periods. Alley suggested that the Pine Island Bay drainage "is probably the most likely ... to experience the onset of dramatic ice-sheet changes. Here, thick, fast-moving ice discharges into relatively warm ocean waters without the protection of a large shelf" (Alley 2002b, 452).

Regardless of some ice sheets' thickening, marked decreases in salinity measured in the Ross Sea during the last four decades indicate that the West Antarctic Ice Sheet is melting as a whole. According to S. S. Jacobs and colleagues, writing in *Science*, "These changes have been accompanied by atmospheric warming on Ross Island, ocean warming at depths of more than 300 meters north of the continental shelf ... and thinning of the southeast Pacific ice shelves" (Jacobs, Giulivi, and Mele 2002, 386). The freshening of water in the Ross Sea (its decline in salinity), according to this study, "appears to have resulted from a combination of factors, including increased precipitation, reduced sea-ice production, and increased melting of the West Antarctic Ice Sheet" (Jacobs, Giulivi, and Mele 2002, 386).

Is Some Antarctic Sea Ice Expanding?

According to a 2002 report by the Environment News Service, satellite records of sea ice around Antarctica indicated that Southern Hemisphere ice cover had increased since the late 1970s, at the same time that Arctic sea ice has declined. Continued decreases or increases could have substantial impacts on polar climates, because sea ice spreading over large areas increases albedo, reflecting solar radiation away from the Earth's surface ("Antarctic Sea Ice" 2002).

Claire Parkinson of the Goddard Space Flight Center analyzed the length of the sea ice season throughout the Southern Ocean to obtain trends in sea ice coverage. Parkinson examined 21 years (1979–1999) of Antarctic sea ice satellite records and found that, on average, the area where southern sea ice seasons have lengthened by at least one day per year is about twice as large as the area where sea ice seasons have shortened by at least one day per year. One day per year equals three weeks over the 21-year period. Parkinson also reported that the area of sea ice in the Arctic was decreasing by 13,000 square miles a year—an ice pack the size of Maryland and Delaware (Russell 2002, A-4).

FURTHER READING

Alley, Richard B. "On Thickening Ice?" *Science* 295 (January 18, 2002b):451–452.

de Angelis, Hernán, and Pedro Skvarca. "Glacier Surge after Ice Shelf Collapse." *Science* 299 (March 7, 2003):1560–1562.

"Antarctic Sea Ice Has Increased." Environment News Service, August 23, 2002. http://ens-news.com/ens/aug2002/2002-08-23-09.asp#anchor5.

Jacobs, S. S., C. F. Giulivi, and P. A. Mele. "Freshening of the Ross Sea during the Late 20th Century." *Science* 297 (July 19, 2002):386–389.

Joughin, Ian, and Slawek Tulaczyk. "Positive Mass Balance of the Ross Ice Streams, West Antarctica." *Science* 295 (January 18, 2002):476–480.

Kaiser, Jocelyn. "Glaciology: Warmer Ocean Could Threaten Antarctic Ice Shelves." *Science* 302 (October 31, 2003):759.

Radford, Tim. "Antarctic Ice Cap Is Getting Thinner: Scientists' Worries That the South Polar Ice Sheet Is Melting May Be Confirmed by the Dramatic Retreat of the Region's Biggest Glacier." *London Guardian*, February 2, 2001a, 9.

Revkin, Andrew C. "A Chilling Effect on the Great Global Melt." The *New York Times*, January 18, 2002a, A-17.

Revkin, Andrew C. "Study of Antarctic Points to Rising Sea Levels." *New York Times*, March 7, 2003, A-8.

Rignot, Eric, and Stanley S. Jacobs. "Rapid Bottom Melting Widespread near Antarctic Ice Sheet Grounding Lines." *Science* 296 (June 14, 2002): 2020–2023.

Russell, Sabin. "Glaciers on Thin Ice: Expert Says Melting to Be Faster Than Expected." *San Francisco Chronicle*, February 17, 2002, A-4.

Shepherd, Andrew, Duncan J. Wingham, Justin A. D. Mansley, and Hugh F. J. Corr. "Inland Thinning of Pine Island Glacier, West Antarctica." *Science* 291 (February 2, 2001):862–864.

Shepherd, Andrew, Duncan Wingham, Tony Payne, and Pedro Skvarca. "Larsen Ice Shelf Has Progressively Thinned." *Science* 302 (October 31, 2003):856–859.

Toner, Mike. "Huge Ice Chunk Breaks off Antarctica." *Atlanta Journal-Constitution*, March 20, 2002a, A-1.

Infectious Diseases among Wildlife

Warming temperatures have been contributing to epidemics of infectious diseases in wildlife around the world, ecologists and epidemiologists have warned. A scientific team's two-year study was the first to comprehensively analyze worldwide epidemics across entire plant and animal systems, on land and in the oceans.

C. Drew Harvell of Cornell University's ecology and evolutionary biology department, lead author of the study, said, "What is most surprising is the fact that climate-sensitive outbreaks are happening with so many different types of pathogens—viruses, bacteria, fungi and parasites—as well as in such a wide range of hosts, including corals, oysters, terrestrial plants, birds, and humans" (Cookson 2002, 4). The spread of diseases, in most cases, is supported by increased vigor associated with warming and increased moisture that the team attributed to various "vectors," organisms that spread diseases, such as mosquitoes, ticks, and rodents. "The accumulation of evidence has us extremely worried," said epidemiologist Andrew Dobson of Princeton University. "We share diseases with some of these species. The risk for humans is going up" (Cookson 2002, 4).

Population Declines or Species Extinctions

Harvell and colleagues found that "[i]nfectious diseases can cause rapid population declines or species extinctions. Many pathogens of terrestrial and marine taxa are sensitive to temperature, rainfall, and humidity, creating synergisms that could affect biodiversity." Climate warming can increase pathogen development and survival rates, disease transmission, and host susceptibility (Harvell et al. 2002, 2,158). Climate changes associated with El Niño events that may be related to global warming have had a "detectable influence" on pathogens, including coral diseases, oyster pathogens, crop pathogens, Rift Valley fever, and human cholera (Harvell et al. 2002, 2158).

Harvell and colleagues reviewed potential consequences of temperature changes on infectious diseases and considered the hypothesis that climate warming will affect host-pathogen interactions by (1) increasing pathogen development rates, transmission, and number of generations per year; (2) relaxing overwintering restrictions on pathogen life cycles; and (3) modifying host susceptibility to infection (Harvell et al. 2002, 2158).

The scientists anticipated that greater overwintering success of pathogens could increase the severity of many diseases. Because temperatures are increasing more rapidly during winter than in other seasons, this "population bottleneck" may be removed for many pathogens. "Several plant diseases are more severe after mild winters or during warmer temperatures, which suggests that directional climate warming will alter plant disease severity," they wrote (Harvell et al. 2002, 2159). Harvell and colleagues cited an example of fungi infecting Mediterranean oak and the Dutch elm disease. "Warming," they wrote, "can decrease plant resistance to both fungi and viruses. Plant species that have faster growth rates in warmer climates also may experience increased disease severity, because higher host density increases the transmission of many pathogens" (Harvell et al. 2002, 2159).

Tropical Diseases Spreading

Harvell and colleagues said that vectorborne human pathogens, such as malaria, Africa trypanosomiasis, Lyme disease, tick-borne encephalitis, yellow fever, plague, and dengue fever, have increased in incidence or geographic range in recent decades (Harvell et al. 2002, 2160). Similar expansions have been noted in animal diseases, according to Harvell and colleagues.

Warmer summers and milder winters are encouraging diseases that threaten lions, cranes, vultures, and ferrets. "The global warming is also helping to spread tropical diseases to human habitations previously unaffected by such illnesses," according to Harvell and colleagues (Radford 2002b, 7). "The number of similar increases in disease incidence is astonishing," said Richard Ostfeld, of the Institute for Ecosystem Studies, in Millbrook, New York (Radford 2002b, 7). Warming is shrinking areas where pests are controlled by frosts, and cool periods can cut insect, parasite, and fungus pest populations by up to 99 percent. According to one account, "In Hawaii, avian malaria has already wiped out native song birds living below an altitude of 1,400 meters (4,500 feet)" (Radford 2002b, 7).

The team reserved special attention for El Niño events. The scientists looked at the massive die-off of corals during the unusually warm El Niño year of 1998. Much of the coral had died from fungal and other diseases thriving in warmer seas. Damage was found to be worldwide. Oysters in Maine have been blighted by parasites that heretofore had been restricted to more southerly waters. Lions in the Serengeti

suffered canine distemper, and cranes, vultures, and even wild American ferrets had been hit by disease outbreaks. The monarch butterfly has come under pressure from an exploding parasite population.

Scientists fear that the spread of tropical infections to temperate zones could bring calamity. Compared with other geographic regions, the tropics have a richer variety of species and fewer individuals in each species; diversity acts as a buffer against the spread of disease. The temperate zones have a smaller range of species, with greater numbers in each, so pathogens moving into the temperate zones could affect a few common and abundant susceptible species (Radford 2002b, 7). "Human destruction of biodiversity makes this a double whammy. It means we are exacerbating the problem," said Dobson. "We have to get serious about global change. It is not only going to be a warmer world, it is going to be a sicker world" (Radford 2002b, 7).

Taking the Climate Signal Seriously

Almost everywhere the authors looked—on land and sea, in tropical climes and temperate ones—they found examples of plants and animals falling prey to disease as temperatures rose (McFarling 2002a, A-7). "When you see the same pattern across so many organisms, you need to start taking the climate signal seriously," said Richard S. Ostfeld, an ecologist at the Institute of Ecosystem Studies in Millbrook, New York. "We don't want to be alarmist, but we are alarmed" (McFarling 2002a, A-7). For example, "Malaria-carrying mosquitoes invade higher elevations, leading to more malaria deaths in mountain-dwelling Hawaiian songbirds. Ligurian Sea sponges that suffer heat stress in warmer waters are more vulnerable to fatal infection" (McFarling 2002a, A-7).

The authors of this study included some examples, paradoxically, in which warmer temperatures could lead to less disease. These include a number of diseases in fish, insects, and amphibians that spread only in cool conditions. Harvell said that her team could not yet make specific forecasts on how warming might alter ecosystems, but she said there was concern because "diseases are capable of reshaping whole communities very rapidly." She pointed to Caribbean corals, where many species that existed for 4,000 years have been wiped out by disease during the last 15 years (McFarling 2002a, A-7).

"This is a truly disturbing panorama," Paul Epstein of Harvard Medical School's Center for Health and the Global Environment told Usha Lee McFarling of the *Los Angeles Times*. "I'm afraid we've underestimated the true costs of what climate change is doing to our environment and our society" (McFarling 2002a, A-7).

FURTHER READING

Cookson, Clive. "Global Warming Triggers Epidemics in Wildlife." *London Financial Times*, June 21, 2002, 4.

Harvell, C. Drew, Charles E. Mitchell, Jessica R. Ward, Sonia Altizer, Andrew P. Dobson, Richard S. Ostfeld, and Michael D. Samuel. "Climate Warming and Disease Risks for Terrestrial and Marine Biota." *Science* 296 (June 21, 2002):2158–2162.

McFarling, Usha Lee. "Study Links Warming to Epidemics: The Survey Lists Species Hit by Outbreaks and Suggests That Humans Are Also in Peril." *Los Angeles Times*, June 21, 2002a, A-7.

Radford, Tim. "World Sickens as Heat Rises: Infections in Wildlife Spread as Pests Thrive in Climate Change." *London Guardian*, June 21, 2002b, 7.

Insects and Other Pests, New Ranges in Midlatitudes

Warming climate has favored the spread of many insects that previously were kept under control by seasonal frosts and freezes. Spreading insect infestations have been notable in many midlatitude urban areas, among them London, England, and Boston, Massachusetts. According to James Meek, writing in the *London Guardian*:

> They're chomping in Chelsea, dining *al fresco* in Fulham and more than pecking at their food in Pimlico. Despite their fancy taste in London addresses they are neither posh nor particularly fussy. They are alien vine weevils, and they want to eat your plants. Two species of vine weevil previously unable to survive Britain's cold winters have been discovered in southwest London. One of the species has also been detected in Surrey and Cardiff and as far north as Edinburgh. Two new species, *otiorhynchus armadillo* and *otiorhynchus salicicola*, not previously known north of Switzerland, will add to gardeners' woes. (Meek 2002b, 6)

"This is probably the most serious new garden pest in recent memory," said Max Barclay, curator of beetles at the Natural History Museum in London, who has been following waves of insect infestations novel to Britain. Classed as beetles, the immigrating species are now prevalent in Chelsea, Victoria, Pimlico, and Fulham, causing significant damage to gardens, in some cases defoliating garden plants almost entirely. "It's very likely these weevils have been introduced to Britain through imported ornamental plants from Italy," said Barclay. "It looks like they're here to stay" (Meek 2002b, 6). Earlier springs and later, milder winters have drawn other new pests to Britain. "Of course, the fact [that] Mediterranean species are doing well in Britain invites speculation about climate amelioration," said Barclay (Meek 2002b, 6).

Urban Wildlife in Boston

After one of Boston's mildest winters on record in 2001–2002, residents and exterminators reported "one of the largest explosions of urban wildlife in recent memory" (Wilmsen 2002, B-1). Rats, ants, wasps, snakes, squirrels, rats, skunks, and other animals used the mild winter to procreate at times when the rigors of a New England winter usually would have killed many and kept the rest sexually dormant. Many small mammals had litters early and probably would have one or more extra litters of young during the warm months of early 2002. (Many of the same creatures experienced the traditional perils of a New England winter during 2002–2003 and 2003–2004, however, when snow and cold returned with a vengeance.)

"I expect sharp increases in rat populations all over New England," said Bruce Colvin, a Lynnfield, Massachusetts, ecologist who is an expert on rats. "When you have a mild winter, there's more food and warm places for them to live, which means less stress. Instead of killing each other in the competition for food, they're all getting along and reproducing," he explained. Rat pups born during the winter were already mature enough by early spring to have litters of their own (Wilmsen 2002, B-1). Larger-than-usual numbers of carpenter ants and other insects, including bees and other wasps, were emerging in Boston by late March 2002, well ahead of their usual seasonal introduction in June.

By the first week of June 2002, also in Boston, pollen levels were at levels not usually observed until the end of the month. Writing in the *Annals of Allergy, Asthma, and Immunology*, as reported in the *Boston Globe*, Epstein said that if carbon-dioxide levels double, ragweed pollen levels will rise 61 percent, and he anticipates that level will be reached by 2050 (Smith 2002, A-1). The record mild weather of the 2001–2002 winter was followed in New England by two severe winters that played a role in reducing pest populations to more normal levels.

Rats, Spiders, and Scorpions Thrive in Great Britain

Great Britain's rat population rose by nearly a third between 1998 and 2002. According to a report described in London's *Daily Mail*, by 2002, an estimated 60 million brown rats inhabited Great Britain's buildings, streets, sewers, and waterways. The newspaper reported that "[t]he rodents narrowly outnumber the human population" (Utton 2003).

Even Buckingham Palace called in exterminators after a rat infestation in its kitchens in 2002. The 29 percent four-year rise in rat populations was blamed on milder, wetter winters as well as fast-food litter on which the rats feast. The amount of rubbish on London streets had grown by an estimated 80 percent in 35 years, according to the report. Rats breed quickly, reaching sexual maturity in eight weeks. They can have sex 20 times a day, and can give birth every four weeks. A single pair of rats can produce as many as 2,000 offspring a year (Utton 2003).

Barrie Sheard of the British National Pest Technicians' Association, which conducted the study, pointed to a rapid increase in rat numbers during the summer months, with warming a probable factor. Other factors also are at work. For example, many sewer systems built during the late nineteenth century have decayed, allowing more rats to reach the surface. To save money, many British town councils have started billing private property owners for rat control, which has caused many to avoid reporting infestations.

"Until these last four to five years, our summer months were always recognized as a period of the year when rat complaints were always at a far-reduced level," Sheard said (Utton 2003). Peter Gibson, speaking for Keep Britain Tidy, said, "The worrying thing is not just their numbers, but their behavior. Rats are not just in the sewers any more. They are coming out into the open, on to the streets, because they know that's where the most food is" (Utton 2003).

Spiders, wasps, and similar species have been moving northward through Britain at unprecedented rates as temperatures rise. Insects with historic ranges in the northern reaches of England are moving into Scotland. Those previously residing in southern England have arrived in the north, as European spiders are increasingly invading Britain from the south and west. Global warming is being held responsible (Browne and Simons 2002, 8).

With no natural predators, scorpions have been spreading in England as temperatures rise. Two-inch-long, European yellow-tailed scorpions probably were brought to England from Italy or southern France during 1860 on ships docking at Sheerness in Kent. For many decades, a colony of about 1,000 hid in the brickwork of the dockyard. As temperatures warmed after the late 1980s, the scorpion colony grew to about 3,000 and spread into neighboring residential areas, where people have been advised to follow practices used in tropical areas: turn shoes upside down and shake them before wearing. Paul Hillyard, curator of arachnids at the Natural History Museum in London, said, "In the warmer temperatures these scorpions can flourish and people could start transporting them throughout Britain" (Ingham 2004, 40). European yellow-tailed scorpions also have been sighted in Harwich and Pinner and at the Ongar underground station in Essex.

According to one English observer, "The magnificent yellow and black wasp spider, *Argiope bruennichi*, which grows up to 50 millimeters (two inches) long, arrived in Hampshire from the Continent and now occurs as far north as Derbyshire" (Browne and Simons 2002, 8). Peter Harvey, head of the Spider Recording Scheme for the British Arachnological Society, said that the wasp spider had spread rapidly during the five years ending in 2002. "It is almost certainly because of warmer and longer summers and autumns. It lays its eggs in autumn, so cold autumns are very bad for it" (Browne and Simons 2002, 8).

Examples of insect migrations are plentiful in the British Isles:

> The bee wolf, a wasp that feeds honeybees to its offspring, had until recently been confined to the Isle of Wight and East Anglia. However, it has spread rapidly in recent years and is now found as far north as Yorkshire. Arachnophobes in Scotland will be anxious to learn that the giant house spider, *Tegenaria gigantea*, which was once confined to England, has now reached Ullapool in the

northwest Highlands. A species of jumping spider entirely new to Britain was recently discovered in London, having arrived from the Continent. A survey of Mile End Park in the East End of London has revealed a thriving colony of *Macaroeris nidicolens*, which is usually found in continental Europe, where it lives on small trees and bushes. Exactly how the spider arrived in Britain is unclear, although the young spiderlings could have crossed the Channel using thread of gossamer blown on the wind. (Browne and Simons 2002, 8)

FURTHER READING

Browne, Anthony, and Paul Simons. "Euro-spiders Invade as Temperature Creeps up." *London Times*, December 24, 2002, 8.

Ingham, John. "Stingers Thrive as the Country Gets Warmer: Invasion of the Scorpions." *London Express*, June 18, 2004, 40.

Meek, James. "Global Warming Gives Pests Taste for Life in London." *London Guardian*, October 8, 2002b, 6.

Smith, Stephen, "Comin' Ah-choo: Tepid Temperatures Speeding Allergy Season." *Boston Globe*, April 10, 2002, A-1.

Utton, Tim. "The Rat Rampage." *London Daily Mail*, January 22, 2003, n.p. (LEXIS)

Wilmsen, Steven. "Critters Enjoy a Baby Boom: Mild Winter's Downside Is Proliferation of Vermin." *Boston Globe*, March 30, 2002, B-1.

Intergovernmental Panel on Climate Change: The Politics of Climate Consensus

The Intergovernmental Panel on Climate Change (IPCC) 2007 assessment forecast a global average worldwide temperature rise of 2° to 11.5°F by 2100, a wider range than the 2001 assessment's 2.5° to 10.4°F. However, the panel also said a temperature rise of 3.2° to 7.1°F was most likely. The 2007 IPCC assessment placed the blame for global warming on human activities with near certainty (90 to 99 percent), compared with its likelihood (66 percent) in the 2001 report (Rosenthal and Revkin 2007). If greenhouse gases continue to increase at even a moderate pace, temperatures by the end of the century could match those last seen 125,000 years ago, a time when seas were 12 to 20 feet higher than today. The difference is thermal inertia—melting that is "in the pipeline," but not yet realized.

The 2007 assessment anticipates that, by the year 2020, 75 million to 250 million people in Africa will suffer water shortages, and residents of Asia's large cities will be at great risk of river and coastal flooding, Europeans can expect extensive species loss, and North Americans will experience longer and hotter heat waves and greater competition for water.

The 2007 assessment also forecast intensifying drought in southern Europe and the Middle East, Sub-Saharan Africa, the U.S. Southwest, and Mexico, coupled with ocean flooding that could imperil low-lying islands and densely populated river deltas of southern Asia, affecting the world's poorest peoples most dramatically.

"It's the poorest of the poor in the world, and this includes poor people even in prosperous societies, who are going to be the worst hit," said Rajendra K. Pachauri, chair of the panel. "People who are poor are least-equipped to be able to adapt to the impacts of climate change, and therefore in some sense this does become a global responsibility in my view" (Kanter and Revkin 2007b).

Widespread Effects Already Measurable

Without swift action to bring down greenhouse-gas levels in the atmosphere, the IPCC said that a worldwide temperature rise of 3° to 5°F over the next century would inundate coasts and islands inhabited by hundreds of millions of people. The panel concluded with 90 percent certainty that humans have been the main cause of worldwide warming since 1950. Martin Parry, the co-chair of the team that wrote the new report, said widespread effects were already measurable, with much more to come. "We're no longer arm-waving with models," he said. "This is empirical information on the ground" (Kanter and Revkin 2007b).

Some effects of warming may be beneficial. The IPCC assessment said that some regions could benefit from increased rainfall, longer growing seasons, open Arctic seaways, and fewer deaths from cold weather. However, most will be harmful. "The warnings are clear about the scale of the projected changes to the planet," said Bill Hare, an author of the report and a visiting scientist at the Potsdam Institute for Climate Impact Research in Germany. "Essentially there's going to be a mass extinction within the next 100 years unless climate change is limited," added Hare. "These impacts have been known for many years, and are now seen with greater clarity in this report," he said. "That clarity is perhaps the last warning we're going to get before we actually have to report in the next IPCC review that we're seeing the disaster unfolding" (Kanter and Revkin 2007b).

Three Choices: Mitigation, Adaptation, and Suffering

Sea-level rise at the end of the century is estimated by the IPCC to be 7–23 inches. An additional 3.9–7.8 inches is possible, according to this forecast, if recent, accelerating melting of polar ice sheets continues. The report anticipates that sea levels will continue to rise for a thousand years after that because of thermal inertia. Many experts said those estimates are probably too low because the panel had failed to consider a growing body of data on melting glaciers and inland ice sheets, which are major contributors to sea-level rise (Dean 2007b). Climate experts have "a great deal of confidence" in observations that sea-level rise is accelerating, said Laury Miller, an oceanographer at the National Oceanographic and Atmospheric Administration who was a reviewer for part of the report (Dean 2007b).

Drew Shindell, a climate expert at the NASA Goddard Institute for Space Studies, described the difficulty of forming firm scientific conclusions in a field in which research develops so quickly: "The melting of Greenland has been accelerating so incredibly rapidly that the IPCC report will already be out of date in predicting sea level rise, which will probably be much worse than is predicted in the IPCC report" (Dean 2007b).

"We basically have three choices: mitigation, adaptation and suffering," said John Holdren, the president of the American Association for the Advancement of Science and an energy and climate expert at Harvard University. "We're going to do some of each. The question is what the mix is going to be. The more mitigation we do, the less adaptation will be required and the less suffering there will be" (Kanter and Revkin 2007a).

The report said no matter how much civilization slows or reduces its greenhouse-gas emissions, global warming and sea-level rise will continue for centuries. "This is just not something you can stop. We're just going to have to live with it," said co-author Kenneth Trenberth. "We're creating a different planet. If you were to come up back in 100 years time, we'll have a different climate" ("Panel Says" 2007).

Political Pressures on the IPCC

IPCC reports often indicate as much about the political atmosphere surrounding global warming as the science. These documents are approved by politicians as well as scientists, often after considerable wrangling. Scientific assessments in 2007, for example, passed review (and censorship) of officials from China, the United States, and other countries.

For example, U.S. negotiators eliminated language in one section that called for cuts in greenhouse-gas emissions, said Patricia Romero Lankao, a scientist at the National Center for Atmospheric Research (NCAR) in Boulder, Colorado, who was one of the report's lead authors (Eilperin 2007a, A-5). Because of political meddling, the final IPCC document was "much less quantified and much vaguer and much less striking than it could have been," said Stéphane Hallegatte, a participant from France's International Center for Research on the Environment and Development (Kanter and Revkin 2007b).

Juliet Eilperin of the *Washington Post* described how politics shaped the report:

China objected to wording that said "based on observed evidence, there is very high confidence that many natural systems, on all continents and in most oceans, are being affected by regional climate changes, particularly temperature increases." The term "very high confidence" means researchers are at least 90 percent sure of their findings. When China asked that the word "very" be stricken, three scientific authors balked, and the deadlock was broken only by a compromise to delete any reference to confidence levels. "That was a really hard discussion," said Romero Lankao, who participated in the talks. But the scientist, a lead author on the chapter examining the effect of global warming on industry, settlements and society, said the panel's overall message remained clear: "No one on Earth will escape the impacts of a warming planet." (Eilperin 2007a, A-5)

Pressure from Russia, China, and Saudi Arabia softened sections on coral damage and discussions of tropical storms that appeared in the summary. The nations also succeeded in getting the authors to drop parts of an illustration showing how different emissions policies might limit damage.

Scientific Consensus

While media reports of the public-policy debate regarding global warming has often conveyed an impression that scientists are hopelessly divided over the issue of whether human activities are warming the lower atmosphere. In actuality, a high degree of agreement has existed since the IPCC's *First Assessment* was published in 1990. The IPCC's first major report forecast widely varying temperature increases by region with an assumed doubling of carbon dioxide in the atmosphere. The largest increases (6° to 7°C) were forecast in the interiors of northern North America and Asia during the winter; increases in the summer for the same regions were forecast at between 3° and 4°C. The largest summer temperature increase (4.8°C) was forecast for the interior of southern Asia. The smallest increases year-round were forecast for the tropics, especially areas near large bodies of water.

An IPCC conference during November 1990 in Geneva, Switzerland, issued a "ministerial declaration" representing 137 countries that agreed that, although the climate had varied in the past, "the rate of climate change predicted by the IPCC to occur over the next century [due to greenhouse warming] is unprecedented." The ministers declared that "climate change is a global problem of unique character" (Jager and Ferguson 1991, 525). The ministers also declared that the eventual goal should be "to stabilize greenhouse gas concentrations at a level that would prevent dangerous anthropogenic interference with climate" (Jager and Ferguson 1991, 536).

The question of whether the Earth is becoming unnaturally warmer because of human activities was largely settled in scientific circles by 1995, with publication of the IPCC's *Second Assessment*, a worldwide group of about 2,500 experts. The panel concluded that the Earth's temperature had increased between 0.5° and 1.1°F (0.3° to 0.6°C) since reliable worldwide records became available between 1850 and 1900. The IPCC noted that warming accelerated as measurements approached the present day.

The IPCC's *Second Assessment* concluded that human activity—increased generation of carbon dioxide and other greenhouse gases—was at least partially responsible for the accelerating rise in global temperatures. The amount of carbon dioxide in the atmosphere has been rising nearly every year because of increased use of fossil fuels by ever-larger human populations experiencing higher living standards. The IPCC's *Second Assessment*, according to one observer,

> [M]akes an unprecedented, though qualified, attribution of the observed climate change to human causes. Though the human signal is still building and somewhat masked within natural variation, and while there are key uncertainties to be resolved, the Panel concludes that "the balance of evidence suggests that there is a discernible human influence on global climate."

In its *Second Assessment*, the IPCC declared:

> During the past few decades, two important factors regarding the relationship between humans and the Earth's climate have become apparent. First, human activities, including the burning of fossil fuels, land-use change and agriculture, are increasing the atmospheric concentrations of greenhouse gases (which tend to warm the atmosphere) and, in some regions, aerosols (microscopic airborne particles, which tend to cool the atmosphere). These changes in greenhouse gases and aerosols, taken together, are projected to change regional and global climate and climate-related parameters such as temperature, precipitation, soil moisture and sea level. Second, some human communities have become more vulnerable to hazards such as storms, floods and droughts as a result of increasing population density in sensitive areas such as river basins and coastal plains. Potentially serious changes have been identified, including an increase in some regions in the incidence of extreme high-temperature events, floods and droughts, with resultant consequences for fires, pest outbreaks, and ecosystem composition, structure and functioning, including primary productivity. (Bolin et al. 1995)

The *Second Assessment* indicated that rising atmospheric concentrations of greenhouse gases would cause interference with the climate system to grow in magnitude, as "the likelihood of adverse impacts from climate change that could be judged dangerous will become greater" (Bolin et al. 1995). Because of these dangers, the IPCC called on the governments of the worlds to "take precautionary measures to anticipate, prevent or minimize the causes of climate change and mitigate its adverse effects" (Bolin et al. 1995).

The IPCC linked expected climate changes to rising levels of several greenhouse gases in the atmosphere following the rapid spread of fossil-fueled industry, including carbon dioxide, (from about 280 to almost 360 parts per million [ppm] as of 1992), methane (from 700 to 1,720 parts per billion [ppb]), and nitrous oxide (from about 275 to about 310 ppb). While aerosols may have a short-term impact in some areas, the IPCC's *Second Assessment* said that their short-lived nature does little to mitigate the effects of the greenhouse gases, which often reside in the atmosphere for hundreds of years.

FURTHER READING

Bolin, Bert, John T. Houghton, Gylvan Meira Filho, Robert T. Watson, M. C. Zinyowera, James Bruce, Hoesung Lee, Bruce Callander, Richard Moss, Erik Haites, Roberto Acosta Moreno, Tariq Banuri, Zhou Dadi, Bronson Gardner, J. Goldemberg, Jean-Charles Hourcade, Michael Jefferson, Jerry Melillo, Irving Mintzer, Richard Odingo, Martin Parry, Martha Perdomo, Cornelia Quennet-Thielen, Pier Vellinga, and Narasimhan Sundararaman. *Intergovernmental Panel on Climate Change. Second Assessment Synthesis of Scientific-Technical Information Relevant to Interpreting Article 2 of the United Nations Framework Convention on Climate Change.* Approved by the IPCC at its eleventh session, Rome, December 11–15, 1995. http://www.unep.ch/ipcc/pub/sarsyn.htm.

Dean, Cornelia. "Even Before Its Release, World Climate Report Is Criticized as Too Optimistic." *New York Times*, February 2, 2007b. http://www.nytimes.com/2007/02/02/science/02oceans.html.

Eilperin, Juliet. "U.S., China Got Climate Warnings Toned Down." *Washington Post*, April 7, 2007a, A-5. http://www.washingtonpost.com/wp-dyn/content/article/2007/04/06/AR2007040600291_pf.html.

Jager, J., and H. L. Ferguson. *Climate Change: Science, Impacts, and Policy. Proceedings of the Second World Climate Conference.* Cambridge: Cambridge University Press, 1991.

Kanter, James and Andrew Revkin. "World Scientists Near Consensus on Warming." *New York Times*, January 30, 2007a. http://www.nytimes.com/2007/01/30/world/30climate.htm.

Kanter, James, and Andrew C. Revkin. "Scientists Detail Climate Changes, Poles to Tropics." *New York Times*, April 7, 2007b. http://www.nytimes.com/2007/04/07/science/earth/07climate.html.

"Panel Says Humans 'Very Likely' Cause of Global Warming." Associated Press in *New York Times*, February 3, 2007. http://www.nytimes.com/aponline/science/AP-France-Climate.

Rosenthal, Elisabeth, and Andrew C. Revkin, "Climate Panel Issues Urgent Warning to Curb Gases," *New York Times*, February 2, 2007. http://www.nytimes.com/2007/02/02/science/earth/02cnd-climate.htm.

Ireland, Flora and Fauna in

In Ireland, by 2001, swallows were arriving earlier and the growing season for trees was steadily increasing, according to scientists at Trinity College of Dublin and National University of Ireland at Maynooth who have been examining natural responses to temperature changes. Alison Donnelly of the Climate Change Research Centre, School of Botany, Trinity College, found that the beginning of the growing season, defined by the unfolding of leaves, has been occurring earlier in the spring. The scientists also noted that the end of the growing season, the time when the leaves fall, occurred later in autumn, both of which point to an influence of rising global temperatures.

Donnelly said that records of the dates that swallows arrive on the east coast of Ireland only extend back to the 1980s, but nevertheless provided some evidence of the swallows' changing patterns of migration (Fahy 2001). Ireland's Environmental Protection Agency issued a report in late November 2002 titled "Climate Change Indicators for Ireland," which said that temperatures there have been rising 0.25°C per decade, with increases accelerating during the 1990s. The number of frost days had declined (Hogan 2002).

Hummingbird hawk moths, usually residents in southern Europe, have started appearing in Ireland, according to broadcaster and naturalist Eanna Ni Lamhna. Often mistaken for hummingbirds, the hawk moths hover as they swig nectar from flowers (Meagher 2004). Bird watchers have noticed an increase in the number of exotic birds. The little egret, a smaller version of the heron, has taken to nesting along the coast of Cork and Waterford. These birds, which usually are found in lands bordering the Mediterranean Sea, first started nesting in Ireland in 1997, and there are thought to be 50 pairs residing in the country. Other migrant species from southern climates that have recently appeared in Ireland include the pied flycatcher, the bearded tit, and the Mediterranean gull (Meagher 2004).

In the oceans surrounding Ireland, jellyfish have thrived in warmer sea temperatures, and in some enclosed bays, cockles expired from the warming waters. Karin Dubsky, the environmentalist who runs Coastwatch Ireland, said that warming ocean temperatures may cause native oysters to be outbred by Pacific oysters being farmed along the coast. "We are worried that

Pacific oysters might thrive here as a result of the increase in sea temperatures. They could out-breed the Irish oyster in the same way that the grey squirrel knocked out the red squirrel" (Meagher 2004).

FURTHER READING

Fahy, Declan. "Nature Charts its Own Change: Irish Researchers Are Finding Signs of Climate Change in Trees and Bird Species." *Irish Times*, September 13, 2001, n.p. (LEXIS)
Hogan, Treacy. "Still Raining in Costa del Ireland." *Belfast Telegram*, November 26, 2002, n.p. (LEXIS)
Meagher, John. "Look What the Changing Climate Dragged in ..." *Irish Independent*, July 9, 2004, n.p. (LEXIS)

Iron Fertilization of the Oceans

Should the oceans be seeded with large amounts of iron ore that will stimulate the growth of carbon-dioxide-consuming phytoplankton? The idea has attracted some support among corporations and foundations looking for ways to minimize the effects of carbon dioxide without changing the world's basic energy-generation mix. The idea is simple on its face: iron stimulates the growth of phytoplanktonic algae that are believed to be responsible for about half of the world's biologic absorption of carbon dioxide.

Nearly half of the Earth's photosynthesis is performed by phytoplankton in the world's seas and oceans. This fact has led to proposals to "geo-engineer" a carbon-dioxide sink using the same "biological pump" that is believed to have driven at least some of the Earth's past climate cycles (Chisholm 2000, 685). Atsushi Tsuda and colleagues have studied iron fertilization and have found that, under some circumstances, iron fertilization can dramatically increase phytoplankton mass (Tsuda et al. 2003, 958–961).

Ulf Riebesell, a marine biologist at the Alfred Wegener Institute for Polar and Marine Research in Bremerhaven, Germany, believes that ambitious iron seeding of the oceans could remove three to five billion tons of carbon dioxide per year, or about 10 to 20 percent of human-generated emissions, and be on the order of at least 10 times cheaper than planting forests to do the same thing (Schiermeier 2003a, 110). Patents have been issued for ocean fertilization, and

demonstration projects have been undertaken. One such project was described and evaluated in *Nature* (Watson et al. 2000, 730–733).

The chemistry of the oceans varies widely with regard to the amount of iron necessary to prime this pump and substantially increase carbon-dioxide sequestration. In the equatorial Pacific and Southern ocean, wrote Sallie W. Chisholm, a marine biologist at the Massachusetts Institute of Technology, it "is possible to stimulate the productivity of hundreds of square kilometers of ocean with a few barrels of fertilizer" (Chisholm 2000, 686).

In an experiment conducted between Tasmania and Antarctica, researchers confirmed that vast stretches of the world's southern oceans are primed to explode with photosynthesis, but lack only iron. The researchers said that it is too soon to start large-scale iron seeding because the new experiment raised as many questions as it answered. At best, they said, iron seeding would absorb only a small amount of the carbon dioxide in the atmosphere. They also said that their experimental bloom of plankton was not tracked long enough to determine whether the carbon harvested from the air sank into the deep sea or was again released into the environment as carbon-dioxide gas.

Weighing Costs and Benefits

"There are still fundamental scientific questions that need to be addressed before anyone can responsibly promote iron fertilization as a climate control tactic," said Kenneth H. Coale, an oceanographer who has helped design studies of iron's effects in the tropical Pacific (Revkin 2000c, A-18). This seemingly simple proposition has some potential problems, however. First, there exists no way to measure the amount of carbon taken up by phytoplankton. Additionally, the algae produce dimethylsulfide, which plays a role in cloud formation.

Additionally, phytoplankton increase the amount of sunlight absorbed by ocean water, as well as heat energy. They also produce compounds such as methyl halides, which play a role in stratospheric ozone depletion. The iron could promote the growth of toxic algae, which may kill other marine life and change the chemistry of ocean water by removing oxygen. "The oceans are a tightly linked system, one part of which cannot be changed without resonating through the whole system," said Chisholm. "There is no free lunch" (Schiermeier 2003a, 110).

So much iron may be required to produce the desired effect that fertilization of this type will never be commercially possible. "The experiments enabled us to make an initial determination about the amount of iron that would be required and the size of the area to be fertilized," said Ken O. Buesseler of the Woods Hole Oceanographic Institution, who co-authored a study of the idea. "Based on the studies to date, the amount of iron needed and [the] area of ocean that would be impacted is too large to support the commercial application of iron to the ocean as a solution to our greenhouse gas problem," he explained ("Iron Link" 2003). "It may not be an inexpensive or practical option if what we have seen to date is true in further experiments on larger scales over longer time spans," Buesseler said. "The oceans are already naturally taking up human-produced carbon dioxide, so the changes to the system are already underway," he said. "We need to first ask will it work and then what are the environmental consequences?" ("Iron Link" 2003).

Limits on Fertilization

One study asserted that "[t]o assess whether iron fertilization has potential as an effective sequestration strategy, we need to measure the ratio of iron added to the amount of carbon sequestered in the form of particulate organic carbon to the deep ocean in field studies" (Buesseler and Boyd 2003, 67). To date, wrote Buesseler and Boyd, experiments of this type have "produced notable increases in biomass and associated decreases in dissolved inorganic carbon and macronutrients. However, evidence of sinking carbon particle carrying P.O.C. [particulate organic carbon] to the deep ocean was limited" (Buesseler and Boyd 2003, 67).

Given the limits of present technology, this study estimates that an area larger than the Southern Ocean (waters southerly of 50 degrees south latitude) would have to be fertilized to remove 30 percent of the carbon that human activity presently injects into the atmosphere. Thus, according to this study, "ocean iron fertilization may not be a cheap and attractive option if impacts on carbon export and sequestration are as low as observed to date" (Buesseler and Boyd 2003, 68).

Despite prospective problems, iron fertilization is considered possibly viable in some quarters. Scientists who fed tons of iron into the Southern Ocean reported evidence during 2004 that stimulating the growth of phytoplankton in this way may strengthen the oceans' use as a carbon sink. In a report in the April 16, 2004, edition of *Science*, ocean biologists and chemists from more than 20 research centers said they triggered two huge blooms of phytoplankton that turned the ocean green for weeks and consumed hundreds, perhaps thousands of tons of carbon dioxide.

"These findings would be encouraging to those considering iron fertilization as a global geoengineering strategy," said Ken Coale, a chief scientist at the Moss Landing Marine Laboratories. "But the scientists involved in this experiment realize that this looked only skin deep at the functioning of ocean ecosystems and much more needs to be understood before we recommend such a strategy on a global scale" (Hoffman 2004). "From my work, I don't think this could solve a significant fraction of our greenhouse-gas problem while causing unknown ecological consequences," said Buesseler (Hoffman 2004).

A scientific team led by Stéphane Blain reported observations of a phytoplankton bloom induced by natural iron fertilization:

> We found that a large phytoplankton bloom over the Kerguelen plateau in the Southern Ocean was sustained by the supply of iron and major nutrients to surface waters from iron-rich deep water below. The efficiency of fertilization, defined as the ratio of the carbon export to the amount of iron supplied, was at least ten times higher than previous estimates from short-term blooms induced by iron-addition experiments. This result sheds new light on the effect of long-term fertilization by iron and macronutrients on carbon sequestration, suggesting that changes in iron supply from below—as invoked in some palaeoclimatic and future climate change scenarios—may have a more significant effect on atmospheric carbon dioxide concentrations than previously thought. (Blain et al. 2007, 1070)

A Moratorium on Iron Fertilization

The United Nations Convention on Biological Diversity, meeting in Berlin during late May 2008, issued a moratorium, signed by representatives of 191 nations, on large-scale commercial projects to fertilize the oceans in pursuit of climate-change mitigation. The ban will hold until scientists better understand the effects of such mitigation on the ocean food chain. Iron fertilization could increase ocean acidity and reduce oxygen levels.

On October 31, 2008, delegates from 85 nations, which include those countries that signed the London Convention Treaty (created in 1972 by the United Nations International Maritime Organization), limited iron fertilization to "legitimate scientific research." The move was made to inhibit premature commercial application that exploits awards of carbon credits under diplomatic agreements such as the Kyoto Protocol. The delegates are charged with regulating pollution in international waters through such compacts as the London Protocol. The group acted after announcements of several plans to commercialize iron fertilization. Any research must consider carbon flux (such as emissions of nitrous oxides), impacts on the food web and oxygen levels, and the possibility that toxic species may grow (Kintisch 2008, 835).

FURTHER READING

Blain, Stéphane, Bernard Quéguiner, Leanne Armand, Sauveur Belviso, Bruno Bombled, Laurent Bopp, Andrew Bowie, Christian Brunet, Corina Brussaard, François Carlotti, Urania Christaki, Antoine Corbière, Isabelle Durand, Frederike Ebersbach, Jean-Luc Fuda, Nicole Garcia, Loes Gerringa, Brian Griffiths, Catherine Guigue, Christophe Guillerm, Stéphanie Jacquet, Catherine Jeandel, Patrick Laan, Dominique Lefèvre, Claire Lo Monaco, Andrea Malits, Julie Mosseri, Ingrid Obernosterer, Young-Hyang Park, Marc Picheral, Philippe Pondaven, Thomas Remenyi, Valérie Sandroni, Géraldine Sarthou, Nicolas Savoye, Lionel Scouarnec, Marc Souhaut, Doris Thuiller, Klaas Timmermans, Thomas Trull, Julia Uitz, Pieter van Beek, Marcel Veldhuis, Dorothée Vincent, Eric Viollier, Lilita Vong, and Thibaut Wagener. "Effect of Natural Iron Fertilization on Carbon Sequestration in the Southern Ocean." *Nature* 446 (April 26, 2007):1070–1074.

Buesseler, Ken O., and Philip W. Boyd. "Will Ocean Fertilization Work?" *Science* 300 (April 4, 2003):67–68.

Chisholm, Sallie W. "Stirring Times in the Southern Ocean." *Nature* 407 (October 12, 2000):685–686.

Hoffman, Ian. "Iron Curtain over Global Warming: Ocean Experiment Suggests Phytoplankton May Cool Climate." *Daily Review* (Hayward, CA), April 17, 2004, n.p. (LEXIS)

"Iron Link to CO$_2$ Reductions Weakened." Environment News Service, April 10, 2003. http://ens-news.com/ens/apr2003/2003-04-10-09.asp#anchor8.

Kintisch, Eli. "Rules for Ocean Fertilization Could Repel Companies." *Science* 322 (November 7, 2008):835.

Revkin, Andrew C. "Antarctic Test Raises Hope On a Global-Warming Gas." *New York Times*, October 12, 2000c, A-18.

Schiermeier, Quirin. "The Oresmen." *Nature* 421 (January 9, 2003a):109–110.

Tollefson, Jeff. "UN Decision Puts Brakes on Ocean Fertilization." *Nature* 453 (June 5, 2008):704.

Tsuda, Atsushi, Shigenobu Takeda, Hiroaki Saito, Jun Nishioka, Yukihiro Nojiri, Isao Kudo, Hiroshi Kiyosawa, Akihiro Shiomoto, Keiri Imai, Tsuneo Ono, Akifumi Shimamoto, Daisuke Tsumune, Takeshi Yoshimura, Tatsuo Aono, Akira Hinuma, Masatoshi Kinugasa, Koji Suzuki, Yoshiki Sohrin, Yoshifumi Noiri, Heihachiro Tani, Yuji Deguchi, Nobuo Tsurushima, Hiroshi Ogawa, Kimio Fukami, Kenshi Kuma, and Toshiro Saino. "A Mesoscale Iron Enrichment in the Western Subarctic Pacific Induces a Large Centric Diatom Bloom." *Science* 300 (May 9, 2003):958–961.

Watson, A. J., D. C. E. Bakker, A. J. Ridgwell, P. W. Boyd, and C. S. Law. "Effect of Iron Supply on Southern Ocean CO$_2$ Uptake and Implications for Glacial Atmospheric CO$_2$." *Nature* 407 (October 12, 2000):730–733.

Island Nations and Sea-Level Rise

On many small islands around the world, rising seas are everyday news, and global warming is on everyone's lips. Nicholas Kristof of the *New York Times* described how Teunaia Abeta, a resident of Kiribati Island, in the Pacific, watched in horror, as

> A high tide came rolling in from the turquoise lagoon and did not stop. There was no typhoon, no rain, no wind, just an eerie rising tide that lapped higher and higher, swallowing up Abeta's thatched-roof home and scores of others in this Pacific island nation. "This had never happened before," said Abeta, 73, who wore only his colorful *lava-lava*, a skirt-like garment, as he sat on the raised platform of his home fingering a home-rolled cigarette. "It was never like this when I was a boy." (Kristof 1997)

The Alliance of Small Island States (AOSIS), including the Philippines, Jamaica, the Marshall Islands, the Bahamas, Samoa, and others, has been one of few voices at climate talks favoring swift, worldwide reductions in greenhouse-gas emissions. Their self-interest is evident: with warming already "forced," but not yet fully worked into the world's temperature equilibrium, many small island states will lose substantial territory and economic base within the next few decades. For every other country on Earth, competing economic interests drive climate negotiations. For the small island nations, the driving force is survival.

David Losia, 21, cuts coconuts for the family pigs in the front of his family house flooded by the rising tides where cinder blocks are used as a makeshift walkway in Funafuti, Tuvalu, March 23, 2004. (AP/Wide World Photos)

"For us, it's a matter of death and life, whereas in terms of the citizens of the industrialized countries, [global warming] will affect their lifestyles basically, but not to the extent they will be disappearing," said Bikenibeu Paeniu, Tuvalu's prime minister (Webb 1998b).

An Urgent Issue in Kiribati and Other Islands

In no other place is global warming a more urgent issue, in a practical sense, than in equatorial Kiribati (pronounced "Kirabas"), a chain of coral atolls in the Pacific, near the Marshall Islands. Kiribati is a republic, a member of the United Nations, and home to about 79,000 people. Its land area, at any point, is no more than two meters above sea level. Additionally, the living corals that have raised the islands above sea level are threatened by warmer ocean waters. Ierimea Tabai, who became president of Kiribati after its independence from Britain in 1979, has said that "[i]f the greenhouse effect raises sea

levels by one meter, it will eventually do away with Kiribati. In fifty or sixty years, my country will not be here" (Webb 1998b).

Kiribati's 33 islands, comprising 277 square miles, are scattered across 5.2 million square kilometers (two million square miles) of ocean, including three groups of islands: 17 Gilbert Islands, eight Line Islands, and eight Phoenix Islands. Kiribati includes Kiritimati (formerly Christmas Island), the world's largest coral atoll (150 square miles). The island of Kiribati, located roughly halfway between California and Australia, more than doubles its surface area at low tide.

During 1997, the area was devastated by an El Niño episode, which brought heavy rainfall, a half-meter rise in sea level, and extensive flooding. About 40 percent of the atolls' coral were killed by overheated water, and nearly all of Kiritimati Island's roughly 14 million birds died or deserted the island.

Sea-level rise provoked by global warming could imperil more than 3,000 small isolated islands, grouped into 24 political entities in the Pacific Ocean, with a population of about five million people in 800 distinct cultures. Many coral atolls also contain permanent areas of fresh water (lagoons), which are vulnerable to salinization as sea levels rise.

The Ocean Is No Longer Their Friend

Central lagoons of the small islands are social centers and a source of food. According to Kristof,

> Children play in the water from infancy. Many adults fish or sail for a living. The Kiribati men in their loincloths go out in outrigger canoes each day to catch tuna, and the women wade out on the coral reef to dive for shellfish and net smaller fish for dinner. (Kristof 1997)

"People here think of the ocean as their source of livelihood, as their friend," said Ross Terubea, a Kiribati radio reporter. "It's hard to think that it would destroy us" (Kristof 1997).

By 1998, rising sea levels already were swallowing some small Pacific islands and contaminating drinking water on others. The rising sea has endangered sacred sites and drowned some small islands near Kiribati and Tuvalu, including the islet of Tebua Tarawa, once a landmark for Tuvalu fishermen. Kiribati already has moved some roads inland on its main island as the rising Pacific Ocean eats into its shores.

Rising sea levels are seeping into soils on some islands, as water tables rise. Soils in some areas are becoming too salty for growing most vegetables. In Tuvalu, according to a dispatch from Reuters, farmers "are beginning to grow their taro crops in tin containers filled with compost instead of traditional pits" (Webb 1998b). Soil contamination is also an issue in the Bahamas, where fear has been expressed that the limestone that underlies the soil on many islands will absorb saline ocean water like sponges.

Given expectations that global temperatures will rise in coming decades, the sea-level rise of the twentieth century is expected to be dwarfed by the rise during the twenty-first century. Scientists who attend to this issue typically estimate sea-level rise possibilities in wide ranges, because no one knows how much ice will melt given a certain rise in temperatures over the poles. A team led by J. J. Wells Hoffman, for example, in 1983 estimated that sea levels in the year 2100 would be between 58 and 368 centimeters higher

than they were in the 1980s. R. Thomas, in 1986, put the estimated range at 56 to 345 centimeters (Thomas 1986). The estimates vary by a factor of roughly seven, an indication of the uncertainty that plagues forecasts of how much the sea may rise under various assumptions about global warming.

Patrick Barkham of the *London Guardian* reported from the South Pacific that islands widely imaged as paradises are falling victim to rising seas:

> The dazzling white sand and dark green coconut palms of Tepuka Savilivili were much like those on dozens of other small islets within sight of Funafuti, the atoll capital of Tuvalu. But shortly after cyclones Gavin, Hina and Kelly had paid the tiny Pacific nation a visit, islanders looked across Funafuti's coral lagoon and noticed a gap on the horizon. Tepuka Savilivili had vanished. Fifty hectares of Tuvalu disappeared into the sea during the 1997 storms. The tiny country's precious 10 square miles of land were starting to disappear. (Barkham 2002, 24)

Nothing living remains on the few square meters of sand that remains of the island, "only several odd flip-flops and a rusty tin." Vasuaafua, another small island, by 2002 had been reduced by the rising ocean to nine coconut palms on a narrow ribbon of sand. Several years ago, the island had a sandy beach several hundred yards long (Barkham 2002, 24). During the highest tides of 2001, the island's meteorological office, near the airstrip on Funafuti, was swamped by seawater (Barkham 2002, 24). Before the mid-1980s, flooding tides usually arrived only in February. By 2002, however, floods became likely at any time between November and March.

Seas Rise as Atolls Sink

Many of these islands face two compounding problems: the sea is rising while the land itself is slowly sinking, as 65-million-year-old coral atolls reach the end of their life spans. The atolls were formed as formerly volcanic peaks sank below the ocean's surface, leaving rings of coral. For five years, the government of Tuvalu has noticed many such troubling changes on its nine inhabited islands and concluded that, as one of the smallest and lowest-lying countries in the world, it is destined to become among the first nations to be sunk by a combination of global warming-provoked sea-level rise and slow erosion. According to Barkham's report, the evidence before their eyes, including forecasts for a rise in sea level as much as 88 centimeters during the twenty-first century by international scientists, has convinced most of Tuvalu's 10,500 inhabitants that rising seas and more frequent violent storms are certain to make life unlivable on the islands, if not for them, then for their children (Barkham 2002, 24).

Residents of the islands have been seeking higher ground, often in other countries. The number of Tuvalu's residents living in New Zealand, for example, doubled from about 900 in 1996 to 2,000 in 2001, many of them fleeing the rising seas on their home islands. A sizable Tuvaluan community has grown up in West Auckland (Gregory 2003). The highest point on Tuvalu is only about three meters above sea level. "From the air," wrote Barkham, "Its islands are thin slashes of green against the aquamarine water. From a few miles out at sea, the nation's numerous tiny uninhabited islets look smaller than a container ship and soon slip below the horizon" (Barkham 2002, 24).

"As the vast expanse of the Pacific Ocean creeps up on to Tuvalu's doorstep, the evacuation and shutting down of a nation has begun," Barkham wrote. With the curtains closed against the tropical glare, the prime minister, Koloa Talake, works in a flimsy Portakabin at the lagoon's edge on Funafuti. Talake, "who sits at his desk wearing flip-flops and bears a passing resemblance to Nelson Mandela," likens his task to the captain of a ship: "The skipper of the boat is always the last man to leave a sinking ship or goes down with the ship. If that happens to Tuvalu, the prime minister will be the last person to leave the island" (Barkham 2002, 24).

Many Pacific island farmers report that their crops of swamp taro (*pulaka*), a staple food, are dying because of rising soil salinity (Barkham 2002, 24). Another staple food, breadfruit (*artocarpus altilis*), also is threatened by saltwater inundation. The breadfruit are harvested from large evergreen trees with smooth bark and large, thick leaves that reach a height of 20 meters (about 60 feet). The fruit, which is large and starchy, reaches edible maturity once or twice a year, depending on type of tree. Some varieties bear fruit all year. Breadfruit probably originated in the Moluccas, Philippines, and New Guinea (Field 2002). Diana Ragone, director of science at the National Tropical Botanical Garden in Hawaii, said that "[o]n some atolls now the people are seeing total die-back of the breadfruit trees, one hundred percent" (Field 2002). Shallow-rooted breadfruit trees are especially vulnerable to increasing numbers of tropical storms, Ragone said. Two cyclones, Val and Ofa, wiped out breadfruit trees in parts in Samoa during 2002.

Living within a few meters of ocean level, island residents share a special fear of typhoons that roil the seas and drive storm surges through their villages. Two decades ago, one or two serious tropical storms usually hit the islands each decade; during the 1990s, seven such storms ravaged them. Now, more frequent El Niños (possibly spurred by global warming) seem to be making storms more common.

FURTHER READING

Barkham, Patrick. "Going Down: Tuvalu, a Nation of Nine Islands—Specks in the South Pacific—Is in Danger of Vanishing, a Victim of Global Warming. As Their Homeland Is Battered by Ferocious Cyclones and Slowly Submerges under the Encroaching Sea, What Will Become of the Islanders?" *London Guardian*, February 16, 2002, 24.

Gregory, Angela. "Fear of Rising Seas Drives More Tuvaluans to New Zealand." *New Zealand Herald*, February 19, 2003, n.p. (LEXIS)

Kristof, Nicholas. "For Pacific Islanders, Global Warming Is No Idle Threat." *New York Times*, March 2, 1997. http://sierraactivist.org/library/990629/islanders.html.

Thomas, R. "Future Sea-level Rise and its Early Detection by Satellite Remote Sensing." In *Effects of Changes in Atmospheric Ozone and Global Climate*, Vol. 4. New York: United Nations Environment Programme/United States Environmental Protection Agency, 1986.

Webb, Jason. "Small Islands Say Global Warming Hurting Them Now." Reuters, 1998b. http://bonanza.lter.uaf.edu/~davev/nrm304/glbxnews.htm.

J

Jatropha: An Alternative to Corn Ethanol?

Jatropha, proposed first as a biofuel crop in India, was, by 2008, being raised on plantations in East Africa. Two Australian states have banned it as an invasive species. Jatropha, which is poisonous, could overgrow farmland or pastures, and damage food supplies in Africa and elsewhere.

The seeds of the Jatropha fruit will grow almost anywhere—deserts, trash dumps, rock piles. Until it began to be taken seriously as an energy crop in India, Jatropha was dismissed as a weed. Like many weeds, it needs little water or fertilizer, and no one eats it. The plant was imported to India by Portuguese traders, and had been used there by native peoples to light lamps.

India has been taking a close look at Jatropha as an energy crop because the nation has a great deal of land that is too dry or infertile to grow other crops. An energy industry centered on Jatropha has been eyed by BP (formerly British Petroleum), which has joined with the British firm D1 Oils in a $90 million joint partnership to develop biodiesel fuel from Jatropha. Australia-based Mission Biofuels invested $80 million in the same idea by 2007, and had 80,000 acres under cultivation in India, aiming at 250,000. India's state railway uses the fuel to power some of its locomotives, and has planted the bushes along its tracks. The estimated cost of producing a barrel of biodiesel from Jatropha is $43, half that of corn and a third the cost for rapeseed. This estimate is based on a small amount of experience, however, and may vary widely (Barta 2007c, A-1, A-12; Fairless 2007b, 652).

Jatropha may soon be blended into jet fuel. On December 30, 2008, Air New Zealand ran one of four engines of a Boeing 747-400 passenger jet on a 50 percent biofuel blend made from jatropha during a two-hour test flight out of Auckland. Air New Zealand chief executive Rob Fyfe called the flight a milestone for the airline and commercial aviation. "Today we stand at the earliest stages of sustainable fuel development and an important moment in aviation history," said Fyfe ("Air New Zealand" 2008). Air New Zealand plans to use biofuel for 10 percent of its fuel by 2013. On January 7, 2009, Continental Airlines conducted a test flight with a fuel blend that included algae and jatropha; on January 30, Japan Airlines flight-tested fuel based on camelina oilseed.

FURTHER READING

"Air New Zealand Jet Flies on Jatropha Biofuel." Environment News Service, December 30, 2008. http://www.ens-newswire.com/ens/dec2008/2008-12-30-02.asp.

Barta, Patrick. "Jatropha Plant Gains Steam in Global Race for Biofuels." *Wall Street Journal,* August 24, 2007c, A-1, A-12.

Fairless, Daemon. "Biofuel: The Little Shrub that Could—Maybe." *Nature* 449 (October 11, 2007b): 652–655.

Rosenthal, Elisabeth. "New Trend in Biofuels Has New Risks." *New York Times,* May 21, 2008c. http://www.nytimes.com/2008/05/21/science/earth/21biofuels.html.

Jellyfish Populations and Potency

Some sea species thrive on conditions that kill others. For example, consider jellyfish, which seem to have a biological affinity for some forms of human pollution as well as warmer habitats. Jellyfish populations have been increasing rapidly in many parts of the world. In some areas, the increase appears to be part of a natural cycle (jellyfish populations also are declining in a few other areas) (Pohl 2002, F-3). Jellyfish have been called "the cockroaches of the open waters" and are maritime survivors that thrive in damaged environments (Rosenthal 2008g). They breed more quickly as waters warm.

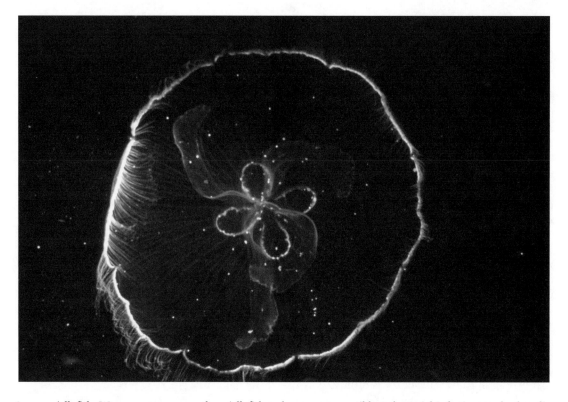

A moon jellyfish. Warmer waters cause these jellyfish to be more susceptible to bacterial infections and other diseases. (National Oceanic and Atmospheric Administration Photo Library)

By the summer of 2004, reports indicated that jellyfish populations were on the rise in Puget Sound, the Bering Strait, and the harbors of Tokyo and Boston. "Smacks" or swarms of jellyfish shut down fisheries in Narragansett Bay, parts of the Gulf of Alaska, and sections of the Black Sea. In the Philippines, 50 tons of jellyfish shut down a power plant, provoking blackouts, when they were sucked into its cooling system (Carpenter 2004, 68). During late July 2003, thousands of barrel jellyfish and moon jellyfish washed up on the coast of southern Wales.

Jellyfish Thrive Worldwide

Near Barcelona, Spain, fishermen found their nets slimed with jellyfish as more than 300 beachgoers per day were treated for stings during the summer of 2008. All over the world, jellyfish have been proliferating, a sign of overfishing of their predators (tuna, sharks, and jellyfish), increasing pollution that depletes oxygen levels, and gradually rising water temperatures. These problems are especially severe in the landlocked Mediterranean Sea.

During 2008, beaches have closed because of jellyfish swarms at Cote d'Azur in France, the Great Barrier Reef of Australia, and at Waikiki and Virginia Beach in the United States. In Australia, where the deadly Irukandji jellyfish is expanding its range in warming waters, the number of jellyfish stings rose to 30,000 treated cases in 2007, double 2005's total (Rosenthal 2008g).

"Human-caused stresses, including global warming and over-fishing, are encouraging jellyfish surpluses in many tourist destinations and productive fisheries," said the U.S. National Science Foundation, which issued a report on the explosion of jellyfish populations during the fall of 2008. The report lists Australia, the Gulf of Mexico, Hawaii, the Black Sea, Namibia, coastal Britain, the Mediterranean, the Sea of Japan, and the Yangtze estuary as problem areas (Rosenthal 2008g).

A jellyfish smack 16 kilometers square at a depth of 35 feet late in 2007 invaded the Northern Salmon Company, Northern Ireland's only salmon farm, killing more than 100,000 fish, at a cost of more than $2 million. Billions of *Pelagia nocticula* (Mauve Stingers) swarmed into the cages holding the fish about a mile into the Irish Sea, off Glenarm Bay and Cushendun. Managing director John Russell said he had never seen anything similar during three decades in the business. "The sea was red with these jellyfish and there was nothing we could do about, it, absolutely nothing," he said ("Jellyfish Attack" 2007). The jellyfish had not been sighted so far north until recent years (their usual range is in the Mediterranean Sea), so warming waters probably played a role. Observers likened the attack to a swarm of locusts on land.

Jellyfish as Indicator of Stressed Ecosystems

"Jellies are a pretty good group of animals to track coastal ecosystems," said Monty Graham, a scientist at the University of South Alabama. "When you start to see jellyfish numbers grow and grow, that usually indicates a stressed system" (Pohl 2002, F-3). Those stresses include increased water temperature, a rise in nutrients (from fertilizers and sewage), and depleted stocks of other fish, often caused by overfishing, which removes the jellyfish's competitors. All of these changes usually are caused by humans, according to Graham.

Otto Pohl of the *New York Times* described increasing numbers and toxic potency of jellyfish in some areas of Australia that are popular with swimmers:

> When Robert King climbed back on the boat after snorkeling off the Great Barrier Reef ... on March 31 [2002], he knew something was wrong. "I don't feel so good," he said, rubbing his chest. He had been stung by a jellyfish, and his condition deteriorated rapidly. By the time [an] emergency helicopter arrived, he was screaming in agony; a few hours later he was in a coma, eyes frozen wide, bleeding into his brain. He never regained consciousness. King, 44, from Columbus, Ohio, was the second person in Australia to die this year from the sting of a species of jellyfish, *Carukia barnesi*, found only in Australia and never before known to be fatal. More than 200 other victims went to hospitals, several times the number in a normal summer season here.... Jellyfish release millions of microscopic harpoons when touched, shooting tiny hypodermic needles into a victim's skin. They are lined with barbs and filled with venom, and they often linger painfully in the skin for months after the toxin has worn off. (Pohl 2002, F-3)

In Australia, regarding jellyfish stings, "This year [2002] [was] incredibly abnormal," said Jamie Seymour, a jellyfish expert at James Cook University (Pohl 2002, F-3). Seymour believes that strong, unusual wind patterns helped to blow the jellyfish toward the shore, where they flourished in unseasonably warm waters. Seymour, who has analyzed the venom from each sting that receives hospital treatment in the Barrier Reef region for years, had never seen the type of venom that killed the two tourists in 2002. In the Gulf of Mexico, according to a report in the *New York Times*, shrimp fishermen are struggling with a rising numbers of jellyfish that fill their nets with slimy gelatin, ruining their catch (Pohl 2002, F-3).

Smackdown: Hawaii to New England

On Waikiki Beach in Hawaii a lifeguard, Landy Blair, counted jellyfish stings on more than 900 people during a single day in 2002, about 1 percent of which sent victims to hospitals. Blair has been keeping track of jellyfish populations near the beach since 1991. The problem has grown steadily worse, he said, adding, "[w]e have see the highest numbers ever over the past year" (Pohl 2002, F-3). On the beaches near Auckland, New Zealand, half a dozen sting victims have required hospitalization, said Robert Ferguson of Surf Life Saving New Zealand, a lifeguards' organization. "It's the first time I've ever heard of victims needing hospital care," Ferguson said. "This is a new type of jellyfish with stings that are much more severe, much harsher" (Pohl 2002, F-3).

At about the same time, a report appeared in the *Boston Globe* describing a massive infestation of jellyfish in Narragansett Bay and Long Island Sound. A group of fishermen who expected "an array of marine life in their nets ... got jellyfish, nothing but jellyfish; jellyfish so plentiful that the gelatinous organisms came up dangling through the net like slimy icicles. And with each haul came more" (Arnold 2002, C-1).

"Eventually it seemed that our deck was coated with Vaseline," said Captain Eric Pfirrmann, who works for Save the Bay, a group whose members engage in environmental issues related to Rhode Island's Narragansett Bay. He piloted a research vessel that had taken several high-school teachers on a marine field trip. "I've seen blooms like this before," Pfirrmann said, "but never so early in the summer." The culprit is a nonstinging invertebrate about the size and shape of a tulip blossom and commonly known as the combjelly. These jellyfish, along with sea squirts (an entirely different organism) were taking over Long Island Sound, thriving in large part because water temperatures have risen about 3°F over the past two decades, according to scientists (Arnold 2002, C-1).

The rapid increase in jellyfish has nearly wiped out the winter flounder population in Narragansett Bay at the same time that the non-native sea squirt had overrun local oysters and blue mussels. Both are favored by warmer water temperatures and human pollution. "There is evidence of jellyfish explosions around the world that appear related to the adverse impact of human activities, and those include global warming," said Sarah Chasis, the senior attorney for the New York City-based Natural Resources Defense Council (Arnold 2002, C-1).

The *Boston Globe* reported that

> Historically, Narragansett Bay was the northern limit of the combjelly, whose domain extends as far south as Argentina. During November of 2000 combjellies were documented for the first time in Boston Harbor, although in numbers too sparse at that time to affect the harbor's ecology. In Narragansett Bay, according to Barbara Sullivan, an oceanographer from the University of Rhode Island who has been studying the jellyfish infestation, warmer water is changing the rules of who eats whom. (Arnold 2002, C-1)

The combjelly's reproductive cycle adjusts to the warmth of the water in which it lives. Populations usually explode during the warmth of late summer and early autumn. Narragansett Bay has warmed an average of 3.4°F during the past 20 years, while between the late 1970s to 2001, the average temperature of Long Island Sound during for the first three months of each year has increased about 8 percent, from 37.4°F to 40.2°F (Arnold 2002, C-1).

As waters warmed, the combjelly has reproduced and "bloomed" four months earlier, which enables it to gobble up the eggs and larvae deposited in the spring by spawning fish to the point of nearly replacing them (Arnold 2002, C-1). "We have seen areas of the bay where these things have cleaned out everything edible floating in the water column," Chasis said.

The winter flounder population has nearly vanished. In 1982, approximately 4,200 metric tons of flounder were landed in Rhode Island.

Within 20 years, the harvest fell to approximately 600 tons (Arnold 2002, C-1). "Combjellies do indeed eat many winter flounder larvae," said Tim Lynch, a marine biologist with the state Division of Fish and Wildlife, "but I do not believe they are the sole cause of the population decline in Narragansett Bay." Factors such as overfishing and water quality probably also play a role, Lynch said (Arnold 2002, C-1).

By 2008, the National Science Foundation reported that swarms of stinging jellyfish and jellyfish-like animals were turning many fisheries and vacation spots into "jellytoriums." Jellyfish populations were exploding not only because of warming waters but also because of non-native species, overfishing, and the presence of artificial structures, such as oil and gas rigs. In addition to damaging fisheries, jellyfish swarms were inflicting an increasing toll on fish farms, ocean mining, and desalination plants, as well as nuclear power plants, by clogging intake pipes. In the Gulf of Mexico, some swarms had reached 100 jellyfish in a cubic meter of water ("Jellyfish Swarms" 2008).

Monty Graham of Alabama's Dauphin Island Sea Lab, on a barrier island in the Gulf of Mexico, says the abnormally large, dense, or frequent jellyfish swarms are "a symptom of an ecosystem that has been tipped off balance by environmental stresses" ("Jellyfish Swarms" 2008). Jellyfish often swarm in dead zones, where they face few predators or competitors. Rising water temperatures accelerate their growth and reproduction. An NSF report, "Jellyfish Gone Wild: Environmental Change and Jellyfish Swarms," says that marine turtles, which eat jellyfish, have declined in many areas. One jellyfish-like creature, Australia's *Chironex fleckeri*, the world's most venomous animal, can kill a person in less than three minutes.

FURTHER READING

Arnold, David. "Global Warming Lends Power to a Jellyfish in Narragansett Bay and Long Island Sound: Non-native Species Are Taking Over." *Boston Globe*, July 2, 2002, C-1.

Carpenter, Betsy. "Feeling the Sting: Warming Oceans, Depleted Fish Stocks, Dirty Water—They Set the Stage for a Jellyfish Invasion." *U.S. News and World Report*, August 16, 2004, 68–69.

"Jellyfish Attack Destroys Salmon." British Broadcasting Corporation, November 21, 2007. http://news.bbc.co.uk/2/hi/uk_news/northern_ireland/7106631.stm.

"Jellyfish Swarms Invade Ecosystems Out of Balance." *Environment News Service*, December 16, 2008. http://www.ens-newswire.com/ens/dec2008/2008-12-16-01.asp.

Pohl, Otto. "New Jellyfish Problem Means Jellyfish Are Not the Only Problem." *New York Times*, May 21, 2002, F-3.

Rosenthal, Elisabeth. "Stinging Tentacles Offer Hint of Oceans' Decline." *New York Times*, August 3, 2008g. http://www.nytimes.com/2008/08/03/science/earth/03jellyfish.html.

Jet Contrails, Role in Climate Change

A study published in the *Journal of Climate* estimated that increasing coverage of cirrus clouds over the United States (to which air traffic is a major contribution) could raise tropospheric temperatures 0.2° to 0.3°C per decade (Minnis et al. 2004, 1671). The study, by researchers at a NASA facility in Langley, Virginia, concluded that clouds from aircraft exhaust, or contrails, contributed to a 0.27°C per decade warming trend in the United States between 1975 and 1994.

The NASA study was the first time weather observations had been used to document temperature change relating to contrails, said Patrick Minnis, a senior research scientist at Langley. "Cirrus clouds can have a net warming or net cooling effect on the Earth, depending on how thick they are," Minnis said. Contrails may stretch 1,600 kilometers and widen to 60 kilometers, depending on the weather. "Cirrus clouds from contrails tend to be thin, and the effect of thin clouds tends to be warming," he said (Schleck 2004, D-6).

Contrails from high-flying jets also may be helping to narrow the diurnal range of high and low temperatures, making days slightly cooler and nights slightly warmer, according to measurements taken during the three days after the September 11, 2001, terrorist attacks on New York City and Washington, D.C., when nearly all air travel in the United States was grounded, providing scientists with a rare view of nearly contrail-free skies for the first time in almost half a century ("Contrails Linked" 2002).

David Travis, a climatologist at the University of Wisconsin-Whitewater, led a study that offered some of the first evidence for the climate-changing effects of contrails. Travis and colleagues, including Pennsylvania State University geographer Andrew Carleton and Wisconsin-Whitewater undergraduate Ryan Lauritsen, used satellite images to compare cloud cover from those three days to 30 years of data for mid-September. Then they reviewed daytime and night-time surface-air temperatures across North America collected from 4,000 weather stations.

Results of this study indicated that jet contrails, like other forms of cloudiness, are compressing diurnal temperature ranges in some regions of North America. "Scientists have been noticing unusual changes in diurnal temperatures for quite some time, but can't explain why," said Travis. "We're providing one possible explanation here. Maybe jet contrail coverage is one of the reasons for this shrinking temperature range" ("Contrails Linked" 2002; Travis, Carleton, and Lauritsen 2002, 601).

Travis said the findings of this study may complicate the global warming debate because in some regions contrails may offset some of the temperature increases anticipated by global-warming models. The study also underscores the point that not all influences on climate are global. Factors such as contrails can make a difference on a regional or local scale. *See also:* Air Travel

FURTHER READING

"Contrails Linked to Temperature Changes." *Environment News Service*, August 8, 2002. http://ens-news.com/ens/aug2002/2002-08-08-09.asp#anchor4.

Minnis, Patrick, J. Kirk Ayres, Rabindra Palikonda, and Dung Phan. "Contrails, Cirrus Trends, and Climate." *Journal of Climate* (April 5, 2004):1671–1685.

Schleck, Dave. "High Fliers May be Creating Clouds: Global Warming May Be Worsened by Contrails from Aircraft." *Montreal Gazette*, July 18, 2004, D-6.

Travis, Davis J., Andrew M. Carleton, and Ryan G. Lauritsen. "Climatology: Contrails Reduce Daily Temperature Range." *Nature* 418 (August 8, 2002): 593–594.

Jet Streams

Jet streams, high-altitude winds that steer weather, have been shifting, and global warming may be a factor. Scientists at the Carnegie Institution made a case that between 1979 and 2001 the jet streams around the world have risen in altitude and moved toward the poles. The Northern Hemisphere's jet stream weakened, all of which conforms to climate models anticipating effects of a warming climate.

Cristina Archer and Ken Caldeira of the Carnegie Institution's Department of Global Ecology tracked changes in the average position and strength of jet streams using records compiled by the European Centre for Medium-Range Weather Forecasts, the National Centers for Environmental Protection, and the National Center for Atmospheric Research. Data included outputs from weather prediction models, conventional observations from weather balloons and surface instruments, and remote observations from satellites. The results were published in the April 18, 2008, edition of *Geophysical Research Letters*.

The poleward shift of jet streams thus far is small, about 19 kilometers (12 miles) per decade in the Northern Hemisphere, compared with their daily and seasonal changes, but cumulative changes could provoke changes in climate. Storm paths in North America may move northward with the migrating jet streams. Hurricanes' development can be inhibited by jet streams, and their high-altitude wind shear, so they could become more powerful and more numerous as jet streams move away from subtropical ocean regions.

"At this point we can't say for sure that this is the result of global warming, but I think it is," said Caldeira. "I would bet that the trend in the jet streams' positions will continue. It is something I'd put my money on" ("Changing Jet Streams" 2008).

A team of atmospheric scientists has associated the general mildness of winters in the Northern Hemisphere during recent decades with variations in the Northern Hemisphere Annular Mode (NAMS), part of the high-level jet stream. When NAMS is in its "positive" phase, winds strengthen over the eastern Atlantic, bringing soggy gales to Britain while also keeping continental Arctic air locked to the north and east most of the time. The "negative" phase of NAMS brings fewer storms to the British Isles, but allows more frequent intrusion of cold air.

A newspaper account (Henderson 2001a) stated that NAMS, rather than global warming, is responsible for a string of recent mild winters in England, but the scientists quoted in the piece did not make this distinction. For example, David Thompson, an atmospheric scientist at Colorado State University (one of the leaders of the cited research), was quoted as saying, "It is conceivable that the behavior of the Arctic Oscillation could be linked to the build-up of greenhouse gases in the atmosphere" (Henderson 2001a). Thompson and Wallace's own words read: "If this trend proves to be anthropomorphic, as suggested by recent climate modeling experiments, statistics like those presented here may prove useful in making projections of what winters will be like later in this century" (2001, 88).

FURTHER READING

Archer, C. L., and K. Caldeira. "Historical Trends in the Jet Streams." *Geophysical Research Letters* 35 (April 18, 2008), L08803, doi: 10.1029/2008 GL033614.

"Changing Jet Streams May Alter Paths of Storms and Hurricanes." NASA Earth Observatory, April 16, 2008. http://earthobservatory.nasa.gov/Newsroom/MediaAlerts/2008/2008041626579.html.

Henderson, Mark. "Positive Winds Keeping Arctic Winters at Bay." *London Times*, July 6, 2001a, n.p. (LEXIS)

Thompson, David W. J., and John M. Wallace. "Regional Climate Impacts of the Northern Hemisphere Annular Mode." *Science* 293 (July 6, 2001):85–89.

K

Keeling, Charles D. ("Keeling Curve")

Until the late 1950s, scientists had no reliable records of carbon dioxide and other greenhouse-gas levels in the atmosphere, nor even an instrument to accurately measure the amount of CO_2 in the air. At that time, Charles David Keeling began to measure CO_2 levels and assemble documentation indicating that the worldwide level of carbon dioxide had risen to about 315 parts per million (ppm), compared with about 280 ppm at the end of the previous century. Keeling's calculations also indicated that the level was continuing to rise. It reached 385 ppm by the year 2008. The historic graph of Earth's atmospheric carbon-dioxide level came to be called "The Keeling Curve."

When Keeling decided to measure the concentration of carbon dioxide in the atmosphere, he first had to construct a machine to obtain readings in parts per million. No such machine existed at the time. Keeling worked on his "manometer" for a year, building the machine from an old blueprint at the California Institute of Technology in Pasadena. Keeling's first readings, on the roof of a Caltech laboratory, showed an atmospheric concentration of 310 ppm.

Keeling next took his manometer on family vacations, recording readings. What he found

Dr. Charles D. Keeling. (Scripps Institution of Oceanography)

contradicted scientific assumptions of the time, which held that carbon-dioxide levels would vary widely, depending on local sources of the gas. Instead, he found that carbon-dioxide readings were very similar in different places.

"I decided all the data in the literature was wrong," Keeling later told William K. Stevens, a science reporter for the *New York Times* (Stevens 1999, 140). After Keeling and his associates recorded a number of readings for more than a year, Keeling discovered that the readings on his manometer were rising steadily, year by year.

Keeling also was discovering that atmospheric carbon-dioxide levels vary annually, with lower readings in the spring and summer (when plants are respiring oxygen on the large landmasses of the Northern Hemisphere), and higher levels during fall and winter, when many plants in the same regions are dormant and more carbon is being released because of vegetative decay. This annual cycle can vary by as much as 3 percent in the Northern Hemisphere, compared with 1 percent in the Southern Hemisphere, where the dominance of oceans mutes seasonal cycles and restricts seasonal variability.

Until Keeling showed that CO_2 levels were rising, most scientists who studied carbon-dioxide levels in the atmosphere had assumed that the oceans absorbed all of the extra carbon that human activities were injecting into the air. The readings of Keeling and his associates showed that human activity was steadily raising the proportion of carbon dioxide in the atmosphere more quickly than the oceans or other "sinks" of the gas could absorb it.

FURTHER READING

Keeling, Charles D., and Timothy P. Whorf. "The 1,800-year Oceanic Tidal Cycle: A Possible Cause of Rapid Climate Change." *Proceedings of the National Academy of Sciences of the United States of America* 97, no. 8 (April 11, 2000):3814–3819.

Stevens, William K. *The Change in the Weather: People, Weather, and the Science of Climate.* New York: Delacorte Press, 1999.

Kidney Stones

Scientists writing in the *Proceedings of the National Academy of Sciences,* July 15, 2008, proposed a connection between kidney stones and a warming environment. Tom H. Brikowski, Margaret S. Pearle, and Yair Lotan wrote that patients in the U.S. South have a 50 percent higher incidence of kidney stones than in the Northeast. The researchers expect that the fraction of the U.S. population living in high-risk stone zones is predicted to grow from 40 percent in 2000 to 56 percent by 2050, provoking as much as a 30 percent increase (Brikowski, Lotan, and Pearle 2008, 9841), resulting in 1.6 million new cases in the United States by 2050.

Incidence of kidney stones increases with dehydration, which aggravates the formation of stones from salts that crystallize in the kidneys. Because dehydration increases in warmer climates, the scientists asserted that similar conditions may also increase kidney stone incidence in large parts of Africa, the Middle East, and South Asia.

Pearle, of the University of Texas Southwestern Medical School in Dallas, said, "We see a relationship between kidney stones and temperatures everywhere. Even in places with air conditioning, warmer temperatures mean more stones" (Vergano 2008, A-1). By 2007, kidney stones were afflicting 12 percent of men and 7 percent of women in the United States.

FURTHER READING

Brikowski, Tom H., Yair Lotan, and Margaret S. Pearle. "Climate-related Increase in the Prevalence of Urolithiasis in the United States." *Proceedings of the National Academy of Sciences* 105, no. 28 (July 15, 2008):9841–9846.

"Global Warming Linked to Increase in Kidney Stones." Environment News Service, May 19, 2008. http://www.ens-newswire.com/ens/may2008/2008-05-19-091.asp.

Vergano, Dan. "Kidney Stone Cases Could Heat Up." *USA Today,* July 15, 2008, A-1.

Kilimanjaro, Snows of

Mount Kilimanjaro has lost 90 percent of its glacial mass during the last 90 years. A third of the ice cap that Ernest Hemingway celebrated melted during the 12 years ending in 2006 (Braasch 2007, 38). By 2002 Mount Kilimanjaro had lost 82 percent of its ice cap's volume since it was first carefully measured in 1912, according to Glaciologist Lonnie Thompson. Kilimanjaro's ice field shrank from 12 square kilometers in 1912 to only 2.6 square kilometers in 2000, reducing the height of the mountain by several meters. The ice covering the 19,330-foot peak "will be gone by about 2020," said Thompson (Arthur 2002, 7).

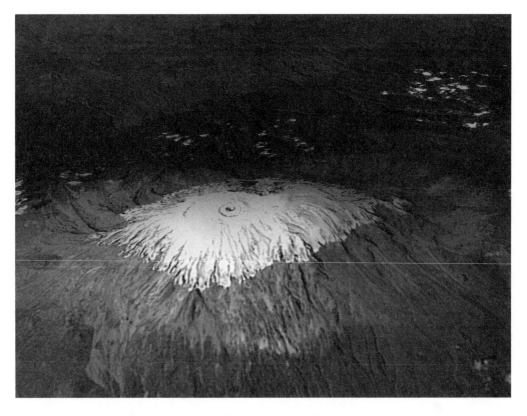

Aerial view of Mount Kilimanjaro, showing the ice cap, 1993. (NASA)

A Glacial and Literary History

Global warming may not be the only culprit in the demise of Kilimanjaro's ice cap; natural climate changes also have been blamed, along with deforestation on the mountain's slopes that sucks moisture out of rising winds that once coated the upper elevations of the mountain with snow. Euan Nisbet of Zimbabwe's Royal Holloway College has suggested, in all seriousness, that plastic tarps be draped across thee remaining ice fields to extend their life (Morton 2003).

The demise of Kilimanjaro's ice cap could imperil Tanzania's economy, which relies on tourism driven by the attraction of the mountain. In the Hemingway short story "The Snows of Kilimanjaro," a disillusioned writer, Harry Street, reflects on his life while lying injured in an African campsite. The short story was made into a film starring Gregory Peck in 1952. "Kilimanjaro is the number-one foreign currency earner for the government of Tanzania," said Thompson. "It has its own international airport and some 20,000 tourists every year" (Arthur 2002, 7).

The Kilimanjaro Ice Cap, which once was 50 meters deep, originated during an extremely wet period about 11,700 years ago, according to ice cores examined by Thompson and colleagues (Arthur 2002, 7). Thompson and colleagues' ice-core research indicates that Kilimanjaro has lost its ice cap before, during catastrophic droughts 8,300, 5,200, and 4,000 years ago. The most recent drought, which lasted 300 years, also threatened the rule of the Egyptian pharaohs. Professor Thompson said, "Writings on tombs 'talk' about sand dunes moving across the Nile and people migrating. Some have called this the Earth's first dark age" (Arthur 2002, 7). "Whatever happened to cause these dramatic climate changes could certainly occur again," Thompson said (Arthur 2002, 7).

Drought Is a Factor

Some scientists assert that the melting snows of Kilimanjaro are a result mainly of declining moisture rather than rising temperatures. A study published in the *International Journal of Climatology* found no evidence of recent temperature rise in the region. Researchers led by Georg Kaser, professor of tropical glaciology at the University of Innsbruck, Austria, and Douglas Hardy

Aerial view of Mount Kilimanjaro, showing a reduction in the ice cap, 2000. (NASA)

of Massachusetts University has pinpointed a sharp drop in atmospheric moisture in the 1880s and a subsequent prolonged dry period in the region as being behind the ice cap's decline (Kaser et al. 2004, 329).

When Kaser's findings were pumped up by climate contrarians as "proof" that global warming has not contributing to the demise of Kilimanjaro's snows, however, he repudiated their assumption. "We are entirely against the black-and-white picture that says it is either global warming or not global warming," Kaser told the *New York Times.* "We have a mere 2.5 years of actual field measurements from Kilimanjaro glaciers … so our understanding of their relationship with climate … is just beginning to develop," said Douglas R. Hardy, a co-author of the paper (Revkin 2004a).

Some other researchers also deny that the diminishing snows of Kilimanjaro are related to warming temperatures. Philip Mote of the University of Washington, for example, said that most of the ice loss on Kilimanjaro occurred before the 1950s, when warming temperatures were not a dominant factor. Reduced snowfall is

an important factor, according to Mote, as well as sublimation, which converts ice to water vapor at below-freezing temperatures without turning it to water in between (Mote and Kaser 2007).

Rapid melting of glaciers often blesses farmers downstream with a large amount of water in the short term. Farmers near Mount Kilimanjaro recently found the water supply so bountiful that they have been able grow a surplus of crops for sale. They have begun to raise water-hungry crops, such as tulips, for export to Europe. The water supply is a temporary blessing, of course. After the glaciers have melted, however, the surplus of water will be replaced by severe drought. "More water now means more agriculture," Graumlich said. "But what will they do when there is much less water later on?" (McFarling 2002b, A-4)

FURTHER READING

Arthur, Charles. "Snows of Kilimanjaro Will Disappear by 2020, Threatening World-wide Drought." *London Independent*, October 18, 2002, 7.

Braasch, Gary. *Earth Under Fire: How Global Warming Is Changing the World*. Berkeley: University of California Press, 2007.

Kaser, Georg, Douglas R. Hardy, Thomas M.A. Org, Raymond S. Bradley, and Tharsis M. Hyera. "Modern Glacier Retreat on Kilimanjaro as Evidence of Climate Change: Observations and Facts." *International Journal of Climatology* 24 (2004):329–339.

McFarling, Usha Lee. "Glacial Melting Takes Human Toll: Avalanche in Russia and Other Disasters Show That Global Warming Is Beginning to Affect Areas Much Closer to Home." *Los Angeles Times*, September 25, 2002b, A-4.

Morton, Oliver. "The Tarps of Kilimanjaro." *New York Times*, November 17, 2003. http://www.nytimes.com/2003/11/17/opinion/17MORT.html.

Mote, Philip, and Georg Kaser. "The Shrinking Glaciers of Kilimanjaro: Can Global Warming Be Blamed?" *American Scientist*, July-August 2007. http://www.americanscientist.org/template/AssetDetail/assetid/55553.

Revkin, Andrew C. "Climate Debate Gets Its Icon: Mt. Kilimanjaro." *New York Times*, March 23, 2004a.

Kyoto Protocol

By 2007, all major nations in the world had signed onto the Kyoto Protocol to reduce greenhouse-gas emissions except the United States; a total of 176 national governments. The United States did sign onto the original document, drafted in Kyoto, Japan, with representatives of 160 other countries, during December 1997, under President Bill Clinton (Al Gore was the country's chief negotiator), but President George W. Bush repudiated the pact after he took office in 2001.

The Kyoto Protocol places binding limits on each industrial country's combined emissions of the six principal categories of greenhouse gases: carbon dioxide, nitrous oxide, methane, sulfur hexafluoride, perfluorocarbons, and hydrofluorocarbons. These limits apply to the 38 so-called Annex I (industrial) countries. Under the Kyoto Protocol, each industrial country's baseline for calculating reductions is its 1990 emissions of carbon dioxide, methane, and nitrous oxides and its choice of 1990 or 1995 levels in the other three categories of gases. The United States agreed to a target of 7 percent below this baseline by the period between 2008 and 2012.

Diplomacy Has Had Little Practical Effect

The Kyoto Climate Convention and refinements negotiated at Buenos Aires during 1998 have not done much to curtail global greenhouse-gas emissions because some industrial countries (most notably the United States) will not accept cuts in their own emissions if developing nations are not forced to reduce their fossil-fuel consumption as well. This debate became sharper 10 years after the Kyoto Protocol was negotiated as China and India became major global drivers in carbon-dioxide increases with a boom powered mainly by coal-fired electricity, the dirtiest of fossil fuels.

Many developing countries reply that holding high per capita consumption and low-consumption countries to the same reductions amounts to energy colonialism, which locks all players in place over time. An across-the-board cut in allowable emissions would, for example, lock Bangladesh at roughly 1 percent of the per capita greenhouse effluent of the United States.

Anil Agarwal and Sunita Narain, in *Global Warming in an Unequal World: A Case of Environmental Colonialism* quoted the Worldwatch Institute on the nature of this problem:

> There remains the extraordinarily difficult question of whether carbon emissions should be limited in developing countries, and, if so, at what level. It is a simple fact of atmospheric science that the planet will never be able to support a population of 10 billion people emitting carbon at, say, the rate of Western Europe today. This would imply carbon emissions of four times the current level, or as high as 23 billion tons per year. (Agarwal and Narain 1991, 1)

One very large problem in climate diplomacy is the reluctance of people representing the diverse interests of many of the world's nation-states to place those interests aside in pursuit of a common goal of reducing greenhouse-gas emissions. National rather than global interests dominated climate diplomacy throughout the 1990s (despite the rhetoric of many national leaders, cited above), as atmospheric greenhouse gases continued to rise.

Author Ross Gelbspan called the results of climate diplomacy at the Kyoto conference "a yawning disjunction between what may be politically feasible and the natural requirements of the planet's inflamed atmosphere" (Gelbspan 1997a).

> All of this, in the end, leaves the resolution of perhaps the most ominous environmental problem ever confronting humanity to the mercy of the global marketplace.... Meanwhile, the world's glaciers are melting, the world's oceans are warming, plants and insects are migrating northward, the

Campaigners from the World Wildlife Fund on June 6, 2001, dressed up as polar bears demanding from European Union Environment Ministers an unequivocal commitment to finalize the Kyoto Protocol irrespective of the stance of the United States. (AP/Wide World Photos)

zooplankton are dying in the Pacific Ocean, the Antarctic ice shelves are breaking up, and the planet continues to heat at a faster rate than at any time in the last ten thousand years. (Gelbspan 1997a)

John Prescott, the U.K. deputy prime minister, in April 2000, challenged the United States to tackle the threat posed by global warming, warning that "no country, however big" can afford to ignore its consequences (Newman 2000, 17). While British emissions of greenhouse gases by the year 2007 had fallen between 12 and 14 percent compared with the Kyoto Protocol 1990 targets, Prescott noted that emissions in the

United States rose 11 percent between 1990 and 1998 (they continued to rise after that as well) Canada's greenhouse-gas emissions rose 13 percent during the 1990s, while several European countries (including Britain) have made progress toward meeting the goals of the Kyoto Protocol.

A Band-Aid or a Dead Letter?

Global greenhouse-gas emissions are rising and evidence of a warming planet are developing much more quickly than world diplomacy has been able to address them. The snail's pace nature of consultative diplomacy combines with the fact that we feel the results of fossil-fuel effluvia perhaps 50 years after the fact (through thermal inertia) to create a trap in which human responses to global warming take place several decades after nature requires them. Given these circumstances, the Kyoto Protocol may be a climatic dead letter.

Ten years after its negotiation in 1997, however, the only sizable countries that have came close to meeting Kyoto Protocol target emission reductions have been Great Britain and Germany. Most other signatories have not met their goals, and most developing countries (India and China among them) are not bound by its provisions. By 2007, India and China were responsible for 70 percent of the world's increase in energy demand, and both were supplying it mainly with coal, the fossil fuel that is highest in greenhouse-gas emissions.

Thus, the Kyoto Protocol has become more of a political rallying cry than a serious challenge to global warming. Even if the protocol is fully implemented, a projected temperature rise of 2°C by 2050 would be shaved only by 0.07°C, according to calculations by atmospheric scientist Thomas M. L. Wigley. In other words, the Kyoto goals are only a small fraction of the reduction in emissions required if worldwide temperature levels are to be stabilized during the twenty-first century and afterward (Wolf 2000, 27).

As governments around the world dickered over climate diplomacy (and the United States, which produces about one-fifth of the world's greenhouse gases, ignored the Kyoto Protocol) global emissions of carbon dioxide from fossil-fuel combustion increased by about 25 percent above 1990 levels by the year 2007. According to statistics compiled by the International Energy Agency, emissions mainly increased because of large pollution increases in developing nations and substantial growth in the United States and other western industrial nations.

The increase would have been higher, except for the collapse of former state socialist economies in Russia and Eastern European nations during the period (Holly 2002).

Emissions rose in all major economic sectors, including energy, transport, industry, and agriculture. One exception was waste management, where emissions declined slightly. The figures did not include emissions and removals from land-use change and forestry ("Rich Countries" 2003).

After the Kyoto Protocol

By 2007, constituent parties to the Kyoto Protocol were meeting to plan the next diplomatic framework to govern world climate-change policy after it expires in 2012. One such meeting, in Vienna, Austria, during August 2007, drew 1,000 participants from more than 150 governments, as well as business and industry, environmental organizations, and research institutions. United Nations Secretary-General Ban Ki-moon has said that countries must agree on a plan to follow Kyoto in 2009 to allow three years for transition.

The Kyoto Protocol is merely a first step in part because it does not include such major emissions sources as the United States, China, India, Brazil, Mexico, and South Africa. Its goals also are modest compared with the 80 percent reduction in greenhouse gases by 2050 that many scientists believe is necessary to avert threatening changes in the environment.

Delegates to the meeting in Vienna discussed measures required to limit global temperature rise to 2°C, above which effects on biosystems could be catastrophic. Delegates also examined a financial report indicating that such measures by 2030 would require investments of up to US$210 billion a year. Josef Pröll, Austria's federal minister for agriculture, forestry, environment, and water management, host of the five-day meeting, said that "[e]ach year without mitigation measures is a year which drives the human and financial cost of adaptation steeply upwards" ("Building Consensus" 2007).

The Vienna meetings broke down amid disagreement along the usual lines, with various countries protecting their own interests. Some industrial countries, including the United States, refused to adopt measures to support a reduction of greenhouse-gas emissions by 25 to

40 percent below 1990 levels by 2020. The various parties agreed that this target would act as a starting point for future discussion. The Group of 77 (developing nations) said that industrial countries should target an 80 percent reduction in greenhouse gases by 2020. German Chancellor Angela Merkel favored per capita cuts to bring industrial countries' emissions in line with those of developing countries (Anderson 2007c, A-20).

FURTHER READING

Agarwal, Anil, and Sunita Narain. *Global Warming in an Unequal World: A Case of Environmental Colonialism.* New Delhi, India: Centre for Science and Environment, 1991.

Anderson, John Ward. "U.N. Climate Talks End in Cloud of Discord Industrialized, Developing Nations Still at Odds Over How and When to Cut Emissions." *Washington Post,* September 1, 2007c, A-20. http://www.washingtonpost.com/wp-dyn/content/article/2007/08/31/AR2007083102052_pf.html.

"Building Consensus to Keep the Earth Cool after 2012." Environment News Service, August 28, 2007. http://www.ens-newswire.com/ens/aug2007/2007-08-28-01.asp.

Gelbspan, Ross. "A Global Warming." *The American Prospect* 31 (March/April 1997a). http://www.prospect.org/archives/31/31gelbfs.html.

Holly, Chris. "World CO_2 Emissions Up 13 Percent from 1990-2000." *Energy Daily* 30, no, 206 (October 25, 2002):n.p. (LEXIS)

Newman, Cathy. "Prescott Warns U.S. over Climate." *The Financial Times* (London), April 27, 2000, 7.

"Rich Countries' Greenhouse Gas Emissions Ballooning." Environment News Service, June 9, 2003. http://ens-news.com/ens/jun2003/2003-06-09-02.asp.

Wolf, Martin. "Hot Air about Global Warming." London *Financial Times,* November 29, 2000, 27.

$\mathcal{L}$

Land-Use Changes and the Amazon's Carbon Flux

University of Missouri scientist Deborah Clark, who works at the La Selva Biological Station in Costa Rica, told the Association for Tropical Biology annual meeting in Panama City, Panama, during the first week of August 2002 that tropical forests may soon become carbon-dioxide generators rather than carbon sinks. She said that data from La Selva "show a strong negative correlation between tree growth and higher temperatures. Temperatures experienced by canopy leaves may be close to the point at which respiration exceeds photosynthesis so that net production of carbon dioxide results," she said ("Global Warming Is Changing" 2002). According to a report by the Environment News Service, Clark continued, "Positive feedback between higher temperatures and CO_2 production by tropical forests could be catastrophic by resulting in accelerated increase in global CO_2 levels" ("Global Warming Is Changing" 2002).

The devastation of tropical forests is not being caused by climate change alone. Deforestation also plays a major role. James (Bud) Alcock, professor of environmental sciences at Pennsylvania State University, has developed a mathematical model to study the effects of human-driven deforestation that forecasts the demise of the Amazon rainforest within 40 to 50 years, given present-day deforestation rates of about 1 percent per year. Alcock said his model showed that without immediate and aggressive action to change current agricultural, mining, and logging practices, the 2 million square mile Amazon rainforest could pass "the point of no return" in 10 to 15 years ("Deforestation Could Push" 2001). The model indicates that the rainforest could essentially disappear within 40 to 50 years, while other estimates give the area 75 to 100 years, given "business as usual."

"Because of the way tropical rainforests work, they are dependent on trees to return water to the air," said Alcock, noting that the size of the Amazon River Basin's forests already has been reduced by about 25 percent. "This interdependence of climate and forest means risks to the forests are much closer at hand than what we might expect" ("Deforestation Could Push" 2001). A healthy forest holds a large proportion of precipitation and returns it to the atmosphere so it can be recycled in a process called "evapotranspiration." Without a healthy base of vegetation, water runoff occurs at a higher rate and creates the potential for destruction of rainforest ecosystems. Less rain also could mean more frequent forest fires, further threatening the balance of the rainforest, as well as the animal life it supports, Alcock said ("Deforestation Could Push" 2001).

The Amazon: Carbon Sink to Source

As early as the early 1990s, Brazilian government figures (publication of which was delayed almost a decade because of political pressures) indicated that human activity was turning the Amazon Valley from the "lungs of the world" into a net source of greenhouse-gas emissions. A report by Claudio Dantas, "Brazil: A Major Air Polluter," published by the Brazilian newspaper *Correio Braziliense,* indicated that the main factor in this change was burning initiated by settlers turning forests into farming areas ("Amazon Deforestation" 2004). Largely because of deforestation, Brazil ranks among the world's top 10 greenhouse-gas sources. Deforestation is so pervasive in Brazil that it represents 70 percent of that nation's greenhouse gases, dwarfing the contributions of industry and transportation ("Amazon Deforestation" 2004). The county's government is so permeated with corruption that in 2008 its own agencies were found to be the main culprits in continuing deforestation of the Amazon Valley.

Marcelo Furtado, campaign director at environmentalist organization Greenpeace, said: "No government had the ability and political will to combat deforestation in Brazil. Considering that data from 1994 already showed a worrisome situation, it is obvious that things are much worse today." According to Furtado, deforestation reportedly increased 40 percent during the past decade. "We are reaching the point of 25,000 to 30,000 square kilometers of forest destruction. Urgent action must be taken," Furtado warned ("Amazon Deforestation" 2004). According to the Brazilian government, deforestation increased by 28 percent from 2002 to 2003 alone.

Mark Maslin and Stephen J. Burns assembled a "moisture history" of the Amazon basin, which indicated that the area has been much drier in some past epochs than today:

> [T]he Amazon River outflow history for the past 14,000 years.... shows that the Amazon Basin was extremely dry during the Younger Dryas, with the discharge reduced by at least 40 percent compared with that of today. After the Younger Dryas, a melt-water-driven discharge event was followed by a steady increase in the Amazon Basin effective moisture throughout the Holocene. (Maslin and Burns 2000, 2285)

F. Siegert and colleagues, writing in *Nature*, surveyed damage to tropical rainforests worldwide by fires during the 1997–1998 El Niño (ENSO) episode. They found that selective logging "may lead to an increased susceptibility of forests to fire" (Siegert et al. 2001, 437). This scientific team tested this hypothesis on East Kalimantan (Borneo), part of Indonesia, which suffered the worst fires, and found that "[f]orest fires primarily affected recently logged forests; primary forests or those logged long ago were less affected" (Siegert et al. 2001, 437). Recently logged forests contain an abundance of flammable dead wood, causing fires to spread more rapidly.

Effects of Rising Carbon-Dioxide Levels on Amazonian Forests

In addition, a rising level of carbon dioxide is causing a dramatic change in the composition of tree species in the Amazonian forest, and some tree species are growing faster and dominating forests. In turn, other species are being forced into decline. The rate at which trees absorb

carbon dioxide and convert it into growth varies by species. These factors help determine which species will come to dominate in the forest, and how useful the forest may be as a "sink" that can soak up the pollution ("Carbon Pollution" 2004).

A U.S. and Brazilian team of biologists, led by William Laurance of the Smithsonian Tropical Research Institute in Panama and the National Institute for Amazonian Research in Manaus, Brazil, described the long-term effects of rising carbon-dioxide levels on virgin Amazonian forests. They marked out 18 one-hectare (2.5-acre) plots in central Amazonia, tagged nearly 13,700 trees with a trunk diameter of more than 10 centimeters (4.5 inches), and monitored the growth and each species' population for 20 years.

Of the 115 most abundant species, 27 showed spectacular changes in population density and "basal area" (the amount of land occupied by the trunk of that species), which is a reliable indicator of biomass. Thirteen species gained in population density and 14 declined; 14 species occupied a greater portion of the land, and 13 species retreated ("Carbon Pollution" 2004).

The scientists found that "[g]enera of faster-growing trees, including many canopy and emergent species, are increasing in dominance or density, whereas genera of slower-growing trees, including many sub-canopy species, are declining. Rising atmospheric CO_2 concentrations may explain these changes" (Laurance et al. 2004, 171). "These changes could have both local and global consequences. Undisturbed Amazonian forests appear to be functioning as an important carbon sink, helping to slow global warming, but pervasive changes in tree communities could modify this effect," they continued (Laurance et al. 2004, 174).

The big winners as carbon-dioxide levels rise have been spindly canopy-level trees and shrubs, such as the manbarklak, sclerobium, and parkia, which are fast-growing, with light-density wood. The losers have been slow-growing, dense tropical hardwoods, such as the croton and oenocarpus, which live in the dark forest interior. These slow-growing trees are by far the biggest absorbers of carbon. They are the species that make the Amazon Valley a vital carbon sink ("Carbon Pollution" 2004).

Human-initiated land-use changes have been accelerating the Amazon's transition away from its role as a major carbon sink. As jungle has been razed for cattle pasture, crops, logging, highways, and human settlements at an increasingly faster rate, what was once a large carbon sink is turning into a net source. Scientists at the National Institute for Amazon Research in Manaus have estimated that carbon emissions in Brazil may have risen by as much as 50 percent since 1990. According to their calculations, land-use changes by 2002 produced annual emissions of 400 million tons of greenhouse gases, much more than the 90 million tons generated annually by use of fossil fuels in Brazil (Rohter 2003).

FURTHER READING

"Amazon Deforestation Causing Global Warming, Brazilian Government Says." British Broadcasting Corporation International reports, December 10, 2004. (LEXIS)

"Carbon Pollution Wreaking Havoc with Amazonian Forest." Agence France Presse, March 10, 2004. (LEXIS)

"Deforestation Could Push Amazon Rainforest to Its End." Unisci: Daily University Science News, July 3, 2001. http://www.Unisci.com.

"Global Warming Is Changing Tropical Forests." Environment News Service, August 7, 2002. http://ens-news.com/ens/aug2002/2002-08-07-01.asp.

Laurance, William F., Alexandre A. Oliveira, Susan G. Laurance, Richard Condit, Henrique E. M. Nascimento, Ana G. Sanchez-Torin, Thomas E. Lovejoy, Ana Andrade, Sammya D'Angelo, Jose E. Ribeiro, and Christopher W. Dick. "Pervasive Alteration of Tree Communities in Undisturbed Amazonian Forests." *Nature* 428 (March 11, 2004):171–175.

Maslin, Mark, and Stephen J. Burns. "Reconstruction of the Amazon Basin Effective Moisture Availability over the Past 14,000 Years." *Science* 290 (December 22, 2000):2285–2287.

Rohter, Larry. "Deep in the Amazon Forest, Vast Questions about Global Climate Change." *New York Times*, November 4, 2003. http://nytimes.com/2003/11/04/science/earth/04AMAZ.html.

Siegert, F., G. Ruecker, A. Hinrichs, and A. A. Hoffmann. "Increased Damage from Fires in Logged Forests during Droughts Caused by El Niño." *Nature* 414 (November 22, 2001):437–440.

Land-Use Patterns May Aggravate Warming

The growth of cities and industrial-scale agriculture may be responsible for as much as half of recent rises in temperature across the United States, according to a study by scientists at the University of Maryland. Meteorologists Eugenia Kalnay and Ming Cai found evidence that a temperature increase of 0.13°C (0.234°F) over 50 years may be attributed to changes in land use ("Half U.S." 2003).

Kalnay and Cai's research utilized records from 1,982 surface stations located below 500 meters elevation in the 48 contiguous U.S. states during 50 years (1950–1999), as well as trends based on data from satellite and weather balloons. They reported:

The most important anthropogenic influences on climate are the emission of greenhouse gases and changes in land use, such as urbanization and agriculture. But it has been difficult to separate these two influences because both tend to increase the daily mean surface temperature.… Our results suggest that half of the observed decrease in diurnal temperature range is due to urban and other land-use changes. Moreover, our estimate of 0.27°C mean surface warming per century due to land-use changes is at least twice as high as previous estimates based on urbanization alone. (Kalnay and Cai 2003, 528–529)

Land-Use Changes: Rural and Urban

A comparison of urban and rural weather stations, without including agricultural effects, would underestimate the total impact of land-use changes, Kalnay and Cai asserted. The urban heat-island effect actually takes place mostly at night, the two scientists wrote, "when buildings and streets release the solar heating absorbed during the day" ("Half U.S." 2003). At the time of maximum temperature, the urban effect is one of slight cooling due to shading, aerosols, and thermal-inertia differences between city and country that presently are not well understood, they believe ("Half U.S." 2003).

Urban sprawl, deforestation, and agricultural practices can alter temperatures and rainfall patterns in ways that sometimes augment the effects of increased greenhouse gases. "Our work suggests that the impacts of human-caused land-cover changes on climate are at least as important, and quite possibly more important[,] than those of carbon dioxide," said Roger Pielke, an atmospheric scientist at Colorado State University. "Through land-cover changes over the last 300 years, we may have already altered the climate more than would occur … [from] the radiative effect of a doubling of carbon dioxide," added Pielke, lead author of a study published in the August 2002 issue of *Philosophical Transactions: Mathematical, Physical and Engineering*

Sciences, a journal published by The Royal Society of London (Lazaroff 2002c; Pielke 2002).

Pielke and his colleagues asserted that if carbon-dioxide emissions continue to rise at recent rates, the level of carbon dioxide in the atmosphere will double from preindustrial levels by 2050. At the same time, land-surface uses will continue to change. According to a report by the Environment News Service,

> Different land surfaces influence how the sun's energy is distributed back to the atmosphere. For example, if a rainforest is removed and replaced with crops, there is less transpiration, or evaporation of water from leaves. Less transpiration leads to warmer temperatures in that area. (Lazaroff 2002c)

Land Use and Temporary Cooling

Land-use changes can produce temporary regional cooling as well as warming. If farmland is irrigated, for example, more water transpires and evaporates from moist soils, which cools and moistens the atmosphere, changing local precipitation and cloudiness. Forests may influence the climate in more complicated ways than previously thought, the authors contend. According to the Environment News Service report,

> In regions with heavy snowfall, reforestation or the growth of new forests would cause the land to reflect less sunlight, meaning that more heat would be absorbed. This could result in a net warming effect, even though the new trees would remove carbon dioxide from the atmosphere through photosynthesis during the growing season. (Lazaroff 2002c)

Reforestation also could increase transpiration in an area, putting more water vapor into the air. Water vapor in the troposphere, the lowest and densest part of the Earth's atmosphere, is the biggest contributor to greenhouse-gas warming, the researchers said.

Australian researchers also believe that they have found strong evidence that clearing of land may trigger major climatic changes. A Macquarie University team said that its findings indicate that the climate can respond suddenly and dramatically to centuries of environmental abuse. The researchers used one of Australia's most powerful supercomputers to model changing rainfall patterns after the mid-1970s in the southwestern corner of Western Australia.

Some parts of the region have suffered declines of up to 15 and 20 percent in winter rainfall, threatening Perth's water supply. Research by Macquarie University atmospheric science researchers Andy Pitman, Neil Holbrook, and Gemma Narisma (working with Pielke) suggested the clearing of land may be responsible for about half of the rain shortfall (Macey 2004, 2). Pitman said the results of the modeling closely reflected the rainfall changes observed since the mid-1970s, suggesting that land clearing may exert 5 to 10 times the influence previously expected. "We didn't expect to find that," he said. "It scared the hell out of us" (Macey 2004, 2).

Forests, now cleared for farming, once slowed moist winds blowing in from the Indian Ocean, he said.

> This slowing of the atmosphere causes turbulence, which in turn generates rainfall. Without the tree cover, the water in the atmosphere flows across the landscape and is deposited elsewhere. Our results suggest, and observations indicate, that it's falling farther inland, outside of the catchments that provide the Perth water supply.... It may be another argument why we shouldn't be logging old-growth forests. (Macey 2004, 2)

These findings, to be published in the *Journal of Geophysical Research*, also showed that climatic changes could appear decades or centuries after humans began interfering with the environment. "It may be that when the effects of deforestation suddenly exceed a threshold the climate is likely to respond in a dramatic way" (Macey 2004, 2).

FURTHER READING

"Half U.S. Climate Warming Due to Land Use Changes." Environment News Service, May 28, 2003. http://ens-news.com/ens/may2003/2003-05-28-01.asp.

Kalnay, Eugenia, and Ming Cai. "Impact of Urbanization and Land-use Change on Climate." *Nature* 423 (May 29, 2003):528–531.

Lazaroff, Cat. "Land Use Rivals Greenhouse Gases in Changing Climate." Environment News Service, October 2, 2002c. http://ens-news.com/ens/oct2002/2002-10-02-06.asp.

Macey, Richard. "Climate Change Link to Clearing." *Sydney Morning Herald*, June 29, 2004, 2.

Pielke, Roger, 2002, "Land Use Changes and Climate Change." *Philosophical Transactions: Mathematical, Physical and Engineering Sciences* (Journal of The Royal Society of London), August 2002.

Legal Liability and Global Warming

The debate over global warming now turns more frequently to legal liability: Who or what groups are responsible, and how should laws be changed to make them accountable? In 2008, *Law and Policy*, a legal journal, published a special issue on global warming, covering topics related to redress through the civil law and measures for redress for vulnerable island nations.

The phrase "law of the air" was first used by Margaret Mead during the early 1970s at a conference titled "The Atmosphere: Endangered and Endangering," in a Fogarty Conference report co-edited with William Kellogg (Mead and Kellogg 1980). Stephen Schneider picked up the idea in his works during the 1980s. "Of course, we now have a 'law of the air' known as the Kyoto Protocol," commented Schneider (Schneider 2000c). Thus, we now have a *design* for a Law of the Air. We will have a *real* Law of the Air when countries and companies can be taken to court, convicted, and penalized for illegal emission of greenhouse gases. Such an idea started to emerge in legal actions after the year 2000.

Specific Lawsuits Emerge

The Inuit Circumpolar Conference (ICC), which represents 155,000 people inside the Arctic Circle, assembled a human-rights case against the United States (specifically the George W. Bush administration) in 2005 because global warming is threatening their way of life.

The ICC's 167-page petition to a subagency of the Organization of American States (OAS) alleged violations of fundamental human rights among Arctic peoples who ring the Arctic Ocean from Nunavut (a semi-sovereign Inuit province of Canada), to Greenland, the Russian Federation, and Alaska. The petition, which was compiled in defense of Inuit rights as a people within the context of international human-rights law, seeks a declaration from the commission that emissions of greenhouse gases from the United States—the source of more than 25 percent of the world's cumulative greenhouse gases during the last century—are violating Inuit human rights as outlined in the 1948 American Declaration on the Rights and Duties of Man.

Late in September 2007, a federal judge threw out of federal court in San Francisco a suit filed by the California's government against the world's six largest automakers. The judge, Martin Jenkins, ruled that it was impossible, legally, to separate automobiles' responsibility in regard to emissions of carbon dioxide from that of natural sources and other human activities. The suit, filed by California attorney general Jerry Brown, sought a determination of damages caused by floods and other natural disasters. The court held that the extent to which these disasters are caused by global warming also was impossible to determine. Therefore, liability could not be established. Although the suit failed, it asked questions that had never before been placed before a court.

At about the same time, Ceres, an alliance of environmentally active investors, joined Environmental Defense, several investment managers who supervise $1.5 trillion in assets, and the financial officers of 10 states and New York City to ask the Securities and Exchange Commission to require companies to disclose the risks of climate change to their financial situations.

An Inupiat Village Sues

Early in 2008, lawyers for the Inupiat village of Kivalina, and its 400 Alaskan native residents, north of Nome, filed suit in Federal District Court, San Francisco, against five oil companies, 14 electric utilities, and the largest U.S. coal-mining company, alleging that they were partially responsible for the climatic circumstances that are forcing the villagers to leave their ancestral location on a barrier reef between the Chukchi Sea and two rivers.

Among the companies named in the suit were Exxon Mobil, American Electric Power, BP America, Chevron, Peabody Energy, Duke Energy, Conoco Phillips, and Southern Company. The suit accused Exxon, Conoco, and AEP of conspiracy, "a long campaign by power, coal, and oil companies to mislead the public about the science of global warming" (Barringer 2008b, A-16). The village is being threatened with inundation by seas whipped by storms that in colder times would have been sealed under ice much of the year. The seas are encroaching to a point at which sea walls and sandbags are of little help.

The Inuit filing occurred the same week that New York State's attorney general, Andrew M. Cuomo, initiated an inquiry into the affairs of five energy companies regarding whether they had adequately disclosed financial risks associated with coal-fired power plants that they own or manage. Mindy S. Lubber, president of Ceres, said in an interview, "We need this because right now more than half of the S&P [Standard & Poor's] 500 are not disclosing their climate risk, which we would consider a material risk in this day and age" (Barringer 2007c). *See also:* Arctic Warming and Native Peoples; Human Rights, Global Warming and the Arctic

FURTHER READING

Barringer, Felicity. "Effort to Get Companies to Disclose Climate Risk." *New York Times*, September 18, 2007c. http://www.nytimes.com/2007/09/18/business/18disclose.html.

Barringer, Felicity. "Flooded Village Files Suit, Citing Corporate Link to Climate Change." *New York Times*, February 27, 2008b, A-16.

Mead, Margaret, and William Kellogg, eds. *The Atmosphere: Endangered and Endangering.* Tunbridge Wells, England: Castle House, 1980.

Schneider, Steven H. Personal communication, March 18, 2000c.

Light-Emitting Diodes (LEDs)

For Christmas and other ceremonial lighting, light-emitting diodes (LEDs) have become more popular. While a 26-light string of old-fashioned incandescent light bulbs burns at 125 watts and lasts about 1,000 hours. The same size string of LEDs burns at 0.66 watts and lasts about 20,000 hours. Like compact fluorescents, LEDs are more expensive, about twice as much per string, but they make it up in bulb life and energy savings. The flashing globe that heralded in the New Year in 2008 in New York City's Times Square was, for the first time, composed of LEDs.

While incandescent lights work by channeling an electrical current through a fire filament (that eventually burns out), LEDs are small semiconductor chips (or diodes) that use a stream of electrons to put out light as they use energy.

Lobster Catches Decline in Warmer Water

During 2002, record high temperatures in New England were a factor in lighter-than-usual lobster catches off shore. Marine scientists attributed warmer-than-usual waters to seasonal aberrations, including a mild winter, a short spring, and a hot summer (Healy 2002, B-1). Bob Glenn, a biologist at the state Division of Marine Fisheries, said he took a temperature reading of 77°F in 20 feet of water off the coast

of Bourne, Massachusetts, during the third week of August. The average reading is near 70°F. "I'm not ready to jump on the global warming bandwagon, but this has been an exceptionally warm summer in many places off the Cape," Glenn said. "And this kind of water has a real impact on just about everything" (Healy 2002, B-1).

An account in the *Boston Globe* said that

> The movements of lobsters, blue-fin tuna, and cod have become difficult to predict this summer [2002] because they are moving out of warmer water, several fishermen said. Even on the ocean side of the Cape, which typically has colder waters than its bay, lobsters haven't been appearing in some of the usual streams this month, and some lobstermen's businesses have taken a beating as a result. "I've lost $8,000 in two weeks because my traps aren't seeing nearly as many lobsters," said Billy Souza, who has 700 traps from Provincetown to Wellfleet and sells his catch out of his garage here. (Healy 2002, B-1)

Leaning against his truck after delivering about 150 pounds of lobster to his wife and chief saleswoman, Cheryl, Souza declared the day his best in weeks. His haul fell as low as 25 or 30 pounds on some days, when he mostly netted "eggers"—lobsters bearing eggs that must be thrown back—and found himself baffled by the lobsters' movements. "I've been lobstering for 20 years, and from my own system these lobsters like it in 42-degree water," Souza said. "More and more of my traps are coming up totally empty" (Healy 2002, B-1).

Massachusetts State biologists associated the warm water temperatures during the summer of 2002 with temperatures in Cape Cod Bay and Buzzards Bay the previous December, which were the warmest in 12 years, according to the state Division of Marine Fisheries. In one reading in 65 feet of water off Plymouth, the mean temperature was 46°F in December, compared with the usual mean of 42°F degrees. In 75 feet of water off Buzzards Bay, December's monthly mean was 50°F, compared with the normal reading of 45°F (Healy 2002, B-1). Temperatures in New England returned to average or below during the winters of 2002–2003 and 2003–2004, as pervasive cold and snow returned.

FURTHER READING

Healy, Patrick. "Warming Waters: Lobstermen on Cape Cod Blame Light Hauls on Higher Ocean Temperatures." *Boston Globe*, August 30, 2002, B-1.

Lodging and Greenhouse-Gas Emissions

By 2007, many business guests were demanding environmental consciousness from their hotels as owners discovered that "green" decisions could save money. Light bulbs that use less energy and bathroom fixtures that limit water flow save money in the long run. "Environmental issues are one of the hottest issues within the travel industry right now," said Bill Connors, the executive director of the National Business Travel Association, which made eco-friendly elements in hotel design and operations a focus (for the first time) at its annual convention in July 2007 (White 2007).

Increasing numbers of hotels have been registering for certification under the U.S. Green Building Council's Leadership in Energy and Environmental Design (LEED) program. Compact fluorescent bulbs, policies that extend the use of towels between washings, green roofs, use of nontoxic cleaning agents, and recycling bins in rooms have become more popular. Marriott Hotels plans to reduce energy consumption in its hotels 20 percent between 2000 and 2010.

Key card systems are now being used in many hotels that control lights. Room lights turn on when a guest inserts a card next to the door, and turn off when it is removed. It is, therefore, impossible to leave lights on in an empty room. Such systems are becoming common in several countries. Along similar lines, some hotels now have infrared sensors that turn off air conditioning or heat when a guest leaves a room. The cost of such a system usually can be recouped in a year or two from energy savings.

The Gaia Napa Valley Hotel and Spa in California greets guests with a monitor above the front desk that measures how much water and power the hotel uses. In each of its 133 rooms, a copy of Al Gore's *An Inconvenient Truth* has been placed beside the usual Bible (Peterka 2007, 20).

The King Pacific Lodge, a luxury resort in the Great Bear Rainforest along the British Columbia coast (where a three-night stay costs about $5,000 per person), has turned down the temperature of its showers and fills its twin-engine boats (usually used for salmon fishing) to capacity and restricts their speed with engine governors, all to curtail greenhouse-gas emissions. The lodge is acting on a plan to cut its carbon footprint by half in five years (Ball 2007c, P-1). The lodge's carbon footprint has been

enormous—1.7 tons per guest per stay (3.5 days), or about as much as an average U.S. citizen generates in a month. The resort's biggest carbon producers are two 110-kilowatt diesel generators for electricity. Propane is being substituted in the kitchen.

The most carbon-intensive part of a vacation in a remote area is usually the air travel required to reach it. *The Wall Street Journal* carried an account of one couple who installed a solar hot water heater in their home and, feeling environmentally correct, took flights around the world. Soneva Fushi, a 65-villa resort on the Maldives, a chain of islands in the Indian Ocean (often mentioned as one of the first locations that will drown in rising seas generated by melting ice) announced plans to go carbon-neutral by 2010, partially by using coconut oil in its diesel generator. Devices also were installed to cut off air conditioning if a room's door was left open for more than a minute (Ball 2007c, P-5). The resort also added a fee to guests' bills to offset the massive amount of carbon dioxide required to get there by air from anywhere else.

Guests who drive hybrid vehicles to the Fairmont hotels in California and British Columbia receive free parking. The Habitat Suites in Austin, Texas, installed a solar hot-water heating system that cut natural gas use 60 percent. The Lenox Hotel of Boston composts 120 tons a year of restaurant waste (Bly 2007, D-2). At San Francisco's Kimpton Hotel Triton, guests can book on a designated "Eco Floor," with low-flow toilets and nontoxic cleaning supplies. Shampoo and conditioner are dispensed from wall-mounted canisters, instead of individual plastic bottles.

At Greenhouse 26, which opened during 2008 on Manhattan Island, the elevator captures energy generated when it stops, much as a hybrid car recycles energy released when the brakes operate. Heating and cooling are provided geothermally. Water from sinks and showers is recycled for use in toilets (White 2007).

FURTHER READING

Ball, Jeffrey. "The Carbon Neutral Vacation." *Wall Street Journal*, July 30, 2007c, P-1, P-4-5.

Bly, Laura. "How Green Is Your Valet and the Rest?" *USA Today*, July 12, 2007, D-1, D-2.

Peterka, Amanda E. "Sustainable Hospitality." *EJ* [Environmental Journalism], Michigan State University Knight Center for Environmental Journalism 7, no. 1(Fall 2007):20–21, 38.

White, Martha C. "Enjoy Your Green Stay." *New York Times*, June 26, 2007. http://www.nytimes.com/2007/06/26/business/26green.html.

Loggerhead Sea Turtles: Warmth Alters Gender Ratio

Climate change threatens loggerhead turtles nesting on Florida beaches, according to British researchers' work. Anticipated temperature increases could destroy male North American loggerhead populations, and thereby portend extinction for the species. Researchers analyzed 26 years of loggerhead turtle nesting and climate data and compared the findings with models for future temperatures. Warmer temperatures during incubation produce female loggerhead turtles and cooler conditions produce males.

Increasing incubation temperatures 1°C eliminates male turtles in some locales, the research team said. An increase of 3°C "would lead to extreme levels of infant mortality and declines in nesting beaches across the United States" ("Climate Change Threatens" 2007). Loggerheads already are listed as threatened under the U.S. Endangered Species Act. Loggerhead nest counts declined 22.3 percent between 1989 and 2005. In addition to the threat of rising temperatures, loggerheads are being killed before they breed, drowning in fishing trawlers' nets.

Published in the journal *Global Change Biology*, the research was conducted in partnership with the Bald Head Island Conservancy and the North Carolina Wildlife Resources. "We are stunned by these results and what they could mean for the species in the future," said Brendan Godley of the University of Exeter's School of Biosciences. "In particular, we are concerned that populations that are already predominantly female could become 100 percent female if temperatures increase by just one degree. This is a major issue for nesting populations further south, in Florida, for example, where males are already in short supply," Godley explained ("Climate Change Threatens" 2007).

In Florida, which presently hosts more than 90 percent of loggerhead nests in the United States, 90 percent of hatchlings are female. By

Loggerhead turtle. (Courtesy of Getty Images/Digital Vision)

contrast, in North Carolina, where nesting sites are generally cooler, 42 percent are male. Males from more northerly locations may be migrating southward to help balance the gender ratio there. The turtles, which also nest in Oman and South Africa, reach maturity in 20 to 30 years, and can weigh as much as 400 pounds. In South Africa nesting populations have been increasing, according to Richard Penn Sawers, head of the World Wildlife Fund/Green Trust Turtle Monitoring and Community Development Project. These turtles breed almost entirely within the Greater St. Lucia Wetland Park, which is a designated marine protected area and World Heritage site ("Climate Change Threatens" 2007).

Loggerhead nesting by 2007 reached the lowest level since Florida began keeping official records in the 1980s. The turtles were being endangered not only by rising sand temperatures that skew their sex ratio, but also by sea walls and coastal development, which invade their nesting areas, and by collisions with fishing boats and stranding in nets far at sea. On the Atlantic shore, nestings have declined 50 percent since 1998. Some homeowners along the coast also are installing sand-filled geotubes, 1,000-ton sandbags, to block the encroaching ocean. These bags also destroy nesting areas ("Loggerhead Turtle" 2007).

FURTHER READING

"Climate Change Threatens Loggerhead Turtles." Environment News Service, February 22, 2007. http://www.ens-newswire.com/ens/feb2007/2007-02-22-02.asp.

"Loggerhead Turtle Nesting Down by Half since 1998." Environment News Service, November 8, 2007. http://www.ens-newswire.com/ens/nov2007/2007-11-08-094.asp.

$\mathcal{M}$

Malaria in a Warmer World

In the tropics, elevation has long been used to shield human populations from diseases that are endemic in the lowlands. With global warming, mosquito-borne diseases have been reaching higher altitudes, affecting peoples with little or no immunity. Temperature must exceed 59°F for the malaria parasite to develop inside a host mosquito. According to Pim Martens, "A minor temperature rise will be sufficient to turn the populated African highlands into an area that is suitable for the malaria mosquito and parasite" (Martens 1999, 537).

During 1997, malaria ravaged large areas of Papua New Guinea at an elevation of 2,100 meters, notably higher than the 1,200 to 2,000 meters that heretofore had provided a barrier to the disease in different parts of Central and Southern Africa. In northwestern Pakistan, according to Martens, a rise of about 0.5°C in the mean temperature was a factor in increasing the incidence of malaria from a few hundred cases a year in the early 1980s to 25,000 in 1990 (Martens 1999, 537). While most strains of malaria can be controlled, drug-resistant strains were proliferating late in the twentieth century.

Writing in the March 1998 *Bulletin of the American Meteorological Society*, Paul Epstein and seven co-authors described the spread of malaria and dengue fever to higher altitudes in tropical areas of the Earth because of warmer temperatures. Rising winter temperatures also have allowed disease-bearing insects to survive in areas previously closed to them. According to Epstein, frequent flooding, which is associated with warmer temperatures, also promotes the growth of fungus and provides excellent breeding grounds for large numbers of mosquitoes. The flooding caused by Hurricane Floyd and other storms in North Carolina during 1999 are cited by some as a real-world example of global warming that is promoting conditions ideal for the spread of diseases imported from the tropics (Epstein et al. 1998).

Temperatures Rise, Malaria Spreads

According to the Intergovernmental Panel on Climate Change's (IPCC) projections for human health, a rise in global average temperatures of 3° to 5°C by 2100 could lead to 50 to 80 million additional cases of malaria per year worldwide, primarily in tropical, subtropical, and less well-protected temperate-zone populations. Britain's Hadley Climate Center expects the same disease to reach the Baltic states by 2050. In parts of the world where malaria is now unknown, most people have no immunity. The World Health Organization also projected that warmer weather will cause tens of millions additional cases of malaria and other infectious diseases. The Dutch health ministry anticipates that more than a million people may die annually as a result of the impact of global warming on malaria transmission in North America and Northern Europe (Epstein 1999, 7).

Malaria could return to Britain as an endemic disease, scientists at the University of Durham warned as they announced a plan to produce a "risk map" showing which areas were most likely to suffer an outbreak (Connor 2001c, 14). With millions of tourists visiting malaria-infested regions of the world, the risk of the disease making a comeback is further increased by global warming, which expands mosquito habitats in the United Kingdom, said Rob Hutchinson, an entomologist at the university, at the annual meeting of the Royal Entomological Society in Aberdeen. He said that of the 25 million overseas visitors who came to Britain in 1999, about 260,000 came from Turkey and the countries of the former Soviet Union, where vivax malaria was endemic and health care was poor (Connor 2001c, 14).

By the late 1990s, malaria had been transmitted by mosquitoes as far north as Toronto, Canada, according to Epstein. "The extreme events we are seeing today in Nicaragua and Honduras [as a result of Hurricane Mitch in 1998] are spawning outbreaks of cholera and dengue fever with new breeding sites for mosquitoes and increased water-born diseases," Epstein said (Epstein 1998). In northerly latitudes, night-time and winter temperatures have warmed twice as fast as overall global temperatures since 1950, Epstein said, meaning that fewer pests are being killed by frost in the southern reaches of the temperate zones. Humidity also has increased in many regions, including much of the eastern United States, helping mosquitoes to breed. Disease-carrying mosquitoes usually required a certain level of temperature *and* humidity to survive.

In May 1995, researchers in the Netherlands and in England estimated the increase in malaria's geographic range, which could occur if the IPCC's projections for global warming prove correct. These researchers concluded that, in tropical regions, the epidemic potential of the mosquito population would double. In temperate climates, according to these projections, the epidemic potential could increase a hundred-fold. Furthermore, this study said, "There is a real risk of reintroducing malaria into non-malarial areas, including parts of Australia, the United States, and southern Europe" ("Rachel's No. 466" 1995).

Several Factors at Work

By the late 1970s, dwindling investments in public health programs, growing resistance to insecticides among some species, and prevalent environmental changes (such as deforestation) also contributed to a widespread resurgence of malaria, according to Epstein. "By the late 1980s." he reported, large epidemics of malaria were being associated with warm, wet weather. Between 1993 and 1998, reported Epstein, worldwide incidence of malaria quadrupled.

Malaria is now found in higher-elevations in central Africa and could threaten cities such as Nairobi, Kenya (at about 5,000 feet, roughly the elevation of Denver, Colorado), as freezing levels have shifted higher in the mountains. In the summer of 1997, for example, malaria took the lives of hundreds of people in the Kenyan highlands, where populations had previously been unexposed. (Epstein 1998)

Between 1970 and 1995, the lowest level at which freezing occurs had climbed about 160 meters higher in mountain ranges from 30 degrees north to 30 degrees south latitude, based on radiosonde data analyzed at NOAA's Environmental Research Laboratory. This shift corresponds to a warming at these elevations of about 1°C (almost 2°F), which is nearly twice the average warming that had been documented over the Earth as a whole by the end of the twentieth century (Epstein 1998). As higher elevations warm, mosquito-vector diseases are ascending tropical mountainsides around the world. Bill Weinburg, in *Native Americas*, described changes provoked by warming in the mountains of Mexico:

> Juan Blechen Nieto, a Cuernavaca physician, traveled through the Sierra del Sur on a survey of local health conditions in November 1998, and found an alarming incidence of dengue fever and malaria. "These are diseases that are traditionally associated with lowland coastal regions, and are now appearing in the Sierra del Sur," he told me.... Indians in highland Oaxaca communities tell me they have mosquitoes now, for the first time. This has to do with deforestation impacting local and regional climate. It gets hotter, and the undergrowth that comes up after forests are destroyed provides a habitat for pests. (Weinberg 1999, 58–59)

See also: Human Health, World Survey

FURTHER READING

Connor, Steve. "Malaria Could Become Endemic Disease in U.K." *London Independent*, September 12, 2001c, 14.

Epstein, Paul R. "Climate, Ecology, and Human Health." December 18, 1998. http://www.iitap.iastate.edu/gccourse/issues/health/health.html.

Epstein, Paul R. "Profound Consequences: Climate Disruption, Contagious Disease, and Public Health." *Native Americas* 16, no. 3/4 (Fall/Winter 1999):64–67.

Epstein, Paul R., Henry F. Diaz, Scott Elias, Georg Grabherr, Nicholas E. Graham, Willem J. M. Martens, Ellen Mosley-Thompson, and Joel Susskind. "Biological and Physical Signs of Climate Change: Focus on Mosquito-borne Diseases." *Bulletin of the American Meteorological Society* 79, Part 1 (1998): 409–417.

Martens, Pim. "How Will Climate Change Affect Human Health?" *American Scientist* 87, no. 6 (November/December 1999):534–541.

Martens, Willem J. M., Theo H. Jetten, and Dana A. Focks. "Sensitivity of Malaria, Schistosomiasis, and Dengue to Global Warming." *Climatic Change* 35 (1997):145–156.

"Rachel's No. 466: Warming and Infectious Diseases." November 2, 1995. Environmental Research Foundation, Annapolis, Maryland. http://www.igc.apc. org/awea/wew/othersources/rachel466.html.

Weinburg, Bill. "Hurricane Mitch, Indigenous Peoples and Mesoamerica's Climate Disaster." *Native Americas* 16, no. 3/4 (Fall/Winter 1999):50–59. http:// nativeamericas.aip.cornell.edu/fall99/fall99weinberg. html.

"Managed Realignment" in Great Britain

Parts of England's coastline are afflicted by the same problems as the U.S. East and Gulf of Mexico coasts. The land is subsiding, as ice melt and thermal expansion slowly raise sea levels. The U.K. Climate Impact Programme, a government-funded program at Oxford University, forecast that the sea could be as much as a meter higher by late in the twenty-first century. In addition to climate change, "isostatic rebound," the rise of Scotland's coast following the last ice age, is contributing to a subsiding coastline southward along the English coast. As the sea rises three millimeters a year abreast of Essex, the land itself is sinking half as rapidly. By 2003, the rising waters were threatening the closed Bradwell Nuclear Power Station ("Rising Tide" 2003, 4). The marks of climate change are everywhere. Cornwall farmers, for example, have adapted to warmer, wetter weather by growing jalapeño peppers.

Anthony Browne commented in the *London Times*:

> A thousand years after King Canute showed that man could not hold back the tide, the Government has come to the same conclusion. The Environment Agency, the government body responsible for flood defenses, is planning a strategic withdrawal from large parts of the English coastline because it believes that it can no longer defend them from rising sea levels, the result of global warming. (Browne 2002b, 8)

The Sea Floods Low-Lying Farmland

The new strategy, officially called "managed realignment," will allow the sea to flood low-lying farmland rather than attempting to fend off the invading waters by building ever-higher defenses. The policy, which will allow the encroaching sea to submerge several thousand acres of land, has been welcomed by environmentalists. Farmers, however, said that the strategy was "unviable" and have demanded more concrete flood defenses. The area of coast affected ranges from the Humber Estuary, around East Anglia, to the Thames estuary and west to the Solent. Strategic withdrawal has been planned for sections of the Severn Estuary (Browne 2002b, 8). The first site surrendered to the sea was in Lincolnshire. About 200 acres of farmland were flooded by seawater at Freiston Shore after diggers broke through the flood defense banks to create a salt-marsh bird reserve (Browne 2002b, 8).

"Managed realignment is in its infancy ... but it has been an emotive issue; a lot of people are concerned about it. But you have to get the message across that we are defending property," said Brian Empson, flood defense policy manager for the Environment Agency (Browne 2002b, 8). The land that was returned to the sea in 2002 had been reclaimed 150 years earlier and protected by man-made banks. A grass-covered wall in Abbots Hall farm country on the east coast of Essex that had held back the sea for almost 400 years was breached intentionally in 2002 to surrender the area to the sea. The use of breakwaters made of clay, bricks, and, finally, large blocks of concrete, had not stopped rising waters.

Until recently, the Thames River barrier, built to protect London and surrounding areas from unusually high tides and storm surges, closed an average of two or three times a year. Between November 2001 and March 2002, however, the barrier was raised 23 times. A British report released in September 2002 said that 59,000 square miles (home to 750,000 people) in and around London were vulnerable to flooding because they were below high-tide levels, some by as much as 12 feet.

Local Land Losses

Late in 2006, 50 feet of Roger Middleditch's sugar-beet field fell into the North Sea, Middleditch, a tenant farmer, once tilled 23 acres of potatoes, but 20 acres have been reclaimed by the sea over several years. "We've lost so much these last few years," he said. "You plant, and by harvest it's fallen into the water" (Rosenthal 2007a). While the area has been sinking slowly for at least 100 years, the pace has accelerated during the last few years. The government no longer maintains local seawalls, part of its policy of "managed realignment" or "managed retreat." Homes in Happisburgh, an English village, fall into the sea each year, collapsing over a crumbling cliff. The North Norfolk District Council and Coastal Concern Action Limited have started to shore up Happisburgh's cliff with rocks, financed in part by an Internet campaign, "Buy a Rock for Happisburgh." According to David Viner, a climate expert at the University of East Anglia, "The U.K. won't let London flood, but the national government's not going to worry about an odd village or farm" (Rosenthal 2007a).

Land loss at Benacre "has accelerated dramatically," said Mark Venmore-Roland, the estate's manager. "At first it was like a chap losing his hair—bit by bit, so you'd get used to it." But in the past few years, he said, "it's been really frightening" (Rosenthal 2007a). Elisabeth Rosenthal of the *New York Times* described how the shoreline of eastern England is changing:

Walkers and birders who frequent these famous Broads, or salt marshes, will find that the hiking path through Benacre that once gently declined from a low grassy plateau toward the beach, now ends in a precipitous drop of 16 feet to the water; the rest fell into the sea in February. The 6,000-acre Benacre Estate is losing swaths of land 30 feet wide along its entire two miles of coastline each year. Inland trees that were once sold for timber are

dying or no longer commercially valuable, because the proximity to the salty sea air has left them stunted. (Rosenthal 2007a)

FURTHER READING

Browne, Anthony. "Canute Was Right! Time to Give up the Coast." *London Times*, October 11, 2002b, 8.
"Rising Tide: Who Needs Essex Anyway." *London Guardian*, June 12, 2003, 4.
Rosenthal, Elisabeth. "As the Climate Changes, Bits of England's Coast Crumble." *New York Times*, May 4, 2007a. http://www.nytimes.com/2007/05/04/world/europe/04erode.html.

Maple Syrup Wanes and Other Changes in New England

New England's maple trees require cold weather to yield the sap that becomes maple syrup; they yield less sap in warmer winters. An analysis of syrup production between 1920 and 2000 indicated a decline in every New England state except Maine (Donn 2002). At the same time, titmice, red belly woodpeckers, northern cardinals, and mockingbirds are being observed more often at bird feeders in Vermont. All of these birds have migrated from more southerly latitudes as temperatures have warmed.

Sap buckets hang on maple trees in East Montpelier, Vermont, April 8, 2005. (AP/Wide World Photos)

University of New Hampshire forester Rock Barrett, who supervised the survey, said that pervasive warming already may have doomed New England's maple-syrup industry. "I think the sugar maple industry is on its way out, and there isn't much you can do about that," he said. Even in 2002, however, roughly one in four Vermont trees still was a sugar maple. Vermonters made almost 60 percent of New England's 850,000 gallons of syrup that year, according to federal farm data (Donn 2002).

A newspaper account described the industry that Barrett said is endangered:

> Every year, as winter begins to melt away, Vermont's sugarhouses come back to life. They puff thick, white smoke from stainless steel evaporators that boil sap down to syrup. Punctuated by vacuum gauges, sap-carrying plastic tubing—a technology that is replacing suspended buckets—snakes through thick woods to collection tubs. Vermont kids, as always, freeze maple treats in the snow. (Donn 2002)

Oak and Hickory Move in, Maples Out

Much of New England could lose its maple forests during the twenty-first century in favor of the oak and hickory that are dominant farther south. Already, during recent decades, most expansion in syrup production has occurred to the north, in colder Quebec. During a decade ending in 2002, yearly production there has doubled to satisfy a booming market, which by the year 2000 surpassed the United States fivefold, according to the North American Maple Syrup Council (Donn 2002). Over the last 80 years, New England's typical syrup output has dropped by more than half, from more than 1.6 million gallons a year to less than 800,000. Syrup has dwindled to a $22 million annual regional business, according to the U.S. Department of Agriculture (Donn 2002). The syrup-making season in many parts of New England has advanced from March to February during recent decades, another sign that the weather is warming. "The winter weather has been interrupted, with spring … in between," said Toni Pease, a sugar maker in Oxford, New Hampshire, who complained of reduced sap flows from such fluctuations (Donn 2002).

As Temperatures Rise, the Environment Changes

The average annual temperature in Vermont and far-north New York State climbed nearly 1.6°F during the twentieth century (Donn 2002). A shift away from an agrarian society and increasing popularity of syrup substitutes has pressured the maple-syrup industry. Maine, by contrast, has cooled, slipping a yearly average of nearly a 0.5°F. Meanwhile, Maine's syrup production has exploded from less than 12,000 gallons a year through most of the 1980s, to 230,000 gallons by 2002. The boom is not entirely due to climate change; a large part of it stems partly from an invasion of Quebec sugar makers working Maine forests (Donn 2002). Cold winters give New England more than syrup; they discourage forest insect pests, such as the maple-eating pear thrip. Researchers also say that warming promotes tree-damaging weather extremes like storms and drought (Donn 2002).

Maple syrup could be only one traditional casualty of rapid climate change in the northeastern regions of the United States. According to the first comprehensive U.S. federal study summarizing possible effects of global warming on the U.S. Northeast,

> New England's maple trees stop producing sap. The Long Island and Cape Cod beaches shrink and shift, and disappear in places. Cases of heatstroke triple. And every 10 years or so, a winter storm floods portions of Lower Manhattan, Jersey City and Coney Island with seawater. The Northeast of recent historical memory could disappear this century, replaced by a hotter and more flood-prone region where New York could have the climate of Miami and Boston could become as sticky as Atlanta…. In the most optimistic projection, we still end up with a 6- to 9-degree increase in temperature," said George Hurtt, a University of New Hampshire scientist and co-author of the study on the New England region. "That's the greatest increase in temperature at any time since the last Ice Age. (Powell 2001, A-3)

Scientists who compiled this study concluded that significant warming already was taking place in New England by the year 2000. They noted that, on average, temperatures in the region rose 2°F during the twentieth century. The scientists added that temperature rise during the twenty-first century "will be significantly larger than in the 20th century." One widely used climate model cited in the report predicted a 6°F increase, the other 10°F (Powell 2001, A-3).

A report by the U.S. Office of Science and Technology Policy said that expected climate changes within the present century are likely to shift the ideal range for some North American forest species northward by as much as 300 miles, exceeding the species' ability to migrate naturally. If emissions of greenhouse gases continue to increase, the report projects that maples will recede from all areas of the continental United States except the northern tip of Maine.

The Sierra Club also has described the possible effects of global warming on sugar maples:

> Regions dependent on forests for various commodities, such as the maple-syrup-producing states of New England and the Midwest, could also face devastating losses. EPA projects that by 2050 the range for the sugar maple could shift north of all but the northernmost tip of New England. This possibility has serious implications for the maple syrup industry, which currently provides up to $40 million annually to the regional economy. (Glick 1998)

The same Sierra Club report traced increasing temperatures and insect infestations in boreal forests to a rise in temperatures that began in 1976. With rising temperatures have come larger and more frequent forest fires, according to Greenpeace. "Unless atmospheric concentrations of greenhouse gases are quickly stabilized," the report stated, "climate-vegetation models predict that large areas of boreal forest will be reduced to patchy open woodland and grassland, resulting in lowered biological diversity and a reduced ability to store carbon" (Jardine 1994).

"We can't rely on tradition like we used to," said Burr Morse, 58, who once routinely began the sugaring season by inserting taps into trees around Town Meeting Day, the first Tuesday in March, and collecting sap to boil into syrup up until about six weeks later. The maple's biological clock is set by the timing of cold weather (Belluck 2007). For at least 10 years, some farmers have tapped maple trees earlier in the season. Some of them now tap in February and still miss up to 30 percent of the sap. During the 2007 season, Morse tapped in February and still missed 300 of his usual 1,000 gallons of syrup. "But the way I feel, we get too much warm [weather]. How many winters are we going to go with Decembers turning into short-sleeve weather, before the maple trees say, 'I don't like it here any more'?" (Belluck 2007).

Some farmers insist that the warm weather is a natural cycle that will correct itself, but the string of early springs has turned into a trend. After 1971, according to the National Oceanic and Atmospheric Administration, winter temperatures in the Northeast have increased by 2.8°F. The season has begun earlier than average for more than 30 years. Some farmers have adapted technology to maintain production, but that is a stopgap

remedy. At some point, given a continuation of the trend, the weather will simply become too warm to nurture maple syrup in Vermont. "In the 1950s and 1960s, 80 percent of world's maple syrup came from the United States, and 20 percent came from Canada," said Barrett N. Rock, a professor of natural resources at the University of New Hampshire. "Today it's exactly the opposite. The climate that we used to have here in New England has moved north to the point where it's now in Quebec" (Belluck 2007).

Rising Oceans a Complex Threat

Rising ocean waters present a more complicated threat. The seas around New York City have risen 15 to 18 inches in the past century (including subsidence of the land), and scientists forecast that by 2050, waters could rise an additional 10 to 20 inches (Powell 2001, A-3). By 2080, storms with 25-foot surges could hit New York every three or four years, inundating the Hudson River tunnels and flooding the edges of the financial district, causing billions of dollars in damage (Powell 2001, A-3).

"This clearly is untenable," said Klaus Jacob, a senior research scientist with Columbia University's Lamont-Doherty Earth Observatory. According to a *Washington Post* summary of the report,

> Sea-level rise could reshape the entire Northeast coastline, turning the summer retreats of the Hamptons and Cape Cod into landscapes defined by dikes and houses on stilts. Should this come to pass, government would have to decide whether to allow nature to have its way, or to spend vast sums of money to replenish beaches and dunes. Complicating the issue is the fact that some wealthy coastal communities exclude nonresident taxpayers from their beaches. (Powell 2001, A-3)

The dates of peak river flow in New England (from melting snow) have advanced about two weeks in a century, according to Glenn Hodgkins, a lead researcher with the U.S. Geological Survey in Augusta, Maine. Hodgkins also has determined that the average date at which ice melts advanced by nine days in northern New England and 16 days in southern New England between 1850 and 2000 (Sharp 2003).

FURTHER READING

Belluck, Pam. "Warm Winters Upset Rhythms of Maple Sugar." *New York Times*, March 3, 2007. http://www.nytimes.com/2007/03/03/us/03maple.htm.

Donn, Jeff. "New England's Brilliant Autumn Sugar Maples—and Their Syrup—Threatened by Warmth." Associated Press, September 23, 2002. (LEXIS)

Glick, Patricia. *Global Warming: The High Costs of Inaction.* San Francisco: Sierra Club, 1998. http://www.sierraclub.org/global-warming/inaction.html.

Jardine, Kevin. "The Carbon Bomb: Climate Change and the Fate of the Northern Boreal Forests." Ontario, Canada: Greenpeace International, 1994. http://dieoff.org/page129.htm.

Powell, Michael. "Northeast Seen Getting Balmier: Studies Forecast Altered Scenery, Coast." *Washington Post*, December 17, 2001, A-3.

Sharp, David. "Study: New England's Winters Not What They Used to Be." Associated Press State and Regional News Feed, July 23, 2003. (LEXIS)

Medieval Warm Period, Debate over Temperatures

During 2003, a media stir was created by a study promoting the point of view that temperatures had been warmer during the "medieval warm period" from 900 to 1300 C.E. than during the twentieth century. Willie Soon, a physicist and astronomer, wrote the study with four co-authors: Sallie Baliunas; Sherwood Idso and his son Craig Idso, who are the former and current president of the Center for the Study of Carbon Dioxide and Global Change; and David Legates, a climate researcher at the University of Delaware.

This study was promoted as a product of Harvard University (two of the authors had affiliations with Harvard), but most of it was supported by nonprofit groups with ties to the oil industry. The paper provoked a global storm of e-mail among scientists, some of whom proposed a boycott of a journal that published the study. "Energy interests paid for the study and help finance the groups promoting it. The study illustrates a strategy adopted in the late 1980s to attack the credibility of climate science," said John Topping, president of the Climate Institute. "They saw early on that what they had to do was keep the science at issue," said Topping, a former Republican congressional staff member who founded the institute in 1986 (Nesmith 2003a).

The study, "Reconstructing Climatic and Environmental Changes of the Past 1,000 Years: A Reappraisal," which purported to be an analysis of data from more than 200 other studies, was underwritten by the American Petroleum Institute, an advocacy association for the world's largest oil companies. Two of the five authors received support from the ExxonMobil Foundation. Two others, affiliated with the Harvard-Smithsonian Center for Astrophysics, also were listed as "senior scientists" with a Washington-based organization supported by several right-wing foundations and ExxonMobil. One of these organizations, the George T. Marshall Institute, is headed by William O'Keefe, a former executive of the American Petroleum Institute. O'Keefe also was past president of the Global Climate Coalition, a defunct group created by oil and coal interests to lobby against U.S. participation in treaties such as the Kyoto Protocol (Nesmith 2003a).

Raymond F. Bradley, Malcolm K. Hughes, and Henry F. Diaz, writing in *Science*, asserted that while comparisons of the medieval period with the present are difficult because surviving records are sparse (especially on a worldwide scale), "[t]emperatures from 1000 to 1200 C.E. (or 1000 to 1100 C.E.) were … 0.03°C cooler than the period from 1901 to 1970 C.E. The latter period was on average about 0.35°C cooler than the last 30 years of the twentieth century" (Bradley, Hughes, and Diaz et al. 2003, 405). These authors also asserted that the period 1100 C.E. to 1260 C.E. was characterized by high levels of explosive volcanism, which in our time have been associated with warmer-than-usual winters in northern Europe and northwestern Russia that may have helped support Viking colonization of Iceland and Greenland. Most of the weather records of the time also come from Europe, which may make the data a poor proxy for worldwide averages.

FURTHER READING

Bradley, Raymond S., Malcolm K. Hughes, and Henry F. Diaz. "Climate in Medieval Time." *Science* 302 (October 17, 2003):404–405.

Nesmith, Jeff. "Is the Earth too Hot? A New Study Says No, but Then It Was Funded by Big Oil Companies." *Atlanta Journal-Constitution* in *Hamilton Spectator* (Ontario, Canada), May 30, 2003a, n.p. (LEXIS)

Methane as a Greenhouse Gas

Methane's concentration in the atmosphere has increased 250 percent since 1800, a faster rate than carbon dioxide, yielding a radiative forcing by itself of almost 0.5 watt per square meter, about half that of carbon dioxide. Methane, in other words, is responsible for about half the heat retention in the air that carbon dioxide provides.

A landfill. (Courtesy of Getty Images/PhotoDisc)

In addition, methane is chemically reactive, leading to complications. For example, according to one analysis,

Rising methane concentrations can cause increases in ozone and stratospheric water vapor concentrations [water vapor itself is a greenhouse gas]. A recent emissions-based view of methane's radiative forcing that included these additional effects estimated that methane's true contribution was nearer 0.9 watt per square meter, equivalent to more than half the radiative forcing caused by CO_2. (Shine and Sturges 2007, 1805)

Methane Levels on the Rise

Atmospheric methane levels are on the rise, after stagnating during the 1990s, and could accelerate global warming. Research indicates that drought in recent years reduced methane emissions from natural sources and masked the impact of methane increases from human activities. "The bad news is that the slowdown in global methane emissions in the past few decades was only temporary," said Jos Lelieveld, director of the Max Planck Institute for Chemistry in Germany ("Methane Emissions" 2006).

"Had it not been for this reduction in methane emissions from wetlands, atmospheric levels of methane would most likely have continued rising," said study co-author Paul Steele, a scientist with the Commonwealth Scientific and Industrial Research Organization (CSIRO) in Australia. "This suggests that, if the drying trend is reversed and emissions from wetlands return to normal, atmospheric methane levels may increase again, worsening the problem of climate change" ("Methane Emissions" 2006).

Unlike carbon-dioxide emissions, which have continued to rise steadily, the proportion of methane in the atmosphere stagnated during the 1990s, following a peak about 1980. In addition to drought, methane levels probably stopped rising because of concerted efforts by governments around the world, according to a report from the Goddard Institute for Space Studies (Hansen and Sato 2001, 14,778; "Limiting Methane" 2002). By 2003, methane emissions were poised to decline further, in large part because of more efficient oil production in the Russian Federation. During 2003, scientists with Australia's CSIRO announced that methane levels in the Earth's atmosphere had been stable since 1999, after a 15 percent rise during the preceding

20 years, and a 150 percent spike since preindustrial times ("Australian Scientists" 2003).

China's methane emissions, 0.2 million tons in 1950, rose to a world-leading 5.6 million by 1988; the United States produced 2.3 million tons of methane in 1950 and 2.8 million in 1988, third in the world behind the former Soviet Union's 4.1 million tons and China. The United Kingdom cut its methane production in half during the 1950–1988 period, from 1.6 million tons (then third in the world) to 0.7 million, mainly due to decreased burning of coal for heating (Hariss, Bensel, and Blaha 1993, 343).

Using data from tracking stations taken between 1984 and 2003, as well as computer simulations, researchers have determined that methane levels fell from an annual growth rate of 12 parts per billion (ppb) in the 1980s to 4 ppb in the 1990s because industry cut emissions and natural gas was used more efficiently. After that, however, levels have risen again, due in large part to rising emissions from Asian fossil-fuel consumption, mainly coal in China (Bousquet et al. 2006, 439; "Methane Emissions" 2006).

Natural emissions of methane are also on the rise. During December 2008, *National Geographic News* reported that areas of the East Siberian Sea were bubbling with methane released from underwater reserves that had been detected on a recent expedition, a possible sign that underwater permafrost has been thawing. Acceleration of such emissions could cause "really serious climate consequences," said expedition member Igor Semiletov of the University of Alaska, Fairbanks (Inman 2008). Semiletov and colleagues traveled along the Siberian coast monitoring methane. "According to our data, more than 50 percent of the Arctic Siberian shelf is serving as a source of methane to the atmosphere," Semiletov said (Inman 2008).

Methane and Dams

Methane also has been found to aggravate global warming in unexpected ways—for example, as a by-product of hydroelectric power in the tropics.

Hydropower has gained a reputation as an alternative source of electric power because it uses no fossil fuels. However, the flooding of vegetated land behind dams can increase emissions of methane. During the 1980s, for example, about 2,500 square kilometers of Amazonian forest were flooded behind the Balbina dam to supply the Brazilian city of Manaus. More than 80 percent of Brazil's domestic energy is produced by hydropower. While hydropower uses no fossil fuels directly, "the global-warming impact of hydropower plants can often outweigh that of comparable fossil-fuel power stations" (Giles 2006c, 524).

Large hydroelectric dams release methane as trees and other plants settle to the bottom and decompose without oxygen when the reservoir is flooded. The methane is released as water passes through the dam's turbines, said Patrick McCully, executive director of International Rivers Network. McCully said that Lima's calculations imply that the world's 52,000 large dams contribute more than 4 percent of the total warming impact of human activities. Hydropower reservoirs in the can have a much higher warming impact than even the dirtiest fossil-fuel plants generating similar quantities of electricity, he said. "It is unfortunate that Lima's study has come too late to be included in the recent reports from the Intergovernmental Panel on Climate Change (IPCC)," McCully said. "Partly because of the influence of the hydro industry and its government backers, climate policymakers have largely overlooked the importance of dam-generated methane" ("Methane from Dams" 2007).

Large amounts of organic matter become trapped behind a dam when land is flooded, and more is flushed in after that. Methane also is released when water is released from the dams. This is especially notable in warm climates, where organic matter quickly decays, giving off methane and carbon dioxide. The Balbina dam has been analyzed and found to be worse for emissions than a fossil-fuel plant. In some cases, hydropower can release four times as much methane as fossil-fuel plants per amount of energy generated. That ratio is open to debate (Giles 2006c, 524). These debates bear on implementation of the Kyoto Protocol, which allows Clean Development Mechanism credit for some hydroelectric development.

Research by scientists affiliated with the National Institute for Space Research (INPE) in Brazil suggested that large dams contribute to global warming by releasing methane into the atmosphere. The authors proposed to solve this problem by capturing the methane emissions and using them to generate electricity. Ivan Lima and colleagues said that, worldwide, large dams annually release about 104 million metric tons of methane to the atmosphere through reservoir surfaces, turbines, and spillways. "If we can generate electricity from the huge amounts of methane produced by existing tropical dams we can avoid the need to build new dams with their associated human and environmental costs," Lima said ("Methane from Dams" 2007).

Philip Fearnside from Brazil's National Institute for Research in the Amazon in Manaus published a study in 2002 indicating that that the greenhouse effect of emissions from the Curuá-Una dam in Pará, Brazil in 1990, was more than 3.5 times what would have been produced by generating the same amount of electricity from oil ("Methane from Dams" 2007).

Methane Management

Methane has been easier to reduce than carbon dioxide. For example, cattle respiration may account for nearly 20 percent of the methane gas released into the atmosphere. Researchers at the University of Nebraska are developing a food additive for cattle that may reduce the amount of methane they generate. "The reason we're focusing on methane is because it's a short-lived, highly potent greenhouse gas that needs be reduced," said University of Nebraska at Lincoln biochemistry professor Stephen Ragsdale (Thiessen 2003).

Methane is released from a cow not only via burping and flatulence, but also through ordinary respiration. The methane is produced in the cow's rumen (the first of a cow's four stomachs), enters the bloodstream, and exits through the lungs. Ken Olson, a range livestock nutritionist at Utah State University, conducted a six-year study that found that better range management practices, such as providing higher-quality forage, can make a small difference in the amount of methane released by cattle.

Ragsdale and fellow researchers James Takacs and Jess Miner conceived the idea of reducing methane by blocking enzymes in the cow's rumen that are required to produce methane. They have patented an additive to block enzymes, having tested more than 200 compounds in an attempt to find the right formula that blocks methane but does not harm beneficial microbes in the cow's rumen. Such a product might be tested on smaller animals before cattle—sheep or even termites, Ragsdale said. The Nebraska researchers said they don't believe methane does anything beneficial for cattle. In fact, the researchers believe that up to 16 percent of the feed given to cattle is wasted because it is used to produce methane. If methane production is removed, producers would have to feed the cattle less, because more of the animal's energy would go to its own production of proteins, amino acids, and fat (Thiessen 2003).

According to James E. Hansen, director of NASA's Institute for Space Studies:

The non-CO_2 portion of the Alternative scenario is essential for success in climate management. A CO_2 amount as large as 475 ppm can result in global warming less than 1°C only if some non-CO_2 forcings decrease in absolute magnitude. Actual non-CO_2 forcings during 2000–2005 have come close to matching the Alternative scenario, and their growth has been notably slower than in IPCC Business-as-Usual scenarios. The most important of these forcing agents, methane (CH_4), has been almost stable in abundance for several years, for reasons that are not well understood. One candidate reason for the slowdown is reduction of methane loss (release to the atmosphere) during the mining of fossil fuels. Capture of methane at landfills and waste management facilities also may have contributed to the slowdown in methane growth, although the number of landfills continues to increase. (Hansen 2006b, 27)

Pigs Make Power

In Sterksel, the Netherlands, farmers cook manure from 3,000 pigs and capture its methane to generate electricity on the local power grid. The pigs' offal is combined with other waste materials, then pumped into digesting tanks where bacteria remove natural gas that can be burned to generate heat and electricity. This one farm also saves $190,000 a year in disposal fees. In Denmark, farmers are legally required to inject manure under soil, which adds richness, reduces odors, and prevents emissions from escaping, at least for a time. This is a start on a large problem. The IPCC estimates that several billion farm animals worldwide generate 18 percent of greenhouse-gas emissions—more than cars, buses and airplanes (Rosenthal 2008i). In 2009, Sweden began affixing labels on food products listing greenhouse-gas emissions.

Consumption of red meat has risen 33 percent in China, India, and Brazil during the last decade. Every step of meat's production creates greenhouse-gas emissions. Production of beef creates 11 times as much greenhouse-gas emissions as a similar amount of chicken and 100 times that of carrots, according to Lantmannen, the Swedish group. *See also:* Feedback Loops; Permafrost and Climate Change

FURTHER READING

"Australian Scientists Announce Good News at Last on Global Warming." Agence France Presse, November 25, 2003, n.p. (LEXIS)

Bousquet, P., P. Ciais, J. B. Miller, E. J. Dlugokencky, D. A. Hauglustaine, C. Prigent, G. R. Van der Werf, P. Peylin, E.-G. Brunke, C. Carouge, R. L. Langenfelds, J. Lathière, F. Papa, M. Ramonet, M. Schmidt, L. P. Steele, S. C. Tyler, and J. White. "Contribution of Anthropogenic and Natural Sources to Atmospheric Methane Variability." *Nature* 443 (September 228, 2006):439–443.

Giles, Jim. "Methane Quashes Green Credentials of Hydropower." *Nature* 444 (November 30, 2006c): 524–525.

Hansen, James E., and Makiko Sato. "Trends of Measured Climate Forcing Agents." *Proceedings of the National Academy of Sciences* 98 (December 18, 2001): 14,778–14,783.

Hariss, Robert C., Terry Bensel, and Denise Blaha. "Methane Emissions to the Global Atmosphere from Coal Mining." In *A Global Warming Forum: Scientific, Economic, and Legal Overview,* ed. Richard A. Geyer, 339–346. Boca Raton, Florida: CRC Press, 1993.

Inman, Mason. "Methane Bubbling Up from Undersea Permafrost?" *National Geographic News,* December 19, 2008. http://news.nationalgeographic.com/news/2008/12/081219-methane-siberia.html.

"Limiting Methane, Soot Could Quickly Curb Global Warming." Environment News Service, January 16, 2002. http://ens-news.com/ens/jan2002/2002L-01-16-01.html.

"Methane Emissions Increasing and Could Hasten Global Warming." Environment News Service, September 28, 2006. http://www.ens-newswire.com/ens/sep2006/2006-09-28-01.asp.

"Methane from Dams: Greenhouse Gas to Power Source." Environment News Service, May 9, 2007. http://www.ens-newswire.com/ens/may2007/2007-05-09-04.asp.

Rosenthal, Elisabeth. "As More Eat Meat, a Bid to Cut Emissions." *New York Times,* December 4, 2008i. http://www.nytimes.com/2008/12/04/science/earth/04meat.html.

Shine, Keith P. and William T. Sturges. "CO_2 Is Not the Only Gas." *Science* 315 (March 30, 2007):1804–1805.

Thiessen, Mark. "Researchers Fighting 'Bad' Breath in Cattle." Associated Press, June 8, 2003. (LEXIS)

Methane Burp (or Clathrate Gun Hypothesis)

Warming oceans eventually could cause intense eruptions of methane from the seafloor, accelerating global warming caused by humankind's industries and transportation. About 10,000 billion tons of solid methane clathrates, described by Fred Pearce as "a lattice of ice crystals rather like a honeycomb" (2007b, 91), are stored beneath the ocean and on the continents. By comparison, the contribution of humans to the atmosphere's inventory of greenhouse gases via burning of fossil fuels has amounted to about 200 billion tons of carbon. If even a small portion of the oceans' stored methane were to escape into the atmosphere, greenhouse warming could greatly accelerate. Among scientists, this mechanism has come to be called the "methane burp" (or "clathrate gun hypothesis").

Past as Precedent

The study of methane clathrates has become more popular in recent years, as evidence accumulates that their release, especially from oceans, may occasionally have been a major, sudden driving force in the Earth's past climate cycles. James P. Kennett and colleagues studied climate records for the last 60,000 years off Santa Barbara, California, and parts of Greenland, finding that "surface and bottom temperatures change in concert" (Blunier 2000, 68; Kennett et al. 2000). This finding supports assertions by E. G. Nisbet that the massive release of oceanic methane from clathrates has played a significant role in rapid warmings during the past, even without added provocations by human industry (Nisbet 1990, 148).

Scientists have yet to reach any sort of consensus on causes of the Earth's "methane burps." No one yet knows why a trillion tons of methane may have been released so suddenly from solid methane hydrates around the world. This chemical reaction provoked a sudden (in geologic time) global warming of 4° to 8°C. James Cook and Gerald Dickens theorized that "[t]he methane probably oxidized to form carbon dioxide which eventually reached the atmosphere, driving greenhouse warming" (Kerr 1999d, 1465).

The sediment cores drilled by Miriam E. Katz and colleagues contained remnants of small marine organisms called *foraminifera*, which preserve a record in their shells of carbon levels in the ocean. The shells tell a story of an extreme warming (possibly more than 10°F) in the ocean over a short time, which killed more than half of the *foraminifera.* The sediment core also contained evidence of an underwater landslide, which scientists believed took place as melting methane clathrates "warmed dramatically, breaking apart into water and methane gas, and bubbl[ed] ferociously out of the sea floor" (Witze 2000, 4).

This line of reasoning was supported by Richard Norris of the Woods Hole Research Center and Ursula Rohl of Germany's University of Bremen. Norris and Bremen wrote in *Nature* that the "methane burp" occurred when an as-yet-unknown natural provocation pumped greenhouse gases into the atmosphere, causing a sudden bout of global warming. "Our results suggest that large natural perturbations to the global carbon cycle have occurred in the past ... at rates that are similar to those induced today by human activity," they wrote (Norris and Rohl 1999, 775). Miriam E. Katz, Dorothy K. Pak, Gerald R. Dickens, and Kenneth G. Miller asserted that this surge in global temperatures may have played a crucial role in the evolution of warm-blooded mammals as the Earth's dominant species 10 million years after a cataclysmic event (probably the impact, on the Earth, of a very large asteroid) ended dominance by the dinosaurs. Katz and colleagues contended that "elevated temperatures quickly opened high latitude migration routes for the widespread dispersal of mammals" (Katz et al. 1999, 1531).

Linking Methane Releases and the Carbon Cycle

Writing in *Science*, Kai-Uwe Hinrichs and colleagues Laura Hmelo and Sean Sylva of the Woods Hole Oceanographic Institution provided a direct link between methane reservoirs in coastal marine sediments and the global carbon cycle, an indicator of global warming and cooling (Hinrichs, Hmelo, and Sylva 2003, 1214–1217). Molecular fossils of methane-consuming bacteria found in the Santa Barbara basin off California deposited during the last glacial period (70,000 to 12,000 years ago) indicate that large quantities of methane were emitted from the seafloor during warmer phases of Earth's climate in the recent past. Preserved molecular remnants found by the Woods Hole team result from bacteria that fed exclusively on methane and indicate that large quantities of this powerful greenhouse gas were present in coastal waters off California. The team studied samples that were deposited 37,000 to 44,000 years ago.

"For the first time, we are able to clearly establish a connection between distinct isotopic depletions in forams and high concentrations of methane in the fossil record," said Hinrichs, an assistant scientist in the Woods Hole Institution's Geology and Geophysics Department. "The large amounts of methane presumably released during one event about 44,000 years ago suggest a mechanism different from those underlying the emissions at warmer periods, i.e. slow decomposition of methane hydrate triggered by warming of bottom waters," Hinrichs continued. "The sudden release of these enormous quantities of methane was probably caused by landslides and melting of the methane hydrate" (Hinrichs, Hmelo, and Sylva 2003, 1214–1217).

A "Methane Pulse" from the Ocean Floors

Stephen P. Hesselbo and colleagues reported that roughly 140 to 200 million years ago large quantities of methane were liberated from ocean floors, possibly because of warming global temperatures. This "methane pulse"—a "voluminous and extremely rapid release of methane from gas hydrate contained in marine continental-margin sediments" (Hesselbo et al. 2000, 392)—combined with oxygen in the oceans to form carbon dioxide, accelerating the worldwide warming. Along the way, a large proportion of oceanic animal life (perhaps 80 percent) died for lack of oxygen. "One of the important questions that is debated a lot today is the stability of this methane hydrate reservoir, and how easy it is to release the methane," said Stephen P. Hesselbo, lead author of the paper. "The extinction and the association with lack of oxygen has been fairly well established, but the association with methane release is something that hadn't been realized before," he said ("Prehistoric" 2000, 9).

During past periods of rapid warming, methane in gaseous form has been released from the seafloor in intense eruptions. An explosive rise in temperatures on the order of about 8°C during a few thousand years accompanied a methane release 55 million years ago, called the Paleocene-Eocene Thermal Maximum. Evidence abounds that "[j]ust a small amount of warming could kick-start a positive feedback loop between hydrate release and further warming, sending global temperatures soaring" (Schiermeier 2003, 681). Following the temperature spike 55 million years ago, the Earth eventually recovered its temperature equilibrium—but it took roughly 100,000 to 200,000 years to do so.

This period has assumed a crucial position in climate studies because atmospheric concentrations of carbon dioxide are believed to have reached 800 to 1,000 parts per million at its peak—the level that will become prevalent by the end of the twenty-first century at present rates of increase. The warming triggered 55 million years ago was quite spectacular. Scientists drilling in the Arctic have discovered that temperatures in the Arctic Ocean rose suddenly to as high as 68°F (20°C). "What no one expected was how much warmer it really was. That is a huge surprise," said Andy Kingdon of the British Geological Survey ("North Pole" 2004). Scientists discovered fossils of marine plants and animals that died quickly because they could not cope with the surge in temperatures.

Signatures on the Geological Record

Scientists have known for some time about the massive release of methane about 55 million years ago, because they can detect a distinct chemical signature in the geological record. Until recently, however, the cause of this spike in greenhouse gases has eluded them. A European team, led by Henrik Svensen at the University of Oslo, discovered ancient "escape conduits" on the floor of the Norwegian Sea. Svensen and several colleagues suggested that eruption of carbon-rich sedimentary strata from these vents under the present-day Norwegian Sea may have provided the trigger.

The researchers suggested in *Nature* that such conduits were common throughout the Northeast Atlantic 55 million years ago as the seafloor literally ripped apart between Greenland and northern Europe (Munro 2004, A-10). The European scientists suggested that upwellings of molten rock from deep in the Earth heated and cooked organic material in sea sediments, producing excessive amounts of methane gas that bubbled out into the atmosphere. "Similar volcanic and metamorphic processes may explain climatic events associated with other large igneous provinces such as the Siberian Traps (about 250 million years ago)," they wrote in *Nature* (Svensen et al. 2004, 542). Believing that carbon-dioxide release alone was insufficient to explain the magnitude of warming, a case has been made by Gabriel J. Bowen and colleagues that rapid increases in humidity (water vapor being a greenhouse gas) occurred at the same time (Bowen et al. 2004, 495).

An Analogue to Human-Provoked Warming?

Some scientists who have studied past releases of methane clathrates believe that they could prove a dramatic analogue for what might happen if human transport and industry continue to pump massive amounts of carbon dioxide and other greenhouse gases into the air. The rise in temperature that has been associated with this event is, indeed, very close to what the Intergovernmental Panel on Climate Change (IPCC) projects (given "business as usual" conduct) during the lives of our great-grandchildren.

"The case is getting stronger that global warming is associated with large hydrocarbon releases, and here's an example where it seems to have happened in the past," said Roy Hyndman, a senior scientist at the Pacific Geoscience Centre of the Geological Survey of Canada (Munro 2004, A-10). "If you put a lot of methane or other hydrocarbons into the atmosphere[,] you can get abrupt and very large global warming and that's what we're doing now," said Hyndman. "The amount we are putting in now is huge in the geological context, very large" (Munro 2004, A-10).

Kennett and colleagues (2003) asserted that the conversion of clathrates was much more important in driving climate change than increases in wetland emissions, a common explanation. However, wrote Gerald R. Dickens, an earth scientist at Rice University in Houston,

> There are ... two gaping problems: the authors provide no compelling evidence that methane released from the sea floor passed through the water column or that atmospheric methane initiated climate change. Until these major holes are plugged, their full hypothesis should rightfully be considered highly speculative. (Dickens 2002, 1017)

Dickens added, "It's the first really nice evidence that hydrocarbons were coming out of the sea floor at this time" (Munro 2004, A-10). Dickens and Hyndman said that the warming spike 55 million years ago is an intriguing, if not yet completely solid, analogue for what humans are doing to the atmosphere today. "The question is, will it just warm steadily as we add more and more carbon, or is there some point where it will warm a lot? There is a danger of a sudden or catastrophic big change," said Hyndman (Munro 2004, A-10).

Hot magma was delivered into organic-rich sediments throughout the North Atlantic at this time, initiating a massive hydrocarbon discharge from the seafloor. The composition of the expelled fluids and the timing of their release has not been explicitly defined, however, leading to questions about how many gigatons of carbon were released. Some calculations put that amount at about 3,000 gigatons, a "stupendous amount of hydrocarbons," considering that the entire world's deposits of oil and gas today total about 5,000 gigatons (Dickens 2004, 514). While Dickens's explanation for the warming spike 55 million years ago needs refining, he asserted that if cause and effect can be proved here, these events may become "an intriguing but imperfect analogue of current fossil-fuel emissions" (Dickens 2004, 515).

"It is the first time geology has isolated an individual methane release event in the distant past," said Santo Bains of Oxford University's department of earth sciences. "Now we can see just how it played its part in the global warming process of that era" (Keys 2001). Bains led geologists during three years of research through the badlands of Wyoming, parts of Antarctica, and off the east coast of Florida. "By studying that event, we may well be able to understand the effect of future global warming on the Arctic," said paleoclimatologist Professor Euan Nisbet of Royal Holloway College, University of London.

Crocodiles and Palm Trees in the Arctic

Researchers who have drilled into sediment layers near the east coast of Florida found evidence that melting methane clathrates thawed suddenly (over the course of a few thousand years) about 55 million years ago, initiating a sudden episode of global warming that ended with crocodiles and palm trees in the Arctic. At the peak of this episode, greenhouse-gas levels in the atmosphere were between two and six times as high as at present. Lisa Sloan, a paleoclimatologist at the University of California–Santa Cruz, and Gerald Dickens, a paleoceanographer at James Cook University in Queensland, Australia (two of several scientists who conducted the study), reported the findings at a meeting of the American Geophysical Union late in 1999.

Could the warming episode about 55 million years ago have been ignited by different methane releases in different parts of the world? T. L. Hudson and L. B. Magoon have advanced a theory that, 52 to 58 million years ago in the Gulf of Alaska, "large amounts of sediment, eroded from freshly uplifted mountains, were deposited in deep water off the 2,200-kilometer coast of Alaska" (Clift and Bice 2002, 130).

The oceanic tectonic plate, subducting below the continental plate, generates heat, along with the production of hydrocarbons (including liquid

methane), from organic matter in the sediment. According to Peter Clift and Karen Bice, writing in *Nature*, the methane may subsequently have turned to gas and bubbled into the atmosphere, "increasing greenhouse-gas concentrations [that] might [help to] explain the higher global temperatures 58 to 52 million years ago" (Clift and Bice 2002, 130). The same two authors asserted that while this process may have contributed significantly to a prolonged period of sudden atmospheric warming at that time, it was probably too geographically isolated to have been the sole cause of worldwide warming (Clift and Bice 2002, 130).

FURTHER READING

Blunier, Thomas. "'Frozen' Methane Escapes from the Sea Floor." *Science* 288 (April 7, 2000):68–69.

Bowen, Gabriel J., David J. Beerling, Paul L. Koch, James C. Zachos, and Thomas Quattlebaum. "A Humid Climate State during the Palaeocene/Eocene Thermal Maximum." *Nature* 432 (November 25, 2004):495–499.

Clift, Peter, and Karen Bice. "Earth Science: Baked Alaska." *Nature* 419 (September 12, 2002):129–130.

Dickens, Gerald R. "A Methane Trigger for Rapid Warming?" *Science* 299 (February 14, 2003):1017. Review of James P. Kennett, Kevin G. Cannariato, Ingrid L. Hendy, and Richard J. Behl. *Methane Hydrates in Quaternary Climate Change: The Clathrate Gun Hypothesis.* Washington, D.C.: American Geophysical Union, 2002.

Dickens, Gerald R. "Global Change: Hydrocarbon-driven Warming." *Nature* 429 (June 3, 2004):513–515.

Hesselbo, Stephen P., Darren R. Grocke, Hugh C. Jenkyns, Christian J. Bjerrum, Paul Farrimond, Helen S. Morgans Bell, and Owen R. Green. "Massive Dissociation of Gas Hydrate during a Jurassic Oceanic Anoxic Event." *Nature* 406 (July 27, 2000):392–395.

Hinrichs, Kai-Uwe, Laura R. Hmelo, and Sean P. Sylva. "Molecular Fossil Record of Elevated Methane Levels in Late Pleistocene Coastal Waters." *Science* 299 (February 21, 2003):1214–1217.

Katz, Miriam E., Dorothy K. Pak, Gerald R. Dickens, and Kenneth G. Miller. "The Source and Fate of Massive Carbon Input during the Latest Paleocene Thermal Maximum." *Science* 286 (November 19, 1999):1531–1533.

Kennett, James P., Kevin G. Cannariato, Ingrid L. Hendy, and Richard J. Behl. "Carbon Isotopic Evidence for Methane Hydrate Instability During Quaternary Interstadials." *Science* 288 (April 7, 2000):128–133.

Kennett, James P., Kevin G. Cannariato, Ingrid L. Hendy, and Richard J. Behl. *Methane Hydrates in Quaternary Climate Change: The Clathrate Gun*

Hypothesis. Washington, D.C.: American Geophysical Union, 2003.

Kerr, Richard A. "A Smoking Gun for an Ancient Methane Discharge." *Science* 286 (November 19, 1999d):1465.

Keys, David. "Global Warming: Methane Threatens to Repeat Ice-age Meltdown." *London Independent*, June 16, 2001, n.p. (LEXIS)

Munro, Margaret. "Earth's 'Big Burp' Triggered Warming: Prehistoric Release of Methane a Cautionary Tale for Today." *Edmonton Journal*, June 3, 2004, A-10.

Nisbet, E. G. "The End of the Ice Age." *Canadian Journal of Earth Sciences* 27 (1990):148–157.

Norris, Richard D., and Ursula Röhl. "Carbon Cycling and Chronology of Climate Warming during the Palaeocene/Eocene Transition." *Nature* 401 (October 21, 1999):775–778.

"North Pole Had Sub-tropical Seas because of Global Warming." Agence France Presse, September 7, 2004. (LEXIS)

Pearce, Fred. *With Speech and Violence: Why Scientists Fear Tipping Points in Climate Change*. Boston: Beacon Press, 2007b.

"Prehistoric Extinction Linked to Methane." Associated Press in *Omaha World-Herald*, July 27, 2000, 9.

Schiermeier, Quirin. "Gas Leak: Global Warming Isn't a New Phenomenon—Sea-bed Emissions of Methane Caused Temperatures to Soar in Our Geological Past, but No One Is Sure What Triggered the Release." *Nature* 423 (June 12, 2003):681–682.

Svensen, Henrik, Sverre Planke, Anders Malthe-Sorenssen, Bjorn Jamtveit, Reidun Myklebust, Torfinn Rasmussen Eidem, and Sebastin S. Rey. "Release of Methane from a Volcanic Basin as a Mechanism for Initial Eocene Global Warming." *Nature* 429 (June 3, 2004):542–545.

Wignall, Paul B., John M. McArthur, Crispin T. S. Little, and Anthony Hallam. "Methane Release in the Early Jurassic Period." *Nature* 441 (June 1, 22006):E5. [Arising from: D. B. Kemp, A. L. Coe, A. S. Cohen, and L. Schwark. *Nature* 437 (2005):396–399.]

Witze, Alexandra. "Evidence Supports Warming Theory." Dallas *Morning News* in Omaha *World-Herald* (Metro Extra), January 12, 2000, 1.

Monsoon Precipitation Patterns

David M. Anderson and colleagues reconstructed wind speeds of the Asian Southwest Monsoon for the last 1,000 years, using fossil *Globigerina bulloides* (sediment) abundance in box cores from the Arabian Sea. They found that "monsoon wind speed increased during the past four centuries as the Northern hemisphere warmed" (Anderson et al. 2002, 596). They inferred that "the observed link between Eurasian snow cover and the southwest monsoon persists on a centennial scale"

(Anderson et al. 2002, 596). Alternately, they wrote in *Science*,

> The forcing implicated in the warming trend (volcanic aerosols, solar output, and greenhouse gases) may directly affect the monsoon. Either interpretation is consistent with the hypothesis that the southwest monsoon strength will increase during the coming century as greenhouse gases continue to rise and northern latitudes continue to warm. (Anderson, Overpeck, and Gupta 2002, 596)

The Asian Southwest Monsoon affects nearly half the world's population; its effects range from the Sahara to Japan. In addition to greenhouse gases, the strength of the monsoon is governed by 10,000- to more than 100,000-year cycles in solar insolation related to the Earth's orbit around the sun. The strength of the monsoon reached a 10,000-year low about C.E. 1600, coinciding with the Maunder Minimum, a period of reduced solar activity, as well as cold temperatures during the Little Ice Age (Black 2002, 528). Evaluating the work of Anderson and colleagues, David M. Anderson, writing in *Science*, said, "This argument has critical implications in the face of global warming" (Black 2002, 528). Monsoon intensity was most notable during the twentieth century, as the pace of temperature increases accelerated.

Indian Ocean Temperatures and African Drought

Warming in the Indian Ocean is probably linked to drought in parts of eastern and southern Africa, where rainfall has declined about 15 percent since the 1980s during the March through May monsoon season. The reduction is caused by irregularities in moisture transport between the ocean and land caused by warming in the ocean, according to research published in the August 5, 2008, *Proceedings of the National Academy of Sciences*. Irregularities in the African monsoon affect marginal farmers' ability to feed themselves, according to the U.S. Agency for International Development's Famine Early Warning Systems Network (Funk et al. 2008, 11,081).

"The last 10 to 15 years have seen particularly dangerous declines in rainfall in sensitive ecosystems in East Africa, such as Somalia and eastern Ethiopia," said Molly Brown of NASA's Goddard Space Flight Center, Greenbelt, Maryland, a co-author of the study. "We wanted to know if the trend would continue or if it would start getting wetter" ("NASA Data" 2008).

Lead author Chris Funk of the University of California–Santa Barbara and colleagues showed that, paradoxically, a warmer ocean caused more precipitation off shore, while drying the atmosphere over the land. The team then ran 11 models to forecast future trends, and 10 of them agreed that the trend will continue as the ocean warms, depriving Africa's eastern seaboard of rainfall. "We can be quite certain that the decline in rainfall has been substantial and will continue to be," Funk said. "This 15 percent decrease every 20–25 years is likely to continue" ("NASA Data" 2008).

FURTHER READING

Anderson, David M., Jonathan T. Overpeck, and Anil K. Gupta. "Increase in the Asian Southwest Monsoon During the Past Four Centuries." *Science* 297 (July 26, 2002):596–599.

Black, David E. "The Rains May Be A-comin.'" *Science* 297 (July 26, 2002):528–529.

Funk, Chris, Michael D. Dettinger, Joel C. Michaelsen, James P. Verdin, Molly E. Brown, Mathew Barlow, and Andrew Hoell. "Warming of the Indian Ocean Threatens Eastern and Southern African Food Security but Could be Mitigated by Agricultural Development." *Proceedings of the National Academy of Sciences* 105 (August 12, 2008): 11,081–11,086.

"NASA Data Show Some African Drought Linked to Warmer Indian Ocean." NASA Earth Observatory, August 5, 2008. http://earthobservatory.nasa.gov/Newsroom/NasaNews/2008/2008080527314.html.

Mountain Glaciers, Slip-Sliding Away

Ice-core samples taken from high elevations in the Himalayas indicate that the 1990s were the warmest decade of the last 1,000 years at "the roof of the world." A team of scientists drilled holes at roughly 500-foot elevation intervals in the ice on a flank of Xixabangma, which rises to 26,293 feet. "This is the highest climate record ever retrieved, and it clearly shows a serious warming during the late twentieth century, one that was caused, at least in part, by human activity," said Lonnie Thompson, a professor of geological sciences at Ohio State University, who led the study.

The ice cores also contain evidence of major droughts, including one that began in 1790, which caused the deaths of more than 600,000 people in India, where monsoon rains did not arrive for six years. The research team also found evidence that levels of atmospheric dust have quadrupled over

the Himalayas during the twentieth century. Concentrations of chlorine have doubled during the same period (Thompson et al. 2000).

Rising Snow Levels, Glacial Retreat

Global warming will raise snow levels and generally cause seasonal snow packs to melt more quickly. M. Zimmermann and W. Haeberli, who have studied mountain debris flow in their native Switzerland, assert that

> Glaciers usually protect subglacial sediments from instability due to increased normal pressure and sheet strength.... [G]lacier retreat [has] caused remarkable changes in the cryosphere of the Alps, which resulted in an extension of zones prone to debris-flow initiation. The melting of small glaciers in particular exposes steep … moraines. In addition, many small lakes [have] formed behind … retreating glacier margins. Breaking of such natural dams can cause disastrous debris flows.... Continued glacier shrinkage would expose even more morainic material and, hence, further expand the area of potential debris-flow initiation. The reduction of summer runoff due to decreasing glacier surface is not likely to compensate for this negative effect. (Zimmermann and Haeberli 1992, 69)

Switzerland's Grindewald Glacier has retreated about 1.5 kilometers since 1850, with about half the retreat since 1940 (Gribben 1990, 3). The mean surface temperature of alpine permafrost in Switzerland rose between 1° and 1.5°C from 1880 to 1950, then continued to be more or less stable from 1950 to 1980, rising again after that (Haeberli 1992, 28). Warming mountain temperatures have caused "upward displacement" of plant species on 26 alpine peaks, according to one study (Grabbher et al. 1994), and 30 peaks, according to another (Pauli, Gottfried, and Grabherr 1996).

As glacial ice melted in the higher elevations of the Alps during the summer of 2000, long-lost debris and human remains surfaced in unusually large amounts. On 15,800-foot Mont Blanc, the highest elevation in the Alps, fragments from an Indian Airlines Boeing 707, including human bones, surfaced at the foot of the Bossons Glacier, near the ski resort of Chamonix. The airliner crashed in 1966. The remains of a Dijon woman who disappeared during October 1977 were found at the 9,000-foot level on the Peclet-Polset Glacier. According to an account in the *London Times*, "Bones, climbing equipment and parts from the Boeing, and another Indian airliner that crashed in 1950, had been appearing for years on the glaciers, but the heat wave had accelerated the process" (Bremner, Owen, and Henderson 2000).

Ice cores drilled in the Tibetan plateau indicate that the last half-century probably was the warmest period in the last 12,000 years in that area (Gelbspan 1997b, 139). Other ice-core records indicate that higher elevations in the tropics are warmer than they have been in at least 2,000 to 3,000 years, causing glaciers on the highest peaks to retreat. The edge of the Qori Kalis Glacier in the Peruvian Andes retreated at an annual rate of 13 feet a year from 1963 to 1978, but by 1995, the annual rate of retreat was 99 feet a year (*Understanding* 1997, 3).

In the meantime, sea levels have risen 4 to 10 inches during the last century, more than they had changed during the previous thousand years. The sea level is now at its highest level in 5,000 years, and rising (*Understanding* 1997, 5). In Peru, the retreat of glaciers in the Andes was raising concern during the late 1990s for irrigation water, which usually flows from the mountains to the coastal desert. The same rivers also are dammed for hydroelectric power.

Ice Melt and Expected Flooding

Indian researchers have warned that glaciers in the Himalayas are melting at an alarming rate and could cause a catastrophe if meltwater lakes overflow into surrounding valleys. "All the glaciers in the middle Himalayas are retreating," Syed Hasnain, of Jawaharlal Nehru University in Delhi, told *New Scientist* ("Aircraft Pollution" 1999). The Gangorti Glacier at the head of the Ganges River is receding 100 feet a year, according to Hasnain's study. If this trend continues, all the glaciers in the central and eastern Himalayas could disappear by 2035, Hasnain said.

Glaciologists Lonnie G. Thompson and Ellen Mosley-Thompson of Ohio State University have measured the retreat of the largest glaciers in the Peruvian Andes. They calculated a rate of retreat for 1963 to 1983, and found that the rate had tripled between 1983 and 1991. The loss in volume of ice increased sevenfold in less than half a human lifetime (Gelbspan 1997b, 139). Between 1963 and 1987, the glacial mass on Mount Kenya in tropical east Africa decreased 40 percent. The

Thompsons commented, "The loss of these valuable hydrological stores may result in major economic and social disruptions in those areas dependent upon the glaciers for hydrologic power and fresh water" (Gelbspan 1997b, 140). *See also:* Ice Melt, World Survey

FURTHER READING

"Aircraft Pollution Linked to Global Warming: Himalayan Glaciers Are Melting, with Possibly Disastrous Consequences." Reuters in *Baltimore Sun*, June 13, 1999, 13-A.

Bremner, Charles, Richard Owen, and Mark Henderson "Heat Wave Uncovers the Grim Secrets of the Snows." *London Times*, August 26, 2000, n.p. (LEXIS)

Gelbspan, Ross. *The Heat Is On: The High Stakes Battle Over Earth's Threatened Climate.* Reading, MA: Addison-Wesley Publishing Co., 1997b.

Grabherr, G., N. Gottfried, and H. Pauli. "Climate Effects on Mountain Plants." *Nature* 369 (1994): 448–451.

Gribben, John. *Hothouse Earth: The Greenhouse Effect and Gaia.* London: Bantam Press, 1990.

Haeberli, W. "Possible Effects of Climatic Change on the Evolution of Alpine Permafrost." In *Greenhouse-Impact on Cold-Climate Ecosystems and Landscapes,* ed. M. Boer and E. Koster, 23–35. *Selected Papers of a European Conference on Landscape Ecological Impact of Climatic Change,* Lunteren, the Netherlands, December 3–7, 1989. CARTENA Supplement 22. Cremlingen, Germany: Catena Verlag, 1992.

Pauli, H., M. Gottfried, and M. Grabherr. "Effects of Climate Change on Mountain Ecosystems—Upward Shifting of Alpine Plants." *World Resources Review* 8 (1996):382–390.

Thompson, L. G., T. Yao, E. Mosley-Thompson, M. E. Davis, K. A. Henderson, and P.-N. Lin. "A High-Resolution Millennial Record of the South Asian Monsoon from Himalayan Ice Cores." *Science* 289 (September 15, 2000):1916–1919.

Understanding the Science of Global Climate Change. Washington, D.C.: Environmental Media Services, 1997.

Zimmermann, M., and W. Haeberli. "Climatic Change and Debris-flow Activity in High Mountain Areas: A Case Study in the Swiss Alps." In *Greenhouse-Impact on Cold-Climate Ecosystems and Landscapes,* ed. M. Boer and E. Koster, 59–72. *Selected Papers of a European Conference on Landscape Ecological Impact of Climatic Change,* Lunteren, the Netherlands, December 3–7, 1989. CARTENA Supplement 22. Cremlingen, Germany: Catena Verlag, 1992.

$\mathcal{N}$

National Security and Global Warming

By 2007, even the George W. Bush administration, which long had dismissed global warming as a hoax, was beginning to parse rising ocean levels, droughts, and violent weather as national-security issues. "Unlike the problems that we are used to dealing with, these will come upon us extremely slowly, but come they will, and they will be grinding and inexorable," said Richard J. Truly, a retired U.S. Navy vice admiral and former NASA administrator (Revkin and Williams 2007). Consultants' reports to the government warned that effects of global warming could provoke large-scale migrations, increased tensions across borders, spread of diseases, and intensifying conflicts over food and water. All could involve U.S. military forces.

These reports urged that the possible outcomes of global warming should become part of U.S. security strategies and that the country "should commit to a stronger national and international role to help stabilize climate changes at levels that will avoid significant disruption to global security and stability" (Revkin and Williams 2007).

The Military Studies Global Warming

One such report, titled "National Security and the Threat of Climate Change," was commissioned by the Center for Naval Analyses, financed by the federal government, and authored by several retired generals and admirals on a Military Advisory Board. During March 2007, a similar report from the Global Business Network (which advises intelligence agencies and the Pentagon) asserted that rising oceans and intensifying storms could provoke social and political problems as many poor regions are threatened. Bangladesh was mentioned as an example. These and other studies are beginning to trace environmental roots of wars and other conflicts in places such as Afghanistan, Nepal, and Sudan.

"Just look at Somalia in the early 1990s," said one of the authors of the report, Peter Schwartz. "You had disruption driven by drought, leading to the collapse of a society, humanitarian relief efforts, and then disastrous U.S. military intervention. That event is prototypical of the future.... Picture that in Central America or the Caribbean, which are just as likely," he said. "This is not distant, this is now. And we need to be preparing" (Revkin and Williams 2007). "The evidence is fairly clear that sharp downward deviations from normal rainfall in fragile societies elevate the risk of major conflict," said Marc Levy of the Earth Institute at Columbia University, which published a study on the relationship between climate and civil war (Revkin and Williams 2007).

Two additional U.S. federal government studies (one from the Energy Information Administration and the other provided to the House of Representatives' Permanent Select Committee on Intelligence, by the National Intelligence Council) released late in June 2008 pointed out official nervousness about trends that have been concerning scientists for years. One report on energy trends emphasized the world's voracious appetite for fossil fuels, which is raising prices for oil and driving increasing consumption of coal as well. The other report characterized global warming as a national-security issue that will intensify conflict around the world.

Climate Change and U.S. Security

The studies indicated that world energy consumption will rise 50 percent from 2005 to 2030, most of the increase coming from coal and oil, with solar, wind, nuclear, and others increasing as well, but lagging in scope, driven by growth in China, India, and other industrializing nations. Coal accounted for 24 percent of total world energy use in 2002 and 27 percent in 2005, and probably will increase further, along with added emissions (Revkin 2008f). China's coal consumption nearly doubled from 2000 to 2007.

Thomas Fingar, the chairman of the National Intelligence Council, said:

> We judge global climate change will have wide-ranging implications for U.S. national security interests over the next 20 years. Although the United States will be less affected and is better equipped than most nations to deal with climate change, and may even see a benefit owing to increases in agriculture productivity, infrastructure repair and replacement will be costly. We judge that the most significant impact for the United States will be indirect and result from climate-driven effects on many other countries and their potential to seriously affect U.S. national security interests. We assess that climate change alone is unlikely to trigger state failure in any state out to 2030, but the impacts will worsen existing problems—such as poverty, social tensions, environmental degradation, ineffectual leadership, and weak political institutions. Climate change could threaten domestic stability in some states, potentially contributing to intra- or, less likely, interstate conflict, particularly over access to increasingly scarce water resources. We judge that economic migrants will perceive additional reasons to migrate because of harsh climates, both within nations and from disadvantaged to richer countries. (Revkin 2008f)

FURTHER READING

Revkin, Andrew C. "Reports: Energy Thirst Still Topping Climate Risks." Dot.Earth, *New York Times*, June 25, 2008f. http://dotearth.blogs.nytimes.com/ 2008/06/25/reports-energy-thirst-still-topping-climate-risks/index.html.

Revkin, Andrew C., and Timothy Williams. "Global Warming Called Security Threat." *New York Times*, April 15, 2007. http://www.nytimes.com/2007/04/ 15/us/15warm.html.

Netherlands, Sea Levels

The Dutch fear that rising storm surges could inundate much of the Netherlands, large areas of which have been reclaimed from the sea. Fears have been expressed that the country's western provinces may flood. The Hague, for example, may become uninhabitable as low-lying suburbs of Amsterdam return to marshland or open water.

The Dutch already have been forced to anticipate surrender of 200,000 hectares of farmland to river floodplains. A major construction program of floating homes has started. Pieter van Geel, the Dutch environment secretary, said,

> Half of our country is below sea level and so beyond a certain level, it is not possible to build dikes anymore. If we have a sea level rise of two meters, we have no control, no possibility of solving that. It's unthinkable.... I fear the problem is going to catch us within 25 years, and that is a very short period. (Evans-Pritchard 2004, A-6).

During mid-2008, a Netherlands commission recommended spending $144 billion to reinforce the country's sea defenses through the year 2100 as a precaution against sea-level rise. The measures include widening dunes facing the North Sea and raising the height of dikes along the coastline and rivers.

Overview of part of the storm-surge barrier, in Burgh-Haamstede, Netherlands, September 27, 2006. The storm-surge barrier is part of the Delta Works that protect the Netherlands against the sea. (AP/Wide World Photos)

In the Netherlands, the threat of sea-level rise is being met with amphibious homes. Anne van der Molen's two-bedroom, two-story house in Maasbommel, which cost about $420,000,

> rest[s] on land but built to rise with the water level. It sits on a hollow concrete foundation and is attached to six iron posts sunk into the lake bottom. Should the river swell, as it often does in the rain, the house will float up as much as 18 feet, held in place by two horizontal mooring posts that connect it to the neighboring house, and then float back down as the water subsides. (Lyall 2007a)

"Dutch people have always had to fight against the water," she said. "This is another way of thinking about it. This is a way to enjoy the water, to work with it instead of against it" (Lyall 2007a).

Anne van der Molen's home is one of 46 built in a single development that is designed to anticipate relentlessly rising sea levels and a heightened chance of flooding rains because of climate change, said Steven de Boer, a concept developer at Dura Vermeer, the company that developed the project. In 1995, local rivers flooded and forced 250,000 people to evacuate. Local dikes have since been raised to counter a similar flood, but since the Netherlands is in essence a large river delta, caution is the byword for the future.

"All the universities are united in one big program with the government; we have a team of some 500 people working on climate-proofing the Netherlands," said Pier Vellinga, a professor of climate change at the University of Amsterdam. "Whatever happens—Greenland melting or tropical storms surging on the Atlantic—we are here to stay. That is becoming our national slogan" (Lyall 2007a).

FURTHER READING

Evans-Pritchard, Ambrose. "Dutch Have Only Years before Rising Seas Reclaim Land: Dikes No Match against Global Warming Effects." *London Daily Telegraph* in *Ottawa Citizen*, September 8, 2004, A-6.

Lyall, Sarah. "At Risk from Floods, but Looking Ahead with Floating Houses." *New York Times*, April 3, 2007a. http://www.nytimes.com/2007/04/03/science/earth/03clim.html.

New Jersey and Long Island: Sea-Level Rise

Rising waters are nibbling at the coastlines of New Jersey and Long Island. Some beachfront vacation homes on Long Island have been raised on stilts as tides have risen. Hoboken subway stations also have been reinforced with "tide guards" (Nussbaum 2002, A-1). Vivian Gornitz, a sea-rise specialist at NASA's Goddard Institute of Space Studies in New York City, said that the amount of land lost to the sea could accelerate to between two and five times the present rate by the end of the twenty-first century. She said that seas in the area could rise between 4 and 12 inches by 2020, 7 inches to 2 feet by 2050, and 9.5 inches to almost 4 feet by 2080.

Jim Titus, project manager for sea-level rise at the U.S. Environmental Protection Agency, said that such a rise could imperil all beaches in the area. "The data doesn't lie," he said. "The sea is rising. The shore is eroding because the sea is rising. There's no doubt about that." At Long Beach Island, said Titus, sea levels already have risen a foot in 70 years. Two days a month, high tides at full moon (called "spring tides") often cause flooding in the area. Homes are now built on stilts and garbage cans are anchored to prevent them from floating away on the tides (Nussbaum 2002, A-1). Local officials said that future storms could flood transportation infrastructure, including the Newark-Liberty International Airport, which lies less than 10 feet above present sea levels.

FURTHER READING

Nussbaum, Alex. "The Coming Tide: Rise in Sea Level Likely to Increase N.J. Floods." *Bergen County Record* (New Jersey), September 4, 2002, A-1.

El Niño, La Niña (ENSO), and Climate Change

No single variable feature of Earth's oceans has a more profound influence on climate than the El Niño Southern Oscillation (ENSO) cycle, or "El Niño," which warms a broad swath of the Pacific Ocean near the Equator, westward from South America, in irregular cycles of roughly three to seven years. Intervening periods of cooler water are called "La Niña."

The El Niño pattern causes dramatic warming of equatorial waters adjacent to the coast of South America, an area that is usually cooler than surrounding waters. It can affect air-circulation patterns (and therefore weather) in North and South America, as well as across the Pacific Ocean into southern Asia. Most notably, these shifting

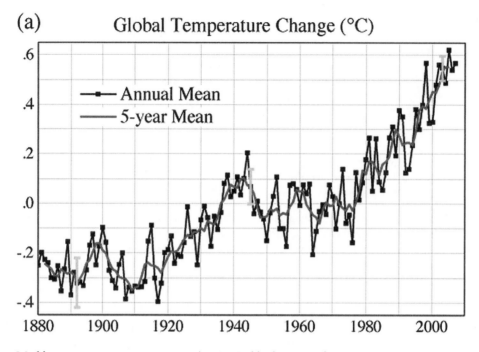

World temperature mean, 1880–2008. (NASA/Goddard Institute for Space Studies)

patterns can turn the western coast of South America, which is usually a desert, into an area plagued by torrential rains, while changing the Amazon Valley, which is usually wet, into a much drier region. This change takes place as the direction of winds across the Andes changes from generally east-to-west to west-to-east. The El Niño pattern also restricts the supply of moisture to Asia, hobbling the usual monsoon pattern, with widespread effects for hundreds of millions of people from India, through Indonesia, to Australia.

The ENSO cycle can be so powerful that climate-change contrarians mistake a change in phrase for the coming of a new ice age. In 2008, for example, a strong La Niña sent temperatures to below average in large parts of the world for the first time in more than 20 years, except for the Mount Pinatubo eruption in 1991, which produced an overload of cooling sulfur in the atmosphere for several months.

While ENSO "is the dominant year-to-year climate signal on Earth," according to Michael J. McPhaden and colleagues, writing in *Science* (McPhaden, Zebiak, and Glantz 2006, 1740), its frequency may not be related to global warming itself. Recent behavior of the El Niño/La Niña cycle "is most likely not outside the range

expected for natural climate variability.... There is no definitive evidence from the instrumental record at present for changes in ENSO behavior in response to greenhouse-gas forcing" (McPhaden, Zebiak, and Glantz 2006, 1744).

While future influence of global warming on ENSO is "open to debate," these authors argued that "[t]he consensus outlook from the current generation of global climate models suggest no significant change in ENSO characteristics under various greenhouse-gas emission scenarios that presume a doubling of atmospheric CO_2 from pre-industrial levels over the next 100 years" (McPhaden, Zebiak, and Glantz 2006, 1744). The authors also wrote, "We cannot say with confidence at present how global warming will affect either ENSO variability or the background state on which it is superimposed" (McPhaden, Zebiak, and Glantz 2006, 1744).

More Frequent El Niños in a Warmer World?

McPhaden's position has been debated by other scientists, who believe that the frequency and potency of El Niño episodes may be related to a warming climate. As global temperatures have risen since 1980, the ENSO cycle has occurred more frequently than at any other time in the century and a half during which detailed worldwide records have been kept. Climate modeling by A. Timmermann and colleagues tends to support the idea that further warming will enhance the prominence of El Niño–type events in world oceanic and terrestrial circulation and climate patterns. The forecasts of Timmermann and colleagues come with an added twist: the La Niña side of this oscillation may be rarer, but possibly colder as well. In brief, their models indicate that weather may become generally wilder as well as warmer as levels of greenhouse gases increase. The researchers believe that in a warmer world the climate will change in several ways.

> The climatic effects will be threefold. First, the mean climate in the tropical Pacific region will change toward a state corresponding to present-day El Niño conditions. It is therefore likely that events typical of El Niño will also become more frequent. Second, a stronger inter-annual variability will be superimposed on the changes in the mean state, so year-to-year variations may become more extreme under enhanced greenhouse conditions. Third, the inter-annual variability will be more strongly skewed, with cold events (relative to the warmer mean state) becoming more frequent. (Timmermann et al. 1999, 696)

Alexander W. Tudhope and colleagues found evidence that El Niño episodes have occurred for at least 130,000 years, under both glacial and interglacial conditions, but that ENSO episodes usually increase with warmer weather.

This study used coral populations off the New Guinea coast as a climate proxy. Rapid tectonic lift in the area has exposed corals that are much older than most available samples. "At their site," reported Julia Cole in *Science*, "modern (late nineteenth century to today) corals show the highest amplitude of inter-annual ENSO variance of all samples over the last 130,000 years" (Cole 2001, 1497). The relationship between warming and more frequent El Niños is complicated by other factors, however. Variations in the Earth's orbit (and changing intensities on incoming sunlight) also may play a role.

A Hothouse Climate and El Niño

Matthew Huber and Rodrigo Caballero looked at the warm climate of the Eocene, 55 to 35 million years ago, as a testing ground for interaction between a "hothouse" climate and frequency of El Niño cycles. Comparing climate simulations of that period with variability records preserved in

lake sediments, they found little connection. "The simulations show Pacific deep-ocean and high-latitude surface warming of about 10°C but little change in the tropical thermocline structure, atmosphere-ocean dynamics, and ENSO [El Niño], in agreement with proxies," they wrote in *Science*. "This result contrasts with theories linking past and future 'hothouse' climates with a shift toward a permanent El Niño–like state" (Huber and Caballero 2003, 877).

The most definitive support for the idea that a warmer Earth promotes El Niño conditions was published in *Science* during the summer of 2005. Michael W. Wara and colleagues reported that during the warm early Pliocene, between 3 million and 4.5 million years ago (the most recent interval in global climate markedly warmer than the present), El Niño–type conditions were predominant in the Pacific Ocean. "Sustained El Niño–type conditions … could be a consequence of, and play an important role in determining, global warmth," they wrote (Wara, Ravelo, and Delaney 2005, 758). This study contradicted another, which was compiled from much less data, which made a case for a persistent cool La Niña state in the eastern Pacific during the early-mid Pliocene (Kerr 2005b, 687; Rickaby and Halloran 2005, 1948–1952).

El Niños in Cool Periods

El Niño conditions do not always occur in warm weather, however. The lead author of an article in *Science*, Tammy Rittenour, said that traces of El Niño–like conditions have been traced to the last ice age, 13,500 to 17,500 years ago (Rittenour, Brigham-Grette, and Mann 2000, 1039–1042). The work of Rittenour and colleagues suggests that temperature may be only one of several factors affecting the ebb and flow of several El Niño cycles (2000, 1039).

Rittenour was surprised by indications that her climate research was pointing to El Niño–type events during a glacial period. She had begun by investigating the rate at which melting drained glacial Lake Hitchcock, in present-day southern New England. "It kind of threw me back," Rittenour told a reporter. "Coming up with El Niño was kind of a shock" (Anderson 2000, 12-B).

Amazon Valley May Become a Desert

Mat Collins, a senior research fellow at the Meteorological Office in Reading, England, has raised a possibility that persistent El Niño conditions accelerated by global warming could change prevailing winds in South America in such a way that the Amazon Valley may become a desert by the end of the twenty-first century. "In our model, 50 years from now the Amazon dries up and dies," he told the British Association at the University of Salford (Arthur 2003a, 6). Whatever its causes, El Niño is a complex phenomenon; its frequency and severity also has been related to atmospheric changes provoked by major volcanic eruptions (Adams, Mann, and Amman 2003, 274–278).

If and when such a change takes place, Arthur said that it may compound global warming: "There would be a reinforcing effect because, as the rainforest dried up, the carbon that is presently locked in its vegetation would be released into the atmosphere" (Arthur 2003a, 6). Collins rated the chance of such a scenario at between 10 and 20 percent, which he called a preliminary estimate. Desert conditions in the Amazon could be initiated, said Collins, by a "super El Niño" that would cause prevailing winds in equatorial South America to surge over the Andes from the west, causing floods there, as is usual during El Niño episodes. Collins believes that warming will increase the number, intensity, and duration of El Niño events, which cause desiccating westerly winds to tumble down the Andes into the Amazon Valley, reversing the usual up-slope flow that helps cause copious rainfall.

FURTHER READING

Adams, J. Brad, Michael E. Mann, and Casper M. Amman. "Proxy Evidence for an El Niño-like Response to Volcanic Forcing." *Nature* 426 (November 20, 2003):274–278.

Anderson, Julie. "UNL Student Helps Shed New Light on El Niño." *Omaha World-Herald*, May 14, 2000, 12-B.

Arthur, Charles. "Super El Niño Could Turn Amazon into Dustbowl: British Association for the Advancement of Science." *London Independent*, September 9, 2003a, 6.

Cole, Julia. "A Slow Dance for El Niño." *Science* 291 (February 23, 2001):1496–1497.

Huber, Matthew, and Rodrigo Caballero. "Eocene El Niño: Evidence for Robust Tropical Dynamics in the 'Hothouse.'" *Science* 299 (February 7, 2003): 877–881.

Kerr, Richard A. "El Niño or La Niña? The Past Hints at the Future." *Science* 309 (July 29, 2005b):687.

McPhaden, Michael J., Stephen E. Zebiak, and Michael H. Glantz. "ENSO as an Integrating Concept in Earth Science." *Science* 314 (December 15, 2006): 1740–1745.

Rickaby, R. E. M., and P. Halloran. "Cool El Nina During the Warmth of the Pliocene?" *Science* 307 (March 25, 2005):1948–1952.

Rittenour, Tammy M., Julie Brigham-Grette, and Michael E. Mann. "El Niño-Like Climate Teleconnections in New England during the Late Pleistocene." *Science* 288 (May 12, 2000):1039–1042.

Timmermann, A., J. Oberhuber, A. Bacher, M. Esch, M. Latif, and E. Roeckner. "Increased El Niño Frequency in a Climate Model Forced by Future Greenhouse Warming." *Nature* 398 (April 22, 1999):694–696.

Wara, Michael W., Ana Christina Ravelo, and Margaret L. Delaney. "Permanent El Niño-Like Conditions during the Pliocene Warm Period." *Science* 309 (July 29, 2005):758–761.

Nitrogen Cycle and Warming Seas

Some research has indicated that warming waters may change the nitrogen cycle in many coastal waters in ways that increase acidification and endanger sea life by depriving waters of oxygen. Changes in the way that seawater handles nitrogen also have been aggravated by other factors, such as fertilizer runoff from farming fields. Oxygen is less soluble in warmer waters, a fact that comes into play as nitrogen-rich runoff from fertilizer contributes to "dead zones" in the oceans where large rivers reach the sea. Both nitrates and sulfates in these polluted zones cause water to become more acid, making it even more hostile to sea life as waters warm.

Problems have been noted in Narragansett Bay, Rhode Island, which for many years acted as a sink for reactive nitrogen that human activity dumped into it. By 2006, however, the overload of nitrates and rising water temperatures caused the bay to switch, in one year, from a nitrate sink to a nitrate source (Lane 2007, 778). Phytoplankton production in the bay, a basis of the maritime food chain, has been failing rapidly,

"If Narragansett is typical of other bays," wrote Nick Lane in *Nature*,

[I]t could be a harbinger of a new threat [from warming seas]. Shifting the effect of anthropogenic nitrogen loading beyond the immediate coastal zone could destabilize ocean ecosystems by acidifying the waters, exacerbating harmful algal blooms, killing fish and shellfish, or perhaps even powering a vicious new cycle of global warming. (Lane 2007, 778)

Don Canfield, a geochemist at the University of South Denmark, believes that the problems in Narragansett Bay are symptoms of a global problem, and that "nitrate export" into the open oceans could rise quickly within a few years (Lane 2007, 780). A by-product of this process is nitrous oxide (laughing gas), which, as a greenhouse gas, is 200 to 300 times as potent per molecule as carbon dioxide.

FURTHER READING

Lane, Nick. "Climate Change: What's in the Rising Tide?" *Nature* 449 (October 18, 2007):778–780.

Northeast United States, Anticipated Weather in 2100

A study by the Union of Concerned Scientists sketched a picture of life in the U.S. Northeast at the end of the twenty-first century if "business as usual" greenhouse-gas emissions continue (see http://ucsaction.org/ct/Sdwb-cd1ZR7s). It isn't pretty at all.

East coast cities such as Philadelphia and Washington, D.C., are projected to have temperatures of 100°F or more for 30 days during an average summer (compared with two a year presently), as well as more frequent droughts and more extreme rainstorms. Spruce and hemlock trees may retreat across the Canadian border, and coastal communities (Boston, New York, Atlantic City, and others) could face regular flooding, all products of summer temperatures that are 6° to 14°F warmer and winters that are 8° to 12°F warmer.

The area would realize a shorter list of benefits, said Peter Frumhoff, one of the report's lead authors, who was lead author of the impact and mitigation report by the Intergovernmental Panel on Climate Change. Some farmers would enjoy a longer growing season, and heating bills would decline substantially in winter (Kaufman 2007d, A-4).

The report said that "[t]he Northeast's climate is changing; spring arrives sooner, with hotter summers and milder, less snowy winters. Annual temperatures across the nine states of the Northeast have risen more than 1.5°F between 1970 and 2006. Winter temperatures rose 3.8°F during the same period" ("Global Warming Will" 2006). "The very notion of the Northeast as we know it is at stake," said Dr. Cameron Wake, a research professor at the University of New Hampshire's Climate Change Research Center and co-author of the report. "The near-term emissions choices we make in the Northeast and throughout the world will help determine the climate and quality of life our children and grandchildren experience," according to the report ("Global Warming Will" 2006).

The report describes impacts of the two emissions scenarios over 30-year increments: 2010–2039, 2040–2069, and 2070–2099. If "business-as-usual" emissions continue, according to its analysis, "summer in upstate New York could feel like present-day South Carolina by the end of the century. Average annual temperatures in the Northeast will likely rise 6.5° to 12.5°F by the end of the century absent change," the report said, compared with an increase of 3.5° to 6.5°F if emissions are sharply reduced ("Global Warming Will" 2006).

FURTHER READING

"Global Warming Will Dramatically Alter U.S. Northeast." Environment News Service, October 4, 2006. http://www.ens-newswire.com/ens/oct2006/2006-10-04-03.asp.

Kaufman, Marc. "Report Warns of a Much Warmer Northeast Effects Could Be Disastrous, Says Two-Year Study." *Washington Post*, July 12, 2007d, A-4. http://www.washingtonpost.com/wp-dyn/content/article/2007/07/11/AR2007071100942_pf.html.

North Sea Ecological Meltdown

Global warming is contributing to changes that some scientists describe as an "ecological meltdown," with devastating implications for fisheries and wildlife. The "meltdown" begins at the base of the food chain, as increasing sea temperatures reduce plankton populations. The devastation of the plankton then ripples up the food chain as fish stocks and sea-bird populations decline as well.

Scientists at the Sir Alistair Hardy Foundation for Ocean Science in Plymouth, England, which has been monitoring plankton growth in the North Sea for more than 70 years, have said that unprecedented warming of the North Sea has driven plankton hundreds of miles to the north. They have been replaced by smaller, warm-water species that are less nutritious (Sadler and Lean 2003, 12). Overfishing of cod and other species has played a role, but fish stocks have not recovered after cuts in fishing quotas. The number of salmon returning British waters is now half of what it was 20 years previously. A decline of plankton stocks is a major factor in this decline.

"A regime shift has taken place and the whole ecology of the North Sea has changed quite dramatically," said Chris Reid, the foundation's director.

> We are seeing a collapse in the system as we knew it. Catches of salmon and cod are already down and we are getting smaller fish. We are seeing visual evidence of climate change on a large-scale ecosystem. We are likely to see even greater warming, with temperatures becoming more like those off the Atlantic coast of Spain or further south, bringing a complete change of ecology. (Sadler and Lean 2003, 12)

According to a report by Richard Sadler and Geoffrey Lean in the *London Independent*, research by the British Royal Society for the Protection of Birds (RSPB) has determined that seabird colonies off the Yorkshire coast and the Shetlands during 2003 "suffered their worst breeding season since records began, with many simply abandoning nesting sites" (Sadler and Lean 2003, 12). The sea birds are declining because sand eels are declining. The sand eels feed on plankton, which have declined as water temperatures have risen. This survey concentrated on kittiwakes, one breed of sea birds, but other species that feed on the eels, including puffins and razorbills, also have been seriously affected. Euan Dunn of the RSPB commented: "We know that sand eel populations fluctuate and you do get bad years. But there is a suggestion that we are getting a series of bad years, and that suggests something more sinister is happening" (Sadler and Lean 2003, 12).

Sand eels also account for a third to half of the North Sea catch, by weight. They have heretofore been caught in huge quantities by Danish factory ships, which turn them into food pellets for pigs and fish. During the summer of 2003, the Danish fleet caught only 300,000 English tons of its 950,000-ton quota, a record low (Sadler and Lean 2003, 12). The situation is "unprecedented in terms of its scale and the number of species it's affecting," said ornithologist Eric Meek of the RSPB (Kaiser 2004, 1090).

FURTHER READING

Kaiser, Jocelyn. "Reproductive Failure Threatens Bird Colonies on North Sea Coast." *Science* 305 (August 20, 2004):1090.

Sadler, Richard, and Geoffrey Lean. "North Sea Faces Collapse of its Ecosystem." *London Independent*, October 19, 2003, 12.

Northwest and Northeast Passages

During late August 2007, for the first time, a Northwest Passage from Baffin Bay to Northern Alaska opened during a season of record ice melt for the Arctic Ice Cap. During September 2008, shipping passages opened briefly on both sides of the Arctic Ice Cap, a Northwest *and* a Northeast passage. A shipping channel through the Arctic could cut nearly 8,000 miles off the trade route between Europe and Asia.

According to a report released during 2002, Gary Brass, director of the U.S. Arctic Research Commission, said that within a decade both the Northwest Passage and the Northern Sea Route could be open to vessels lacking reinforcement against the ice for at least a month in the summer, assuming that recent trends reducing ice coverage continue. By 2050, both routes may be open for the entire summer, according to this report (Kerr 2002b, 1490). German explorers asserted on October 10, 2002, that global warming and unusual wind patterns had enabled them to become the first navigators to sail unaided in a yacht through the usually icebound route along Russia's Arctic coast. A 12-man team, led by the explorer Arved Fuchs, 49, made the trip on its fourth west-to-east attempt during the summer in a 60-foot sail-driven, wooden trawler that they had converted into a yacht.

European mariners have been seeking and failing to find a Northwest Passage since 1497, when English King Henry VII sent Italian explorer John Cabot to look for a route from Europe to the Orient that would avoid the southern tip of Africa. Many explorers failed at the task, including Sir Francis Drake and Captain James Cook. NASA's Advanced Microwave Scanning Radiometer aboard the Aqua Satellite observed open water along nearly the entire route on August 22, 2007. "Although nearly open, the Northwest Passage was not necessarily easy to navigate in August 2007," NASA noted. "Located 800 kilometers (500 miles) north of the Arctic Circle and less than 1,930 kilometers (1,200 miles) from the North Pole, this sea route remains a significant challenge, best met with a strong icebreaker ship backed by a good insurance policy" ("Northwest Passage" 2007).

Some industrialists are gleefully anticipating the melting of polar ice so that they can open new shipping lanes and ports, drill for oil, or engage in other profitable activities. Bad news for polar bears, for example, may be good for

Pat Broe, a Denver entrepreneur, who bought the derelict port of Churchill, Manitoba, on Hudson's Bay, from the Canadian government in 1997 for about $10 Canadian. By Broe's calculations, once the Arctic Ice Cap melts, Churchill could bring in as much as $100 million a year as a port on Arctic shipping lanes between Europe, Asia, and the Americas—the fabled Northwest Passage—shorter by thousands of miles than present southerly routes (Krauss et al. 2005).

Oil companies have been pushing into the frigid Barents Sea seeking undersea oil and gas fields made accessible not only by melting ice but also by advances in technology. But now, as thinning ice stands to simplify construction of drilling rigs, exploration is likely to move even farther north (Krauss et al. 2005). In 2004, scientists found evidence of oil in samples taken from the floor of the Arctic Ocean only 200 miles from the North Pole. All told, one-quarter of the world's undiscovered oil and gas resources lies in the Arctic, according to the U.S. Geological Survey (Krauss et al. 2005).

Tourists also have been swarming to the Arctic. One day during the summer of 2005, residents of Pangnirtung, on the east coast of Baffin Island, were greeted by a surprise: a 400-foot European cruise ship, which had dropped anchor unannounced and sent several hundred tourists ashore in small boats.

FURTHER READING

Kerr, Richard A. "A Warmer Arctic Means Change for All." *Science* 297 (August 30, 2002b):1490–1492.

Kraus, Clifford, Steven Lee Myers, Andrew C. Revkin, and Simon Romero. "As Polar Ice Turns to Water, Dreams of Treasure Abound." *New York Times*, October 10, 2005. http://www.nytimes.com/2005/10/10/science/10 arctic.html.

"Northwest Passage Nearly Open." NASA Earth Observatory, August 27, 2007. http://earthobservatory. nasa.gov/Newsroom/NewImages/images.php3?img_ id=17752.

Nuclear Power as "Clean" Energy

James Lovelock, who pioneered measurement of trace gases in the atmosphere and developed the idea of the Gaia hypothesis, has become a staunch advocate of nuclear power to bridge the "gap" between fossil fuels and other sources of power. Gaia, named after the Greek goddess of the Earth,

has been defined by Lovelock as a view that the planet acts as a living organism to maintain "life on Earth [that] actively keeps the surface conditions always favorable for whatever is the contemporary ensemble of organisms" (Volk 2006, 869). Lovelock, faced with scientific criticism, has since reformulated his school of thought as more abstract theory. Lovelock now asserts that human manipulation of greenhouse-gas levels in the atmosphere has stirred Gaia to declare war on humanity in which she "now threatens us with the ultimate penalty of extinction" (Volk 2006, 869). Such language strikes many other scientists as metaphorical and anthropomorphic. Pressed, Lovelock agrees that the idea is a metaphor, with limited literal meaning.

In *The Revenge of Gaia*, Lovelock asserted that solar, wind, or biomass will take too much room (a quarter million wind turbines, for example) to provide the power needs of the United Kingdom, and that nuclear fission should be used as a bridge to nuclear fusion and more efficient renewables. He sees nuclear waste as a small price to pay for its value as a carbon-free, proven source of power. Fears of radiation are overblown, Lovelock contended. (Lovelock also considered huge-scale engineering solutions, such as sun-blocking reflectors in space, when and if warming goes into a feed-back powered runaway mode.) Tyler Volk, reviewing the book in *Nature,* concluded: "Read this book for its thoughtful sections on global energy and climate, but steer clear of its web of Old Testament-like prophecy" (Volk 2006, 870).

Barry Commoner, one of the founders of the modern environmental movement, opposed Lovelock's position dramatically:

> This is a good example of shortsighted environmentalism. It superficially makes sense to say, "Here's a way of producing energy without carbon dioxide." But every activity that increases the amount of radioactivity to which we are exposed is idiotic. There has to be a life-and-death reason to do it. I mean, we haven't solved the problem of waste yet. We still have used fuel sitting all over the place. I think the fact that some people who have established a reputation as environmentalists have adopted this is appalling. (Vinciguerra 2007)

"Breeder" Reactors

Tom Blees, in *Prescription for the Planet*, asserted that fourth-generation nuclear power (formally called the Integral Fast Reactor [IFR] or "breeder reactors") eliminates moderating materials that are used in thermal reactors, using less than 1 percent of fissionable material in uranium ore, very gradually burning all the uranium. This type of reactor virtually eliminates the cruelties of uranium mining (at present-day rates of use, 1,000 years of supply are on hand). It also solves the problem of spent fuel. What fuel is consumed can be recycled on-site into new rods. Ashes left over from this process cannot be used as weapons, and their radioactive half-life is relatively short, a few hundred years for reduction to background levels of uranium ore.

These reactors also can burn the waste of thermal reactors. Furthermore, Blees argued that "IFR design can be practically failsafe, relying on physical properties of reactor components to shut down in even the most adverse situations, thus avoiding coolant problems of Chernobyl and Three Mile Island, as well as the earthquake problem" (Hansen 2008c). Blees asserted that governmental tests of this technology were canceled for political reasons in 1994 by the Clinton administration. Research into "fast" reactors continues in other countries, according to Blees. Blees pointed out that coal-fired power plants "are exposing the population to more than 100 times more radioactive material than nuclear power plants—some of it spewed out the smokestacks, but much of it in slag heaps of coal ash" (Blees 2008; Hansen 2008c).

China's Nuclear Construction Campaign

As environmental philosophers debate the role of nuclear power in relation to global warming, by 2007, China had begun construction on dozens of new nuclear-power plants to address part of its growing global-warming burden from low-quality coal. China plans to spend $50 billion to build 32 nuclear plants by 2020. Some experts believe that China may build 300 more such plants by about 2050, to power what will be, by then, the largest national economy in the world. By that time, China may have half the nuclear-power capacity in the world (Eunjung Cha 2007, D-1). China also is building the world's largest repository for spent nuclear fuel in its western desert, amid the Beishan Mountains, an area that is nearly bereft of human habitation. The nuclear-construction binge represents a major change for China, where only 2.3 percent of electricity was generated from nuclear power in 2006, compared with about 20 percent in the United States and almost 80 percent in France. In 2007, China had only nine nuclear-power plants.

In part because of Chinese demand, the price of processed uranium ore increased from $10 to $120 a pound between 2003 and 2007. Expecting that China will be one of its major customers, Japan's Toshiba paid $5.4 billion during 2006 to acquire a U.S. company, Westinghouse Electric, which specializes in construction of nuclear plants. The Chinese government, emphasizing safety, has been using companies such as Westinghouse to instruct its engineers who will build and operate new nuclear plants.

Sweden and Nuclear Power

During the 1970s, energy interests in Sweden pitched nuclear power as an antidote to oil. Seven new nuclear reactors came into operation between 1972 and 1980. Because of environmental considerations, high production costs, and low world market prices, Sweden's substantial uranium reserves (250,000 to 300,000 tons, or about 20 percent of known world reserves) have not been exploited. By 2006, 45 percent of Sweden's electricity still was being generated by nuclear power, and 8 percent from fossil fuels.

Swedish nuclear power sustained a blow during 2006 because of a near-meltdown at one of the country's reactors. An electricity failure on July 25, 2006, led to the immediate shutdown of the Forsmark 1 reactor after two of four backup generators, which supply power to the reactor's cooling system, malfunctioned for about 20 minutes.

This near-accident revived memories of the Ukraine's Chernobyl, which showered radioactive fallout over much of Sweden in April 1986. By 2010, all 12 of Sweden's nuclear plants will be shut down, to be replaced by natural gas, a fossil fuel, until other sources are available.

FURTHER READING

Blees, Tom. *Prescription for the Planet: The Painless Remedy for Our Energy and Environmental Crises.* Self-published, 2008. http://www.prescriptionfortheplanet.com.

Eunjung Cha, Ariana. "China Embraces Nuclear Future: Optimism Mixes with Concern as Dozens of Plants Go Up." *Washington Post*, May 29, 2007.

Hansen, James E. "Trip Report." August 5, 2008c. http://www.columbia.edu/~jeh1/mailings/20080804_TripReport.pdf.

Lovelock, James. *The Revenge of Gaia: Why the Earth Is Fighting Back—and How We Can Still Save Humanity.* London: Allen Lane, 2006.

Volk, Tyler. "Real Concerns, False Gods: Invoking a Wrathful Biosphere Won't Help Us Deal with the Problems of Climate Change." Review, James Lovelock, *The Revenge of Gaia,* 2006. *Nature* 440 (April 13, 2006):869–870.

Vinciguerra, Thomas. "At 90, an Environmentalist from the '70s Still Has Hope." *New York Times,* June 19, 2007. http://www.nytimes.com/2007/06/19/science/earth/19conv.html.

O

Ocean Circulation, Worldwide

Warming seas could affect patterns of oceanic circulation around the world (called thermohaline circulation) that replenish the world's oceans with oxygen, leading to the possible extinction of several sea creatures. A debate also has developed regarding whether changes in ocean circulation could cause some areas of the world (most notably, Western Europe) to experience a marked cooling trend by suppressing the warm waters of the Gulf Stream, which help to supply that area with a winter climate that is unusually warm for such a northerly latitude.

By 2007, however, scientific consensus was leaning away from this possibility, as changes earlier taken to be long term were being regarded, instead, as part of cyclical variation that probably is not related to a warming climate (Schiermeier 2007b, 844–845). Stuart Cunningham of the National Oceanography Centre at Southampton, United Kingdom, and colleagues found that the thermohaline circulation varied by 25 percent in one year.

Warming and Inhibition of Thermohaline Circulation

The term "thermohaline" is used because "water density in the ocean is determined by both temperature and salinity" (Rahmstorf 2003, 699). According to Peter U. Clark and colleagues, writing in *Nature*, most, but not all, coupled general circulation model projections of the twenty-first century's climate anticipate a reduction in the strength of the Atlantic overturning circulation with warming seas that are provoked by increasing concentrations of greenhouse gases (Clark et al. 2002, 863–868). This flow is part of what marine scientists called the Global Conveyor, a vast submarine flow of water south from the Arctic. The Gulf Stream keeps Britain 5°C warmer than expected for such a northerly latitude and delivers 27,000 times more heat to British shores than all of that nation's power stations (Radford 2001b, 3).

A report presented at the annual meeting of the American Association for the Advancement of Science in February 2005 by Ruth Curry, a scientist at the Woods Hole Oceanographic Institute, indicated that massive amounts of fresh water from melting Arctic ice are seeping into the Atlantic Ocean. According to Curry's research, between 1965 and 1995, about 4,800 cubic miles of fresh water (more water than Lake Superior, Lake Erie, Lake Ontario, and Lake Huron combined) melted from the Arctic region and poured into the northern Atlantic. Curry projected that if this pattern continues at current rates, the thermohaline circulation may begin to shut down in about two decades.

Furthermore, said Curry, Greenland's ice, which to date had not been melting quickly, now is thawing at more rapid rates. "We are taking the first steps," Curry said in a news conference. "The system is moving in that direction" (Borenstein 2005, A-6). In the longer range, according to calculations by Curry and colleagues, "at the observed rate, it would take about a century to accumulate enough fresh water (e.g., 9,000 cubic thousand cubic meters) to substantially affect the ocean exchanges across the Greenland-Scotland Ridge, and nearly two centuries of continuous dilution to stop them" (Curry et al. 2005, 1774).

Ocean Salinity and Circulation

The salinity of the North Atlantic is closely related to world ocean circulation. If the North Atlantic becomes too fresh (due, most likely, to melting Arctic ice), its waters could stop sinking, and the Conveyor could slow, even perhaps stop. Spencer R. Weart, in *The Discovery of Global Warming*, provided a capsule description of possible changes in the ocean's thermohaline circulation under the influence of sustained, substantial global warming:

> If the North Atlantic around Iceland should become less salty—for example, if melting ice sheets diluted the upper ocean layer with fresh water—the surface layer would no longer be dense enough to sink. The entire circulation that drove cold water south along the bottom could lurch to a halt. Without the vast compensating drift of tropical waters northward, a new glacial period could begin. (Weart 2003, 64)

While atmospheric winds move heat quickly across large distances, fundamental changes in ocean temperatures and circulation may take centuries. Thermohaline circulation involves warm surface currents that distribute tropical heat, as well as deeper currents that carry cold water back toward the Equator. Together, these currents form a system that circulates ocean water, heat, oxygen, and nutrients through the North and South Atlantic, then into the Indian Ocean and the Pacific. The North Atlantic is one of only two areas in the world's oceans where the presence of ice aids the formation of salty, dense water, which then sinks and helps to drive ocean circulation, carrying oxygen with it. The second is the Weddell Sea in the Antarctic.

According to one expert observer, "If the warming is strong enough and sustained long enough, a complete collapse [of thermohaline circulation] cannot be excluded" (Clark et. al. 2002, 863). "What is fairly clear is that if the ocean circulation patterns that now warm much of the North Atlantic were to slow or stop, the consequences could be quite severe," Clark said. "This might also happen much quicker than many people appreciate. At some point the question becomes how much risk do we want to take?" ("Slowing Ocean" 2002). This circulation is vital to the circulation of oxygen and nutrients in much of the Atlantic Ocean—and, thus, necessary to marine life as we know it. Similar circulation systems perform the same function in other oceans around the world. Without the thermohaline circulation, the oceans could become largely stagnant with large areas bereft of life.

Thomas Stocker and colleagues also argue that these changes may be nonlinear: "they may have large amplitudes and may occur as surprises." Inherent in such changes is their "reduced predictability" (Stocker et al. 2001, 277). "Among such non-linear changes," they wrote, "are the collapse of large Antarctic ice masses and rapid sea-level rise, the desertification of entire land regions, the thawing of permafrost and associated release of large amounts of radiatively

active gases, and the collapse of the large-scale Atlantic Thermohaline (i.e., the temperature and salinity driven) circulation" (Stocker et al. 2001, 277).

Carsten Ruhlmann and colleagues wrote in *Nature* that the Earth's emergence from its last glaciation was punctuated by several short, sharp periods during which glacial conditions returned. These variations seem to have been triggered by the type of changes in the North Atlantic thermohaline circulation that are considered possible in a world warmed by increasing emissions of greenhouse gases. "The thermohaline circulation was the important trigger for these rapid climate changes," wrote Ruhlmann and colleagues (1999, 512). Climate changes that diminish thermohaline circulation tend to divert the Gulf Stream southward, cooling the North Atlantic Ocean (as well as Greenland and Western Europe), but "warming ... the western tropical North Atlantic ... and most of the Southern Hemisphere" (Ruhlmann et al. 1999, 512).

A study of North Atlantic ocean circulation published in late 2005 reported a 30 percent reduction in the Meridianal Overturning (Thermohaline) Circulation at 26.5 degrees north latitude, based on readings taken during 1957, 1981, 1992, and 1998. Harry Bryden at the Southampton Oceanography Centre in the United Kingdom, whose group carried out the analysis, said he was not sure whether the change was temporary or part of a long-term trend. This analysis is, however, "the first observational evidence that such a decrease of the oceanic overturning circulation is well underway" (Quadfasel 2005, 565).

"We don't want to say the circulation will shut down," Bryden told the *New Scientist*. "But we are nervous about our findings. They have come as quite a surprise" (Pearce 2005). Bryden's team measured north-south heat flow during 2004 using a set of instruments placed in various locations in the Atlantic between the Canary Islands and the Bahamas, and then compared results with surveys taken in 1957, 1981, and 1992. They calculated that the amount of water flowing north had fallen by around 30 percent (Bryden, Longworth, and Cunningham 2005, 655). The area that was surveyed was limited and thus could be influenced by regional variations.

Worldwide Effects of Circulation Changes

Oceanographers have found similar trends in parts of the oceans bordering Antarctica. Marine geochemist Wallace Broecker of Columbia

University's Lamont-Doherty Earth Observatory and colleagues have found that "[t]he renewal of deep waters by sinking surface waters near Antarctica has slowed to only one-third of its flow a century or two ago" (Kerr 1999c, 1062). The work of Broecker, Stewart Sutherland, and Tsung-Hung Peng suggested that the slowing of deep-water formation may be related to warming that accompanied the end of Europe's Little Ice Age (roughly 1400 to 1800 A.D.). Furthermore, they posited that "[a] see-sawing of deep water production between the northern Atlantic and the Southern oceans may lie at the heart of the 1,500-year ice-rafting cycle" (Broecker, Sutherland, and Peng 1999, 1132).

Variations in the thermohaline circulation of the North Atlantic can have marked climatic consequences in Mesoamerica as well as in Europe. In *The Great Maya Droughts* (2000), for example, Richardson B. Gill postulated that a southward shift in the Gulf Stream, forced by cold deep-water flow from the north, cooled the climate in northwestern Europe between 800 and 1000 A.D. The same shift is said by Gill to have displaced the North Atlantic high-pressure system southwestward, causing many years of severe drought in regions populated by the classic Maya civilization in Mesoamerica. Gill makes a case that these droughts played a major role in collapse of the classic Maya civilization. *See also:* Thermohaline Circulation

FURTHER READING

Borenstein, Seth. "Scientists Worry about Evidence of Melting Arctic Ice." Knight-Ridder News Service in *Seattle Times*, February 18, 2005, A-6.

Broecker, Wallace S. "Are We Headed for a Thermohaline Catastrophe?" In *Geological Perspectives of Global Climate Change*, ed. Lee C. Gerhard, William E. Harrison, and Bernold M. Hanson, 83–95. AAPG [American Association of Petroleum Geologists] Studies in Geology No. 17. Tulsa, OK: AAPG, 2001.

Broecker, Wallace S., Stewart Sutherland, and Tsung-Hung Peng. "A Possible 20th-Century Slowdown of Southern Ocean Deep Water Formation." *Science* 286 (November 5, 1999):1132–1135.

Bryden, Harry L., Hannah R. Longworth, and Stuart A. Cunningham. "Slowing of the Atlantic Meridional Overturning Circulation at 25° North." *Nature* 438 (December 1, 2005):655–657.

Clark, P. U., N. G. Pisias, T. F. Stocker, and A. J. Weaver. "The Role of the Thermohaline Circulation in Abrupt Climate Change." *Nature* 415 (February 21, 2002):863–868.

Curry, Ruth and Cecilie Mauritzen. "Dilution of the Northern North Atlantic Ocean in Recent Decades." *Science* 308 (June 17, 2005):1772–1774.

Gill, Richardson Benedict. *The Great Maya Droughts: Water, Life, and Death.* Albuquerque: University of New Mexico Press, 2000.

Kerr, Richard A. "Oceanography: Has a Great River in the Sea Slowed Down?" *Science* 286 (November 5, 1999c):1061–1062.

Quadfasel, Detlef. "The Atlantic Heat Conveyor Slows." *Nature* 438 (December 1, 2005):565–566.

Pearce, Fred. "Failing Ocean Current Raises Fears of Mini Ice Age." NewScientist.com, November 30, 2005. http://www.newscientist.com/article.ns?id=dn8398.

Radford, Tim. "As the World Gets Hotter, Will Britain Get Colder? Plunging Temperatures Feared after Scientists Find Gulf Stream Changes." *London Guardian*, June 21, 2001b, 3.

Rahmstorf, Stefan. "Thermohaline Circulation: The Current Climate." *Nature* 421 (February 13, 2003):699.

Ruhlemann, Carsten, Stefan Mulitza, Peter J. Muller, Gerold Wefer, and Rainer Zahn. "Warming of the Tropical Atlantic Ocean and Slowdown of Thermohaline Circulation during the Last Deglaciation." *Nature* 402 (December 2, 1999):511–514.

Schiermeier, Quirin. "Ocean Circulation Noisy, Not Stalling." *Nature* 448 (August 23, 2007b):844–845.

"Slowing Ocean Currents Could Freeze Europe." Environment News Service, February 21, 2002. http://ens-news.com/ens/feb2002/2002L-02-21-09.html.

Stocker, Thomas F., Reto Knutti, and Gian-Kasper Plattner. "The Future of the Thermohaline Circulation—A Perspective." In *The Oceans and Rapid Climate Change: Past, Present, and Future*, ed. Dan Seidov, Bernd J. Haupt, and Mark Maslin, 277–293. Washington, D.C.: American Geophysical Union, 2001.

Weart, Spencer R. *The Discovery of Global Warming.* Cambridge, MA: Harvard University Press, 2003.

Ocean Food Web

Warming temperatures will force adaptation or extinction of many plant and animal species in the world's oceans. Great sea turtles, for example, will come ashore at their usual sites to lay their eggs on sand and rocks that have grown warmer. The temperature determines the sex ratio of the turtles, with female births declining as the environment warms. Scientists do not yet know how much warmth will be required before the last sea turtle mother will die (Bates and Project Plenty 1990, 100). According to the World Commission on Environment and Development, one plant or animal species is going extinct somewhere in the world every minute, due mainly to human encroachment on (and pollution of) natural habitats (Bates and Project Plenty 1990, 101).

Warming seas distort marine ecosystems in other ways. According to a report by Climate Solutions of Olympia, Washington, a dramatic ocean-temperature increase near North America's west coast began about 1977. Warming sea-surface temperatures may interfere with phytoplankton production, with impacts rippling through the food web. Cooler, upwelling ocean water will break through warm surface waters less frequently, reducing nutrients available for plants and animals living in the oceans. By the 1990s, such decreases in productivity were detected near the California coast, where scientists have documented a measurable decrease in the abundance of zooplankton, the second level in the food web. By the 1990s, the abundance of zooplankton was 70 percent lower than it had been during the 1950s.

The decline of zooplankton raises fears for the survival of several fish species that feed on it. Water temperatures in some areas near the U.S. west coast have increased 1° to 2°C during the last four decades (Gelbspan 1995). Ocean seabirds in the California Current have declined 90 percent since 1987, one indication of food-web collapse. Reduced primary phytoplankton production usually means less overall fertility in marine ecosystems, including reduced fisheries, according to a report compiled by the World Wildlife Fund (Mathews-Amos and Berntson 1999).

Harvard University Epidemiologist Paul Epstein has warned that rising temperatures in many ocean areas may be related to increasing outbreaks of algal toxins, bacteria, and viruses that affect large numbers of marine species, as well as shorebirds and mammals. "Of great concern," wrote Epstein, "are diseases that attack coral and sea grasses, essential habitats that sustain mobile aquatic species" (Epstein 1999, 66). During the 1997–1998 El Niño, weak winds and strong sunlight created highly stratified, low-nutrient surface waters in the eastern Bering Sea, resulting in a massive bloom of *coccolithophores*, a type of phytoplankton usually associated with low-nutrient areas. According to the World Wildlife Fund, such a large-scale bloom had never before been documented in that area (Mathews-Amos and Berntson 1999).

Scientists have documented decreased reproduction and increased mortality in seabirds and marine-mammal populations in warming water. The World Wildlife Fund reports that Sooty Shearwaters off the California coast declined 90 percent during the late 1980s and early 1990s, as Cassin's Auklets declined 50 percent.

Zooplankton populations declined markedly at the same time. In Alaska, a severe decline in Shearwaters from 1997 to 1998 "was clearly due to starvation," according to the World Wildlife Fund (Mathews-Amos and Berntson 1999).

By the late 1990s, the Mediterranean Sea had become home to at least 110 species of tropical fish for the first time in recorded history. The tropical fish, some of which are crowding out native species, have followed warming water into the Mediterranean through the Suez Canal and the Strait of Gibraltar. Barracuda and other tropical fish are migrating to the Mediterranean as water temperatures warm.

Eighty-five species previously unknown on a significant scale in the waters of Italy or the Levant have set up home there over the past ten years, according to research carried out by Professor Franco Andolaro and his team at the Palermo office of the Italian Marine Research Institute. They are competing for food with 550 indigenous species. "What is in danger," Professor Andolaro said, "is the ecological balance of the Mediterranean basin. Tropical species now make up almost 20 percent of the fauna and the colonization is continuing constantly" (Phillips 2000). The Italian Environment Ministry said that water temperatures in the Mediterranean have increased about 2°C since 1970. Fifty-five of the new species are from the Red Sea and the remainder from the Atlantic. Ten of the Red Sea migrants are so numerous that they are now being fished commercially in the Mediterranean, including gold-band goatfish, striped-fin goatfish, Haifa grouper, streamlined spinefoot, the wahoo (a small tuna), banded barracuda, and Brazilian lizard fish.

News of the influx has caused considerable excitement in Italy, where tourism to exotic destinations is a burgeoning business. "We won't have to go as far as the Seychelles, Sharm el-Sheikh or the Maldives to enjoy the most fascinating marine beauty," Il Giornale of Milan commented. "Today we can admire these curious and highly coloured tropical fish by simply taking a swim in Liguria or Sardinia" (Phillips 2000).

Another potential problem with warming seas is an expected reduction in the pH factor, indicating an increase in relative acidity of the water. Wilfrid Bach wrote:

A CO_2 doubling in the atmosphere might reduce the pH of surface seawater from the usual 8.1 to 7.6. It has been shown that a reduction of the pH to 7.6 would increase the copper-ion activity by ten-fold, a change to which the marine phytoplankton would react dramatically ... a pH of 7.7 would cause the death of fish larvae. (Bach 1984, 183)

FURTHER READING

Bach, Wilfrid. *Our Threatened Climate: Ways of Averting the CO₂ Problem though Rational Energy Use*, trans. Jill Jager. Dordrecht, Germany: D. Reidel, 1984.

Bates, Albert K., and Project Plenty. *Climate in Crisis: The Greenhouse Effect and What We Can Do.* Summertown, TN: The Book Publishing Co., 1990.

Epstein, Paul. "Profound Consequences: Climate Disruption, Contagious Disease and Public Health." *Native Americas* 16, no. 3/4 (Fall/Winter 1999): 64–67. http://nativeamericas.aip.cornell.edu/fall99/fall99epstein.html.

Gelbspan, Ross. "The Heat Is On: The Warming of the World's Climate Sparks a Blaze of Denial." *Harper's*, December 1995. http://www.dieoff.com/page82.htm.

Mathews-Amos, Amy, and Ewann A. Berntson. "Turning up the Heat: How Global Warming Threatens Life in the Sea." World Wildlife Fund and Marine Conservation Biology Institute, 1999. http://www.worldwildlife.org/news/pubs/wwf_ocean.htm.

Phillips, John. "Tropical Fish Bask in Med's Hot Spots." *London Times*, July 15, 2000, n.p.

Ocean Life: Whales, Dolphins, and Porpoises

Whales, dolphins, and porpoises are declining in the Arctic and the Antarctic because shrinking sea ice has been destroying species that they eat. Cetaceans (belugas, narwhals, and bowhead whales) have suffered most acutely from the reduction of sea ice, according to a report, *Whales in Hot Water?* published by the Whale and Dolphin Conservation Society and the global wildlife organization (formerly, the World Wildlife Fund [WWF]).

"Whales, dolphins and porpoises have some capacity to adapt to their changing environment," said Mark Simmonds, international director of science at the Whale and Dolphin Conservation Society (WDCS). "But the climate is now changing at such a fast pace that it is unclear to what extent whales and dolphins will be able to adjust, and we believe many populations to be very vulnerable to predicted changes" ("Warming Oceans" 2007).

Populations of tiny, shrimp-like krill, the main source of food for many of the great whales, are declining in many areas as waters warm. Krill's life cycle is dependent on sea ice. In January 2006, a 30-year study published by an international team of scientists showed that El Niño ocean warming events affected the availability of krill in the Southern Ocean. This in turn has affected the number of calves produced by southern right whales in the South Atlantic, as reported by the Environment News Service ("Warming Oceans" 2007).

Other human activities, including chemical contamination, noise pollution, and collisions with ships, are adding to the species problems in addition to a warming environment, according to the report. About 1,000 cetaceans every day are snared in fishing nets. Acidification of the oceans as they absorb growing quantities of carbon dioxide also harms sea animals. Human activities, such as commercial shipping, oil and gas mining exploration and development, and military activities, will increase as sea ice shrinks and the Arctic warms. "This will result in much greater risks from oil and chemical spills, worse acoustic disturbance and more collisions between whales and ships," said the report's lead author Wendy Elliott, from the WWF's Global Species Programme ("Warming Oceans" 2007).

FURTHER READING

"Warming Oceans Put More Stress on Whales." Environment News Service, May 21, 2007. http://www.ens-newswire.com/ens/may2007/2007-05-21-04.asp.

Oceans, Carbon-Dioxide Levels

Because humankind is a narcissistic species, when considering the implications of global warming, our focus usually is fixed on the one-third of the planet that is composed of dry land. While the land warms, the other two-thirds of the Earth will be warming as well, with profound implications for the species that inhabit it, including a holocaust for coral reefs, which already has begun.

According to Peter G. Brewer and colleagues, writing in *Science*, the chemistry of ocean water already is being changed by human-induced greenhouse gases in the atmosphere: "The invading wave of atmospheric CO_2 has already altered the chemistry of surface seawater worldwide, and over much of the ocean, this tracer field has now permeated to a depth of more than one kilometer" (Brewer et al. 1999, 943). Given that the ocean is slowly warming, as its carbon-dioxide

A salmon shark landed in Alaskan waters. (Bruce Wright, Conservation Science Institute)

level rises, its capacity as a carbon sink is prob-ably being compromised.

Ken Caldeira and Philip B. Duffy asserted in *Science* that "uptake," or removal, of human-induced carbon dioxide by the oceans is less than many earlier investigators have assumed, and that, as temperatures warm, carbon uptake will diminish further. Additionally, Caldeira and Duffy contended that absorption of carbon into the oceans of the Southern Hemisphere (the focus of their study) has been diminishing since fossil-fuel effluvia became a factor in the compo-sition of the atmosphere about 1880. "Venti-lation of the deep Southern Ocean was much more vigorous in the period from about 1350 to 1880 than in the recent past," wrote the research-ers (Caldeira and Duffy 2000, 620).

Decline of Phytoplankton

Free-floating, microscopic plants called phyto-plankton, the basis of the oceanic food web, have been declining rapidly in areas where they once nourished rich fisheries, most notably in the temperate areas of the world's oceans in the Northern and Southern hemispheres. Cod and salmon, both cold-water fish, are becoming increasingly scarce in English offshore waters, for example, as tropical species are being sighted in the same areas more frequently. Sharks have been sighted in Alaskan waters.

Around the world, coral reefs have suffered due to warming above their tolerances in tropical oceans. Not all sea life has suffered because of warming in the oceans, however. Jellyfish, for example, thrive in warming waters. They also benefit from some forms of human water pollu-tion, which has led to increasing size and toxicity of their stings. Giant squid also seem to benefit from warming seas.

Oceans Lose Ability to Absorb Carbon Dioxide

The oceans have long been regarded as the Earth's largest carbon sink, but since 1980 their ability to absorb humankind's excess carbon has been decreasing. This decline in the Southern

Ocean has been averaging about 15 percent per decade, according to research ("Antarctic Ocean" 2007). "This is the first time that we've been able to say that climate change itself is responsible for the saturation of the Southern Ocean sink. This is serious," said lead author Corinne Le Quere of the University of East Anglia and the British Antarctic Survey. "All climate models predict that this kind of feedback will continue and intensify during this century," said Le Quere. "The Earth's carbon sinks—of which the Southern Ocean accounts for 15 percent—absorb about half of all human carbon emissions. With the Southern Ocean reaching its saturation point more CO_2 will stay in our atmosphere," she said ("Antarctic Ocean" 2007).

In addition, acidification in the Southern Ocean is likely to reach dangerous levels before 2050, the scientists said. The international team included researchers from Commonwealth Scientific and Industrial Research Organization (CSIRO) in Australia, Max-Planck Institute in Germany, University of East Anglia and British Antarctic Survey in England, Climate Monitoring and Diagnostics Laboratory in the United States, National Institute of Water and Atmospheric Research in New Zealand, South African Weather Service, Laboratoire des Sciences du Climat et de l'Environement/Institut Pierre Simon LaPlace and Centre Nationale de la Recherche Scientifique in France, and Centre for Atmospheric and Oceanic Studies in Japan ("Antarctic Ocean" 2007).

Professor Chris Rapley, director of the British Antarctic Survey, said, "Since the beginning of the industrial revolution, the world's oceans have absorbed about a quarter of the 500 gigatons of carbon emitted into the atmosphere by humans. The possibility that in a warmer world the Southern Ocean—the strongest ocean sink—is weakening is a cause for concern" ("Antarctic Ocean" 2007).

"The researchers found that the Southern Ocean is becoming less efficient at absorbing carbon dioxide due to an increase in wind strength over the ocean, resulting from human-induced climate change," said Paul Fraser, who leads research in atmospheric greenhouse gases at Australia's CSIRO Center for Marine and Atmospheric Research. "The increase in wind strength is due to a combination of higher levels of greenhouse gases in the atmosphere and long-term ozone depletion in the stratosphere, which previous CSIRO research has shown intensifies storms

over the Southern Ocean," he said. The increased winds influence the processes of mixing and upwelling in the ocean, which in turn causes an increased release of carbon dioxide into the atmosphere, reducing the net absorption of carbon dioxide into the ocean, he explained. ("Antarctic Ocean" 2007). *See also:* Acidity and Rising Carbon-Dioxide Levels in the Oceans

FURTHER READING

"Antarctic Ocean Losing Ability to Absorb Carbon Dioxide." Environment News Service, May 18, 2007. http://www.ens-newswire.com/ens/may2007/2007-05-18-04.asp.

Brewer, Peter G., Gernot Friederich, Edward T. Peltzer, and Franklin M. Orr, Jr. "Direct Experiments on the Ocean Disposal of Fossil Fuel CO_2." *Science* 284 (May 7, 1999):943–945.

Caldeira, Ken, and Philip B. Duffy. "The Role of the Southern Ocean in the Uptake and Storage of Anthropogenic Carbon Dioxide." *Science* 287 (January 28, 2000):620–622.

Ocean Sequestration of Carbon Dioxide

Until recently, the oceans were well-known to scientists as a major "sink," or repository, for human-generated atmospheric carbon dioxide and methane. This fact has led to various proposals to remove the nettlesome oversupply of these gases from the atmosphere, and inject at least some of it into the depths of the oceans. Now, however, the oceans are reaching their limits as pollution sponges.

Peter G. Brewer, Gernot Friederich, Edward T. Peltzer, and Franklin M. Orr, Jr. have demonstrated that deep-ocean disposal of carbon dioxide is technologically feasible. They described a series of experiments in *Science* during which carbon dioxide was lowered into several hundred meters of seawater, at different depths. The carbon dioxide, which is a gas at the surface, formed solid hydrates that were expected to remain in the ocean depths for "quite long residence times" (Brewer et al. 1999).

This idea has been dissected and dismissed by Hein J. W. de Baar, of the Netherlands Institute for Ocean Sciences:

The crucial problem with fossil fuel CO_2 is its very rapid introduction within 100 to 200 years into the atmosphere, as opposed to the very slow response of many thousands to millions of years of the deep ocean in absorbing such CO_2. Eventually, the capacity for storage of CO_2 in the deep ocean is very large. Yet, in the meantime, we will witness a transient peak of atmospheric CO_2 which may yield catastrophic changes in the climate. Only after several thousands to millions of years most, but not all, of the fossil fuel CO_2 will be taken up by the oceans. (Baar 1992, 143)

In addition, according to Baar, deep-sea carbon-dioxide injection is suitable only for large stationary energy plants (30 percent of total human emissions) and would raise the cost of generation 30 to 45 percent, while decreasing efficiency by a similar percentage. Much of the carbon dioxide injected into the deep oceans would eventually return to the surface, doubling seawater's acidity, which would be toxic for fish, plankton, and other life in the oceans. "Deep-sea injection is at best a partial, expensive, and temporal [temporary] remedy to the CO_2 problem," wrote Baar (1992, 144). *See also:* Acidity and Rising Carbon-Dioxide Levels in the Oceans

FURTHER READING

Baar, Hein J. W. de, and Michel H.C. Stoll. "Storage of Carbon Dioxide in the Oceans." In *Arctic Ecosystems in a Changing Climate: An Ecophysiological Perspective*, ed. F. Stuart Chapin III, Robert L. Jefferies, James F. Reynolds, Gaius R. Shaver, and Josef Svoboda, 143–177. San Diego: Academic Press, 1992.

Brewer, Peter G., Gernot Friederich, Edward T. Peltzer, and Franklin M. Orr, Jr. "Direct Experiments on the Ocean Disposal of Fossil Fuel CO_2." *Science* 284 (May 7, 1999):943–945.

Oceans Warming: World Survey

The oceans are the final "stop" in global-warming's feedback loop, and potentially one of the most important for human beings—not because we live *in* the oceans, of course, but because more than 100 million people worldwide live within one meter of mean sea level (Meier and Dyurgerov 2002, 350). The situation is particularly acute for island nations. Consider, for example, Indonesia. Jakarta and 69 other sizable cities along Indonesia's coasts probably will be inundated as global warming causes ocean levels to rise during decades to come, according to Indonesia's Secretary of Environment Arief Yuwono ("70 Cities" 2002). Regarding ocean warming, Tom Wigley, a senior scientist at the National Center for Atmospheric Research in

Boulder, Colorado, commented, "We're heading into unknown territory, and we're heading there faster than we ever have before" (Revkin 2001b, A-15).

Some computer models anticipate that the average global sea level will rise one to two feet during the twenty-first century. The oceans are the proverbial "end of the line" of the warming feedback loop. Warming reaches the seas only after a number of decades. Thus, the sea level today may reflect levels of greenhouse gases emitted into the atmosphere 40 to 50 years ago. Even if greenhouse-gas levels were stabilized today at present levels, sea levels might continue to rise for several hundred years. After 500 years, sea-level rise from thermal expansion may have reached only half of its eventual level, "which models suggest may lie within a range of 0.5 to 2.0 meters to 1.0 to 4.0 meters for carbon-dioxide levels at twice and four times pre-industrial [levels] respectively" (Houghton et al. 2001, 77). Similarly, ice sheets will continue to react over thousands of years, even if greenhouse-gas levels stabilize in the atmosphere. An average 100-year warming of 3°C, sustained over a few centuries, could melt most of the Greenland Ice Sheet, eventually raising worldwide sea levels about seven meters.

Warming of Oceans Is "Unmistakable"

A team of oceanographers using a combination of observations and climate models by 2005 had determined that the penetration of human-induced warming in the world's oceans was unmistakable. Furthermore, the same team reported that roughly 84 percent of the extra heat that the entire Earth was absorbing had affected the oceans during the previous 40 years (Barnett 2005, 284). Their study concluded that "little doubt" exists "that there is a human-induced signal in the environment." The record also leaves little doubt that the oceans will continue to warm. "How to respond to the serious problems posed by these predictions is a question that society must decide," they wrote in *Science* (Barnett 2005, 287).

As scientists debated the speed with which the oceans are warming, anecdotal evidence abounds. During the summer of 2002, ocean water was warm enough for swimming along much of Maine's coast, a novelty. A local account said that

Hot days and favorable winds heated up the ocean surface to record temperatures in August. The average daily water temperature here, where the state collects data, reached 67.5°F for the month. That was 3°F warmer than any other August since record keeping began in 1905, and 6.5°F above average. Farther south on beaches in York and Cumberland counties, lifeguards have measured temperatures exceeding 70°F. ("Ocean Temperatures" 2002)

"I've never seen it as warm as it was and for as long as it was," said Joe Doane, a longtime lifeguard at Scarborough Beach. "It was like bathwater" ("Ocean Temperatures" 2002).

Generally, according to studies conducted by Tim Barnett, a marine physicist at the Scripps Institution of Oceanography in San Diego, as much as 90 percent of greenhouse warming ends up in the oceans. Using the oceans' absorption of heat, said Barnett, whose models have estimated the amount of oceanic warming during the past 40 years, "The evidence [of greenhouse warming] really is overwhelming" (von Radowitz 2005).

Earth's Planetary Energy Imbalance

During the spring of 2005, James Hansen and several other scientists published temperature readings from the deep ocean that traced a clear warming trend indicative of the planet's thermal inertia and energy imbalance—the difference between the amount of heat absorbed by Earth and the amount radiated out into space. This thermal imbalance (0.85 watts plus or minus 0.15 per square meter) is evidence of a steadily warming world, raising the odds of an eventual catastrophic sudden change marked by rising seas and melting ice caps (Hansen et al. 2005, 1431).

Hansen and colleagues concluded that the unusual magnitude of the warming trend could not be explained by natural variability, but instead fit precisely with theories suggesting that human activity is the dominant "forcing agent." "This energy imbalance is the 'smoking gun' that we have been looking for," Hansen said in a prepared summary of the study, which was published in the journal *Science*. "The magnitude of the imbalance agrees with what we calculated using known climate forcing agents, which are dominated by increasing human-made greenhouse gases. There can no longer be substantial doubt that human-made gases are the cause of most observed warming" (Hall 2005, A-1).

By Hansen's estimate, 25 to 50 years are required for Earth's surface temperature to reach 60 percent of its equilibrium response—a key

concept when diplomacy and policy usually respond to experience, rather than on expectations of climate change. Regarding those expectations, 0.85 plus or minus 0.15 watts per square meter of excess heat absorbed by the Earth may not seem like much, but multiplied by many square meters over many years, it adds up.

The Earth's planetary energy imbalance did not exceed more than a few tenths of a watt per square meter before the 1960s, according to Hansen and colleagues. Since then, the amount of excess heat absorbed by the atmosphere has grown steadily, except for years after large volcanic eruptions, to a level much above historic averages. Much of this excess heat ends up in the oceans, where it helps to melt ice in the Arctic and Antarctic. In the measured tones with which scientists express themselves, accelerating ice melt could "create the possibility of a climate system in which large sea-level change is practically impossible to avoid" (Hansen et al. 2005, 1434). Hansen and colleagues wrote that continuing the present energy imbalance could lead to a climate that is "out of control" (Hansen et al. 2005, 1434).

"A Definite Sign of Human-Caused Warming"

Tim Barnett, a marine physicist of the Scripps Institution of Oceanography, La Jolla, California, told a meeting of the American Geophysical Union in late December 2000 that the upper 9,000 feet of the world ocean showed "a definite sign of human-caused warming" ("Global Warming Signal" 2000, 4). A team led by Barnett compared measurements of the ocean's heat content with computer models that account for greenhouse-gas increases in the atmosphere, as well as other influences on climate. The model accurately mimicked actual observations, which Barnett interpreted as compelling evidence that humans are responsible for at least part of the oceanic warming trend. Barnett said that its findings are "strong enough to overcome his long skepticism about the models' ability to pinpoint a human influence amid all the naturally chaotic ups and downs of climate" (Revkin 2001b, A-15). According to Barnett and colleagues, the chance that most of the observed ocean warming is caused by human factors is 95 percent (Barnett et al. 2001, 270). "The curves were so good, the data looked faked," said Barnett. "You just don't get it this good the first time around. But there it was" (McFarling 2001a, A-1).

Barnett's study was described in the same issue of *Science* as another article that detailed the results of a study led by Sydney Levitus, director of the National Oceanic and Atmospheric Administration's ocean climate laboratory, in Silver Spring, Maryland. Levitus and colleagues spent 10 years collecting and analyzing 2.3 million temperature measurements in the world's oceans to detect anthropomorphic contributions to their temperature level. The oceans are thought to be a more constant "thermostat" for the Earth than air temperatures, which vary widely day by day. "The oceans are the memory of the climate system," said Levitus (Levitus 2001; McFarling 2001a, A-1). These readings from the world's oceans indicated that human influences have been an important cause of the 0.11° degree C warming that has occurred in the upper two miles of the oceans since 1955. This increase may sound small, but spread over the world's oceans, it contains enough heat energy to meet California's present-day energy requirements for 200,000 years (McFarling 2001a, A-1). The two studies suggested that even if the emission of greenhouse gases were stopped immediately, the effects of warming already present in the atmosphere would double the rate of warming in the oceans within two to four decades. "We're at the very bottom of a sharply accelerating curve," Barnett said (McFarling 2001a, A-1).

Kevin Trenberth, director of the climate analysis section of the National Center for Atmospheric Research, who has written on the use and abuse of climate models, questioned the results of these studies because ocean temperature readings are sparse in some parts of the globe, particularly the southern polar oceans. He added that the two models use different variables. The Parallel Climate Model used by Barnett is not fully accurate, Trenberth said, "because it does not include changes in energy from the sun and the impact of volcanoes, both of which have a cooling effect." Levitus's Geophysical Fluid Dynamics Laboratory model may be a bit too warm, Trenberth added (McFarling 2001a, A-1).

Notable Warming in the Coldest Places

A taste of what's to come in Greenland was provided late in 2004 through a report in *Nature* that described how Greenland's Jakobshavn Glacier's advance toward the sea has accelerated

significantly since 1997. The speed of the glacier nearly doubled between 1997 and 2004 to nearly 50 cubic kilometers a year. This single glacier was responsible for about 4 percent of worldwide sea-level rise during the twentieth century (Joughin, Abdalati, and Fahnestock 2004, 608). The glacier's speed continued to increase after that.

Warming of the oceans has been most notable, relatively speaking, where the sea is usually the coldest, including waters adjacent to Antarctica. A study using seven decades of temperature data indicated that mid-depth water around Antarctica was warming nearly twice as quickly as the oceans' worldwide average (Cowen 2002a, 14). Sarah T. Gille's analysis found that the mid-depth water had warmed 0.17°C since 1950 (Gille 2002, 1275). Gille, of the University of California–San Diego, said, "We thought the ocean between 700 and 1,100 meters (2,300 and 3,600 feet) was pretty well insulated from what's happening at the surface. But these results suggest that the mid-depth Southern Ocean is responding and warming more rapidly than global ocean temperatures [generally]" (Cowen 2002a, 14).

Gille noted that the Southern Ocean "is a very climatically sensitive region" (Cowen 2002a, 14). It is at one end of the conveyor-belt circulation that carries heat toward the poles in upper-level currents and returns cold water back toward the equator at great depths—a key part of the system that maintains Earth's present climate. Any change in Antarctic waters could directly affect circulations in the Atlantic, Indian, and Pacific Oceans (Cowen 2002a, 14). Ocean circulation already may have been affected. Gille, who reported her findings in *Science*, said that her study "suggests that this current has moved closer to the polar continent as the mid-depth water has warmed" (Cowen 2002a, 14).

Many of the world's shorelines already are eroding, only partially because seas are rising. In many areas, shorelines are subsiding from the last ice age, as well as from the removal of underground water and oil reserves. The only major exceptions are areas of high sediment supply, such as along the rims of active delta lobes and regions of glacial outwash (Pilkey and Cooper 2004, 1781). Sea level is rising along midlatitude coastal plain coastlines at a typical rate of 30 to 40 centimeters per century. Typical rates of shoreline retreat average 30 centimeters to one meter per year (Pilkey and Cooper 2004, 1781).

By one estimate, global sea levels rose by roughly 15 to 20 centimeters (six to eight inches) during the twentieth century, more from thermal expansion than from melting ice. The rate at which sea level rose during the twentieth century was 20 times faster than the average rate for the preceding 3,000 years (Smith 2001, 7).

Ocean Temperatures Measured in Detail

Scientists at the National Oceanic and Atmospheric Administration (NOAA) reported during late March 2000 that the oceans had warmed significantly worldwide during the preceding four decades. A broad study of temperature data from the oceans showed that average water temperatures had increased from one-tenth to one-half degree, depending on depth, since the 1950s. The NOAA study uncovered significant warming between 1955 and 1995 as deep as 10,000 feet in the oceans. The greatest warming occurred from the surface to a depth of about 900 feet, where the average heat content increased by 0.56°F. Water as far down as 10,000 feet (in the North Atlantic) was found to have gained an average of 0.11°F.

"We've known the oceans could absorb heat, transport it to subsurface depths and isolate it from the atmosphere. Now we see evidence that this is happening," said Sydney Levitus, chief of NOAA's Ocean Climate Laboratory and principal author of the study. "Our results support climate modeling predictions that show increasing atmospheric greenhouse gases will have a relatively large warming influence on the Earth's atmosphere," said Levitus. According to Levitus, "The whole-Earth system has gone into a relatively warm state" (Levitus et al. 2000; Kerr 2000a, 2126).

The ocean-temperature record compiled by Levitus and colleagues confirmed similar trends in land-surface temperature records, with one notable period of warming between 1920 and 1940, followed by a slight cooling. By the late 1970s, both ocean and land-surface temperatures began to spike upward to new peaks in the observational record. The magnitude of the oceanic warming surprised some experts. Peter Rhines, an oceanographer and atmospheric scientist at the University of Washington in Seattle, said it appeared roughly equivalent to the amount of heat stored by the oceans as a result of seasonal heating in a typical year. "That makes it a big number," he said (Stevens 2000, A-16).

The Environmental Protection Agency has stated that warming seas also may be influenced by changes in the Earth's gravitational field provoked by melting ice at the poles:

> Removal of water from the world's ice sheets would move the earth's center of gravity away from Greenland and Antarctica; the oceans' water would thus be redistributed toward the new center of gravity. Along the coast of the United States, this effect would generally increase sea level rise by less than 10 percent; sea level could actually drop, however, at Cape Horn and along the coast of Iceland.... Climate change could also influences local sea level by changing winds, atmospheric pressure, and ocean currents, but no one has estimated these impacts. (U.S. Environmental Protection Agency 1997)

FURTHER READING

"70 Cities in Indonesia Will Be Inundated." *Antara, the Indonesian National News Agency*, September 25, 2002. (LEXIS)

Barnett, Tim P., David W. Pierce, Krishna M. AchutaRao, Peter J. Gleckler, Benjamin D. Santer, Jonathan M. Gregory, and Warren M. Washington. "Penetration of Human-Induced Warming into the World's Oceans." *Science* 309 (July 8, 2005):284–287.

Cowen, Robert C. "Into the Cold? Slowing Ocean Circulation Could Presage Dramatic—and Chilly—Climate Change." *Christian Science Monitor*, September 26, 2002a, 14.

Gille, Sarah T. "Warming of the Southern Ocean since the 1950s." *Science* 295 (February 15, 2002):1275–1277.

"Global Warming Signal from the Ocean." *Dallas Morning News* in *New Orleans Times-Picayune*, December 31, 2000, 4.

Hall, Carl T. "Ocean Tells the Story: Earth Is Heating Up." *San Francisco Chronicle*, April 29, 2005, A-1.

Hansen, James, Larissa Nazarenko, Reto Ruedy, Makiko Sato, Josh Willis, Anthony Del Genio, Dorothy Koch, Andrew Lacis, Ken Lo, Surabi Menon, Tica Novakov, Judith Perlwitz, Gary Russell, Gavin A. Schmidt, and Nicholas Tausnev. "Earth's Energy Imbalance: Confirmation and Implications." *Science* 308 (June 3, 2005):1431–1435.

Houghton, J. T., Y. Ding, D. J. Griggs, M. Noguer, P. J. van der Linden, X. Dai, K. Maskell, and C. A. Johnson. *Climate Change 2001: The Scientific Basis. Contribution of Working Group I to the Third Assessment Report of the Intergovernmental Panel on Climate Change.* Cambridge: Cambridge University Press, 2001.

Joughin, I., W. Abdalati, and M. Fahnestock. "Large Fluctuations in Speed on Greenland's Jakobshavn Isbrae Glacier." *Nature* 432 (December 2, 2004):608–610.

Kerr, Richard A. "Globe's 'Missing Warming' Found in the Ocean." *Science* 287 (March 24, 2000a):2126–2127.

Levitus, Sydney, John I. Antonov, Timothy P. Boyer, and Cathy Stephens. "Warming of the World Ocean." *Science* 287 (2000):2225–2229.

Levitus, Sydney, John I. Antonov, Julian Wang, Thomas L. Delworth, Keith W. Dixon, and Anthony J. Broccoli. "Anthropogenic Warming of Earth's Climate System." *Science* 292 (April 13, 2001):267–270.

McFarling, Usha Lee. "Studies Point to Human Role in Global Warming." *Los Angeles Times*, April 13, 2001a, A-1.

Meier, Mark F., and Mark B. Dyurgerov. "Sea-level Changes: How Alaska Affects the World." *Science* 297 (July 19, 2002):350–351.

"Ocean Temperatures Reach Record Highs." Associated Press, September 9, 2002. (LEXIS)

Pilkey, Orrin H., and Andrew G. Cooper. "Society and Sea Level Rise." *Science* 303 (March 19, 2004):1781–1782.

Radowitz, John von. "Global Warming 'Smoking Gun' Found in the Oceans." Press Associated Ltd., February 18, 2005. (LEXIS)

Revkin, Andrew C. "Two New Studies Tie Rise in Ocean Heat to Greenhouse Gases." *New York Times*, April 13, 2001b, A-15.

Smith, Craig S. "One Hundred and Fifty Nations Start Groundwork for Global Warming Policies." *New York Times*, January 18, 2001, 7.

Stevens, William K. "The Oceans Absorb Much of Global Warming, Study Confirms." *New York Times*, March 24, 2000, A-16.

U.S. Environmental Protection Agency. *The Cost of Holding Back the Sea*. Washington, D.C.: EPA, 1997. http://users.erols.com/jtitus/Holding/NRJ. html#causes.

Offsets (Carbon): Are They Real?

The market for carbon offsets in the United States grew by 80 percent to $55 million annually in 2006 from 2005, according to a report from New Carbon Finance and Ecosystem Marketplace, despite the fact that no standards exist to define precisely what people are buying other than freedom from guilt. One utility, American Electric Power, agreed to offset about 4.6 million tons of carbon dioxide by paying for projects that reduce methane seeping from farm manure (Fahrenthold and Mufson 2007, A-1).

By 2007, Americans were imitating Europeans by gobbling up carbon-offsets, by which they pay for carbon-creating activities by donating to projects that offset them, such as tree planting, clean-energy projects, or pollution control. Supporters maintained that offsets raise money for good causes and demonstrate the necessity of restricting carbon-producing activities on a personal level. Critics maintain that offsets are "green lite," allowing old-fashioned consumer activities (and their carbon emissions) to continue, meanwhile discharging guilt without actually curtailing greenhouse-gas emissions.

Offsets for the Super-Affluent

"There needs to be more standardization, more verification and more assurances for the consumer that the offsets are real," conceded Ricardo Bayon, director of Ecosystem Marketplace. A number of organizations, including the Center for Resource Solutions in San Francisco and the Climate Group, based in Britain, raced to establish certification standards. "This voluntary stuff is an interim measure," said Judi Greenwald of the Pew Center on Global Climate Change. "But it is certainly better than doing nothing" (Kher 2007).

For those who are seriously rich and *really* feeling guilty about a gigantic carbon footprint, offsets are available for that private jet that uses 15 times the fuel per passenger mile of a commercial flight and about 45 times the amount of an average automobile with one passenger. Mega-mansions can be green-certified (solar-heated swimming pools included). With offsets, no one has to drive a small car, live in a two-room shack, or ride a bicycle. Jets.com, a private jet service, has joined with The Carbon Fund to bill private jet owners for their emissions; V1 Jets International has a similar program. Guilt can be discharged on the cheap: a round-trip flight between Fort Lauderdale, Florida, and Boston, Massachusetts, costs about $20,000, including a $74 carbon offset for the 13 metric tons of carbon dioxide produced (Frank 2007, W-2).

A Carbon-Neutral Airline?

A lengthening list of big businesses, including several large banks, London's taxi fleet, and even a few airline companies also asserted "carbon neutrality." For example, Silverjet, a luxury trans-Atlantic air carrier, promoted itself as the first fully carbon-neutral airline because it donated about $28 from each round-trip ticket to fund such esoteric projects as fertilizing the oceans with iron so that algae can pull more carbon dioxide out of the atmosphere. In theory, said representatives of Silverjet, such projects may eventually eliminate as much carbon dioxide as the airline generates, or about 1.2 tons per passenger per trip.

The cure was on speculation, however, while the carbon emissions remained an everyday reality. An entire consulting industry grew to calculate offsets, billing the companies and handling the transactions with a fee attached. By early 2007, *Business Week* magazine said that the trade in offsets was more than $100 million a year "and growing blazingly fast" (Revkin 2007b). In the meantime, actual greenhouse-gas emissions barely flinched.

"The worst of the carbon-offset programs resemble the Catholic Church's sale of indulgences back before the Reformation," said Denis Hayes, president of the Bullitt Foundation, an environmental grant-making group. "Instead of reducing their carbon footprints, people take private jets and stretch limos, and then think they can buy an indulgence to forgive their sins." Hayes added, "This whole game is badly in need of a modern Martin Luther" (Revkin 2007b)

Michael R. Solomon, a professor at Auburn University who wrote *Consumer Behavior: Buying, Having and Being,* said that he was not surprised by the appeal of offsets. "Consumers are always going to gravitate toward a more parsimonious solution that requires less behavioral change," he said. "We know that new products or ideas are more likely to be adopted if they don't require us to alter our routines very much." He sees danger ahead, "if we become trained to substitute dollars for deeds—kind of an 'I gave at the office' prescription for the environment" (Revkin 2007b).

The Easy Way Out?

Charles Komanoff, an energy economist in New York State, said the commercial market in climate neutrality could have even more harmful effects. According to Komanoff, by suggesting there is an easy way out, it could blunt public support for what really will be needed in the long run: a binding limit on emissions or a tax on the fuels that generate greenhouse gases. "There isn't a single American household above the poverty line that couldn't cut their CO_2 at least 25 percent in six months through a straightforward series of fairly simple and terrifically cost-effective measures," he said (Revkin 2007b).

"There is a very common mind-set right now which holds that all that we're going to need to do to avert the large-scale planetary catastrophes upon us is make slightly different shopping decisions," said Alex Steffen, the executive editor of Worldchanging.com, a Web site devoted to sustainability issues (Williams 2007).

"I was feeling really guilty because I was basically traveling to three continents in the last month: 'I've spent basically six days on an airplane. I've got to fix this,'" said Michael Sheets, 27, who lives in Washington, D.C. Sheets paid $240 to Carbonfund.org. For that $240, he received an e-mail saying that the 52,920 pounds of greenhouse-gas emissions attributable to him for the entire year had been canceled out. This included air travel to Trinidad, Thailand, and Argentina. "I feel much better about it," said Sheets, human resources director for an online-education company in Northern Virginia. "I don't feel as guilty about flying to [Las] Vegas tomorrow for the weekend" (Fahrenthold 2008).

Steffen said that the only valid way to deal with global warming is to actually *reduce* emissions in real time. Instead of jetting off to a second home and pay offsets, own one home and stay there, for example. Paul Hawken, an author and environmental activist, criticized the false promise of chic green consumerism, which he called an oxymoronic phrase. "We turn toward the consumption part because that's where the money is," Hawken said. "We tend not to look at the 'less' part. So you get these anomalies like 10,000-foot 'green' homes being built by a hedge fund manager in Aspen. Or 'green' fashion shows. Fashion is the deliberate inculcation of obsolescence." He added, "The fruit at Whole Foods in winter, flown in from Chile on a 747—it's a complete joke.... The idea that we should have raspberries in January, it doesn't matter if they're organic. It's diabolically stupid" (Williams 2007).

FURTHER READING

Fahrenthold, David A. "There's a Gold Mine in Environmental Guilt: Carbon-Offset Sales Brisk Despite Financial Crisis." *Washington Post,* October 6, 2008, A-1 http://www.washingtonpost.com/wp-dyn/content/article/2008/10/05/AR2008100502518_pf.html.

Fahrenthold, David A., and Steven Mufson. "Cost of Saving the Climate Meets Real-World Hurdles." *Washington Post,* August 16, 2007, A-1. http://www.washingtonpost.com/wp-dyn/content/article/2007/08/15/AR2007081502432_pf.html.

Frank, Robert. "The Wealth Report: Living Large While Being Green," *Wall Street Journal,* August 24, 2007, W-2.

Kher, Unmesh. "Pay for Your Carbon Sins." *TIME,* April 9, 2007. http://www.time.com/time/printout/0,8816,1603737,00.html.

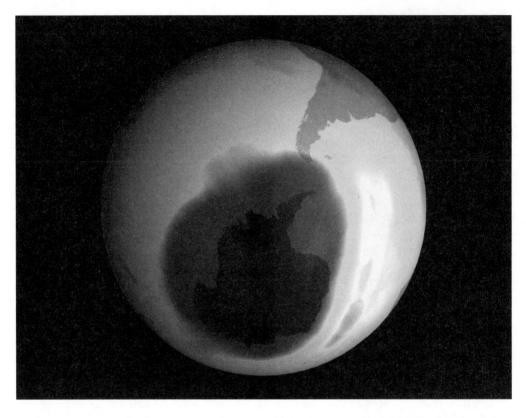

Stratospheric ozone depletion over Antarctica at its greatest extent, September 10, 2000. (NASA)

Revkin, Andrew. "Carbon-neutral Is Hip, but Is It Green? *New York Times*, April 29, 2007b. http://www.nytimes.com/2007/04/29/weekinreview/29revkin.html.

Williams, Alex. "Buying Into the Green Movement." *New York Times*, July 1, 2007. http://www.nytimes.com/2007/07/01/fashion/01green.html.

Ozone Depletion and Global Warming

When chlorofluorocarbons (CFCs) were banned in the late 1980s, most experts expected ozone depletion over the Antarctic to heal. By 2005, ozone depletion there still was a major problem, and an ozone "hole" was beginning to open over the Arctic as well. The nature of science has evolved during those 15 years to explain how the capture of heat near the Earth's surface by greenhouse gases speeds cooling in the stratosphere and plays an important role in continuing ozone depletion at that level. Thus, the healing of the stratospheric ozone layer depends, to some degree, on reduction of greenhouse-gas levels in the lower atmosphere.

The ozone shield protects plant and animal life on land from the sun's ultraviolet rays, which can cause skin cancer, cataracts, and damage to the immune systems of human beings and other animals. Thinning of the ozone layer also may alter the DNA of plants and animals.

While a single atom of chlorine from CFC can destroy more than 100,000 ozone molecules (Flannery 2005, 216), these human-created chemicals do more than destroy stratospheric ozone. They also act as greenhouse gases, with several thousand times the per-molecule greenhouse potential of carbon dioxide.

Ozone depletion over the Antarctic reaches its height in late winter and early spring, as the sun rises after the midwinter night. Solar radiation triggers reactions between ozone in the stratosphere and chemicals containing chlorine or bromine. These chemical reactions occur most quickly on the surface of ice particles in clouds, at temperatures less than $-80°C$ ($-107°F$).

In September 2008, the Antarctic ozone hole reached about 27 million square kilometers, its maximum size for the year, larger than North

America, which is about 25 million square kilometers. Larger than it was in 2007, the 2008 ozone hole was still smaller than the 2006 record ("Ozone Hole" 2008).

Warm at the Surface, Cold Aloft

An increasing level of carbon dioxide near the Earth's surface "acts as a blanket," said NASA research scientist Katja Drdla. "It is trapping the heat. If the heat stays near the surface, it is not getting up to these higher levels" (Borenstein 2000). Deprived of emitted warmth, the stratosphere cools, aggravating depletion of ozone, which protects plants and animals from ultraviolet radiation. Chemical reactions that drive ozone depletion tend to accelerate as the stratosphere cools, retarding the restoration of ozone anticipated after the ban of CFCs under the Montreal Protocol. As levels of greenhouse gases rise, the cooling of the middle and upper atmosphere is expected to continue, with attendant consequences for ozone depletion. Because of this relationship, problems with ozone depletion depend, in a fundamental way, on mitigation of greenhouse warming. In 2006, the Antarctic ozone hole was the largest on record, despite the two-decade ban of CFCs.

"The chemical reactions responsible for stratospheric ozone depletion are extremely sensitive to temperature," Drew T. Shindell wrote in *Nature* (Shindell, Rind, and Lonergan 1998, 589). "Greenhouse gases warm the Earth's surface but cool the stratosphere radiatively, and therefore affect ozone depletion." By the decade 2010 to 2019, Shindell and colleagues expect ozone loses in the Arctic to peak at two-thirds of the "ozone column," or roughly the same ozone loss observed in Antarctica during the early 1990s. "The severity and duration of the Antarctic ozone hole are also expected to increase because of greenhouse-gas-induced stratospheric cooling over the coming decades," they asserted (Shindell, Rind, and Lonergan 1998, 589).

The connection between global warming, a cooling stratosphere, and depletion of stratospheric ozone was confirmed in April 2000, with the release of a lengthy report by more than 300

Paul Crutzen from the Netherlands, director at the Mainz Max Planck Institut für Meteorologie, who, together with Mario Molina and F. Sherwood Rowland, won the 1995 Nobel Prize in Chemistry. (AP/Wide World Photos)

NASA researchers as well as several European, Japanese, and Canadian scientists. The report found that while ozone depletion may have stabilized over the Antarctic, ozone levels north of the Arctic Circle were still falling, in large part because the stratosphere has cooled as the troposphere has warmed. The ozone level over some parts of the Arctic was 60 percent lower during the winter of 2000 than during the winter of 1999, measured year over year.

In addition, scientists learned that as winter ends, ozone-depleted air tends to migrate southward over heavily populated areas of North America and Eurasia. "The largest ultraviolet increases from all of this are predicted to be in the mid-latitudes of the United States," said University of Colorado atmospheric scientist Brian Toon. "It affects us much more than the Antarctic [ozone hole]" (Borenstein 2000).

History of the Ozone Hole

CFCs initially raised no environmental questions when they were first marketed by DuPont Chemical during the 1930s under the trade name Freon. Environmental questions were not often asked at the time. At about the same time, asbestos was being proposed as a high-fashion material for clothing, and radioactive radium was being built into timepieces so that they would glow in the dark.

By 1976, manufacturers in the United States were producing 750 million pounds of CFCs a year, and finding all sorts of uses for them, from propellants in aerosol sprays, to solvents used to clean silicon chips, to automobile air conditioning, and as blowing agents for polystyrene cups, egg cartons, and containers for fast food. "They were amazingly useful," wrote Anita Gordon and David Suzuki. "Cheap to manufacture, non-toxic, non-inflammable, and chemically stable" (Gordon and Suzuki 1991, 24). By the time scientists discovered, during the 1980s, that CFCs were thinning the ozone layer over the Antarctic, they found themselves taking on an industry with sales of $28 billion a year.

By the time their manufacture was banned internationally during the late 1980s, CFCs had been used in roughly 90 million car and truck air conditioners, 100 million refrigerators, 30 million freezers, and 45 million air conditioners in homes and other buildings. Because CFCs remain in the stratosphere for up to 100 years, they will deplete the ozone long after industrial production of the chemicals ceased.

Discovery of the Ozone Hole

During 1985, a team of scientists working with the British Antarctic Survey reported a startling decline in "column ozone values" above an observation station near Halley Bay (Farman, Gardiner, and Shanklin 1985). Ozone depletion had been suspected in theory beginning in the early 1970s, and actual ozone densities had been declining over the Antarctic since 1977. The size of the decline in 1985 was a shocking surprise, however, because theorists had expected stratospheric ozone amounts to fall relatively evenly over the entire Earth.

Mario Molina and Sherwood Rowland, the first scientists to discover the ozone hole had expected a largely uniform decline of 1 to 5 percent (Rowland and Molina 1974, 810–812). Scientists at that time did not realize how ozone depletion related to temperature in the stratosphere. The seasonal variability of the decline was another surprise because existing theoretical models made no allowance for it. Ozone values over Antarctica tended to decline rapidly just as the sun was rising after winter. During the mid-1980s, the cause of dramatic falls in ozone density over the Antarctic was open to debate. Some scientists suspected variability in the sun's radiational output, and others suspected changes in atmospheric circulation. A growing minority began to suspect CFCs. These chemicals were not proven suspects when, in 1987, a majority of the world's national governments signed the Montreal Protocol to eliminate CFCs.

Definite proof of CFCs' role in ozone depletion developed shortly thereafter, as J. G. Anderson and colleagues implicated the chemistry of chlorine and explained a chain of chemical reactions (later broadened to include bromides as a bit player) as the "smoking gun" that explained why ozone depletion was so sharp, and why this depletion was limited to specific geographic areas at a specific time of the year (Anderson, Brune, and Proffitt 1989, 11,465). The temperature of the stratosphere became a key ingredient in the mix—the colder the stratosphere, the more active the chlorine chemistry that devoured ozone. By the year 2000, according to Maureen Christie, ozone depletion was "significantly affecting ozone levels throughout the Southern Hemisphere" (Christie 2001, 86).

Restoration of stratospheric ozone may even become more closely linked to greenhouse warming as temperatures continue to rise near the surface of the Earth. Guy P. Brasseur and colleagues modeled the response of the middle atmosphere to a doubling of carbon-dioxide levels near the surface. Their models indicated that "[a] cooling of about 8°C is predicted at 50 kilometers during summer. During winter, the temperature is reduced up to 14°C at 60 kilometers in the polar region" (Brasseur et al. 2000, 16). Increasing levels of methane also add to this effect. In addition to its properties as a greenhouse gas, "methane oxidation leads to higher water and OH concentrations in the stratosphere and mesosphere, and hence to less ozone at these altitudes" (Brasseur et al. 2000, 16).

During the mid-1990s, scientists were beginning to model a relationship between global warming and ozone depletion. A team led by Drew Shindell at the NASA Goddard Institute for Space Studies created the first atmospheric simulation to include ozone chemistry. The team found that the greenhouse effect was responsible not only for heating the lower atmosphere, but also for cooling the upper atmosphere. The cooling poses problems for ozone molecules, which are most active at low temperatures. Based on the team's model, the buildup of greenhouse gases could chill the high atmosphere near the poles by as much as 8° to 10°C. The model predicted that maximum ozone loss would occur between the years 2010 and 2019 (Shindell, Rind, and Lonergan 1998, 589).

In 1998, the Antarctic ozone hole reached a new record size roughly the size of the continental United States. Some researchers came to the conclusion that, as Richard A. Kerr described in *Science*:

> Unprecedented stratospheric cold is driving the extreme ozone destruction.... Some of the high-altitude chill … may be a counterintuitive effect of the accumulating greenhouse gases that seem to be warming the lower atmosphere. The colder the stratosphere, the greater the destruction of ozone by CFCs. (Kerr 1998c, 391)

Record Antarctic Ozone Depletion

The Antarctic ozone hole formed earlier and endured longer during September and October 2000 than ever before—and by a significant amount. Figures from NASA satellite measurements showed that the hole covered an area of approximately 29 million square kilometers in early September, exceeding the previous record from 1998. These record sizes persisted for several days.

Ozone levels, measured in Dobson units, fell below 100 D.U. for the first time. The area cold enough to produce ozone depletion also grew by 10 to 20 percent more surface area than in any other year. The ozone-depletion zone was coming closer to New Zealand, where usual springtime ozone levels average about 350 D.U. During the spring of 2000, ozone levels reached as low as 260 D.U. when atmospheric circulation patterns nudged the Antarctic zone northward. Scientists usually regard an area of the stratosphere as ozone-depleted when its Dobson-unit level falls below 220.

The decline in stratospheric ozone is striking when viewed on a graph with any sense of historic proportion. As little as a decade or two will do. Until the 1990s, in the Arctic, springtime ozone levels ranged around 500 D.U. By 2001, they were averaging 200 to 300 D.U.; in the Antarctic, in the days before the ozone hole (about 1980) Dobson-unit values ranged from about 250 to 350 D.U.; by the year 2000, they ranged from 100 to 200 D.U. During 2000, the ozone-depleted area over Antarctica grew, at its maximum extent, to a size equal to Africa. While the polar reaches of the Earth have been suffering the most dramatic declines in ozone density, ozone measurements over most of the planet also have declined roughly 15 percent since the mid-1980s.

During early September 2003, indications were that the area of depleted ozone over Antarctica was approaching near-record size again. By the end of the month, the area of severely depleted ozone was the second-largest on record, at about the size of North America. The Antarctic ozone hole swelled to an area of 10 million square kilometers by late-August 2005. Only the ozone holes of 1996 and 2000 had been larger (Schiermeier 2005, 179).

The coupling of global warming near the surface with declining stratospheric temperatures continued in 2006, as the loss of ozone over Antarctica reached a new record, according to scientists with the European Space Agency (ESA). "Such significant ozone loss requires very low temperatures in the stratosphere combined with sunlight," said ESA atmospheric engineer Claus Zehner. "This year's extreme loss of ozone can be explained by the temperatures above Antarctica reaching the lowest recorded in the area since 1979," the beginning of recordkeeping ("Ozone Loss" 2006).

Polar Stratospheric Clouds

Markus P. Rex and colleagues studied climatic conditions in the Arctic and found a surprisingly strong relationship between ozone loss and the number and density of polar stratospheric clouds (PSCs) that form in the stratosphere despite the fact that it is extremely dry. Results were reported in *Geophysical Research Letters* (Rex et al. 2004). By the end of 2001, Michael Proffitt, the World Meteorological Organization's senior scientific officer, said that "[t]he area with temperatures low enough for polar stratospheric clouds that initiate rapid ozone destruction to form during October is double that found during any earlier five-year period" (Tolbert and Toon 2001).

Polar Stratospheric Clouds are not new, having been described as "Nacreous clouds resembling giant abalone shells floating in the sky" (Tolbert and Toon 2001, 61). These clouds form 20 kilometers above the ground during winter and are sometimes called "mother-of-pearl clouds" because they shimmer. Some stratospheric clouds have been reported in Scandinavia for a century, and Edward Wilson noted them on Robert Falcon Scott's 1901 Antarctic expedition. Sometimes, the clouds shine with green and orange shades at sunrise and sunset (Tolbert and Toon 2001, 61). Polar stratospheric clouds remained largely an atmospheric curiosity until the discovery of widespread ozone depletion over the Antarctic during the mid-1980s. Scientists surmised that the ozone loss was occurring in the only place where the stratosphere was cold enough to produce these clouds. The clouds form during the springtime, when sunshine is available.

Daniel Kirk-Davidoff and colleagues wrote in *Geophysical Research Letters* that increasing coverage of polar stratospheric clouds "in a positive feedback loop" have been associated with dramatic polar warming at the surface during periods of high carbon-dioxide levels in the Earth's history, notably the Eocene (55 to 38 million years ago) and the Cretaceous (135 million to 65 million years ago). During these periods, land and surface ocean temperatures in the polar regions are believed to have been much higher than at present, at which time scientists' proxies indicated that tropical temperatures were similar to or only slightly higher than present times (Ball 2004).

Ozone Depletion: "Rocks" in the Stratosphere

As scientists probe the connections between surface warming and atmospheric cooling, they find more potentially dangerous complications.

For example, a team of atmospheric scientists has discovered large particles inside stratospheric clouds over the Arctic that could further delay the healing of the Earth's protective ozone layer. The team found large, nitric acid-containing particles that could delay the recovery and make the ozone layers over both poles more vulnerable to climate change, said atmospheric chemist David Fahey of the National Oceanic and Atmospheric Administration's office in Boulder, Colorado.

Each winter in the stratosphere over the poles, water and nitric acid condense to form polar stratospheric clouds that unleash chlorine and bromine, which degrade the ozone. Later in the winter, nitrogen compounds help shut down the destruction. Fahey's team found previously unknown nitric acid particles that remove nitrogen, allowing the destruction to continue. They nicknamed them "rocks" because they are hundreds of times bigger than other particles in the clouds.

The "rocks" form during the polar winter, when temperatures in the stratosphere decline to below −90°C. If global-warming forecasts become reality, the cooling of the stratosphere compelled by the retention of heat near the surface may cause more "rocks" to form, accelerating ozone depletion. "If it gets colder and you get more 'rocks,' the depletion period is going to last longer. The chlorine can continue to eat ozone," said Paul Newman, an atmospheric physicist at NASA's Goddard Space Flight Center in Maryland (Erickson 2001, 37-A). "What he got is really outstanding," Newman said of the findings by Fahey's team. "This mechanism that we now understand really will help us be able to more precisely predict what's going to happen in the future" (Erickson 2001, 37-A).

Fahey led a team of 27 researchers that included scientists from NOAA in Boulder, the University of Colorado, the National Center for Atmospheric Research in Boulder, and the University of Denver. The scientists described their findings in the February 9, 2001, edition of the journal *Science*.

"It's a major puzzle piece in the process by which ozone comes to be destroyed," Fahey said of the discovery. The discovery of these rocks was made during a January 2000 research flight of a NASA high-flying ER-2 (NASA's version of the U-2 spy plane) over the Arctic near the North Pole (Erickson, 2001, 37-A). A machine on the aircraft that was measuring nitrogen-containing gases "coughed out what looked like disastrous noise" (Kerr 2001, 962). The "noise" turned out to be very large particles (measured against other Arctic cloud mass) containing nitric acid (HNO_3), previously unknown to science. The particles averaged 3,000 times the size of other atmospheric particles in the stratosphere.

These polar stratospheric cloud particles (known in shorthand form as "PSC rocks") remove reactive nitrogen from the atmosphere that would otherwise "tie up chlorine and bromine in inactive, harmless forms" through denitrification (Kerr 2001, 963). The "rocks" also "provide surfaces where chlorine and bromine can be liberated from their inactive forms to enter their ozone-destroying forms" (Kerr 2001, 963). Additionally, the large size of these PSC rocks causes them to fall more quickly than other particles, removing even more nitrogen from the stratosphere. Given all these factors, the PSC rocks "have significant potential to denitrify the lower stratosphere" (Fahey et al. 2001, 1026). Fahey and colleagues concluded:

Arctic ozone abundances will remain vulnerable to increased winter/spring loss in the coming decades as anthropogenic chlorine compounds are gradually removed from the atmosphere, particularly if rising concentrations of greenhouse gases induce cooling in the polar vortex and trends of increasing water vapor continue in the lower stratosphere. Both effects increase the extent of PSC formation and, thereby, denitrification and the lifetime of active chlorine. The role of denitrification in these future scenarios is likely quite important. (Fahey et al. 2001, 1030)

Fahey and his colleagues estimated that ozone depletion in the Arctic stratosphere may not reach its peak until the year 2070, even with a steady decline in chlorine levels. Not everyone is as sanguine as Fahey about the future of stratospheric ozone. The subject is a matter of some rather intense debate. Sherwood Rowland of the University of California at Irvine, who shared the 1995 Nobel Prize for Chemistry for his part in the discovery that CFCs destroy stratospheric ozone, said that the effect of global warming on ozone depletion should be short-lived. "The [ozone depletion] story is approaching closure, and that's very satisfying," Rowland said (Schrope 2000, 627).

Alan O'Neill, a climate modeler at England's University of Redding said that record-breaking ozone holes in the year 2000 are not surprising, and that the ozone holes should heal by the year 2050. He added, however, that "[h]igher concentrations of greenhouse gases … could push that

date back a few decades" (Schrope 2000, 627). O'Neill believes that ozone loses will peak about the year 2005 and then decline. Shindell said that even though scientists are beginning to understand how global warming could delay ozone-shield recovery, "[t]he agreement to limit production [of CFCs] has been an unqualified success. The science was listened to, the policy-makers did something, and it actually worked" (Schrope 2000, 627).

Ozone Depletion in the Arctic

Jonathan Shanklin of the British Antarctic Survey, one of the three scientists credited with discovering severe ozone depletion over Antarctica, has warned that global warming threatens to deplete stratospheric ozone over the Arctic in a manner similar to the ozone hole over the Antarctic. Shanklin told the British Broadcasting Corporation Radio 4's "Costing the Earth" that the hole could ultimately affect the United Kingdom, bathing it in higher levels of cancer-causing ultraviolet radiation. Shanklin also has been quoted as saying that the Arctic's ozone troubles are not over.

Solar flares and frigid stratospheric temperatures during the winter of 2003–2004 provoked the worst depletion of ozone above the Arctic since records have been kept, according to a team of scientists reporting in the March 2, 2005, issue of *Geophysical Research Letters* (Randall et al. 2005). The team reported that levels of nitrous oxides up to four times any previously observed, agitated by solar activity, combined with bitter cold (about −110°F) to drive the depletion of ozone to levels 60 percent below anything previously observed (records reach to 1985). "I don't think we can be confident about whether or not we're seeing an ozone recovery or if we're attributing recovery to the correct causes," said Cora Randall, lead author of the report, who is an atmospheric scientist at the University of Colorado (Human and McGuire 2005, A-8).

The levels of protective ozone over most of Canada will not recover during the twenty-first century and probably will deteriorate during its second half, according to one scientific study. This study contradicted earlier, more optimistic forecasts that ozone levels around the world would begin to recover by mid-century thanks to a ban on synthetic chlorine compounds that destroy ozone. "The more we know, the more we realize we don't know," said Jack McConnell, an atmospheric science professor at York University. Lower ozone levels could mean jumps of up to 10 percent in some skin cancers, which now strike almost 60,000 Canadians a year. The research also found that ozone levels would be lowest in summer, the season of greatest danger. Ozone in the stratosphere—the layer 10 to 40 kilometers above Earth—screens out the ultraviolet rays linked to skin cancers (Calamai 2002a, A-8).

A Bromide-Based Disaster Narrowly Averted

Paul J. Crutzen has asserted that problems with the stratospheric ozone layer could have been much worse if chemists had developed substances based on bromine, which is 100 times as dangerous for ozone atom-to-atom compared with chlorine.

This brings up the nightmarish thought that if the chemical industry had developed organochlorine compounds instead of the CFCs—or, alternatively, if chlorine chemistry had behaved more like that of bromine—then without any preparedness, we would have faced a catastrophic ozone hole everywhere and in all seasons during the 1970s, probably before atmospheric chemists had developed the necessary knowledge to identify the problem. (Crutzen 2001, 10)

Given the fact that no one seemed overly worried about this problem before 1974, wrote Crutzen, "We have been extremely lucky." This shows, he wrote, "[t]hat we should always be on our guard for the potential consequences of the release of new products into the environment … for many years to come" (Crutzen 2001, 10).

Crutzen emphasized the danger inherent in taking chances with the Earth's climate system without understanding its chemistry. The history of atmospheric chemistry during the last few decades, he said, has been one of surprises. "There may be more of these things around the corner," he said (McFarling 2001c, A-1).

FURTHER READING

Anderson, J. G., W. H. Brune, and M. H. Proffitt. "Ozone Destruction by Chlorine Radicals within the Antarctic Vortex: The Spatial and Temporal Evolution of ClO/O3, Anticorrelation Based on In Situ ER-2 Data." *Journal of Geophysical Research* 94 (1989):11,465–11,479.

Ball, Philip. "Climate Change Set to Poke Holes in Ozone: Arctic Clouds Could Make Ozone Depletion Three Times Worse Than Predicted." *Nature* Science Update, March 3, 2004. http://info.nature.com/cgi-bin24/DM/y/eOCB0BfHSK0Ch0JVV0AY.

Borenstein, Seth. "Arctic Lost 60 percent of Ozone Layer: Global Warming Suspected." Knight-Ritter News Service, April 6, 2000. (LEXIS)

Brasseur, Guy P., Anne K. Smith, Rashid Khosravi, Theresa Huang, Stacy Walters, Simon Chabrillat, and Gaston Kockarts. "Natural and Human-induced Pertubations in the Middle Atmosphere: A Short Tutorial." In *Atmospheric Science Across the Stratopause*, David E. Siskind et al. Washington, D.C.: American Geophysical Union, 2000.

Calamai, Peter. "Alert over Shrinking Ozone Layer." *Toronto Star*, March 18, 2002a, A-8.

Christie, Maureen. *The Ozone Layer: A Philosophy of Science Perspective*. Cambridge: Cambridge University Press, 2001.

Crutzen, Paul J. "The Antarctic Ozone Hole, a Human-caused Chemical Instability in the Stratosphere: What Should We Learn from It? In *Geosphere-Biosphere Interactions and Climate*, ed. Lennart O. Bengtsson and Claus U. Hammer, 1–11. Cambridge: Cambridge University Press, 2001.

Erickson, Jim. "Boulder Team Sees Obstacle to Saving Ozone Layer: 'Rocks' in Arctic Clouds Hold Harmful Chemicals." *Rocky Mountain News* (Denver), February 9, 2001, 37-A.

Fahey, D. W., R. S. Gao, K. S. Carslaw, J. Kettleborough, P. J. Popp, M. J. Northway, J. C. Holecek, S. C. Ciciora, R. J. McLaughlin, T. L. Thompson, R. H. Winkler, D. G. Baumgardner, B. Gandrud, P. O. Wennberg, S. Dhaniyala, K. McKinney, T. Peter, R. J. Salawitch, T. P. Bui, J. W. Elkins, C. R. Webster, E. L. Atlas, H. Jost, J. C. Wilson, R. L. Herman, A. Kleinböhl, and M. von König. "The Detection of Large HNO3-Containing Particles in the Winter Arctic Stratosphere." *Science* 291 (February 9, 2001):1026–1031.

Farman, J. C., B. G. Gardiner, and J. D. Shanklin. "Large Losses of Total Ozone Reveal Seasonal ClOx/NOx Interaction." *Nature* 315 (1985):207–210.

Flannery, Tim. *The Weather Makers: How Man Is Changing the Climate and What It Means for Life on Earth*. New York: Atlantic Monthly Press, 2005.

Gordon, Anita, and David Suzuki. *It's a Matter of Survival*. Cambridge, MA: Harvard University Press, 1991.

Human, Katy, and Kim McGuire. "Ozone Decline Stuns Scientists." *Denver Post*, March 2, 2005, A-8.

Kerr, Richard A. "Deep Chill Triggers Record Ozone Hole." *Science* 282 (October 16, 1998c):391.

Kerr, Richard A. "A Variable Sun Paces Millennial Climate." *Science* 294 (November 16, 2001):1431–1433.

Kirk-Davidoff, Daniel, Daniel P. Schrag, and James G. Anderson. "On the Feedback of Stratospheric Clouds on Polar Climate." *Geophysical Research Letters* 29, no. 11 (2002):14,659–14,663.

McFarling, Usha Lee. "Fear Growing over a Sharp Climate Shift." *Los Angeles Times*, July 13, 2001c, A-1.

"Ozone Hole of 2008." NASA Earth Observatory, October 28, 2008. http://earthobservatory.nasa.gov/Newsroom/NewImages/images.php3?img_id=18192.

"Ozone Loss Reaches New Record." Environment News Service, October 2, 2006. http://www.ens-newswire.com/ens/oct2006/2006-10-02-01.asp.

Randall, C. E., V. L. Harvey, G. L. Manney, Y. Orsolini, M. Codrescu, C. Sioris, S. Brohede, C. S. Haley, L. L. Gordley, J. M. Zawodny, and J. M. Russell III. "Stratospheric Effects of Energetic Particle Precipitation in 2003–2004." *Geophysical Research Letters* 32 (March 2, 2005), L05802, doi: 10.1029/2004 GL022003.

Rex, Markus, P. von der Gathen, Alfred Wegener, R. J. Salawitch, N. R. P. Harris, M. P. Chipperfield, and B. Naujokat. "Arctic Ozone Loss and Climate Change." *Geophysical Research Letters* 31 (March 10, 2004). http://www.eurekalert.org/pub_releases/2004-03/agu-ajh031004.php.

Rowland, Sherwood, and Mario Molina. "Stratospheric Sink for Chlorofluoromethanes: Chlorine Atom-Catalyzed Destruction of Ozone." *Nature* 249 (June 28, 1974):810–812.

Schiermeier, Quirin. "Poles Lose Out as Ozone Levels Begin to Recover." *Nature* 437 (September 8, 2005):179.

Schrope, Mark. "Successes in Fight to Save Ozone Layer Could Close Holes by 2050." *Nature* 408 (December 7, 2000):627.

"Scientists Report Large Ozone Loss." *USA Today*, April 6, 2000, 3-A.

Shindell, Drew T., David Rind, and Patrick Lonergan. "Increased Polar Stratospheric Ozone Losses and Delayed Eventual Recovery Owing to Increasing Greenhouse-gas Concentrations." *Nature* 392 (April 9, 1998):589–592.

Tolbert, Margaret A., and Owen B. Toon. "Solving the P[olar] S[tratospheric] C[loud] Mystery." *Science* 292 (April 6, 2001):61–63.

$$\mathcal{P}$$

Palms in Southern Switzerland

Gian-Reto Walther and colleagues tracked the expansion of *Trachycarpus fortunei*, the windmill palm (similar to the palmetto), into southern Switzerland following rising winter minimum temperatures and a lengthening growing season. In addition to *Trachycarpus fortunei* reproducing naturally in the foothills of the southern Alps, they have observed it spreading in seminatural habitats or seeding in gardens and parks as far north as southern coastal England, Brittany, in France, the Netherlands, and coastal southwestern British Columbia, all areas where warmer nights have extended the average annual growing season to well over 300 days a year (Walther et al. 2007).

The palms of Switzerland are being observed about 300 kilometers (more than 200 miles) outside their historical range. Scientists conclude that the spread of these palms is a "significant global bioindicator across continents for present-day climate change and the projected global warming of the near future" (Walther et al. 2007).

FURTHER READING

Walther, Gian-Reto, Emmanuel Gritti, Silje Berger, Thomas Hickler, Zhiyao Tang, and Martin T. Sykes. "Palms Tracking Climate Change." *Global Ecology and Biogeography* (2007), doi: 10.1111/j.1466-8238.2007.00328.x.

Penguins, South African

A century ago, at least 1.5 million African penguins waddled and swam on and along the coasts of Namibia and South Africa. In 2001, the population was about 120,000; by 2007, however, roughly 20,000 survived. With few exceptions, they occupied only a few islands near South Africa's tip, including Robben Island, where Nelson Mandela (and others) were imprisoned. The penguin population has been declining rapidly as the anchovies and sardines they eat migrate southward out of warming water. Some years, the adult population has declined 25 percent.

The African penguin, the continent's only native penguin, is less than two feet tall, "with large eyes set in a field of red and a distinctive black stripe circumnavigating its belly. Its calls range from a delicate trill to a loud night-time bray, which leads some to call it the 'jackass penguin'" (Wines 2007). Rob Crawford, a marine scientist at South Africa's department of marine and coastal management, said that the penguins' decline "has to be [caused by] either fishing or climate or a mixture of both.... but my hunch is that it's environmental" (Wines 2007).

The African penguins can find no land base farther south, so they are limited to food that is available within waters 25 miles out. They have been waiting for the sardines and anchovies to return, and are starving as they wait. Commercial fishing also has depleted sardine stocks. Scientists have concluded the sardines move with changes in the cold, nutrient-rich Benguela Current, which flows from Antarctica along the southwest African coast. In addition to climate change, the sardines may move in a 50-year cycle (Wines 2007).

FURTHER READING

Wines, Michael. "Dinner Disappears, and African Penguins Pay the Price." *New York Times*, June 4, 2007. http://www.nytimes.com/2007/06/04/world/africa/04robben.html.

Permafrost and Climate Change

Given the fact that one-third of the Earth's carbon is stored in far northern latitudes (mainly in tundra and boreal forests) (Mack et al. 2004, 440), the speed with which warming of the ecosystem may release this carbon into the atmosphere is vitally important to forecasts of global warming's pace and effects. The amount of carbon stored in Arctic ecosystems also includes two-thirds of the amount presently found in the atmosphere (Loya and Grogan 2004, 406). Its release into the atmosphere will depend on the pace of temperature rise—and the Arctic, according to several sources, has been the most rapidly warming regions of the Earth.

Most of the permafrost that is capable of releasing methane and carbon dioxide is in the large landmass of Siberia. The amount of carbon trapped in a type of permafrost called yedoma may be 100 times the amount of carbon released into the air each year by the burning of fossil fuels (Borenstein 2006a). Thus, human combustion of fossil fuels acts as a trigger by raising temperatures enough to release even more permafrost from natural sources.

As much as 90 percent of the Northern Hemisphere's permafrost (the top three meters, or 10 feet) could thaw by the end of the twenty-first century given "business-as-usual" increases in greenhouse-gas emissions, according to projections by scientists affiliated with the National Center for Atmospheric Research. Even with major reductions in emissions, the Northern Hemisphere permafrost is projected to shrink from four million to about 1.5 million square miles by 2100.

The Permafrost "Time Bomb"

As temperatures increase, methane and carbon dioxide escapes from the soil. "It's kind of like a slow-motion time bomb," said Ted Schuur, a professor of ecosystem ecology at the University of Florida and co-author of the study in *Science* (Borenstein 2006a). "The effects can be huge," said Katey Walter of the University of Alaska at Fairbanks. "It's coming out a lot and there's a lot more to come out" (Borenstein 2006a; Walter et al. 2006, 71).

"The higher the temperature gets, the more permafrost we melt, the more [likely] it is to become a more vicious cycle," said Chris Field, director of global ecology at the Carnegie Institution of Washington. "That's the thing that is scary about this whole thing. There are lots of mechanisms that tend to be self-perpetuating

and relatively few that tend to shut it off" (Borenstein 2006a).

During 2004, Michelle C. Mack and colleagues presented results in *Nature* of a 20-year fertilization experiment in Alaskan tundra, during which "increased nutrient availability caused a net ecosystem loss [of] almost 2,000 grams of carbon per square meter" (Mack et al. 2004, 440). While aboveground plant production more than doubled under warmer conditions, "losses of carbon and nitrogen from deep soil levels ... were substantial and more than offset the increased carbon and nitrogen storage in plant biomass and litter" (Mack et al. 2004, 440). According to this study, increased releases of carbon to the atmosphere "primed" by increasing decomposition of organic matter could accelerate the rise in atmospheric carbon dioxide—and, therefore, warming.

"Kick-Starting" Greenhouse Gases from Peat

In a series of experiments described in *Nature*, researchers found that the increase in atmospheric carbon dioxide observed in recent decades has had a direct impact on kick-starting the release of carbon from peat (Freeman et al. 2004, 195). "Under elevated carbon-dioxide levels, the proportion of dissolved organic carbon derived from recently assimilated carbon dioxide was ten times higher than that of controlled cases," they wrote. "Concentrations of dissolved organic carbon appear far more sensitive to environmental drivers that affect net primary productivity than those affecting decomposition alone (Freeman et al. 2004, 195). This research is the first to measure and forecast the release of carbon dioxide from peat bogs.

This research was among the first direct physical evidence of a "positive feedback" between carbon dioxide in the atmosphere and the huge stores of carbon locked up on land, with increases in one causing a corresponding increase in the other. Chris Freeman of the University of Wales in Bangor, who led the research team, said, "We've got an enormous carbon store locked up in peat bogs which is equivalent to the entire store of carbon in the atmosphere and yet this store on land appears to have sprung a leak," Freeman said (Connor 2004a, 9). The amount of carbon dioxide being released from peat lands is accelerating roughly at a rate of 6 percent per year, according to the research of Freeman and colleagues. "By 2060 we could see more carbon dioxide being released into the atmosphere than is being released by burning fossil fuel," he said (Connor 2004a, 9).

Tests on peat samples taken from three different sites in Britain show that increasing the amount of carbon dioxide in the air around the samples causes the peat itself to emit up to ten times the amount of carbon dioxide that it would have under usual conditions. Freeman said that peat bogs release carbon in a dissolved organic form. Emissions from peat lands into surrounding rivers and streams has increased by between 65 and 90 percent in the past six years (Connor 2004a, 9).

"The rate of acceleration suggests that we have disturbed something critical that controls the stability of the carbon cycle on our planet," he said. Dissolved organic carbon in rivers and watercourses also may react with the chlorine in water-treatment processes to produce potentially carcinogenic chemicals, Freeman said (Connor 2004a, 9). "We've known for some time that CO_2 levels have been rising and that these could cause global warming. But this new research has enormous implications because it shows that even without global warming, rising CO_2 can damage our environment," he added (Connor 2004a, 9).

Additional evidence that the Earth's colder regions already are releasing additional greenhouse gases into the air has been provided by M. L. Goulden and colleagues who studied boreal forests, finding evidence that "provides clear evidence that carbon dioxide locked into permafrost several hundred to 7,000 years ago is now being given off to the atmosphere as warming climate melts the permafrost" (Davis 2001, 270). These researchers investigated black spruce forest and found that carbon dioxide was being emitted into the atmosphere beneath the biologically active layer containing moss and tree roots (Goulden et al. 1998, 214). The same is true in many cases for methane. According to Neil Davis, author of *Permafrost: A Guide to Frozen Ground in Transition* (2001), this "relict" carbon dioxide

represents a massive source since it is estimated that the carbon dioxide contained in the seasonally and perennially frozen soils of boreal forests is 200 to 500 billion metric tons, enough if all released to increase the atmosphere's concentration of carbon dioxide by 50 percent. Hence, it is possible that the release of carbon dioxide from melting permafrost

during warming, or locking it into newly frozen soil during cooling, may accelerate climate change. (Davis 2001, 270)

In the measured words of science:

We have observed a 65 percent increase in the dissolved organic carbon (DOC) concentration in freshwater draining from upland catchments in the United Kingdom over the past 12 years. Here we show that rising temperatures may drive this process by stimulating the export of DOC from peat lands.... [This process] may increase substantially as a result of global warming. (Freeman et al. 2004, 785)

In simple English, a very dangerous carbon feedback loop has been engaged.

Methane emissions have been increasing from bogs in Sweden, according to an international research team led by the GeoBiosphere Science Centre at Sweden's Lund University. The Abisko region in sub-Arctic Sweden, which the team studied, has long-term records of climate, permafrost, and other environmental variables, which made comparisons possible (Christensen et al. 2004).

"In the present study, airborne infrared images were used to compare the distribution of vegetation in 1970 with that of 2000. Dramatic changes were observed, and the scientists relate them to the climate warming and decreasing extent of permafrost that was observed over the same period" ("Swedish Bogs" 2004). At one site, Stordalen, these researchers estimated an increase in methane emissions between 22 and 66 percent between 1970 and 2000, according to lead researcher Torben R. Christensen of Lund University's GeoBiosphere Science Centre (Christensen et al. 2004; "Swedish Bogs" 2004).

Thaw in Siberia

During August 2005, climate researchers who recently returned from the area said that a large area of western Siberia was undergoing an unprecedented thaw that could dramatically increase the rate of global warming. They said that permafrost across a million square kilometers, an area as large as France and Germany combined, has started to melt. Sergei Kirpotin at Tomsk State University in western Siberia and Judith Marquand at Oxford University reported that was until recently a barren expanse of frozen peat had turned, during the summer, "into a broken landscape of mud and lakes, some more than a kilometer across" (Sample 2005, 1). In the winter, so much methane is being released that its bubbles have kept the surface from freezing.

Kirpotin told the *Manchester Guardian Weekly* that the situation was an "ecological landslide that is probably irreversible and is undoubtedly connected to climatic warming." He added that the thaw had probably begun in the past three or four years (Sample 2005, 1). "When you start messing around with these natural systems, you can end up in situations where it's unstoppable. There are no brakes you can apply," said David Viner, a senior scientist at the Climatic Research Unit at the University of East Anglia. "This is a big deal because you can't put the permafrost back once it's gone. The causal effect is human activity and it will ramp up temperatures even more than our emissions are doing" (Sample 2005, 1).

Euan Nisbet of London's Holloway College, who monitors methane emissions on an international scale, said that methane releases from western Siberia's peat bogs could account for up to 100,000 tons a day, or more than all the human-generated emissions from the United States. Science writer Fred Pearce described the melting:

Numberless lakes stretched to the horizon. From the air, they did not look like lakes that form naturally. They were generally circular, looking more like flooded potholes in a road. The lakes formed individually from small breaches in the permafrost. Wherever ice turned to water, a small pond formed. Then surrounding lumps of frozen peat would slump into the water and the pond would grow in an ever-widening circle, until mile after mile of frozen bog had melted into the mass of lakes. (Pearce 2007b, 79–80)

Melting Permafrost Imperils Alaskan Villages

"I don't want to live in permafrost no more," said Frank Tommy, 47, standing beside gutted geese and seal meat drying on a wooden rack outside his mother's house in Newtok, Alaska, a village that now sinks in permafrost during the warmer months of the year. "It's too muddy. Everything is crooked around here" (Yardley 2007b). As its soil turns to mud, sea erosion has turned Newtok into an island caught between the widening Ninglick River and a slough. Wooden houses that stand beside boardwalks

have been shifted every few months to keep them level. Newtok is one of about 180 Alaskan villages that is being eroded. The Army Corps of Engineers estimates that moving Newtok from its ever-more-perilous location would cost $130 million, or about $413,000 for each of the 315 people who live there (Yardley 2007b).

Newtok's leaders assert that this cost estimate is too high. They say they can move a few people at a time and save money. "They grossly overestimate it, and that's why federal and state agencies are afraid to step in," said Stanley Tom, the current tribal administrator and the brother of Nick Tom Jr. "They don't want to spend that much money" (Yardley 2007b). Newtok has acquired a new site through a land swap with the U.S. Fish and Wildlife Service, above the Ninglick River. They call it Mertarvik, meaning "getting water from the spring" in the Yupik language (Yardley 2007b).

William Yardley described the present village in the *New York Times*:

Excrement dumped from honey buckets is piled on the banks of the slow-flowing Newtok River, not far from wooden shacks where residents take nightly steam baths. An elderly man drains kerosene into a puddle of snowmelt. Children pedal past a walrus skull left to rot, tusks intact, in the mud beside a boardwalk that serves as a main thoroughfare. There are no cars here, just snow machines, boats and all-terrain vehicles that tear up the tundra.

Many men still travel with the seasons to hunt and fish. Some will take boats into Bristol Bay this summer to catch salmon alongside commercial fishermen from out of state. But the waterproof jacket sewn from seal gut that Stanley Tom once wore is now stuffed inside a display case at Newtok School next to other relics. Now Mr. Tom puts on a puffy parka to walk the few hundred feet he travels to work. He checks his e-mail messages to see if there is news from the [Army Corps of Engineers] or from Senator Stevens while his brother, Nick, sketches out a budget proposal for a nonprofit corporation to help manage the relocation, presuming the money arrives. (Yardley 2007b)

Permafrost and Burning Peat Bogs across Russia

Some of the permafrost that covers as much as 65 percent of the Russian Federation has been melting; scientists there expect the permafrost boundary to recede 150 miles northward during the next quarter-century. Already, the diamond-producing town of Mirny, in Yakutia, has evacuated a quarter of its population because their houses melted into the previously frozen soil. Parts of the Trans-Siberian Railway's track has twisted and sunk due to melting of permafrost, causing delays of service of several days at a time.

By 2002, melting permafrost had damaged 300 apartment buildings in the Siberian cities of Norilsk and Yakutsk (Goldman 2002, 1494). "Assuming that the region [Siberia] continues to warm at the modest rate of 0.075° C per year," warned Lev Khrustalev, a geocryologist at Moscow State University, by 2030 "all five-story structures built between 1950 and 1990 in Yakutsk, a city of 193,000 people, could come crashing down unless steps are taken to strengthen them and preserve the permafrost" (Goldman 2002, 1494).

Russian weather experts described Siberia's problems with melting permafrost during an international conference on climate change in Moscow held in early fall 2003. Georgy Golitsyn, director of Moscow's Institute of Atmospheric Physics, said that by the end of the twenty-first century temperature increases in Siberia could be twice the worldwide rise of 1.4° to 5.8° C anticipated by the Intergovernmental Panel on Climate Change (IPCC) (Meuvret 2003). "Extreme weather events might happen more frequently, [with] the melting of permafrost, which is already noticeable, and damage risks to buildings, roads and pipelines. Pipelines are always having some trouble," he said (Meuvret 2003).

"Here in Moscow and in European Russia, really cold episodes are becoming quite rare. And in Siberia, very heavy frosts have almost disappeared. Instead of −40° C or −50° C which were quite frequent, now these occur just occasionally and they usually experience −30° C," Golitsyn said (Meuvret 2003). The disastrous flooding of the Lena River Basin in 2001 gave a foretaste of the problems to come, he noted. "The winter had been normal, but ... the soil was frozen and then in May a heat wave came with temperatures up to 30° [C] when the snow had not melted yet.... This is the type of catastrophe we might have more frequently" (Meuvret 2003).

Golitsyn's warning was supported by Michel Petit, who was, until April 2002, a French representative to the IPCC. Siberia, he said, is "one of the most sensitive regions to climate change on the planet" (Meuvret 2003). While areas that could be farmed would probably expand in Siberia, and maritime transport would be opened at seasons when it is now blocked by ice, melting of the permafrost would be a "real catastrophe," turning large areas of Siberia into swamps. "Major industrial complexes, towns and pipelines would subside. Carbon dioxide and methane would escape, and the greenhouse effect would become even more serious," said Gorgy Gruza, a Russian climate-change expert who opposes the Kyoto Protocol.

Lakes across the Siberian Arctic are shrinking and drying up, according to a comparison of satellite images taken of 10,000 large lakes over a 25-year period. Scientists found that 125 of the lakes disappeared completely and are now revegetated. Researchers at three U.S. universities described their research on lakes spread across more than 200,000 square miles of Siberia, as they asserted that "Arctic warming has accelerated since the 1980s, driving an array of complex physical and ecological changes in the region" (Smith et al. 2005, 1429). After three decades of rising soil and air temperatures in Siberia, "our analysis reveals a widespread decline in lake abundance and area, despite slight precipitation increases" (Smith et al. 2005, 1429). Why? Warming temperatures lead to thinning and eventual "breaching" of permafrost near lakes, greatly facilitating their drainage to the subsurface (Smith et al. 2005, 1429).

Russia has been beset by burning peat bogs during warmer, drier summers. During September 2002, peat fires around Moscow produced so much smoke that motorists drove with headlights on at noon. The fires, which add carbon dioxide to the atmosphere, are especially intense because of a hot, dry summer.

Yukon Caskets Ride Melting Permafrost

Thawing permafrost has been causing Inuvialuit caskets, many more than 80 years old, to the surface on parts of Herschel Island, Yukon Territory. Animals, including caribou, have tampered with some of the caskets, while others have been looted of artifacts, presumably by people. According to DeNeen L. Brown of the *Washington Post*,

Graves are pushing up from the ground as the ice within the carpet of permafrost melts, churning the soil beneath it into a muddy soup, spitting up foreign contents, sending whole hill slopes sliding downward. On a far tip of this island an entire grave site one day got up and slipped into the sea. (Brown 2001, A-27)

The gravesites are all that remains on the island of a once-thriving Inuvialuit community. The island now is mostly deserted, with only the few park rangers as summer residents. Tourists come in by plane to hike, camp, and watch birds.

Brown reported that

> The older generation of Inuvialuit believes that anyone who touches the possessions of the dead after they are buried will be cursed. Some in the younger generation, many years removed from traditional life, are wrestling with both sides of the issue. They want to maintain tradition, but they also do not want to sit back while their ancestors' bones lie uncovered on melting ground. (Brown 2001, A-27)

Wayne Pollard, professor of geography at McGill University in Montreal, who began studying Herschel Island in 1988, said that the island's landscapes are some of the world's most vulnerable to climate change. "If it were simply coastal erosion and the graves dropping into the ocean and disappearing, that wouldn't be a problem," he said. "Because as I understand it, the Inuvialuit are quite comfortable with the idea of bodies being returned to the environment, as part of the natural cycle that they accept in life" (Brown 2001, A-27). Most often, however, the graves do not reach the sea.

The Qinghai-Tibet Railway

Designers and builders of the Qinghai-Tibet Railway in China have found that deteriorating permafrost has presented problems for the world's highest-elevation rail line. Research produced by the Chinese Academy of Sciences indicates that permafrost on the path of the railroad is now five to seven meters thinner than 20 years ago; about 10 percent of the plateau's permafrost has completely melted. Scientists said the climate changes have altered ground temperatures to a depth of at least 40 meters, with greatest changes to a depth of 20 meters. Stabilizing the melting Earth has become a major challenge in design of the railway ("Global Warming Troubles" 2003).

Engineers working on the railway have adopted three special measures to ensure the stability of the roadbed in the permafrost areas, including changing routes, building railway bridges along sections of complex geological conditions, and building soil layers that can insulate the ground from heat created by the railway ("Global Warming Troubles" 2003).

The nearly 2,000-kilometer railway links Xining, the capital of northwest China's Qinghai Province, and Lhasa, the capital of the Tibet Autonomous Region. The section linking Xining and Golmud City in Qinghai was completed in 1984. Construction of the 1,118-kilometer section connecting Golmud with Lhasa began in June 2001 and was completed in 2007 ("Global Warming Troubles" 2003).

FURTHER READING

Borenstein, Seth. "Scientists Find New Global Warming 'Time Bomb' Methane Bubbling up from Permafrost." Associated Press, September 6, 2006a. (LEXIS)

Brown, DeNeen L. "Waking the Dead, Rousing Taboo: In Northwest Canada, Thawing Permafrost Is Unearthing Ancestral Graves." *Washington Post*, October 17, 2001, A-27.

Christensen, Torben R., Torbjörn Johansson, H. Jonas Åkerman, Mihail Mastepanov, Nils Malmer, Thomas Friborg, Patrick Crill, and Bo H. Svensson. "Thawing Sub-arctic Permafrost: Effects on Vegetation and Methane Emissions." *Geophysical Research Letters* 31, no. 4 (February 20, 2004), L04501, doi: 10.1029/2003GL018680.

Connor, Steve. "Peat Bog Gases Accelerate Global Warming." *London Independent*, July 8, 2004a, 9.

Davis, Neil. *Permafrost: A Guide to Frozen Ground in Transition.* Fairbanks: University of Alaska Press, 2001.

Freeman, C., N. Fenner, N. J. Ostle, H. Kang, D. J. Dowrick, B. Reynolds, M. A. Lock, D. Sleep, S. Hughes, and J. Hudson. "Export of Dissolved Carbon from Peatlands under Elevated Carbon Dioxide Levels." *Nature* 430 (July 8, 2004):195–198.

"Global Warming Troubles Qinghai-Tibet Railway Construction." Xinhua (Chinese News Agency), April 30, 2003. (LEXIS)

Goldman, Erica. "Even in the High Arctic, Nothing is Permanent." *Science* 297 (August 30, 2002):1493–1494.

Goulden, M. L., S. C. Wofsy, J. W. Harden, S. E. Trumbore, P. M. Crill, S. T. Gower, T. Fries, B. C. Daube, S.-M. Fan, D. J. Sutton, A. Bazzaz, and J. W. Munger. "Sensitivity of Boreal Forest Carbon Balance to Soil Thaw." *Science* 279 (January 9, 1998):214–217.

Loya, Wendy M., and Paul Grogan. "Carbon Conundrum on the Tundra." *Nature* 431 (September 23, 2004):406–407.

Mack, Michelle C., Edward A. G. Schuur, M. Syndonia Bret-Harte, Gaius R. Shaver, and F. Sturt Chapin III. "Ecosystem Carbon Storage in Arctic Tundra Reduced by Long-term Nutrient Fertilization." *Nature* 432 (September 23, 2004):440–443.

Meuvret, Odile. "Global Warming Could Turn Siberia into Disaster Zone: Expert." Agence France Presse, October 2, 2003.

Pearce, Fred. *With Speech and Violence: Why Scientists Fear Tipping Points in Climate Change.* Boston: Beacon Press, 2007b.

Sample, Ian. "Warming Hits 'Tipping Point: Climate Change Alarm as Siberian Permafrost Melts for First Time since Ice Age." *Manchester Guardian Weekly,* August 18, 2005, 1. http://www.guardian.co.uk/guardianweekly/story/0,12674,1550685,00.html.

Smith, L. C., Y. Sheng, G. M. MacDonald, and L. D. Hinzman. "Disappearing Arctic Lakes." *Science* 308 (June 3, 2005):1429.

"Swedish Bogs Flooding Atmosphere with Methane: Thawing Sub-arctic Permafrost Increases Greenhouse Gas Emissions." American Geophysical Union, February 10, 2004. http://www.scienceblog.com/community/article2366.html.

Walter, K. M. S. A. Zimov, J. P. Chanton, D. Verbyla, and F. S. Chapin, III. "Methane Bubbling from Siberian Thaw Lakes as a Positive Feedback to Climate Warming." *Nature* 414 (September 7, 2006):71–75.

Yardley, William. "Engulfed by Climate Change, Town Seeks Lifeline." *New York Times,* May 27, 2007b. http://www.nytimes.com/2007/05/27/us/27newtok.html.

Phytoplankton Depletion and Warming Seas

Phytoplankton in the world's oceans are a key consumer of excess carbon, but warming of the oceans may be devastating them. Surveys by satellites and ships have confirmed that the productivity of these microscopic plants is declining, most notably from the North Pacific to the high Arctic. Mike Toner wrote in the *Atlanta Journal-Constitution*: "Plankton are also as important to the long-term health of the atmosphere as the world's forests. The photosynthesis of the ocean's tiny green plants account for about half of the carbon dioxide that plants remove from the atmosphere each year" (Toner 2002d, 3-A). "The less phytoplankton you have, the less carbon is taken up by the oceans," said Margarita Conkright, of the National Oceanic and Atmospheric Administration (Toner 2002d, 3-A).

Warmer ocean surface temperatures are related to lower oceanic phytoplankton biomass and productivity, the source of half the photosynthesis ("net primary production") on Earth (and the base of the oceanic food chain), according to a survey of nearly a decade's worth of satellite data compiled by Michael J. Behrenfeld and colleagues in *Nature* (2006, 752–755). They argued that reduction in phytoplankton biomass is a result of changes induced by warming in ocean circulation that reduces supplies of nutrients required for photosynthesis. Many of these nutrients are conducted through the ocean by upwelling of cold, nutrient-rich water.

According to another researcher, "Extrapolating the satellite observations into the future suggests that marine biological productivity in the tropics and mid-latitudes will decline substantially, in agreement with climate-model simulations" (Doney 2006, 696). Productivity probably will increase at higher latitudes.

Zooplankton Move Northward

Gregory Beaugrand and colleagues reported in *Science* that several sea species, notably zooplankton (a major base of the oceanic food chain), moved northward in the Northern Hemisphere between 1960 and 1999 in response to warming water temperatures:

> We provide evidence of large-scale changes in the biogeography of calanoid copepod crustaceans in the Eastern North Atlantic Ocean and European shelf seas.… Strong bio-geographical shifts in all copepod assemblages have occurred with a northward extension of more than 10 degrees latitude of warm-water species associated with a decrease in the number of colder-water species. (Beaugrand et al. 2002, 1692)

The study was based on analysis of 176,778 samples collected by the Continuous Plankton Recorder Survey taken monthly in the North Atlantic since 1946. The scientists wrote:

> The observed bio-geographical shifts may have serious consequences for exploited resources in the North Sea, especially fisheries. If these changes continue, they could lead to substantial modifications in the abundance of fish, with a decline or even a collapse in the stock of boreal species such as cod, which is already weakened by over-fishing. (Beaugrand et al. 2002, 1693–1694)

Studying nitrogen balances, Canadian and American oceanographers found evidence that warmer oceans will lose nitrate, depleting phytoplankton, and boosting atmospheric carbon-dioxide levels (Leggett 2001, 226). Satellite surveys have detected a sharp decline in plankton in several of the world's oceans, a potential threat to the marine food chain, and one that could undercut one of the world's natural buffers to global warming (Toner 2002d, 3-A).

Decline Varies Ocean to Ocean

The mass of phytoplankton can vary by 100 times in various parts of the ocean, depending on local conditions, including the degree of mixing and deposit of windborne iron from the continents. Ocean mixing is inhibited by warming water. Satellites have made possible surveys of plankton biomass over large areas of the world ocean. The amount of plankton biomass is also sensitive to the El Niño/La Niña cycle, decreasing as waters warm, and increasing as they cool.

The greatest decline has been in the Northern Pacific Ocean, where summer levels have dropped by more than 30 percent since the 1980s. Comparing sets of satellite data from early 1980 to data from the late 1990s, researchers reported in *Geophysical Research Letters* that sharp declines in plankton had taken place in *both* the North Pacific and in the North Atlantic, where their abundance decreased by 14 percent. In equatorial regions, plankton levels increased. Worldwide, plankton stocks decreased more than 8 percent (Gregg and Conkright 2002).

The researchers were not certain whether the decline of phytoplankton is part of a natural cycle in the oceans, a reflection of regional changes, a result of a gradual warming of the Earth, or a combination of these factors. They did find, however, a correlation between the decline of plankton and increasing ocean surface temperatures indicating that "[c]limate change could be a cause as well as an effect of plankton declines" (Toner 2002d, 3-A). Plankton need two things to grow: sunlight and nutrients. The researchers said warmer sea-surface temperatures interfere with upwelling of colder water that is rich in nutrients from the oceans' depths.

Scientists in Australia have found similar declines in phytoplankton populations and have asserted that global warming is starving the depths of the Southern Ocean of oxygen. Research scientists working with the Australian climate-research agency Commonwealth Scientific and Industrial Research Organization (CSIRO) said they have found a significant drop in the Southern Ocean's oxygen levels during the past 30 years; they also anticipate that this situation will worsen in the future.

Hobart-based scientist Richard Matear said that oxygen-starved oceans may lead to long-term devastation of marine life. Matear said that the decline in oxygen levels was most likely caused by global warming, as natural variability could not explain the reduction (Barbeliuk 2002). "The interpretation is that less oxygen-rich water is penetrating into the ocean and this in turn gives additional credibility to climate change models," Matear said. Matear said the Southern Ocean was considered by oceanographers to be the "lungs" of the world's oceans, creating 55 percent of the water that regenerates the deep ocean. Any decrease in the Southern Ocean's capacity could have consequences for oceans around the world, he said (Barbeliuk 2002).

Matear said that the Southern Ocean itself is unlikely to be greatly affected by the change, since it is so oxygen-rich. Other oceans that already have limited oxygen could suffer dire consequences, with the variety of marine life greatly reduced, he said. "Most organisms that live in the ocean require oxygen, only a very few don't," Matear said (Barbeliuk 2002). Matear said that warmer waters are unable to carry as much oxygen or sink as deeply into the ocean's depths. The colder the water, the faster it sinks, acting as an oxygen pump. The Southern Ocean's present oxygen level is about 200 micro-moles per kilogram, representing a decline of about 15 micro-moles during 30 years to 2000.

Matear and fellow scientists Tony Hirst (also of the CSIRO), and Ben McNeil, from the Antarctic Co-operative Research Centre, used chemical data gathered during oceanographic research voyages south of Australia to look for changes in the ocean conditions. "Having demonstrated that oxygen is a valuable indicator of climate change in our models, we now have a quantity to monitor to detect future changes," he said (Barbeliuk 2002).

Plankton Decline May be Related to Changes in Thermohaline Circulation

The decline of plankton also may be related to the changing nature of the thermohaline circulation (the Atlantic meridional overturning circulation). Writing in *Nature*, Andreas Schmittner reported on results of models simulating disruption of the Atlantic meridional overturning circulation. The disruption led to a collapse of the North Atlantic plankton stocks to less than half of their initial biomass, "owing to rapid shoaling of winter mixed layers and their associated separation from the deep ocean nutrient reservoir." Schmittner wrote that "[t]hese model results are consistent with the available high-resolution palaeorecord, and suggest that global ocean productivity is sensitive to changes in the Atlantic meridional overturning circulation" (Schmittner 2005, 628).

Adding more support to the idea that phytoplankton populations are declining around the world because of changing ocean circulation, British scientists examining the Atlantic Ocean south of Iceland found that populations of zooplankton, which feed many larger ocean species, have declined by as much as 90 percent in four decades. This may portend population reductions for larger species, from cod and haddock to whales and dolphins. "This is deeply worrying," said marine biologist Dr. Phil Williamson of East Anglia University. "We don't know why zooplankton numbers have plummeted, though global warming looks [like] the best candidate. What is certain is that removing the bottom link from the ocean food chain could have profound and unpleasant results" (McKie 2001, 14; Knutti et al. 2004, 851–854).

Updating a major survey of zooplankton levels in the North Atlantic completed during 1963, a team of British scientists during November 2001 set out in the marine research vessel *Discovery* to measure changes in these levels. Using automated equipment, the scientists sampled concentrations of *Calanus finmarhicus*, the principal type of Atlantic zooplankton. Having carried out 800 samplings in an area 1,000 miles south of Iceland, the scientists found 5,000 to 10,000 zooplankton per square meter, instead of the 50,000 average found in the earlier survey (McKie 2001, 14). The scientists believe that gradually increasing sea temperatures are playing a major role in the zooplankton's decline.

Whether the decline of plankton is directly related to increased ocean temperatures is not yet certain, according to Watson W. Gregg, a NASA biologist at the Goddard Space Flight Center in Greenbelt, Maryland, because other factors also affect the productivity of phytoplankton, such as availability of iron. According to Gregg, the greatest loss of phytoplankton has occurred where ocean temperatures have risen most significantly between the early 1980s and the late 1990s.

In the North Atlantic, during summer, sea-surface temperatures rose about 1.3° F during that period, Gregg said, while in the North Pacific the ocean's surface temperatures rose about 0.7° F (Perlman 2003, A-6). "This research shows that ocean primary productivity is declining, and it may be the result of climate changes such as increased temperatures and decreased iron deposition into parts of the oceans," Gregg said. "This has major implications for the global carbon cycle" (Perlman 2003, A-6).

Warming Aids Phytoplankton in Some Areas

Warming may have aided phytoplankton growth in some areas (unlike northern latitudes of the Atlantic Ocean, where it is declining with rising temperatures). Declining winter and spring snow cover over Eurasia has contributed to an increasing land-ocean temperature gradient that enhances summertime monsoon winds over the Western Arabian Sea, near thee coasts of Somalia, the Republic of Yemen, and Oman. According to one study, increased upwelling in the area contributes to an increase of more than 350 percent in average summertime phytoplankton biomass along the coast and 300 percent offshore, making the aquatic life in the area more productive (Goes et al. 2005, 545).

FURTHER READING

Barbeliuk, Anne. "Warmer Globe Choking Ocean." *The Mercury*, Hobart (Australia), March 16, 2002, n.p. (LEXIS)

Beaugrand, Gregory, Philip C. Reid, Frédéric Ibañez, J. Alistair Lindley, and Martin Edwards. "Reorganization of North Atlantic Marine Copepod Biodiversity and Climate." *Science* 296 (May 31, 2002):1692–1694.

Behrenfeld, Michael J., Robert T. O'Malley, David A. Siegel, Charles R. McClain, Jorger L. Sarmiento, Gene C. Feldman, Allen J. Milligan, Paul G. Falkowski, Ricardo M. Letelier, and Emmanuel S. Boss. "Climate-driven Trends in Contemporary Ocean Productivity." *Nature* 444 (December 7, 2006):752–755.

Doney, Scott C. "Plankton in a Warmer World." *Nature* 444 (December 7, 2006):695–696.

Goes, Joaquim I., Prasad G. Thoppil, Helga do R. Gomes, and John T. Fasullo. "Warming of the Eurasian Landmass is Making the Arabian Sea More Productive." *Science* 308 (April 22, 2005):545–547.

Gregg, Watson, W., and Margarita E. Conkright. "Decadal Changes in Global Ocean Chlorophyll."

Geophysical Research Letters 29, no. 15 (2002), doi: 10.1029/2002GL014689.

Knutti, R. J. Fluckiger, T. F. Stocker, and A. Timmermann. "Strong Hemispheric Coupling of Glacial Climate through Freshwater Discharge and Ocean Circulation." *Nature* 430 (August 19, 2004):851–856.

Leggett, Jeremy. *The Carbon War: Global Warming and the End of the Oil Era.* New York: Routledge, 2001.

McKie, Robin. "Dying Seas Threaten Several Species: Global Warming Could Be Tearing Apart the Delicate Marine Food Chain, Spelling Doom for Everything from Zooplankton to Dolphins." *London Observer*, December 2, 2001, 14.

Perlman, David. "Decline in Oceans' Phytoplankton Alarms Scientists: Experts Pondering Whether Reduction of Marine Plant Life Is Linked to Warming of the Seas." *San Francisco Chronicle*, October 6, 2003, A-6.

Schmittner, Andreas. "Decline of the Marine Ecosystem Caused by a Reduction in the Atlantic Overturning Circulation." *Nature* 434 (March 31, 2005):628–633.

Toner, Mike. "Microscopic Ocean Life in Global Decline: Temperature Shifts a Cause or an Effect?" *Atlanta Journal-Constitution*, August 9, 2002d, 3-A.

Pika Populations Plunge in the Rocky Mountains

Pikas, described by Usha Lee McFarling of the *Los Angeles Times* as "tennis ball-sized critters that whistle at passing hikers and scamper over loose, rocky slopes of the High Sierra and the Rocky Mountains" (McFarling 2003, A-17). A shy, flower-gathering mammal and longtime icon of the West's high peaks an association may fall victim to global warming. By 2003, they already had disappeared from almost 30 percent of the areas where they had been common during the early parts of the twentieth century. Pikas, which resemble hamsters, are biological relatives of rabbits. Over many millennia, they adapted to a degree of intense cold that is becoming scarcer in the mountains. Living above 7,000 feet, the Pikas cannot withstand heat. A comprehensive survey found that sites that have lost pikas were on average drier and warmer than sites where the animals remain, said Erik Beever, a U.S. Geological Survey biologist based in Corvallis, Oregon (McFarling 2003, A-17).

Beever indicated that such factors as cattle grazing and proximity to roads had some effect on the animals. Warmer and drier conditions in recent decades have been a major factor in their rapid disappearance, however. Earlier surveys in the United States found pikas missing from much of their previous range. Observation of pikas in the Yukon revealed that 80 percent of the animals died in some areas after unusually warm winters (McFarling 2003, A-17). Beever said that the demise of the Pika has been rapid. He was most surprised to find groups of animals disappearing over decades, rather than in centuries or millennia, as during climatic swings of the distant past (McFarling 2003, A-17).

FURTHER READING

McFarling, Usha Lee. "A Tiny 'Early Warning' of Global Warming's Effect: The Population of Pikas, Rabbit-like Mountain Dwellers, Is Falling, a Study Finds." *Los Angeles Times*, February 26, 2003, A-17.

Pine Beetles in Canada

The mountain pine beetle has infested an area in British Columbia three times the size of Maryland, devastating swaths of lodgepole pines and reshaping the future of the forest and the communities in it. The beetles are turning trees red, and killing more of them than wildfires. "It's pretty gut-wrenching," said Allan Carroll, a research scientist at the Pacific Forestry Centre in Victoria, whose studies tracked an association between warmer winters and the spread of the beetle. "People say climate change is something for our kids to worry about. No. It's now" (Struck 2006a).

Fires are being set in Alberta, along with traps, as thousands of trees are felled in an attempt to inhibit the beetles' spread across the Rocky Mountains from the West. "This is an all-out battle," said David Coutts, Alberta's minister of sustainable resource development. The Canadian Forest Service calls it the largest known insect infestation in North American history (Struck 2006a). The United States is less vulnerable than Canada because its stands of lodgepole pine are not continuous, and thus do not provide as easy a path of infestation.

Werner A. Kurz, writing in *Nature,* found that mountain pine-beetle infestations have killed enough trees to convert large tracts of Canadian forests, especially in British Columbia, from carbon sinks to sources. By 2007, more than 32 million acres of forests were infested, and the affected area was spreading as warmer summers enable faster beetle reproduction and milder winters fail to kill them (Kurz et al. 2008, 987).

"We are seeing this pine beetle do things that have never been recorded before," said Michael Pelchat, a forestry officer in Quesnel, as he followed moose tracks in the snow to examine a 100-year-old pine killed in one season by the beetle. "They are attacking younger trees, and attacking timber in altitudes they have never been before" (Struck 2006a).

Until now, Canada's lodgepole pine has been protected by the beetles by early, cold winters. For at least a decade, however, winters have not been cold enough to kill the beetles. Scientists with the Canadian Forest Service say the average temperature of winters here has risen by more than 4° F in the last century. "That's not insignificant," said Jim Snetsinger, British Columbia's chief forester. "Global warming is happening. We have to start to account for it" (Struck 2006a).

Doug Struck, writing in the in the *Washington Post,* described how the beetles kill trees:

In an attack played out millions of times over, a female beetle no bigger than a rice grain finds an older lodgepole pine, its favored host, and drills inside the bark. There, it eats a channel straight up the tree, laying eggs as it goes. The tree fights back. It pumps sap toward the bug and the new larvae, enveloping them in a mass of the sticky substance. The tree then tries to eject its captives through a small, crusty chute in the bark.

Countering, the beetle sends out a pheromone call for reinforcements. More beetles arrive, mounting a mass attack. A fungus on the beetle, called the blue stain fungus, works into the living wood, strangling its water flow. The larvae begin eating at right angles to the original up-and-down channel, sometimes girdling the tree, crossing channels made by other beetles. The pine is doomed. As it slowly dies, the larvae remain protected over the winter. In spring, they burrow out of the bark and launch themselves into the wind to their next victims. (Struck 2006a)

At British Columbia's Ministry of Forests and Range in Quesnel, forestry officer Pelchat saw the beetle expansion coming as "a silent forest fire" (Struck 2006a).

Pelchat and his colleagues tried to stall the invasion, hoping for cold temperatures. They searched out beetle-ridden trees, cutting them and burning them. They thinned forests. They set out traps. But the deep freeze never came. "We lost. They built up into an army and came across," Pelchat said. Surveys show the beetle has infested 21 million acres and killed 411 million cubic feet of trees—double the annual take by all the loggers in Canada. In seven years or sooner, the Forest Service predicted, that kill will nearly triple and 80 percent of the pines in the central British Columbia forest will be dead (Struck 2006a). *See also:* Bark Beetles Spread across Western North America; Spruce Bark Beetles in Alaska

FURTHER READING

Kurz, W. A., C. C. Dymond, G. Stinson, G. J. Rampley, E. T. Neilson, A. L. Carroll, T. Ebata, and L. Safranyik. "Mountain Pine Beetle and Forest Carbon Feedback to Climate Change." *Nature* 452 (April 24, 2008):987–991.

Struck, Doug. "'Rapid Warming' Spreads Havoc in Canada's Forests: Tiny Beetles Destroying Pines." *Washington Post,* March 1, 2006a, A-1. http://www.washingtonpost.com/wp-dyn/content/article/2006/02/28/AR2006022801772_pf.html.

Pliocene Paleoclimate

Searching for an analogue to a climate that may result from today's levels of carbon dioxide (after feedbacks take affect) scientists have cast their attention on the Pliocene (three to four million years ago), when atmospheric carbon dioxide was at roughly today's levels, temperatures were about 3° C higher worldwide, but "surface temperatures in the polar regions were so much higher that continental glaciers were absent from the northern hemisphere, and sea level was about 25 meters higher" (Fedorov et al. 2006, 1485).

James Hansen, director of NASA's Goddard Institute for Space Studies, has estimated that a sea-level rise of six meters (the Greenland Ice Cap, the West Antarctic Ice Sheet, or half of each) would displace about 235 million people, based on year 2000 population figures—11 million in the United States, 93 million in China and Taiwan, 46 million in India and Sri Lanka, 24 million in Bangladesh, 23 million in Indonesia and Malaysia, 12 million in Japan, and 26 million in Western Europe.

Raise the sea level by 25 meters (Pliocene level) and the number of displaced people rises to about 700 million, including 47 million in the United States, 224 million in China and Taiwan, 146 million in India and Sri Lanka, 109 million in Bangladesh, 72 million in Indonesia and

Malaysia, 39 million in Japan, and 66 million in Western Europe (Hansen 2005b). A 25 meter sea-level rise would drown most of Florida and sizable U.S. cities on the East and Gulf coasts.

The Pliocene period was not only warm enough to melt much of the world's ice, but it also triggered a semi-permanent "El Niño" condition in the Pacific Ocean, with warm water spanning the Pacific Ocean (Kerr 2005a, 1456). According Hansen,

> If emissions of greenhouse gases continue to increase as in then 'business-as-usual' scenario, the rate of isotherm movement will double in this century to at least 70 miles per decade. If we continue on this path, a large fraction of the species on Earth will go extinct.... For all foreseeable human generations, [Earth] will be a far more desolate planet than the one in which civilization developed and flourished over the past several thousand years. (Hansen 2006a, 12)

FURTHER READING

Fedorov, A. V., P. S. Dekens, M. McCarthy, A. C. Ravelo, P. B. deMenocal, M. Barreiro, R. C. Pacanowski, and S. G. Philander. "The Pliocene Paradox (Mechanisms for a Permanent El Niño). *Science* (June 9, 2006):1485–1489.

Hansen, James E. "Is There Still Time to Avoid 'Dangerous Anthropogenic Interference' with Global Climate? A Tribute to Charles David Keeling." A paper delivered to the American Geophysical Union, San Francisco, December 6, 2005b. http://www.columbia.edu/~jeh1/keeling_talk_and_slides.pdf.

Hansen, James E. "The Threat to the Planet." *New York Review of Books,* July 2006a, 12-16.

Kerr, Richard A. "Looking Back for the World's Climatic Future." *Science* 312 (June 9, 2005a):1456–1457.

Polar Bears under Pressure

Steady melting of Arctic ice threatens the survival of polar bears before the end of this century. Seymour Laxon, senior lecturer in geophysics at the Center for Polar Observation and Modeling in London, said that serious concern exists regarding the long-term survival hopes for polar bears as a species. "To put it bluntly," he said, "No ice means no bears" (Elliott 2003; Laxon et al. 2003, 947).

Andrew Derocher, a professor of biology at the University of Alberta, supported Laxon's beliefs. "If the progress of climate change continues without any intervention, then the prognosis for polar bears would ultimately be extinction," he said ("Expert Fears" 2003, C-8). Although reluctant to give an exact date for polar bears' extinction, Derocher said changes in the northern ecosystem are occurring so quickly that the habitat may be incapable of supporting polar bears within 100 years. "What we're looking at is major changes in terms of the distribution of polar bears," he said. "But there are so many variables to put a number of years on something like extinction of a species. I guess ultimately I'm an optimist that humans will turn the events causing global warming back a bit" ("Expert Fears" 2003, C-8).

The International Union for Conservation of Nature and Natural Resources, a network of 10,000 scientists, forecast in May 2007 that global warming and overhunting could diminish the polar bear population by at least 30 percent in coming decades. "Given what the climate models predict for continuing warming and melting of sea ice, the whole thing leads to an extinction curve," said Peter Ewins, director of the World Wildlife Fund Canada's Arctic Conservation Program. "And it's not a question of if, it's a clear question of when" (Krauss 2006).

During the fall of 2007, after the Arctic Ice Cap had lost almost a quarter of its mass in one summer, warnings became more strident. "Just 10 more years of current global warming pollution trajectories will commit us to enough warming to melt the Arctic and doom the polar bear to extinction," said Kassie Siegel, director of the Center for Biological Diversity's Climate, Air, and Energy Program. "We urgently need to address global warming, not just for the sake of the polar bear but for the sake of people and wildlife around the world" ("Comment Period" 2007).

Melting Ice and Polar Bear Extinction

The offshore ice-based ecosystem is sustained by upwelling nutrients that feed the plankton, shrimp, and other small organisms, which feed the fish. These, in turn, feed the seals, which feed the bears. The Native people of the area also occupy a position in this cycle of life. When the ice is not present, the entire cycle collapses.

The polar bear has become an iconic symbol of global warming's toll in animal life in the Arctic. When the U.S. federal government opened

Two polar bears walking across the ice. (Courtesy of Shutterstock)

comment during one three-month period in 2007 over whether the species should be legally protected under the Endangered Species Act, more than 500,000 people responded (Beinecke 2008, 175).

In 2006, the George W. Bush administration decided to advocate listing the polar bear as threatened under the Endangered Species Act, but not because it wanted to. Environmental groups had backed its lawyers into a legal corner. Natural Resources Defense Council senior attorney Andrew Wetzler, one of the lawyers who filed suit against the administration seeking this action (along with the Center for Biological Diversity and Greenpeace), welcomed the proposal. "It's such a loud recognition that global warming is real," Wetzler said. "It is rapidly threatening the polar bear and, in fact, an entire ecosystem with utter destruction" (Eilperin 2006i, A-1). As part of the U.S. federal government's decision-making process regarding whether to list the polar bear as a threatened species under the Endangered Species Act, the U.S. Geological Survey in 2007 issued a series of reports anticipating that their population would plunge two-thirds by 2050 as Arctic sea ice retreats.

Polar Bears and Legal Liability

On May 14, 2008, about 10,000 polar bears were placed on the U.S. Endangered Species List, the first animal to be placed on that list because of global warming. The move was largely symbolic (it does prohibit importation of hides and trophies from a population of about 15,000 bears in Canada). The declaration did little to modify existing restrictions under the Marine Mammal Protection Act, but it did set the stage for a finely nuanced battle over legal liability for the bears' anticipated demise. For the time being, however, the U.S. Interior Department did little but reaffirm well-known facts: polar bears feed on seals from ice that is rapidly melting during the Arctic's warmer months. Officials then pledged not to use the law to hold any corporations responsible.

The classification was made following a suit filed in 2005 by the Center for Biological Diversity, Greenpeace, and the Natural Resources Defense Council. These groups want to use the declaration as a legal wedge against new coal-fired power plants or other new sources of carbon dioxide that melt the ice that harm the bears. The Interior Department, however, refused to cooperate. Its secretary, Dirk Kempthorne, opposes the Endangered Species Act itself as a legal instrument in any context.

Kempthorne specifically ruled out that possibility, saying, "When the Endangered Species Act was adopted in 1973, I don't think terms like 'climate change' were part of our vernacular." Barton H. Thompson Jr., a law professor and director of the Woods Institute of the Environment at Stanford University, said the decision reflected the administration's view that "there is no way, if your factory emits a greenhouse gas, that we can say there is a causal connection between that emission and an iceberg melting somewhere and a polar bear falling into the ocean" (Barringer 2008e).

Following the polar bears' listing as an endangered species early in September 2008, the American Petroleum Institute and four other business groups joined Alaska Governor Sarah Palin's administration in trying to reverse the listing of the polar bear as a threatened species. The group filed a lawsuit against the Interior Department and U.S. Fish and Wildlife Service, arguing that the bears are not threatened by climate change. The National Association of Manufacturers, the U.S. Chamber of Commerce, the National Mining Association, and the American Iron and Steel Institute joined in the suit.

During 2002, a World Wildlife Fund (WWF) study, "Polar Bears at Risk," reported that the combination of toxic chemicals and global warming could cause extinction of roughly 22,000 surviving polar bears in the wild within 50 years. Lynn Rosentrater, co-author of the report and a climate scientist in the WWF's Arctic program, said:

> Since the sea ice is melting earlier in the spring, polar bears move to land earlier without having developed as much fat reserves to survive the ice free season. They are skinny bears by the end of summer, which in the worst case can affect their ability to reproduce. ("Thin Polar Bears" 2002)

The same report said that increasing carbondioxide emissions have caused Arctic temperatures to rise by 5° C during the past century, and the extent of sea ice has decreased by 6 percent in 20 years. By roughly 2050, scientists believe that 60 percent of today's summer sea ice will be gone, which would more than double the summer ice-free season from 60 to 150 days

("Thin Polar Bears" 2002). Lower body weight reduces female bears' ability to lactate, leading to fewer surviving cubs. Already, fewer than 44 percent of cubs now survive the ice-free season, according to the report ("Thin Polar Bears" 2002).

More pregnant polar bears in Alaska are digging snow dens on land instead of sea ice, and researchers say global warming is the likely cause. From 1985 to 1994, 62 percent of the female polar bears in the study dug dens on sea ice. From 1998 to 2004, just 37 percent gave birth on sea ice. The rest instead dug dens on land, according to the study by three U.S. Geological Survey researchers. Bears that continued to dig dens on ice moved away from ice that was thinner or unstable. "In recent years, Arctic pack ice has formed progressively later, melted earlier, and lost much of its older and thicker multi-year component," said wildlife biologist Anthony Fischbach, the study's lead author ("Polar Bears Shift Dens" 2007, 12).

Less Ice, Thinner Bears

In western Hudson Bay, where warmer temperatures during the 1990s provoked earlier ice melting in the summer and later freezing in the fall, polar bears suffered substantial weight loss. For every week that the ice broke up earlier, bears came ashore 20 to 25 pounds lighter, said zoologist Ian Stirling of the Canadian Wildlife Service in Edmonton (Kerr 2002b, 1492). Stirling has documented a 15 percent decrease in average weight and number of pups born to polar bears in the western Hudson Bay between 1981 and 1998. Such conclusions are amply supported from a number of other sources.

The polar bear population in the Hudson Bay area has dropped from 1,200 in 1989 to 950 in 2004, and the bears that are around are 22 percent smaller than they used to be, she said ("Arctic Ice Melting" 2006). Canadian Wildlife Service scientists reported during December 1998, that polar bears around Hudson Bay are 90 to 220 pounds lighter than 30 years ago, apparently because earlier ice-melting has given them less time to feed on seal pups.

According to Michael Goodyear, director of the Churchill (Manitoba) Northern Studies Center, "For every week a bear has not been hunting, it is 10 kilograms (22 pounds) lighter, which can be dangerous as polar bears need to fatten up for the five months in the summer and fall that they are forced to fast" (Clavel 2002).

Weight loss affects the bears' ability to reproduce, and weakens future generations that are born. Polar bears may soon reach a point at which they have lost enough weight to render them infertile. Lara Hansen, chief scientist at the WWF, told the *London Independent*: "If the population stops reproducing, that's the end of it" (Connor 2004b).

Ringed and bearded seals, the polar bears' usual prey, are one of a number of larger animals that have adapted to life on, in, and under the ice. "Ranging year-round as far as the [North] pole," wrote Kevin Krajick in *Science*, "They never leave the ice pack, keeping breathing holes open all winter, and making lairs under snow mounds. During the spring, the snow lairs camouflage their new pups from polar bears and protect them from cold air" (Krajick 2001a, 425).

Ice in some areas of the Arctic now melts as early as March, when the seals are having their pups. Because the ice breaks up too early, the pups often have not been fully weaned, so many of them starve, or mature in a weakened state. During two decades (1980 to 2002), ice has been breaking up two weeks earlier than previously. Biologist Lois Harwood of the Canadian Department of Fisheries and Oceans said that ice in the Western Arctic broke up three weeks earlier than usual in 1998, dumping many hungry pups into the water before they had been weaned. Adult seals were thinner than usual as well, despite available prey exposed by the early breakup of the ice. "They were starving in the midst of plenty," Harwood said (Krajick 2001a, 425). Peary caribou also have been observed falling through unusually thin ice during their migrations.

Polar Bears as Land Animals

The loss of summer sea ice is pushing polar bears more onto land in northern Canada and Alaska, making people think there are more polar bears when there are not, said NASA scientist Claire Parkinson, who studies the bears ("Arctic Ice Melting" 2006).

Charles Wohlforth, writing in *The Last Polar Bear*, described how the Alaskan North Slope town of Barrow was inundated by landbound polar bears during the early fall of 2002 after sea ice retreated too far offshore to be used as a hunting platform in the usual fashion:

> The strange congregation of dozens of stranded bears ... threw the community into fear and consternation. A 1,100-pound bear—a large male—was shot when it wouldn't leave the school. Visitors were warned not to walk outdoors. Biologists employed by the local government were constantly on the run to keep polar bears off the streets. Polar bears normally don't eat people, but they easily can and, when hungry enough, they occasionally have. Bears on land in the summer fall have been more common all along the coast since the trend of warmer weather shortened the winters and diminished the sea ice. But no one had seen an invasion like this before. Biologists counted more than one hundred bears near town. (Kazlowski et al. 2008, 63)

During the summer of 2004, hunters found half a dozen polar bears that had drowned about 200 miles north of Barrow, on Alaska's northern coast. They had tried to swim for shore after the ice had receded 400 miles. A polar bear can swim 100 miles—but not 400 ("Melting Planet" 2005). As many as 36 other polar bears may have died as they tried to swim to land after the Arctic Ice Shelf along the north shore of Alaska receded (Tidwell 2006, 57).

Without ice, polar bears can become hungry, miserable creatures, especially in unaccustomed warmth. During Iqaluit's weeks of record heat in July 2001, two tourists were hospitalized after they were mauled by a bear in a park south of town. On July 20, a similar confrontation occurred in northern Labrador as a polar bear tried to claw its way into a tent occupied by a group of Dutch tourists. The tourists escaped injury but the bear was shot to death. "The bears are looking for a cooler place," said Ben Kovic, Nunavut's chief wildlife manager (Johansen 2001, 18).

Until recently, polar bears had their own food sources and usually went about their business without trying to steal food from humans. Beset by late freezes and early thaws, hungry polar bears are coming into contact with people more frequently. In Churchill, Manitoba, polar bears waking from their winter's slumber found that the Hudson Bay's ice melted earlier than usual. Instead of making their way onto the ice in search of seals, the bears walk along the coast until they get to Churchill, where they block motor traffic and pillage the town dump.

Churchill now has a holding tank for wayward polar bears that is larger than its jail for people.

TIME magazine described Churchill:

Polar bears that ordinarily emerge from their summer dens and walk north up Cape Churchill before proceeding directly onto the ice now arrive at their customary departure point and find open water. Unable to move forward, the bears turn left and continue walking right into town, arriving emaciated and hungry. To reduce unscheduled encounters between townspeople and the carnivores, natural-resource officer Wade Roberts and his deputies tranquilize the bears with a dart gun, temporarily house them in a concrete-and-steel bear "jail" and move them 10 miles north. In years with a late freeze—most years since the late 1970s—the number of bears captured in or near town sometimes doubles, to more than 100. (Linden 2000)

By 2007, several tour operators were charging $3,000 to $5,000 for two- to four-day "last chance" excursions to Churchill to witness the bears in their natural habitats. In the meantime, the amount of time that polar bears had access to ice for hunting had been reduced by a month in about two decades. Residents of Churchill remarked at wearing flip-flops and shirtsleeves in mid-October, when Hudson Bay once was beginning to freeze. The average weight of female bears had fallen from 600 pounds to a few pounds over 400, inhibiting reproduction. The mortality for bear cubs under five years of age was up 50 percent, according to Robert Buchanan, president of Polar Bears International (Clark 2007, 2-D). The town dump was closed in 2005 after hungry bears made a habit of pillaging it for food scraps. The emaciated bears were tackling garbage trucks before they even reached the dump.

Polar bears face other problems related to warming. For example, some polar bear dens have collapsed because of lightning-sparked brush fires on the tundra, which may be increasing because of global warming, according to Stirling, who said that the Canadian government should consider fighting fires in prime bear denning areas in the north ("Brush Fires" 2002, 18). Some pregnant bears have been digging dens only to have them collapse, Stirling said.

"The fires burn off all of the trees and bushes that are on the upper part of the banks holding the roofs together," he said ("Brush Fires" 2002, 18). "The fires melt the permafrost in the adjacent areas, and in particular the roofs, so there's nothing to hold the roofs together. They just collapse," he explained. The bears then must find another denning site where they can give birth over the winter. "The fires are definitely affecting the ability of the bears to use some of the prime areas," said Stirling. He also said that, after a fire, vegetation may require 70 years to grow back to a density suitable for denning sites ("Brush Fires" 2002, 18). Yet another potential risk to polar bears is the increased chance that rain in the late winter will cause polar bear dens to collapse before females and cubs have departed. Scientists surveying polar bear habitat in Manitoba, Canada, have observed large snow banks used for dens that had collapsed under the weight of wet snow (Stirling and Derocher 1993, 244).

Polar Bears Increasing—Or Just More Visible?

The semisovereign Canadian Inuit territory of Nunavut increased its annual hunting quotas by 29 percent in 2005 to 518 kills, an increase of 115, saying that Inuit hunters were actually seeing an increase in polar bear populations. That impression, some scientists and environmentalists say, is simply a matter of the bear's greater visibility, as shrinking ice pushes them closer to Inuit communities (Krauss 2006). Those experts tick off a list of stresses on the polar bear: Global warming is melting the bear's icy migration routes, critical for breeding and catching seals for food, around the Hudson Bay and Alaska. Poaching is threatening populations in the Russian Federation. Pollution is causing deformities and reproductive failures in Norway (Krauss 2006).

Other experts see a healthier population. They note that there are more than 20,000 polar bears roaming the Arctic, compared with as few as 5,000 some 40 years ago, before Canada, Denmark, Norway, the former Soviet Union, and the United States agreed to strong restrictions on trophy hunting in the 1970s (Krauss 2006).

Mitchell Taylor, manager of wildlife research for the Nunavut government, said warming trends had so far seriously affected only western Hudson Bay, just one of 20 areas where polar bears live. He acknowledged that overhunting could be a problem in Baffin Bay, between Canada and Greenland. "In other areas, polar bears appear to be overabundant," he added. "People have to quit thinking of polar bears as one big continuous mass of animals that are all doing the same thing" (Krauss 2006).

Some Polar Bears Engage in Cannibalism

The survival crisis facing polar bears took a fresh turn for the worse after 2004 when scientists from the United States and Canada found evidence that lack of access to food sources (mainly ringed seals) caused by shrinking ice cover might be forcing some of them to engage in cannibalism in the southern Beaufort Sea north of Alaska and western Canada. The scientists found three examples of polar bears preying on each other between January and April 2004, including killing a female in a den shortly after she gave birth. The bears use the sea ice not only for feeding, but also for mating and giving birth. The predation study was published in an online version of the journal *Polar Biology* (Amstrup et al. 2006).

According to the study's principal author, Steven Amstrup of the U.S. Geological Survey Alaska Science Center, polar bears sometimes kill each other for population regulation, dominance, and reproductive advantage, but killing for food heretofore had been unusual (Joling 2006). "During 24 years of research on polar bears in the southern Beaufort Sea region of northern Alaska and 34 years in northwestern Canada, we have not seen other incidents of polar bears stalking, killing, and eating other polar bears," the scientists said (Joling 2006). The Center for Biological Diversity of Joshua Tree, California, during February 2005, petitioned the federal government to list polar bears as threatened under the federal Endangered Species Act (Joling 2006). "It's very important new information," she said. "It shows in a really graphic way how severe the problem of global warming is for polar bears," said Kassie Siegel, lead author of that petition (Joling 2006).

According to Dan Joling's summary of the predation study for the Associated Press, researchers discovered the first kill in January 2004. A male bear had pounced on a den, killed a female, and dragged it 245 feet away, where it ate part of the carcass. Females are about half the size of males. "In the face of the den's outer wall were deep impressions of where the predatory bear had pounded its forepaws to collapse the den roof, just as polar bears collapse the snow over ringed seal lairs," the paper said,

continuing: "From the tracks, it appeared that the predatory bear broke through the roof of the den, held the female in place while inflicting multiple bites to the head and neck. When the den collapsed, two cubs were buried, and suffocated, in the snow rubble" (Joling 2006).

FURTHER READING

Amstrup, S. C., I. Stirling, T. S. Smith, C. Perham, and G. W. Thiemann. "Recent Observations of Intraspecific Predation and Cannibalism among Polar Bears in the Southern Beaufort Sea." *Polar Biology* 29, no. 11 (2006): 997–1002, doi: 10.1007/s00300-006-0142-5.

"Arctic Ice Melting Rapidly, Study Says." Associated Press in *New York Times*, September 14, 2006. http://www.nytimes.com/aponline/us/AP-Warming-Sea-Ice.html.

Barringer, Felicity. "Polar Bear Is Made a Protected Species." *New York Times*, May 15, 2008e. http://www.nytimes.com/2008/05/15/us/15polar.html.

Beinecke, Frances. "A Climate for Change: Next Steps in Solving Global Warming." In *The Last Polar Bear: Facing the Truth of A Warming World*, Steven Kazlowski, 175–185. Seattle: Braided River, 2008.

"Brush Fires Collapsing Bear Dens." Canadian Press in *Calgary Sun*, November 2, 2002, 18.

Clark, Jayne. "Tours Bear Witness to Earth's Sentinel Species." *USA Today*, November 2, 2007, 1-D, 2-D.

Clavel, Guy. "Global Warming Makes Polar Bears Sweat." Agence France Presse, November 3, 2002.

"Comment Period Extended on Polar Bear Extinction Threat." Environment News Service, October 2, 2007. http://www.ens-newswire.com/ens/oct2007/2007-10-02-091.asp.

Connor, Steve. "Meltdown: Arctic Wildlife is on the Brink of Catastrophe—Polar Bears Could Be Decades from Extinction." *London Independent*, November 11, 2004b, n.p. (LEXIS)

Eilperin, Juliet. "U.S. Wants Polar Bears Listed as Threatened." *Washington Post*, December 27, 2006i, A-1. http://www.washingtonpost.com/wp-dyn/content/article/2006/12/26/AR2006122601034_pf.html.

Elliott, Valerie. "Polar Bears Surviving on Thin Ice." *London Times*, October 30, 2003, n.p. (LEXIS)

"Expert Fears Warming Will Doom Bears." Canadian Press in *Victoria Times-Colonist*, January 5, 2003, C-8.

Johansen, Bruce E. "Arctic Heat Wave." *The Progressive*, October 2001, 18–20.

Joling, Dan. "Study: Polar Bears May Turn to Cannibalism." Associated Press, June 14, 2006. (LEXIS)

Kazlowski, Steven, with Theodore Roosevelt IV, Charles Wohlforth, Daniel Glick, Richard Nelson, Nick Jans, and Frances Beinecke. *The Last Polar Bear: Facing the Truth of a Warming World.* A

Photographic Journey. Seattle: Braided River Books, 2008.

Kerr, Richard A. "A Warmer Arctic Means Change for All." *Science* 297 (August 30, 2002b):1490–1492.

Krajick, Kevin. "Arctic Life, on Thin Ice." *Science* 291 (January 19, 2001a):424–425.

Krauss, Clifford. "Bear Hunting Caught in Global Warming Debate." *New York Times*, May 27, 2006. http://www.nytimes.com/2006/05/27/world/americas/27bears.html.

Laxon, Seymour, Neil Peacock, and Doug Smith. "High Interannual Variability of Sea Ice Thickness in the Arctic Region." *Nature* 425 (October 30, 2003):947–950.

Linden, Eugene. "The Big Meltdown: As the Temperature Rises in the Arctic, It Sends a Chill around the Planet." *TIME* 156, no. 10 (September 4, 2000). http://www.time.com/time/magazine/articles/0,3266,53418,00.html.

"Melting Planet: Species Are Dying Out Faster Than We Have Dared Recognize, Scientists Will Warn This Week." *London Independent*, October 2, 2005. http://news.independent.co.uk/world/environment/article316604.ece.

"Polar Bears Shift Dens." *Townsville Bulletin* (Australia), January 25, 2007, 12.

Stirling, Ian, and Andrew Derocher. "Possible Impacts of Climatic Warming on Polar Bears." *Arctic* 46, no. 3 (September 1993):240–245.

"Thin Polar Bears Called Sign of Global Warming." Environmental News Service, May 16, 2002. http://ens-news.com/ens/may2002/2002L-05-16-07.html.

Tidwell, Mike. *The Ravaging Tide: Strange Weather, Future Katrinas, and the Coming Death of America's Coastal Cities.* New York: Free Press, 2006.

A Polymer That Absorbs Carbon Dioxide at High Temperatures

Early in 2002, scientists revealed details of an inexpensive technology that captures carbon dioxide from industrial processes. According to a report by the Environment News Service, scientists at the Department of Energy's Los Alamos National Laboratory (LANL)

> are developing a new high-temperature polymer membrane to separate and capture carbon dioxide, preventing its escape into the atmosphere. This work is part of the Department of Energy's Carbon Sequestration Program, which is exploring ways to capture carbon dioxide from fossil fuel burning and reduce human impacts on climate. ("Hot Polymer" 2002)

This technology is aimed at sequestering the 30 percent of anthropogenic carbon dioxide that comes from power-producing industries. Present technology is limited to dealing with waste carbon dioxide emitted by such plants up to 150° C; the wastes often reach 375° C. Speaking at an American Geophysical Union conference on May 29, 2002, in Washington D.C., Jennifer Young, principal investigator for LANL's carbon-dioxide membrane separation project, described a new polymeric-metallic membrane that is stable at temperatures as high as 370° C.

"Current technologies for separating carbon dioxide from other gases require that the gas stream be cooled to below 150° C, which reduces energy efficiency and increases the cost of separation and capture," said Young. "By making a membrane which functions at elevated temperatures, we increase the practicality and economic feasibility of using membranes in industrial settings," she added ("Hot Polymer" 2002).

FURTHER READING

"Hot Polymer Catches Carbon Dioxide" Environment News Service, May 29, 2002. http://ens-news.com/ens/may2002/2002-05-29-05.asp.

Populations and Material Affluence

Human transformation of the global environment, including climate change, is driven to an important degree by ever-larger numbers of people. The human population of the Earth reached six billion in 1999, and since then has been increasing by almost 100 million a year. Of all the human beings ever born, *half* were alive in the year 2000. If present projections hold, the Earth by the mid- to late-twenty-first century may be home to at least 10 billion people (Silver and DeFries 1990, 4). The world's population more than doubled in the 45 years between 1955 and 2000, and may double again by about 2050 at current birth rates of roughly 225,000 a day.

Many Factors Increase Emissions

Several factors are at work to enhance greenhouse-gas emissions in addition to a simple increase in human numbers. The average amount of fossil-fuel energy consumed per person is rising gradually worldwide as industrializing nations' middle classes adopt the comforts and conveniences of machine culture, including the basics of a middle-class living standard: an electrified home (often with heat and air conditioning), an automobile, and a variety of appliances, all of which consume electricity. People also live longer (the average life span in the United States did not pass 50 until shortly after the year 1900), providing each individual more time, on the average, to consume fossil-fuel based energy.

Every time human population increases or a tree is felled, the world balance of carbon dioxide changes a little. Compounded millions of times, the change is significant. Settlement of lands previously uninhabited by human beings (such as the rain forests of the Amazon and Indonesia) also introduces their greenhouse-gas-emitting animals and machinery (National Academy of Sciences 1991, 5)

One index of material affluence (and greenhouse-gas production per capita) is the volume of trash and garbage generated. About 1990, North Americans produced roughly four pounds of garbage per person per day, or almost a ton per year. Italians produced 1.5 pounds, and Nigerians produced one pound each per year (Gordon 1991, 184). The amount of garbage produced worldwide is another statistical curve that climbed slowly at first, and then accelerated sharply in our own time, as material affluence has spread around the world. Population is increasing as the effluent per person also rises.

Timothy R. Barber and William M. Sackett found that human-induced greenhouse gases correlate strongly with global population growth, having risen about fivefold between 1950 and 1990. "Anthropogenic fossil sources are especially troublesome because they release previously buried carbon back into the atmosphere," they commented (Barber and Sackett 1993, 220).

Albert K. Bates wrote:

Today [1989], the world must accommodate the equivalent of a new population of the United States and Canada every three years. If North American standards of consumption were universally achieved early in the twenty-first century, world carbon-dioxide production could reach 110 gigatons (billion tons) per year, 20 times present levels. The earth would experience a climate shock of profound proportions, beyond anything we have been able to estimate at the present time. (Bates and Project Plenty 1990, 119)

"Premeditated Waste"

As human populations and per capita resource consumption rise, more products are being designed to be thrown away, as an increasing number of industries use "premeditated waste" to stimulate sales, income, and profit. Premeditated waste involves the manufacture of products designed to be used for only a year or two, until a new fashion or model replaces it. This practice originated in the high-fashion clothing industry and was embraced by automobile makers when they introduced annual model changes. By the end of the twentieth century, the same model of production had been adopted in many other industries. The most visible recent addition to the world garbage heap has been oceans of plastic waste generated by the rapid technological evolution of computer equipment.

According to Anita Gordon and David Suzuki, "Breaking the garbage habit means destroying the cultural myths that we have created as a society—myths that have allowed us to define ourselves by our possessions, that have led us to canonize time and convenience" (Gordon and Suzuki 1991, 186). Such a divorce may be very difficult because consumers love convenience, and industry loves selling them goods that quickly will become obsolete.

During the 1990s, the automobile industry came up with a new variation on an old sales gambit to maximize profits among those drivers who associate their personal identity with the shape, size, and power of their mode of transportation. The sports-utility vehicle allowed automobile companies to sell a larger vehicle (along with more production capacity for greenhouse gases) and thereby increase profits. This strategy reflected a philosophy of "forced consumption," which was outlined during the middle 1950s in the *New York Journal of Retailing*:

Our enormously productive economy demands that we make consumption a way of life, that we convert the buying and use of goods into rituals, that we seek our spiritual satisfactions in consumption.... We need things consumed, burned up, worn out, replaced, and discarded at an ever-growing rate. (Gordon and Suzuki 1991, 186)

Some environmentalists have called upon consumers to boycott such high-pollution forms of personal transportation. These pleas did not do sports-utility vehicles much harm until gasoline prices in the United States passed $4 a gallon in 2008.

Probably the worst example of global warming abuse is the sports utility [vehicle].... We are asking the public not to purchase vehicles that get poor gas mileage and therefore emit large quantities of carbon dioxide.… For example, owing to their poor gas mileage as determined by the Environmental Protection Agency, we are asking people not to buy 23 specific 1998 models of sports utility vehicles, which get 12 to 14 miles to the gallon (city).… A sports-utility [vehicle] that gets 12 miles to the gallon in the city will emit 800 pounds of carbon dioxide over a distance of 500 city miles. (Eco Bridge n.d.)

FURTHER READING

Barber, Timothy R., and William M. Sackett. "Anthropogenic Fossil Carbon Sources of Atmospheric Methane." In *A Global Warming Forum: Scientific, Economic, and Legal Overview*, ed. Richard A. Geyer, 209–223. Boca Raton, FL: CRC Press, 1993.

Bates, Albert K. and Project Plenty. *Climate in Crisis: The Greenhouse Effect and What We Can Do.* Summertown, TN: The Book Publishing Co., 1990.

Eco Bridge. "What Can We Do About Global Warming?" n.d. http://www.ecobridge.org/content/g_wdo.htm (accessed July 22, 2002).

Gordon, Anita and David Suzuki. *It's a Matter of Survival.* Cambridge, MA: Harvard University Press, 1991.

National Academy of Sciences. *Policy Implications of Greenhouse Warming.* Washington, D.C.: National Academy Press, 1991.

Silver, Cheryl Simon, and Ruth S. DeFries. *One Earth, One Future: Our Changing Global Environment.* Washington, D.C.: National Academy Press, 1990.

Poverty and Global Warming

The burden of global warming will fall most suddenly and acutely on poor nations, where a majority of people live close to the land and on the economic and financial margins. The United States, where agriculture is only 4 percent of the economy, will endure climatic stresses more easily than a country like Malawi, in Africa, where 90 percent of the people live in rural areas and about 40 percent of the economy depends on agriculture fed by variable rainfall (Revkin 2007a). Climate change during coming decades will place more than a billion people, most of them now poor, at risk of water stress, including hundreds of millions in harm's way because of sea-level rise, according to the Intergovernmental Panel on Climate Change (IPCC).

Advocates for poorer nations have said that with regard to global warming, the polluter should pay, bearing in mind that carbon dioxide remains in the atmosphere for centuries, so emissions are, to a large degree, cumulative. "We have an obligation to help countries prepare for the climate changes that we are largely responsible for," said Peter H. Gleick, the founder of the Pacific Institute for Studies in Development, Environment and Security in Berkeley, California. "If you drive your car into your neighbor's living room, don't you owe your neighbor something?" Gleick asked. "On this planet, we're driving the climate car into our neighbors' living room, and they don't have insurance and we do" (Revkin 2007a).

"It is the poorest of the poor in the world, and this includes poor people even in prosperous societies, who are going to be the worst hit," IPCC Chairman Rajendra Pachauri told journalists on April 6, 2007, at the release of the panel's summary report for policymakers in Brussels. "This does become a global responsibility in my view," he said ("UN Climate Change" 2007).

By 2020, 75 to 250 million Africans may be exposed to water scarcity and pollution due to climate change, the report said. By the 2080s, large coastal populations that are home to many millions of people may be flooded every year by rising seas, many in the large river deltas of Asia and Africa, and on small, low-lying islands. Temperature increases greater than 2° C also will provoke increases in malnutrition, heat waves, floods, storms, fires, and droughts. Frequency of cardiorespiratory diseases are expected to increase due to higher levels of ground-level ozone, as well as the spread of some infectious diseases ("UN Climate Change" 2007).

"Climate change is having impacts on natural systems—plants, animals, ecosystems and human systems," said Sharon Hays, leader of the U.S. IPCC delegation. "Climate change is clearly a global challenge and we all recognize that it requires global solutions. Not all regions of the world have the same capacity to adapt" ("UN Climate Change" 2007).

"There's no escaping the facts: global warming will bring hunger, floods and water shortages," said Hans Verolme of the global conservation organization World Wildlife Fund (WWF). "Poor countries that bear least responsibility will suffer most—and they have no money to respond—but people should also be aware that even the richer countries risk enormous damage. Doing nothing is not an option," Verolme said. "On the contrary it will have disastrous consequences" ("UN Climate Change" 2007).

"The irritating thing is that we have all the tools at hand to limit climate change and save the world from the worst impacts," said Lara Hansen, chief scientist of the WWF's Global Climate Change Programme. "The IPCC makes it clear that there is a window of opportunity but that it's closing fast. The world needs to use its collective brains to think ahead for the next 10 years and work together to prevent this crisis" ("UN Climate Change" 2007).

Friends of the Earth International's Climate Campaigner, Catherine Pearce, said,

> It is now clear that we are to blame for the last 50 years of warming, and this is already causing adverse changes to our planet. Unless we take action to reduce emissions now, far worse is yet to come, condemning millions in the poorest parts of the world to loss of lives, livelihoods and homes.... Climate change is no longer just an environmental issue.... It is a looming humanitarian catastrophe, threatening ultimately our global security and survival. ("UN Climate Change" 2007)

FURTHER READING

Revkin, Andrew C. "Reports from Four Fronts in the War on Warming." *New York Times*, April 3, 2007a. http://www.nytimes.com/2007/04/03/science/earth/03clim.html.
"UN Climate Change Impact Report: Poor Will Suffer Most." Environment News Service, April 6, 2007. http://www.ens-newswire.com/ens/apr2007/2007-04-06-01.asp.

Public Opinion

Rising temperatures and speculation regarding their effects have made global warming a subject of copious and increasing media discourse during recent years. The amount of debate and discourse has increased markedly with reports of rapidly rising temperatures and serious attempts at legislating greenhouse-gas limitations by a Democratic U.S. Congress. Few scientific issues exceed this one in terms of political and economic gravitas. At stake is not only the future of the Earth as a habitable environment, but the ways in which more than six billion people use energy. The rhetorical temperature has been rising. The *quantity* of attention is not in dispute—but what about the *quality* of this tide of reporting and commentary? To an informed observer, the discourse often displays the tone of a political campaign, a horse race, or a religious revival more than a scientific debate.

Scientists (an important example being NASA Goddard Institute for Space Sciences director James E. Hansen) have become more sensitive to the disconnect between what they know, and what the media are telling the public. Thus, as opinion leaders, they need to creatively "frame" messages that resonate with news organizations. Al Gore has been able to use some capital from his long-term credibility as a spokesman on this issue, his post-2000 role as a counterpoint to the George W. Bush presidency (including its positions regarding global warming), his Hollywood connections, and his ability not take himself too seriously while still being quite serious about the issue. By definition, when Gore speaks on this issue, he is usually newsworthy. Other scientist sources have been required to prove themselves with media to reach media agendas. This involves agenda-setting issues and news media construction of social reality. Ironically, the George W. Bush White House gave Hansen a tremendous boost as a newsworthy subject by introducing an element of conflict in its attempts to censor him.

Global Warming: Basic Rhetorical Issues

When debate broaches the subject of global warming's probability and prospective future effects, people often are asked whether they are "in favor" of global warming or—even more to the political-religious tone of the rhetoric—whether they "believe" in it. Carbon dioxide, methane and other greenhouse gases that retain heat in the lower atmosphere have no morals. Nor do the human activities (such as gasoline-powered transport and coal-generated electrical power) that have been adding human-induced carbon dioxide and methane to the atmosphere at accelerating rates following the advent of the Industrial Revolution two-and-half centuries ago.

News media, however, are created by human beings for human audiences, with appetites for anthropomorphic pathos. Scant attention is often paid in most media coverage to the scientific basics—the fact that carbon-dioxide levels in the atmosphere, for example, are higher than they have been in at least 800,000 years. This figure is being used because ice cores from which scientists can determine accurately the composition of the atmosphere reach back that far as of this writing. As older ice cores are brought to the surface in the future, that date probably will

Greenpeace launches a "Stop Global Warming" balloon over the Mae Moh coal-fired power plant in Lampang, Thailand. (Greenpeace/Unanongrak)

change, given the fact that the interglacial cycle highpoint carbon-dioxide concentration has been about 280 parts per million (ppm). That level is now 380 ppm and rising at 1 to 2 percent per year.

Why should anyone care how high carbon-dioxide and methane levels have risen? What is wrong with warmer temperatures—and just how *much* higher, for how long, and with what effects? At this point, the reporting, and the debate, enters the fog of scientific incomprehension. The popular media are better at pathos than hard science. Not even Al Gore, as he climbs that now-famous ladder in *An Inconvenient Truth*, showing us how high carbon-dioxide levels have risen, does not make the most salient connection.

This connection involves complex scientific concepts such as thermal equilibrium and feedback loops. Today's greenhouse-gas emissions will express themselves in the atmosphere roughly 50 years from now, and in the oceans perhaps a century after that. Today, the wind in our faces carries temperatures a half-century old on global-warming's calendar. Greenhouse-gas levels have risen about 400 percent since roughly 1960. This is the thermodynamic bottom line, and one that has been communicated very poorly by most media reports. Contrarians use this ignorance to downplay the sense of incipient danger posed by the problem.

Increasing Salience of the Issue

Public views have been changing regarding global warming, along with a broader range of environmental issues. A survey conducted at Yale University during 2007 indicated that 83 percent of people questioned in the United States said that global warming was a "serious" problem, compared with 70 percent in 2004. The survey of 1,000 adults released during March 2007 indicated that 63 percent of participants agreed that the United States "is in as much danger from environmental hazards, such as air pollution and global warming, as it is from terrorists" ("Poll: Global Warming" 2007).

The same survey indicated growing concern about dependence on Middle Eastern oil, with 96 percent of interviewees considering this issue a serious problem. The survey also showed strong support for increased use of non-fossil-fuel energy sources, including solar and wind power, as well as increased investment in energy efficiency. Dan Esty, director of the Yale Center for Environmental Law and Policy, which commissioned the survey, said that the United States "is in the midst of a 'revolution,' in which the business community is embracing the profit potential of a burgeoning green consumer movement" ("Poll: Global Warming" 2007).

"There's been a dramatic shift in the business community's attitude toward the environment," said Esty.

Rather than seeing environmental issues as a set of costs to bear, regulations to follow and risks to manage, companies have begun to focus on the upside, recognizing that society's desire for action on climate change, in particular, will create a huge demand for reducing carbon-content products.... It's clear that the public is not waiting for the government to take the lead.... Americans no longer think it's entirely the domain of government to solve environmental problems. They expect companies to step up and address climate change and other concerns. ("Poll: Global Warming" 2007)

The survey indicated that 70 percent of those surveyed believed that former President George W. Bush did not do enough for the environment and should have done more. The results show that many of those questioned want greener products and are ready to spend money to try new technologies that will help reduce greenhouse-gas emissions. Seventy percent indicated a willingness to buy solar panels, and 67 percent would consider buying a hybrid car.

"The coalition supporting action on climate change has broadened considerably," said Gus Speth, dean of Yale's environment school, and author of *Red Sky at Morning: America and the Crisis of the Global Environment* (2004). "With the public ready for carbon controls and business stepping up to the climate change challenge, it is disappointing that our political leadership is lagging so badly on this issue," said Speth ("Poll: Global Warming" 2007).

The Misrepresentations of "Balance"

Reporters and editors at four major U.S. newspapers followed the journalistic custom of balance "at the expense of accurately reporting scientific understanding of the human contributions to global warming," according to an analysis that appeared in *Global Environmental Change* ("Top U.S. Newspapers" 2004). The study, "Balance as Bias: Global Warming and the U.S. Prestige Press," examined coverage of human contributions to global warming in the *New York Times*, the *Washington Post*, the *Los Angeles Times*, and the *Wall Street Journal* from 1988 to 2002.

"By giving equal time to opposing views, these newspapers significantly downplayed scientific understanding of the role humans play in global warming," said Maxwell T. Boykoff, a doctoral candidate in environmental studies at the University of California–Santa Cruz, who co-authored the paper with his brother, Jules M. Boykoff, a visiting assistant professor of politics at Whitman College. "We respect the need to represent multiple viewpoints, but when generally agreed-upon scientific findings are presented side-by-side with the viewpoints of a handful of skeptics, readers are poorly served," added Boykoff. "In this case, it contributed to public confusion and opened the door to political maneuvering" (Boykoff and Boykoff 2004, 125–136; "Top U.S. Newspapers" 2004).

The Boykoffs' study concluded that the journalistic emphasis on balance obscured the scientific consensus in favor of global warming, and created a mistaken impression among many newspaper readers that the sides in the debate were roughly equal. "There's a very small set of people" who question the consensus, said Donald Kennedy, executive editor of *Science*, the leading U.S. scientific journal. "There are a great many thoughtful reporters in the media who believe that in order to produce a balanced story, you've got to pick one commentator from side A and one commentator from side B. I call it the two-card Rolodex problem" (Mooney 2004).

Global-warming contrarians who take advantage of journalistic norms give themselves the appearance of influence in the court of public opinion that they do not possess in scientific discourse. "The journalistic norm of balanced reporting offer[s] a countervailing 'denial discourse'—a voluble minority view [that] argues either that global warming is not scientifically provable, or that it is not a serious threat," according to one observer (Adger et al. 2001, 707). Furthermore, according to the Boykoff brothers, "When it comes to coverage of global warming, balanced reporting can actually be a form of informational bias.... Balanced reporting has allowed a small group of global-warming skeptics to have their views amplified" (Boykoff and Boykoff 2004, 126).

Scientists and Public Media Literacy

Hansen, by repudiating attempts at censorship, engaged in classic media-savvy behavior, involving an important source, an important issue, and a sharp conflict in the public sphere. Hansen is a veteran, having turned attempted censure into media fodder on global warming for almost 30 years. He is among a small number of well-informed scientists willing also to engage the debate as public intellectuals, to raise the salience of an issue that he—like Gore—regards as a planetary emergency. To do this, Hansen endures some heavy flak.

Modes of argumentation differ markedly between proponents of global warming (especially scientists) and several climate contrarians who engage the issue in media with large public audiences. The proponents usually concentrate on scientific arguments, while the contrarians often attack the personalities and motives of their opponents.

The contrarians' scientific case is weak; their argument often boils down to "weather happens," for example, temperatures rise and fall and—so they argue—and no one really knows why. The opponents often ignore the scientific basics that run counter to their argument, such as the temperature record (22 of the 24 warmest years on the instrumental record, or since about 1880, have occurred since 1985), rising levels of greenhouse gases in the atmosphere, and the nature of feedback loops.

Hansen as Public Intellectual

The *Wall Street Journal*, whose editorial writers are among the most trenchant deniers of global warming's validity among major U.S. media, grumbled on March 27, 2007, for example, that Hansen had complained about being silenced while conducting 1,400 interviews "in recent years." With no time axis (and no verifiable source), this statistic is rather useless. Suffice to say that Hansen has become one of the world's better-known figures in the global-warming debate. He has been granting interviews on the subject since at least 1980 and has engaged in the fight against several attempts at censorship by Republican administrations, the most recent being that of George W. Bush (Johansen 2006c).

An editorial cartoon in the January 6, 2006, edition of the *Boston Globe*, by Dan Wasserman, depicted two burley men behind a podium labeled "White House." The men were detaining Hansen—identifiable by his dome forehead and receding hairline—who had been gagged. One of them, with a smug look on his face, is saying, "We are reducing dangerous emissions."

The *Wall Street Journal* editorial made not a single reference to the scientific validity of Hansen's case. Instead, it "swift-boated" his assumed (and rather sloppily assessed) personal motives. The editorial asserted that Hansen had emerged from the conflict with the White House, "remain[ing] on the government payroll, his celebrity vastly burnished in the fable as the man Dick Cheney failed to silence. The question we'd ask is, with all those interviews and public appearances, when does he have time to do his real job?" ("Silenced" 2007).

The editorialist invoked considerable poetic license by never defining Hansen's "real job." On the supposition that Hansen is employed to conduct scientific research (and, as Goddard Institute director, to supervise other research), the editorial could have surveyed Hansen's many scientific papers, in which he engages some of the same debates that shape his public discourse, in prestigious venues such as *Science* and the *Proceedings of the National Academy of Sciences*. Hansen was lead co-author of the first article to use the term "global warming" in a scientific context (Hansen et al. 1981, 957–966).

Hansen has been attacked on such grounds so often that many of his writings in the public realm contain a proviso that he is expressing his First Amendment rights as a private citizen. When he testifies before Congress, he sends his remarks under a private e-mail address from his home on Durham Road, Kintnersville, Pennsylvania. When he speaks out of town on global-warming issues in any forum in which he interjects his own opinions, Hansen often pays his own expenses. While many attempts have been made to denigrate Hansen personally, he has never replied in kind. The most rigorous of scientists, Hansen never loses his rhetorical lunch.

On op-ed pages, Hansen can be uncomfortably direct for fossil-fuel interests as he addresses the central questions of climate change. Hansen laid down the scientific gauntlet in a guest opinion column for the *London Times*:

The climate system has great inertia, responding only slowly to forcings such as the gases that humans are adding to the atmosphere. The inertia is due to the large mass of the ocean and the ice sheets that cover Antarctica and Greenland. This inertia may seem to be a boon, as it reduces near-term change, but it also allows a dangerous level of gases to build up, causing far greater problems in the long term. The gravest threat to humanity, almost surely, lies in the great ice sheets. They can disintegrate rapidly, raising sea level by several metres per century, as we know from the Earth's history....

The best hope for the planet is a grass-roots movement. People concerned about climate change and the legacy that we will leave should consider having a date with the planet. Until the public indicates sufficient interest, and puts pressure on political systems, special interests will continue to rule. (Hansen 2007a)

Engaging the Issues in the Scientific Literature

Personally, Hansen is a rather shy person, rather at odds with his lightning-rod behavior in the public arena. He is also one of few people who engages the issue on both scientific and popular grounds. He often engages the media literacy of other scientists as well as members of the general public. He engaged the following debate in an online forum, *Atmospheric Chemistry and Physics Discussions*, a week after his remarks to Congress in March 2007:

The referee expressed mild concern about terms such as "dangerous anthropogenic interference," "disruptive climate effects," and "tipping points." We appreciate his/her assessment that use of such terms depends on personal preference.... "Tipping point," although objectionable to some scientists, conveys aspects of climate change that have been an impediment to public appreciation of the urgency of addressing human-caused global warming. It is a valid concept: as climate forcing and global warming increase, a point can be reached beyond which part of the climate system changes substantially with only small additional forcing. Examples include loss of Arctic sea ice and ice sheet disintegration.

This phenomenon is made doubly important by the fact that it is difficult to move the public and policymakers to action to address global warming until deleterious effects become obvious. Thus "tipping points" are central to determination of "dangerous" climate change and a legitimate topic for scientific discussion. (Hansen et al. 2007)

Congressional Hearings and Media Literacy

Hansen's stress on addressing the lack of public knowledge about the science of global warming was reflected in his testimony to the Committee on Oversight and Government Reform of the U.S. House of Representatives March 19, 2007. This was one of a number of hearings convened in the House and the Senate to air reports of scientific censorship and gauge public support for limits on greenhouse-gas emissions within months after Democrats took control of both the House and Senate during the November 2006 elections. The hearings provided a small insight into the rhetorical landscape in the United States on global warming at a time when salience and media literacy on the issue was rising.

> Political interference in transmittal of information about climate change science to the public has deleterious effects on the quality of decision-making. Science cannot make decisions for the public. The public and policy makers must consider all factors in making decisions and setting policy. But these other factors should not influence the science itself or the presentation of science to the public. (Hansen 2007b)

Testifying at the hearing, Hansen said editing of scientific studies and efforts to limit scientists' access to the news media and the public amounted to censorship and muddied the public debate over a pressing environmental issue. "If public affairs offices are left under the control of political appointees," he said, "it seems to me that inherently they become offices of propaganda" (Revkin and Wald 2007a).

The grassroots revolution that James Hansen advocates seems to have been ignited with his help as the "Paul Revere of Global Warming" (Johansen 2006c). Like all revolutions, however, the changes in popular perceptions of global warming develop as a function of public opinion, shaped by media literacy, to resolve the disconnect between scientific knowledge and what the public realizes (and acts on) regarding global warming.

An important question at this juncture is: How much time do we have before the force of feedbacks denies humanity any chance at controlling the pace and severity of damage inflicted by rising temperatures? Hansen's analysis allows a decade or two, and 1° to 2° C of global average temperature. Absent basic change in worldwide energy production and consumption within that time, Hansen says that enough inertial warming will be "in the pipeline" by the end of the twenty-first century to raise worldwide sea levels about 25 meters a century or two. The White House, which he has marked precisely on some of his maps forecasting sea-level rise, is 50 to 55 feet above the high-tide line. Thus, public perception, media literacy, technological change, and diplomatic activity find themselves in a race with the amoral forces of geophysics. Feedbacks

delay perception of warming (and, therefore, the will to act on the problem).

James Madison, the primary author of the U.S. Constitution, considered the people's access to information the basic right upon which all other rights depend. "A popular government without popular information, or the means of acquiring it, is but a prologue to a farce or a tragedy, or perhaps both. Knowledge will forever govern ignorance, and a people who mean to be their own governors must arm themselves with the power which knowledge gives," wrote Madison (Johnson 2006). These are words worth heeding, especially at a time when we are setting the environmental course that will determine the habitability of our planet for generations (perhaps even millennia) to come.

FURTHER READING

Adger, W. N., T. A. Benjaminsen, K. Brown, and H. Svarstad. "Advancing a Political Ecology of Global Environmental Discourses. *Development and Change* 32 (2001):681–715.

Boykoff, M. T., and J. M. Boykoff. "Balance as Bias: Global Warming and the U.S. Prestige Press." *Global Environmental Change* 14 (2004):125–136.

Coulter, Ann. "Global Warming Theology." *Pittsburgh Tribune Review*, March 25, 2007, n.p. (LEXIS)

Hansen, James D. "Special Interests Are the One Big Obstacle." *London Times*, March 12, 2007a. http://business.timesonline.co.uk/tol/business/columnists/article1499726.ece.

Hansen, James E. "Political Interference with Government Climate Change Science." Committee on Oversight and Government Reform. United States House of Representatives, March 19, 2007b.

Hansen James, et al. "Dangerous Human-made Interference with Climate: A GISS Model Study." In *Atmospheric Chemistry and Physics Discussions*, March 27, 2007. http://www.cosis.net/copernicus/EGU/acpd/6/S7350/acpd-6-S7350_p.pdf.

Hansen, James, D. Johnson, A. Lacis, S. Lebendeff, D. Rind, and G. Russell. "Climate Impact of Increasing Atmospheric Carbon Dioxide." *Science* 213 (August 28, 1981):957–966.

Johansen, Bruce E. "Media Literacy and 'Weather Wars:' Hard Science and Hardball Politics at NASA." *Studies in Media and Information Literacy Education* (SIMILE) 6, no. 3 (August 2006c). http://www.utpjournals.com/simile/issue23/Johansen6.html.

Johnson, Chalmers. *Nemesis: The Last Days of the American Republic*. New York: Metropolitan Books/Henry Holt, 2006.

Mooney, C. "Blinded by Science: How 'Balanced' Coverage Lets the Scientific Fringe Hijack Reality."

Columbia Journalism Review, November/December 2004. http://www.cjr.org/issues/2004/6/mooney-science.asp.

Pegg, J. R. "Gore Urges Immediate U.S. Freeze on Warming Emissions." Environment News Service, March 21, 2007d. http://www.ens-newswire.com/ens/mar2007/2007-03-21-11.asp.

"Poll: Global Warming as Big a Threat as Terrorism." Environment News Service, March 29, 2007. http://www.ens-newswire.com/ens/mar2007/2007-03-29-09.asp#anchor2.

Revkin, Andrew C., and M. L. Wald. "Material Shows Weakening of Climate Reports." *New York Times*, March 20, 2007a. http://www.nytimes.com/2007/03/20/washington/20climate.html.

"Silenced—1,400 Times." Editorial, *Wall Street Journal*, March 27, 2007, A-18.

"Top U.S. Newspapers' Focus on Balance Led to Skewed Coverage of Global Warming, Analysis Reveals." Ascribe Newsletter. August 25, 2004. (LEXIS)

"What About Us?" Editorial, *New York Times*, July 26, 2006. http://www.nytimes.com/2006/07/28/opinion/28fri2.html.

Public Protests Escalate over Energy-Policy Inertia

By the fall of 2008, Al Gore was calling for civil disobedience to prevent construction of new coal-fired power plants. "I believe we've reached the stage where it is time for civil disobedience to prevent the construction of new coal plants that do not have carbon capture and sequestration," Al Gore declared at the opening session of the Clinton Global Initiative annual meeting this week at the Sheraton New York on September 27, 2008 ("Gore Warns" 2008).

Gore referred to short-sighted economic assumptions that provoked a worldwide financial crisis:

> Well, now is the time to prevent a much worse catastrophe because the world has several trillion dollars in sub-prime carbon assets, based on the assumption that it is perfectly alright to put 70 million tons of global warming pollution into the atmosphere every 24 hours. Since we met here last year, the world has lost ground to the climate crisis. This is a rout.... We are losing badly. ("Gore Warns" 2008)

He continued: "Now, in the midst of this frenetic effort to find a bailout, many are saying we should have prevented this. We should have realized that the short-term greed was overcoming a clear vision of what the risk was" ("Gore Warns" 2008).

Gore ridiculed pledges of "clean coal" technology. "Clean coal is like healthy cigarettes [laughter]. It does not exist. It could theoretically exist. The only demonstration plant was cancelled. How many such plants are there? Zero. How many blueprints? Zero" ("Gore Warns" 2008).

Dumping Coal at the U.S. Capitol

Public protests of energy-policy inertia have become more common in both the United States and Europe. On May 15, 2003, for example, a coalition of environmental and public-interest groups dumped a ton of coal on the lawn of the U.S. Capitol to protest subsidies to that industry. In England, during the third Saturday in May 2002, the world's largest oil conglomerate, Exxon-Mobil, found its 400 service stations besieged by thousands of protesters who protested the company's lack of initiative on global warming and urged motorists to take their business elsewhere.

Farmers posted roadside "Stop Esso" billboards, and the company was condemned at several local music festivals. Using another tactic, about 50 bicyclists on June 12, 2004, rode through downtown Seattle in the nude, wearing only helmets and body paint, to protest society's continued dependence on fossil fuels. Late in August 2007, 60 people posed naked on a melting Swiss glacier. During the fall of 2004, after Florida was devastated by four major hurricanes, a coalition of scientists and environmentalists erected billboards showing a satellite image of a menacing hurricane off the state's coast, saying: "Global warming equals worse hurricanes. George Bush just doesn't get it" (Royse 2004).

Exxon (Esso) Takes Greenpeace to Court

On another front, during 2002, the world's largest oil company tried to silence its biggest critic by taking Greenpeace to court over the use of ExxonMobil logos in a "Stop Esso" boycott campaign. The Texas-based energy group accused Greenpeace of damaging its reputation by doctoring logo letters to resemble the "SS" moniker of the Nazi secret police. Exxon-Mobil demanded that Greenpeace pay it 80,000 euros for damage of its reputation, as well as a further euros 80,000 a day should it continue to use the offending material. Greenpeace said the move was a signal that the world's largest oil company was being hurt by the campaign launched to highlight Exxon's position on global warming (Macalister 2002, 21).

"French law protects your trademark and logo and our employees and customers would not understand if we did not take action to prevent its misuse," said an ExxonMobil spokeswoman in its French office, where the suit had been filed. Exxon's court documents said that the company aimed to prevent publication of symbols on the Stop Esso Internet site. The middle two letters of Esso have been replaced by dollar signs, not the Nazi "SS" logo, said Greenpeace. "Instead of using bully-boy antics to gag free speech, we suggest Esso instead halt its campaign to subvert international action on climate change," said Stephen Tindale, Greenpeace U.K. director. "We simply replaced two letters in Esso's logo with the internationally recognized symbol for the U.S. dollar. We find it ironic that the richest corporation in the world can't recognize the dollar sign and confuses it with a Nazi symbol," he added (Macalister 2002, 21).

ExxonMobil on July 8, 2002, won the first round of the increasingly bad-tempered legal battle designed to bring a halt to Greenpeace's Stop Esso campaign. Exxon was awarded an injunction forcing Greenpeace to cease use of its Esso logo with dollar signs scrawled through it, pending a full hearing. The environmental group was allowed four working days to remove the offending logo from its Web site or face a fine of 5,000 euros.

In the United States, a number of Greenpeace activists on May 27, 2003, blocked the entrance of ExxonMobil's headquarters at Irving, Texas, to protest company inaction regarding global warming. Thirty-six of the protesters were arrested by Irving police and charged with criminal trespassing, a misdemeanor. Many were dressed in tiger costumes, after the old Exxon mascot Tony the Tiger. The protest was staged one day before the company's annual shareholders' meeting in Dallas. At that meeting, environmental activists faced off with pro-Exxon demonstrators who sang "Give Oil a Chance," a parody of John Lennon's "Give Peace a Chance."

Confrontation at an Oil Exchange

Physical confrontations between English Greenpeace activists and the oil industry have become frequent. On February 16, 2005, the day the Kyoto Protocol came into force, 35 Greenpeace activists wearing business suits invaded the International Petroleum Exchange in London, the world's second-largest energy market, shutting down trading for an hour before several were arrested. The exchange, which sets the price for more than half the world's crude oil, was invaded about 2 P.M. by activists blaring foghorns, alarms, and whistles, in an attempt to disrupt "open-cry" floor trading. Some of them attached distress alarms to helium balloons that they lofted above the trading floor. Security guards kicked and punched activists, wounding 12 of them, as three members of Greenpeace scaled the exchange building to hang a banner reading: "Climate Change Kills—Stop Pushing Oil" (Vidal and Macalister 2005, 4).

FURTHER READING

"Gore Warns of Sub-Prime Carbon Catastrophe." *Environment News Service,* September 27, 2008. http://www.ens-newswire.com/ens/sep2008/2008-09-27-01.asp.

Macalister, Terry. "Confused Esso Tries to Silence Green Critics." *London Guardian,* June 25, 2002, 21.

Royse, David. "Scientists: Bush Global Warming Stance Invites Stronger Storms." Associated Press, October 25, 2004. (LEXIS)

Vidal, John, and Terry Macalister. "Kyoto Protests Disrupt Oil Trading." *London Guardian,* February 17, 2005, 4.

$\mathcal{R}$

Railroads: Transport of the Future?

Some of the hottest (which is to say, when it comes to preventing climate change, coolest) forms of transport for the twenty-first century are inventions of the nineteenth century— including the bicycle, the ultimate energy conservation machine, and the passenger railroad. Both have significant room to grow in the United States, where public transportation accounts for only 1 percent of total transport miles. Despite its health benefits (a regular bicycle rider's physiological age is 10 years younger than a person who regularly drives a car), bicycle transport accounts for only two-tenths of 1 percent of travel miles in the United States (one mile of every 500). Walking accounts for 0.7 percent (Hillman, Fawcett, and Rajan 2007, 53–54).

Most U.S. histories of transportation have air travel replacing railroads for all but heavy freight. However, with mounting concern regarding carbon footprints and the decay of the U.S. commercial aviation system (prone to scheduling delays, weather problems, security paralysis, and other dysfunctions), punctual European trains look better every day. Passengers board trains within a few minutes of departure, while airline passengers are advised to arrive hours early to endure long lines at check-in, security, and boarding gates. European trains are clean, fast, spacious, and attended by courteous staff. Remodeled cars include trays for desktops, rooms for small meetings, and Internet access.

Some European trains also are breathtakingly fast. The new TGV train leaves Paris at the speed of a commercial airliner on takeoff—180 miles per hour. Europe's fast trains benefit from a network of "dedicated track," 2,912 miles that allow no freight or slower trains. China has built a magnetic levitation shuttle between the Pudong airport and downtown Shanghai that accelerates to 240 miles an hour during an eight-minute trip. Plans call for a similar line to open between Beijing and Shanghai in 2010 (Finney 2007, 16). Japan has long used 180-mph "bullet trains" between Tokyo and Osaka.

Trains in the United States

Outside the northeast corridor (Boston to Washington) passenger rail barely exists in the United States. The country has long had a skimpy national rail network (Amtrak), but trains are few, schedules are inconvenient, and tracks are often poorly maintained. The system is heavily tax-subsidized, and Congress often threatens to put it out of business. In the biggest eastern cities (Boston, Washington, D.C., and New York City) subways (underground) systems are extensive and widely used. As in London, these trains run on electric rails. In most other cities, subways do not exist. Nearly all intracity and long-distance travel is by car or airline. Trains generally carry freight and meat animals.

Even in the United States, however, trains have been eating into airplanes' share of travel on some routes, notably between Boston, New York City, and Washington, D.C., where Amtrak ridership was up 20 percent year over year ending in July 2007, as airlines experienced scheduling problems and delays, not to mention concern over their carbon footprint. Passenger rail traffic between Chicago and St. Louis surged 53 percent during the same period (Machalaba 2007, B-1). On routes of fewer than 400 miles or so, the airlines' time advantage has disappeared with increasing security screenings, lines, and other problems. Amtrak since 2000 has run Acela locomotives on routes between Boston and Washington at up to 150 miles an hour on part of their routes. During 2007, the trains carried more passengers than airlines. Amtrak attracted 28.7 million riders in the year ended September 30, 2008, up 11 percent in one year. Ticket revenue rose to $1.7 billion, $200 million more than a year earlier ("A Renewed Focus" 2008, A-15).

FURTHER READING

Finney, Paul Burnham. "U.S. Business Travelers Let the Train Take Way the Strain." *International Herald-Tribune*, April 24, 2007, 16.

Hillman, Mayer, Tina Fawcett, and Sudhir Chella Rajan. *The Suicidal Planet: How to Prevent Global Climate Catastrophe.* New York: St. Martin's Press/ Thomas Dunne Books, 2007.

Machalaba, Daniel. "Crowds Heeds Amtrak's 'All Aboard.'" *Wall Street Journal*, August 23, 2007, B-1, B-22.

"A Renewed Focus on Passenger Trains." Associated Press in *New York Times*, November 3, 2008, SA-15.

Red Squirrels' Reproductive Cycle

Birth dates of red squirrels in the Yukon advanced almost three weeks within a decade because of a warming habitat, according to scientists who assert that the change is so profound that it seems to have affected the creature's genetic makeup (Munro 2003a, A-2).

"It's a phenomenal change in such a short period of time," said Stan Boutin, a biologist at the University of Alberta, who tracked hundreds of squirrels in the southwest corner of the Yukon. "We feel we've found some of the first evidence indicating climate change is leading to evolution in animal populations," said Boutin, who worked with Andrew McAdam, also from the University of Alberta, Denis Rèale from McGill University, and Dominique Berteaux from the Universite du Quebec (Ogle 2003, A-12).

Such changes simulate what occurs when animal breeders select for certain traits that are gradually bred into the animals over several generations, he said. In this case, however, climate change is driving the evolutionary change among the squirrels through natural selection. "It's typical Darwinian natural selection," Boutin said. "As the conditions get warmer in spring, it favors those individuals that carry genes that allow them to breed earlier" (Munro 2003a, A-2). Boutin studied four generations of squirrels during a 15-year period at Kluane National Park. The study, published in the March 2003 edition of the *Proceedings of the Royal Society of London* is the first to suggest that warming has triggered genetic change (Rèale et al. 2003).

The squirrels' breeding season has been markedly affected by warmer spring temperatures and an increased food supply from spruce trees, which produce more cones as temperatures warm. Pups are now born as early as March 1. "It's good news in the sense that we are really impressed by the adaptability of these characters and their ability to keep up with the change," said Boutin. "The bad news is that these rates of change are definitely very rapid. Our human activities are affecting organisms in a variety of ways, even to the point where we are causing evolution in them" (Munro 2003a, A-2).

As this study was released during 2003, observers in British Columbia reported that a mild winter there was prompting some animals to procreate several weeks earlier than usual. The first house sparrow hatchlings appeared in the city several weeks earlier than usual, said Stanley Park Ecology Society spokesman Robert Boelens. Boelens expected that other urban critters may be mating earlier than usual—everything from ducks to squirrels to swans to raccoons. "In fact," according to a *Vancouver Sun* report, "Squirrel love is definitely under way.... Several people around the Lower Mainland [near Vancouver's urban area] have reported significant squirrel nest-building activity weeks earlier than usual. The same behaviors have been reported for raccoons, skunks, rabbits, mice, crows, and rats" (Reed 2003, B-1).

FURTHER READING

Munro, Margaret. "Global Warming Affecting Squirrels' Genes, Study Finds: Research in Yukon: 'Phenomenal Change' Seen as Rodents Breed Earlier." *National Post* (Canada), February 12, 2003a, A-2.

Ogle, Andy. "Squirrels Get Squirrelier Earlier: Climate Change to Blame. Breeding Season in Yukon Advances 18 Days in Decade." *Edmonton Journal* (Canwest News Service) in *Montreal Gazette*, February 12, 2003, A-12.

Rèale, D., A. G. McAdam, S. Boutin, and D. Berteaux. "Genetic and Plastic Responses of a Northern Mammal to Climate Change." *Proceedings of the Royal Society of London, Series B* 270 (2003):591–596.

Reed, Nicholas. "Mild Winter Stirs Wildlife to Early Thoughts of Love." *Vancouver Sun*, February 12, 2003, B-1.

Reforestation

Proposals have been made to ameliorate increases in greenhouse-gas emissions through reforestation, the purposeful planting of large forests to absorb some of humankind's surplus carbon dioxide. The 1997 Kyoto Protocol contains mechanisms whereby governments of countries such as the United States, which produce more greenhouse gases per capita than average, may earn credit toward meeting their emissions goals by subsidizing the preservation of forests in poorer nations.

Appreciation of old-growth forests as carbon sinks has grown with recent research. A team of scientists wrote in *Nature* during 2008 that

> Old-growth forests serve as a global carbon dioxide sink, but they are not protected by international treaties, because it is generally thought that aging forests cease to accumulate carbon … literature and databases for forest carbon-flux estimates [indicate] that in forests between 15 and 800 years of age, net ecosystem productivity (the net carbon balance of the forest including soils) is usually positive. Our results demonstrate that old-growth forests can continue to accumulate carbon, contrary to the long-standing view that they are carbon neutral.… Old-growth forests accumulate carbon for centuries and contain large quantities of it. We expect, however, that much of this carbon, even soil carbon, will move back to the atmosphere if these forests are disturbed. (Luyssaert et al. 2008, 213)

Large-Scale Projects

Reforestation could help reduce greenhouse warming, but only if trees are planted on a large scale. For example, if an area of 465 million hectares was planted, the trees on this land area, once mature, would remove almost three billion tons of carbon dioxide from the atmosphere per year, or about 40 percent of the carbon that human beings add to the air. The creation of such a carbon sink would require a land area roughly half the size of the United States (Silver and DeFries 1990, 122–123).

Ecologist George Woodwell estimated that one to two million square kilometers of newly planted trees would remove one billion tons of carbon dioxide (one gigaton) annually, of the roughly seven million gigatons being placed in circulation at the turn of the century. The problem would be finding large tracts of land fertile enough to support trees that are not being used by humans for other purposes.

In England, some private firms have been planting trees to offset their contributions to global warming. The *London Sunday Independent*

conducted a campaign during which readers bought more than 7,000 trees to offset the amount of carbon dioxide created by the manufacture of the newspaper over a year's time. The newspaper itself contributed 750 trees. The Glastonbury arts festival sold 1,333 trees to offset the equivalent amount of carbon dioxide to all the emissions created in the setup, running, and dismantling of the show (Rowe 2000, 5). The trade in trees is coordinated by an organization called Future Forests. Some musical groups, such as The Pet Shops Boys and Neneh Cherry have produced 1.5 million "carbon-neutral" compact discs, meaning they have bought enough trees to offset the carbon emitted by production of their recording as well as movement of their stage materials around the country.

A study of greenhouse-gas reduction potential in Chicago endorsed urban tree-planting projects as a way to reduce air pollution in cities, where reductions are needed the most.

> A strategy focused on tree-planting would have a number of benefits [which] include CO_2 absorption, removal of air pollution, and cooling and sheltering effects. Trees in Chicago have removed significant amounts of pollutants. In 1991, trees in Chicago removed an estimated.… 15 metric tons of carbon monoxide, 84 tons of sulfur dioxide, 89 tons of nitrogen dioxide, and 191 tons of ozone. Heating and cooling costs for buildings can also be reduced by strategically planting trees to shield buildings from wind and to provide shade in the summer months. ("Energy and Equity" n.d.)

Forests May Contribute to Warming

Carbon uptake from reforestation has been proposed even as some models indicate that forests sometimes contribute to global warming, on balance. For example, "The albedo of a forested landscape is generally lower than that of cultivated land, especially when snow is lying.… In many boreal forest areas, the positive forcing induced by decreases in albedo can offset the negative forcing that is expected from carbon sequestration" (Betts 2000, 187). According to this analysis, high-latitude reforestation efforts actually might worsen the greenhouse effect.

Other research indicates that older, wild forests are far better than plantations of young trees at removing carbon dioxide from the atmosphere. One such analysis, published in the journal *Science*, was completed by Dr. Ernst-Detlef Schulze, the director of the Max Planck Institute for Biogeochemistry in Jena, Germany, and two other scientists at the institute. The study provided an important new argument for protecting old-growth forests. The scientists said that their study provided a reminder that the main goal should be to reduce carbon-dioxide emissions at the source, smokestacks and tailpipes.

"In old forests, huge amounts of carbon taken from the air are locked away not only in the tree trunks and branches, but also deep in the soil, where the carbon can stay for many centuries," said Kevin R. Gurney, a research scientist at Colorado State University. When such a forest is cut, he said, almost all of that stored carbon is eventually returned to the air in the form of carbon dioxide. "It took a huge amount of time to get that carbon sequestered in those soils," he said, "So if you release it, even if you plant again, it'll take equally long to get it back" (Revkin 2000b, A-23).

The German study, together with other similar research, has produced a picture of mature forests that differs sharply from long-held notions in forestry, Schulze said. He said that aging forests were long perceived to be in a state of decay that releases as much carbon dioxide as it captures. Soils in undisturbed tropical rain forests, Siberian woods, and some German national parks contain enormous amounts of carbon derived from fallen leaves, twigs, and buried roots that can bind to soil particles and remain stored for 1,000 years or more. When such forests are cut, the trees' roots decay and soil is disrupted, releasing the carbon dioxide (Revkin 2000b, A-23; Schulze, Wirth, and Heimann 2000). "In contrast to the sink management proposed in the Kyoto Protocol, which favors young forest stands, we argue that preservation of natural old-growth forests may have a larger effect on the carbon cycle than promotion of re-growth" (Schulze, Wirth, and Heimann 2000, 2058). Instead of reducing the level of carbon dioxide in the atmosphere, the Kyoto Protocol emphasizes new growth at the expense of established forests, which

will lead to massive carbon losses to the atmosphere mainly by replacing a large pool with a minute pool of re-growth and by reducing the flux into a permanent pool of soil organic matter. Both effects may override the anticipated aim, namely to increase the terrestrial sink capacity by afforestation and reforestation. (Schulze, Wirth, and Heimann 2000, 2059)

FURTHER READING

Betts, Richard A. "Offset of the Potential Carbon Sink from Boreal Forestation by Decreases in Surface Albedo." *Nature* 408 (November 9, 2000):187–190.

"Energy and Equity: The Full Report." Illinois Environmental Protection Agency, n.d. http://www.cnt.org/ce/energy&equity.htm.

Luyssaert, Sebastiaan, E. -Detlef Schulze, Annett Börner, Alexander Knohl, Dominik Hessenmöller, Beverly E. Law, Philippe Ciais, and John Grace. "Old-growth Forests as Global Carbon Sinks." *Nature* 455 (September 11, 2008):213–215.

Revkin, Andrew C. "Planting New Forests Can't Match Saving Old Ones in Cutting Greenhouse Gases, Study Finds." *New York Times*, September 22, 2000b, A-23.

Rowe, Mark. "When the Music's Over ... a Forest will Rise." *London Independent*, June 25, 2000, 5.

Schulze, Ernst-Detlef, Christian Wirth, and Martin Heimann. "Managing Forests after Kyoto." *Science* 289 (September 22, 2000):2058–2059.

Silver, Cheryl Simon, and Ruth S. DeFries. *One Earth, One Future: Our Changing Global Environment.* Washington, D.C.: National Academy Press, 1990.

Refrigerant, Carbon Dioxide's Use as

Several hundred researchers from around the world met at Purdue University on July 12–15, 2004, to discuss innovative air-conditioning and refrigeration technologies, including designs aimed at reducing global warming. At least one of the ideas, the use of carbon dioxide as a refrigerant, is nearly a century old. Carbon dioxide was the first refrigerant used during the early twentieth century but was later replaced with man-made chemicals. Carbon dioxide may be on the verge of a comeback because of technological advances that include the manufacture of extremely thin aluminum tubing called "microchannels" ("Conferences Tackle" 2004).

Hydrofluorocarbons (HFCs), today's most widely used refrigerant, causes about 1,400 times more global warming per molecule than the same quantity of carbon dioxide. The HFCs replaced chlorofluorocarbons (CFCs), also potent greenhouse gases that degrade stratospheric ozone. Tiny quantities of carbon dioxide released from air conditioners would be insignificant compared with the huge amounts produced from burning fossil fuels for energy and transportation, said Eckhard Groll, an associate professor of mechanical engineering at Purdue ("Conferences Tackle" 2004).

Carbon dioxide offers few advantages for large air conditioners, which do not have space restrictions and can use wide-diameter tubes capable of carrying enough of the conventional refrigerants to provide proper cooling capacity. Carbon dioxide, however, may be a promising alternative for systems that must be small and lightweight, such as automotive or portable air conditioners. Various factors, including the high operating pressure required for carbon-dioxide systems, enable the refrigerant to flow through small-diameter tubing, which allows engineers to design more compact air conditioners ("Conferences Tackle" 2004).

In the meantime, Greenpeace asked refrigerator manufacturers to use hydrocarbons such as propane instead of HFCs, and developed its own "Greenfreeze" technology. Unilever became interested in the idea as part of its commitment to reduce the impact of its activities on climate change. While the Greenfreeze technology was adequate for domestic fridges, Unilever carried out lengthy trials to create a system that could be used in larger-scale freezers as well. Unilever agreed to test the product at the 2000 Olympics in Sydney and committed to a global changeover to HFC-free refrigerants.

FURTHER READING

"Conferences Tackle Key Issues in Air Conditioning, Refrigeration." AScribe Newswire, June 23, 2004. (LEXIS)

Revelle, Roger (1909–1991)

In 1957, Roger Revelle and Hans Suess warned, as part of the International Geophysical Year, that

> Human beings are now carrying out a large-scale geophysical experiment of a kind that could not have happened in the past nor be reproduced in the future. Within a few centuries we are returning to the atmosphere and oceans the concentrated organic carbon stored in sedimentary rocks over hundreds of millions of years. (Christianson 1999, 155–156)

Revelle would become known to the world some years later as the mentor of a graduate student, Albert Gore, who, in 1992 (a year after Revelle died), was elected vice president of the United States. The same year, Gore published a book, *Earth in the Balance*, which argued for

mitigation of the greenhouse effect. Gore went on to lose a close contest for the U.S. presidency in 2000, after which he became worldwide publicist regarding global warming, winning a Nobel Peace Prize in 2007 and an Oscar for his documentary film, *An Inconvenient Truth.*

"Grandfather" of the Greenhouse Effect

Revelle, who became known as the "grandfather" of the greenhouse effect, was born in Seattle on March 7, 1909, into a family of Huguenot descent on his father's side and Irish descent on his mother's side. His parents William Roger Revelle, an attorney and schoolteacher, and Ella Robena Dougan Revelle, also a schoolteacher, both had graduated from the University of Washington.

After he was admitted to Pomona College at the age of 16, Revelle entertained thoughts of a career in journalism. However, under the influence of a charismatic professor, Alfred Woodford, Revelle became interested in geology and, after receiving his bachelor's degree in 1929, spent a year of additional study with Woodford. Revelle entered graduate studies at the University of California–Berkeley, in 1930, under the tutelage of geologist George Louderback, who stimulated his interest in marine sedimentation. In 1936 Revelle, completed his doctorate at the University of California.

Called to active duty as an officer in the U.S. Navy six months before the attack on Pearl Harbor, Revelle was assigned to oceanographic research applied to wartime needs. Near the end of World War II, he was assigned to Joint Task Force One, the military command supervising the first postwar atomic test on Bikini Atoll (Operation Crossroads). Revelle organized the Crossroads scientific program, which included a study of the diffusion of radionuclides in the atoll, as well as radioactivity's impact on marine life. What he learned through these studies, and others, made Revelle a life-long opponent of nuclear weapons. After the war, in 1951, Revelle became director of the Scripps Institute of Oceanography.

In 1964, Revelle accepted an appointment as Richard Saltonstall Professor of Population Policy at Harvard University, where he served as director of the Center for Population Studies until 1974. Revelle held the endowed chair at Harvard until 1978. Revelle also was sympathetic to the 1962 book *Silent Spring* by Rachel Carson, which argued that ecosystems were being disrupted by the use of halogenated hydrocarbon pesticides such as dichloro-diphenyl-trichloro-ethane (DDT).

Revelle played a key role in the creation, during 1970, of the Scientific Committee on Problems of the Environment of the International Council of Scientific Unions (ICSU). He also suggested the objective for the ICSU's International Geosphere-Biosphere Program in 1986: "To describe and understand the interaction of the great global physical, chemical, and biological systems regulating planet Earth's favorable environment for life, and the influence of human activity on that environment" (Malone, Goldberg, and Munk n.d.).

Revelle served as president of the American Association for the Advancement of Science during 1974. In 1976, he returned to the University of California–San Diego to become a professor of science and public policy. He received the National Medal of Science in 1991 "for his pioneering work in the areas of carbon dioxide and climate modification, oceanographic exploration presaging plate tectonics, and the biological effects of radiation in the marine environment, and studies for population growth and global food supplies." In response to a reporter asking why he received the medal, Revelle said, "I got it for being the grandfather of the greenhouse effect" (Malone, Goldberg, and Munk n.d.).

FURTHER READING

Christianson, Gale E. *Greenhouse: The 200-Year Story of Global Warming.* New York: Walker and Company, 1999.

Malone, Thomas F., Edward D. Goldberg, and Walter H. Munk. "Roger Randall Dougan Revelle, 1909–1991." *Biographical Memoirs of the National Academy of Sciences.* Cited in Deborah Day. "Roger Randall Dougan Revelle Biography." Accessed January 28, 2009. http://www.repositories.cdlib.org/cgi/viewcontent.cgi?article=1084&context=sio/arch.

Revelle, R., and H. E. Suess. "Carbon Dioxide Exchange between Atmosphere and Ocean and the Question of an Increase of Atmospheric CO_2 during the Past Decades." *Tellus* 9 (1957):18–27.

Rice Yields Decrease with Rising Temperatures

Even modest temperature increases that are anticipated by some climate models could be enough to reduce rice yields significantly over the next century. Researchers at the University of Florida tested several varieties of rice, growing them in chambers that simulated various temperature-change situations. They found that although the rice plants flourished no matter what the temperature, yields of grains declined precipitously as temperatures increased. The more modest temperature increases, the researchers said, could reduce rice yields by 20 to 40 percent by 2100, while the larger increases predicted by more dire forecasts could cut rice production essentially to zero (Fountain 2000, F-5).

In another study, researchers from China, the United States, and the Philippines analyzed weather data from the International Rice Research Institute Farm in Los Banos, Philippines (near Manila), from 1979 to 2003, and compared these records with rice yields at the same location from 1992 to 2003. They found that annual mean maximum and minimum temperature at the location increased by 0.35° C and 1.13° C, respectively, between 1979 and 2003; grain yield was found to decline by 10 percent for each 1° C rise in growing-season minimum temperatures during the dry cropping season (January through April). "This report," they wrote in the *Proceedings of the National Academy of Sciences*, "[p]rovides a direct evidence of decreased rice yields from increased nighttime temperature associated with global warming" (Peng et al. 2004, 9,971). Globally, night-time temperatures have increased faster than daytime readings because greenhouse gases tend to trap more heat radiating from the ground, reducing cooling (Fountain 2004, F-1).

Kenneth G. Cassman of the University of Nebraska, who participated in this study, said that researchers are working to determine the cause of the yield reduction, but they speculate that it is because the warmer nights make the plants work harder just to maintain themselves, diverting energy from growth. "If you think about it, world records for the marathon occur at cooler temperatures because it takes much more energy to maintain yourself when running at high temperatures. A similar phenomenon occurs with plants," he said (Schmid 2004).

Tim L. Setter, a professor of soil, crop, and atmospheric science at Cornell University, who did not take part in this study, commented that higher night-time temperatures "could consume carbohydrates in a nonproductive way, and by

Planting rice seedlings. (Courtesy of Shutterstock)

reducing the reserves of carbohydrates, particularly at time of flowering and early grain filling, would decrease the number of kernels that would be set" (Schmid 2004).

FURTHER READING

Fountain, Henry. "Observatory: Threat to Rice Crops." *New York Times*, December 12, 2000, F-5.
Fountain, Henry. "Observatory: Rice and Warm Weather." *New York Times*, June 29, 2004, F-1.
Peng, Shaobing, Jianliang Huang, John E. Sheehy, Rebecca C. Laza, Romeo M. Visperas, Xuhua Zhong, Grace S. Centeno, Gurdev S. Khush, and Kenneth G. Cassman. "Rice Yields Decline with Higher Night Temperature from Global Warming." *Proceedings of the National Academy of Sciences* 101, no. 27 (July 6, 2004):9971–9975.
Schmid, Randolph E. "Warming Climate Reduces Yield for Rice, One of World's Most Important Crops." Associated Press, June 28, 2004. (LEXIS)

Russia, Glacial Collapse in

During September 20, 2002, a large avalanche in southern Russia may have killed as many as 150 people in North Ossetia, a small republic in the mountains near the Georgian border. According to an account from the scene, "A chunk of a glacier about 500 feet high broke off from beneath a mountain peak and roared down two gorges at more than 62 miles per hour, uprooting trees and accumulating mud and rocks as it went," burying much of the village of Karmadon ("100 Still Missing" 2002, A-5.).

"Within minutes," the account continued,

An immense glacier severed from the mountain near its 15,700-foot peak, roared along the nearby hillside village of Karmadon, burying much of it up to 500 feet of ice and debris.... The collapse of the Maili glacier on the northern edge of the Caucasus Mountains ripped out trees and tossed massive trucks as if they were toys. It left a 20-mile path of

rocky debris, blackened ice and devastation. (Wines 2002, A-11)

The avalanche eroded one-third of the Maili Glacier, about three million tons of ice. A team of experts was in the region to try to discover the cause of the disaster, which an Interior Ministry spokesman said might be connected to global warming. Among the dead was Sergei Bodrov, Jr., one of Russia's best-known young movie stars, who was filming in the area when the avalanche hit.

Russian officials said that the collapse of the glacier seemed at least partly related to climatic change. This is a tricky issue because the collapse of glaciers may be provoked by a variety of causes, including temperature, rainfall, humidity, angle of slope, and even the reflectivity of the glacial ice (McFarling 2002b, A-4). U.S. experts said the Maili Glacier incident followed the pattern of glacier collapse in other areas affected by rising temperatures. "Glaciers tend to [collapse] like that when they're receding, and glaciers are receding all over the world," said Dan Fagre, an ecologist and expert on the ramifications of glacier loss at Glacier National Park in Montana (McFarling 2002b, A-4). Such glacial collapses can be prompted buy accumulation of meltwater, which often pools in cracks of receding glaciers behind walls of sediment and stone, building pressure that eventually causes a slide of ice, mud, and rock (McFarling 2002b, A-4).

FURTHER READING

"100 Still Missing in Russian Avalanche." *Los Angeles Times*, September 24, 2002, A-5.
McFarling, Usha Lee. "Glacial Melting Takes Human Toll: Avalanche in Russia and Other Disasters Show That Global Warming Is Beginning to Affect Areas Much Closer to Home." *Los Angeles Times*, September 25, 2002b, A-4.
Wines, Michael. "Rising Star Lost in Russia's Latest Disaster." *New York Times*, September 24, 2002, A-11.

S

Sachs Harbour, Banks Island, Climate Change

Born in 1954, Rosemarie Kuptana grew up in a traditional Inuit hunting society and spoke only Inuvialuktun (the Western Arctic dialect of the Inuit language) until the age of eight. Her home community of Sachs Harbour is a Banks Island village of about 120 people on the Beaufort Sea, in the Arctic Ocean about 800 miles northeast of Fairbanks, Alaska. Born in an igloo, Kuptana has been an Inuit weather watcher for much of her life (she was 50 years of age in 2004). Her job was to scan the morning clouds and test the wind's direction to help the hunters decide whether to go out and determine what everyone should wear.

"We can't read the weather like we used to," said Kuptana. She said that autumn freezes now occur a month later than they did in her youth; spring thaws come earlier, as well. Residents of Sachs Harbour still suffer through winters that most people from lower latitude would find chilling, with temperatures as low as −40° F. While such temperatures once were commonplace during the winter, however, they now are rare. "The permafrost is melting at an alarming rate," said Kuptana (Johansen 2001, 20). Foundations of homes in Sachs Harbour are cracking and shifting because of the melting permafrost.

Kuptana said that at least three experienced hunters had recently fallen to their deaths through unusually thin ice. Never-before-seen species (including robins, barn swallows, beetles, and sand flies) have appeared on Banks Island. No word exists for robin in Inuktitut, the Inuit language. Growing numbers of Inuit are suffering allergies from white-pine pollen that recently reached Banks Island for the first time. At Sachs Harbour, mosquitoes and beetles are now common sights where they were unknown a generation ago. Sea ice is thinner and now drifts far away during the summer, taking with it the seals and polar bears upon which the village's Inuit residents rely for food. Young seals are starving to death because melting and fracturing sea ice separates them from their mothers.

In the winter, sea ice often is thin and broken, making travel dangerous even for the most experienced hunters. In the fall, storms have become more frequent and more violent, making boating difficult. Thunder and lightning have been seen for the first time at Sachs Harbour, arriving with another type of weather that is new to the area, dousing summer rainstorms. "When I was a child, I never heard thunder or saw lightning, but in the last few years we've had thunder and lightning," Kuptana said (Johansen 2001, 20). "The animals really don't know what to do because they've never experienced this kind of phenomenon. We don't know when to travel on the ice and our food sources are getting further and further away," said Kuptana. "Our way of life is being permanently altered" (Johansen 2001, 20). "We have no other sources of food, the people in my community are completely dependent on hunting, trapping and fishing," said Kuptana. "We have no means of adapting to a different environmental reality, and that is why our situation is so critical" (Johansen 2001, 20).

FURTHER READING

Johansen, Bruce E. "Arctic Heat Wave." *The Progressive*, October 2001, 18–20.

Saint Andrews Golf Course

Golf has been played at St. Andrews since the fifteenth century, but by the year 2000 a warming climate, rising tides, and coastal erosion was allowing the sea to swallow some of the game's oldest and most storied links. According to one observer, "One theory even claims that golfers will be lucky to enjoy another 50 years of competition on the links in Fife before the Old Course becomes just another submerged attraction akin to the lost city of Atlantis" (Aitken 2001, 20).

Peter Mason, external relations manager of the St. Andrews Links Trust, explained how a combination of high tides, large waves, and strong winds have battered the land along the Eden estuary and swept away a coastal path. "Three meters of linksland, just centimeters away from the par-4 eighth hole on the Jubilee Course, were lost to the sea in a single day last year after storms.... It fell away so fast, it frightened the life out of us," Mason recalled (Aitken 2001, 20).

Parts of the St. Andrews "links" once were reclaimed from the ocean. The original courses were called "links" because they were located on strings of dunes and knolls that linked sea and land (Hale 2001, 3-C). Today, all golf courses are called "links" after St. Andrews' origins, whether or not they border water. As satellite surveys indicate that the height of waves striking Britain's coasts has increased 25 percent in 20 years, some of these holes may be relocated for the first time in their two-century history.

Sloping gabions (stone-filled wire cages) were installed along a 100-meter stretch where the path ran before the storms. This defense was backed by 4,400 cubic meters of sand; an additional 12,000 cubic meters of sand was added later. If that does not hold back the sea, more gabions may be needed (Aitken 2001, 20). Roughly 200,000 rounds of golf are played as St. Andrews annually, a major tourist draw.

Golfers who drive off the eighth tee of the Jubilee Course now stare down at 400 yards of rock-filled cages that stand between the course and the ocean. The third hole on the Royal Aberdeen course once was 60 yards from dunes that marked the ocean's edge. By the year 2001, on the eve of the 130th British Open on the course, golfers teed off at this hole two yards from a 40-foot drop into the ocean (Hale 2001, 3-C). Violent winter storms are responsible for much of the erosion.

Other British golf courses also are eroding. At the Royal West Norfolk Golf Club, near Brancaster in England, advancing seas have isolated the clubhouse, forcing use of knee-high rubber boots to play through.

FURTHER READING

Aitken, Mike. "St. Andrews Stymied by Natural Hazard." *The Scotsman*, April 18, 2001, 20.
Hale, Ellen. "Seas Create Real Water Hazard: Changing Climate at Root of Erosion That's Putting Links Courses in Jeopardy." *USA Today*, July 18, 2001, 3-C.

Salmon Decline in Warming Waters

The decline of Pacific salmon runs during the late 1990s suggests that global climate change could devastate fish populations on which millions of people rely for food. In Alaska during 1997 and 1998, few salmon returned from the ocean to spawn at their birthplaces. Those that did return were smaller than average and arrived later than usual. The key factor in the decline of salmon runs appears to have been the fact that water temperatures in 1997 and 1998 were much higher than usual (Kruse 1998).

Salmon Populations Crash

Drastic declines in some western Alaskan salmon populations during 1997 and 1998 led some observers to ask whether rising water temperatures had played a role, because salmon are very sensitive to temperature. According to a report by the World Wildlife Fund,

> While salmon can withstand higher temperatures in summer when food is abundant, in the winter their tolerance drops considerably. As cold-blooded creatures, their metabolism increases in warmer water and keeping up with this high metabolism requires large amounts of food. If sufficient food is not available, salmon can starve. (Mathews-Amos and Berntson 1999)

For Canada's western regions, climate models forecast an increase in precipitation, water runoff, and flooding in winter and a decrease in precipitation and runoff during summer. Higher winter river flows are expected to damage salmon spawning grounds, reduce survival and growth of fish because of increased stream temperatures, and damage Fraser River salmon because of increased predation by warm-water species (Rolfe 1996). In 1995, the Canadian Department of Fisheries and Oceans blamed a collapse of Fraser River salmon runs on predation by mackerel, which invaded the salmon-spawning grounds along with warmer-than-average ocean waters provoked by El Niño conditions. At the same time, and according to the Canadian government, for the same reasons, the Queen Charlotte chinook salmon runs declined about 80 percent.

The salmon catch in Scotland was 35 percent less in 1999 than during 1998, which itself was the second-worst year since comprehensive records began in 1952. The reduced catch, due in part to warming water temperatures, is threatening the livelihood of river proprietors, country hotels and bed and breakfasts, as well as ghillies (guides) who help the angler find his fish. Warmer seawater decreases the population of krill which the salmon eat. Anglers also said that sea lice from salmon reared on fish farms are infesting wild salmon, killing them. The fish farms produce a polluting slurry. In eight of 32 rivers on the west coast of Scotland, salmon were virtually extinct by the year 2000. Seals are killing salmon in some rivers as well (Buxton 2000, 11).

Salmon Fishing Shutdown in British Columbia

Rising temperatures in British Columbia's largest sockeye salmon spawning river, the Fraser, provoked a government shutdown of commercial salmon fishing at the height of the season late in September 1998. According to an Environmental News Service dispatch from Victoria, British Columbia (Thomas 1998), above-average temperatures in the river impaired the salmon's swimming and jumping abilities, and afflicted them with "proliferating pathogens and an invasion of warm-water predators such as squaw fish" (Thomas 1998). The article described fears that seawater temperatures 1.5° C above average in the adjacent Georgia Strait could trigger toxic algae blooms (Thomas 1998).

Between a quarter and two-thirds of salmon returning to various locations in the Fraser River system were dying before they could spawn, many of them from conditions related to rapid warming of their aquatic environment. The water warmed because of below-average snow-pack and above-normal temperatures in the interior of British Columbia, the Fraser's drainage basin. A large aluminum smelter was delivering large amounts of heated water to an upstream tributary of the Fraser, adding yet another human provocation to the salmon's warming

environment. Clear-cutting of upstream forests is also blamed for some of the warming of the river's waters.

Above-average water temperatures in this area also caused salmon to die in unusually large numbers during 1992 and 1994. If the temperature rises 2° C more (from a late-summer peak of 21° C to 23° C), none of the salmon will survive the swim upstream to spawn. At 23° C, most salmon will die of heat prostration. The warming of the river and resulting fishing shutdown in September 1998 brought out angry fishermen, who threatened to block cruise ships in Vancouver's busy harbor.

By late July 2006, the snow pack in the Fraser River watershed of British Columbia, which in most years would be feeding into the river and cooling it, had almost entirely melted. Salmon were threatened by a combination of warm water ands low river levels.

FURTHER READING

Buxton, James. "Suspects in the Mystery of Scotland's Vanishing Salmon: Fish Farms, Seals and Global Warming Are All Blamed for What Some See as a Crisis." *London Financial Times*, June 13, 2000, 11.

Kruse, Gordon H. "Salmon-run Failures in1997–1998: A Link to Anomalous Oceanic Conditions?" *Alaska Fisheries Research Bulletin* 5 (1998):55–63.

Mathews-Amos, Amy, and Ewann A. Berntson. "Turning up the Heat: How Global Warming Threatens Life in the Sea." World Wildlife Fund and Marine Conservation Biology Institute, 1999. http://www.worldwildlife.org/news/pubs/wwf_ocean.htm.

Rolfe, Christopher. "Comments on the British Columbia Greenhouse Gas Action Plan." West Coast Environmental Law Association. A Presentation to the Air and Water Management Association, April 17, 1996. http://www.wcel.org/wcelpub/11026.html.

Thomas, William. "Salmon Dying in Hot Waters." Environment News Service, September 22, 1998. http://www.econet.apc.org/igc/en/hl/9809244985/hl11.html.

Samsø Island, Denmark

The 4,000 residents of largely agricultural Samsø Island, in Denmark, accepted a challenge from the Danish government during the 1990s to convert to a carbon-neutral lifestyle and, by 2008, had largely accomplished the task, with no notable sacrifice of comfort. Farmers grow rapeseed and use the oil to power their machinery;

homegrown straw is used to power centralized home-heating plants; solar panels heat water, and store it for use on cloudy days (of which the 40 square mile island has many); and wind power provides electricity from turbines in which most families on the island have a share of ownership. The island is now building more wind turbines to export power.

During the late 1990s, most of the 4,300 Samsingers' homes were heated with oil imported on tankers, and their electricity was generated from coal, and imported from the Danish mainland on cables. The average resident of the island produced 11 tons of carbon dioxide a year (Kolbert 2008).

In 1997, the Danish Ministry of Environment and Energy sponsored a renewable-energy contest that the city government entered after an engineer who did not live on the island filed a proposal with the mayor's consent. Quite to everyone's surprise, Samsø was picked as Denmark's "renewable-energy island." The designation carried no prize money, so nothing happened for a few months. After that, residents established energy cooperatives and began to tutor themselves in such things as insulation and wind power.

According to Elizabeth Kolbert, writing in *The New Yorker*, "They removed their furnaces and replaced them with heat pumps. By 2001, fossil-fuel use on Samsø had been cut in half. By 2003, instead of importing electricity, the island was exporting it, and by 2005 it was producing from renewable sources more energy than it was using" (Kolbert 2008). "People on Samsø started thinking about energy," said Ingvar Jørgensen, a farmer who heats his house with solar hot water and a straw-burning furnace. "It became a kind of sport" (Kolbert 2008).

Ten years later, Samsø found itself a subject of study by energy tourists, as researchers traveled great distances, burning copious fossil fuels, to observe Samsø's experiment, with its ranks of wind turbines, 11 in all, spinning in a relatively constant wind that makes an ideal wind-power site.

Samsø built district heating plants that run on biomass, mainly bales of locally grown straw, burned to warm water that circulates into local homes, where it provides both heat (the houses are tightly insulated, and the climate is usually mild) and hot water. One district plant, in Nordby, burns wood chips, with solar panels providing hot water when the sun is shining.

Kolbert explained that "[b]urning straw or wood ... produces CO_2 but while fossil fuels release carbon that would have remained sequestered, biomass releases carbon that would have entered the atmosphere through decomposition" (Kolbert 2008).

A few Samsø farmers have converted their cars and tractors to operate on canola oil. Some of them grow their own canola seeds, which also can be used as an ingredient in salad dressing.

FURTHER READING

Kolbert, Elizabeth. "The Island in the Wind: A Danish Community's Victory over Carbon Emissions." *The New Yorker*, July 7, 2008. http://www.newyorker.com/reporting/2008/07/07/080707fa_fact_kolbert.

Satellite Data, Debate Regarding

Climate-change skeptics for nearly two decades have argued that global temperatures as measured by satellites show little or no warming, contrary to most ground-level observations. The skeptics argue that the satellite record is more comprehensive than the ground-level readings, especially over the two-thirds of the Earth that is covered by oceans, where surface measurements are sparse.

By the end of the 1990s, however, older satellite records had been refined and corrected, and the skeptics' case was falling apart. Analysis of satellite data collected between 1979 and 1999 from the lowest few miles of the atmosphere indicated a global temperature rise of about 0.33° F between 1979 and 1999. The results are at odds with previous analyses that showed virtually no warming in the satellite record over the same period. The findings were published by the journal *Science* at its Science Express Web site (at http://www.sciencexpress.org) on May 1, 2003 ("New Look" 2003).

The scientists who compiled the study included Tom Wigley, Gerald Meehl, Caspar Ammann, Julie Arblaster, Thomas Bettge, and Warren Washington, all from the National Center for Atmospheric Research. The lead author of the study was Ben Santer of Lawrence Livermore National Laboratory. "It's undeniable that the agreement with both global climate models and surface data is better for the new analysis than for the old one," said Wigley ("New Look" 2003). Instruments aboard 12 U.S. satellites provide temperature records for the lower strato-sphere. Each sensor intercepts microwaves emitted by various parts of the atmosphere, emissions increasing as temperatures rise. These data are used to infer temperatures ("New Look" 2003).

John Christy, a professor of atmospheric science at Alabama's Global Hydrology and Climate Center, tracked data from U.S. weather satellites, finding in 1998 that changes in their orbits had created errors in their data. When readings were adjusted for their true orbits, evidence of a warming trend emerged.

Dian J. Gaffen, writing in *Nature*, also found that satellite temperature readings of Earth's temperature since 1979 had been distorted by orbital decay. This problem raised others for Gaffen:

> The mere fact that inclusion of the effects of orbital decay on the trend can completely change its sign makes one wonder what other factors may be influencing the apparent trends.... The crux of the matter is that climatologists are relying on systems that were never designed for climate monitoring. (Gaffen 1998, 616)

R. A. Kerr, writing in *Science*, explained that the satellite record was erroneous not only because of orbital decay, but also because of problems with splicing together data sets from many different satellites. Making the corrections he found necessary, Kerr asserted that satellite records show a warming trend of 0.07° 0.12° C in global temperatures per decade between 1979 and 1995. Many climate scientists have concluded that surface thermometers provide the best records, and that these records should be used to measure global warming. This view is supported by F. J. Wentz and M. Schabel, writing in *Nature* (Kerr 1998b, 1948; Wentz and Schabel 1998, 1661).

Contrarians' Use of Data

Since the 1990s, climate-change contrarians have used satellite data to argue the absence of a warming signal. Their figures indicate no warming at higher levels of the atmosphere, contrary to distinct warming at Earth's surface. A 2000 report from the National Research Council concluded that both trends might be correct—in other words, the global atmosphere might be warming more quickly near the ground than higher ("New Look" 2003). "The real issue is the trend in the satellite data from 1979 onward," said Wigley. "If the original analysis of the

satellite data were right, then something must be missing in the models. With the new data set, the agreement with the models is improved, and the agreement with the surface data is quite good" ("New Look" 2003).

As time has passed, the skeptics' case that warming is not supported by satellite date has continued to deteriorate. In 2003, Konstantin Y. Vinnikov and Norman C. Grody reported an analysis of global tropospheric temperatures from 1978 to 2002, using passive microwave sounding data from the NOAA series of polar orbiters and the Earth Observing System Aqua satellite. Their analysis showed a trend of plus 0.22 to 0.26° C per decade, "consistent with the global warming trend derived from surface meteorological stations" (Vinnikov and Grody 2003, 269).

New Evidence in the Debate

A U.S. government study released May 2, 2006, once again undermined the contrarians' argument, denying any statistically significant difference between temperatures measured on the surface of the Earth or via satellites. The U.S. Climate Change Science Program, an interagency body, concluded that the two data sets match. "The bottom line is there are no significant discrepancies in the rates of warming," said Thomas R. Karl, director of the National Climatic Data Center at the National Oceanic and Atmospheric Administration, in a telephone call with reporters. Karl said reconciling the two sets of temperature readings is "really a major step forward" in understanding climate change (Eilperin 2006c, A-3).

Rafe Pomerance, chairman of the Climate Policy Center, a group that advocates mandatory curbs on greenhouse-gas emissions, said the new report settled the scientific debate over humans' role. "This puts the nail in the coffin of [the skeptics'] argument as much as anything I've seen," Pomerance said. "It may not be the first time it's been said, but it's the clearest I've seen it stated coming out of a government agency. Game over" (Eilperin 2006c, A-3).

On the other hand, Senator James M. Inhofe, Oklahoma Republican, who still believes that global warming is a hoax, continued to argue strongly (and often loudly) that no evidence exists of human influence on rising temperatures. He argued that temperatures have risen to the same extent during the past 30 years as 1918 to 1945, when industrial sources were emitting fewer greenhouse gases. "What is clear is that our increased confidence in land-based temperature data in no way implies or supports a conclusion that recent observed warming is due to man instead of natural variability," said Inhofe's spokesman, Matthew Dempsey (Eilperin 2006c, A-3).

FURTHER READING

Eilperin, Juliet. "Study Reconciles Data in Measuring Climate Change." *Washington Post*, May 3, 2006c, A-3.

Gaffen, Dian J. "Falling Satellites, Rising Temperature?" *Nature* 394 (August 13, 1998):615–616.

Kerr, Richard A. "Among Global Thermometers, Warming Still Wins Out." *Science* 281 (September 25, 1998b):1948–1949.

"New Look at Satellite Data Supports Global Warming Trend." AScribe Newswire, April 30, 2003. (LEXIS)

Vinnikov, Konstantin Y., and Norman C. Grody. "Global Warming Trend of Mean Tropospheric Temperature Observed by Satellites." *Science* 302 (October 10, 2003):269–272.

Wentz, F. J., and M. Schabel. "Effects of Orbital Decay on Satellite-Derived Lower Troposheric Temperature Trends." *Nature* 394 (August 13, 1998):661–664.

Scandinavia and Global Warming

Sweden and Norway have some of the highest liquor taxes in the world, which have spawned copious smuggling, mainly from Denmark. Until recently, contraband seized at Malmo, directly across the Oresund Sound from Copenhagen, by Tullverket (Swedish Customs) was poured down the drain. These days, a million bottles a year of illicit liquor is trucked to a new high-tech plant in Linköping (about 80 miles south-southwest of Stockholm) that manufactures biogas fuel for automobiles, as well as fertilizer.

The plant also accepts human and packing-plant waste. Out of this noxious mix, all of which used to be regarded as garbage, comes biofuel to power buses, taxis, garbage trucks, private cars, as well as a methane-propelled "biogas train" that runs between Linköping and Västervik on the southeast coast. The train's boosters (not squeamish vegetarians, from the sound of it) have figured that the entrails from one dead cow, previously wasted, buys four kilometers (2.5 miles) on the biogas train.

Creative Ways to Replace Oil

Sweden and other Scandinavian countries have found many creative ways to replace oil with what used to be waste. People in these countries are making resources of whatever they have in abundance: geothermal resources and wind in Iceland; wood and organic waste in Sweden; wind in Denmark. They are using simple, practical solutions that are available now, at modest cost. Iceland plans by 2050 to power all of its passenger cars and boats with hydrogen made from electricity drawn from local, renewable resources.

Swedish greenhouse-gas emissions fell by 1.7 percent between 2005 and 2006. Emissions have decreased by nearly 9 percent from 1990 levels. This means that Sweden has reduced its emissions by 12.7 percent more than agreed under the Kyoto Protocol. This is shown by data on Swedish greenhouse-gas emissions from 1990 to 2006 that the Swedish Environmental Protection Agency delivered to the government. At the same time, figures from Statistics Sweden show that the economy, in fixed prices, has grown by 44 percent between 1990 and 2006.

In Amsterdam, up to 40 percent of travel in the city was on bicycles by 2007. At many traffic lights, bikes go first. Spain is testing a new solar technology called "Concentrating Solar Power" (CSP) that uses mirrors to focus sunlight with a much greater efficiency than photovoltaic cells. This technology is being tested in Seville, Spain (and, in the United States, by Arizona Power). CSP may open the way for large-scale solar-power plants.

The Linköping plant is not alone, although it is unusual for the number of things it can process. The Danish Crown slaughterhouse uses the fat of 50,000 pigs in an average week to generate biogas. Closer to home, ConocoPhillips and Tyson Foods soon will be making "renewable diesel" from beef, pork, and chicken fat in Texas.

The entire Danish Crown plant has been redesigned with an eye to saving energy, part of a 30-year Danish effort to eliminate waste, conserve energy, and reduce consumption of fossil fuels. Surplus heat from Danish power plants is piped to nearby homes, via insulated pipes, as "cogeneration" or "district heating," which required tearing up streets to install the pipes. This system now heats almost two-thirds of Danish homes. Power plants have been radically reduced in size and built closer to people's homes and offices, to reduce the amount of power lost in transmission, and to encourage the use of formerly wasted heat. In the mid-1980s, Denmark had 15 large power plants; it now has several hundred small ones.

Danish Building codes enacted in 1979 (and tightened several times since) also require thick home insulation and tightly sealed windows. Between 1975 and 2001, Denmark's heating bill fell 20 percent while the amount of heated space rose by 30 percent. Denmark's gross domestic product has doubled on stable energy usage.

Denmark's Energy Sector

The average Dane uses 6,000 kilowatt hours of electricity a year, half of the U.S. average (13,300 kilowatt hours). Much of Denmark's energy sector is owned by nonprofit cooperatives with resident shareholders. Denmark's social-welfare state values high-quality health care, free schools, and pensions over profits, free individual choice, and low taxes.

Sweden and Denmark both were dependent on imported oil during the 1970s, and made a lasting commitment to get off of it. Wind power is an important part of this strategy in Denmark; it also included exploration for oil and gas under the North Sea. By 2007, Denmark was a net exporter of oil, gas, and electricity. Taxes also have been raised to the point that they represent about half an average household's energy bill. Danes buying a large new car pay a registration tax equal roughly to the vehicle's value, as well as higher taxes on fuel than in the United States. For the largest and most wasteful luxury cars, the tax ranges up to 180 percent. Imagine paying $60,000 in taxes to buy a top-of-the-line Hummer, for example. Danish companies pay 43 percent more for electricity than their U.S. counterparts, as an incentive to conserve.

Denmark has become a world leader in wind-turbine technology to the point at which its turbines generate electricity that often is competitive in cost with oil, coal, and nuclear power, meanwhile building infrastructure that provides several thousand jobs.

At the ground level, Sweden and Denmark are no magic alternative energy kingdoms. While progress is being made, and the center of consensus, especially in national government, is more progressive than in the United States, political conflicts do exist. Sweden's carbon tax, for example, contains a jungle of exemptions for various industries.

However, even with a center-right coalition in power, small signs of attention to energy use are visible at street level in Stockholm. Signs in the Riksdag (Parliament) cafeteria advertise that waste food goes partly to the manufacture of biogas. The moving sidewalks in the Riksdag stop when no one uses them. Renovated hotel rooms include a door-side slot for key cards, which must be activated to turn on lights. When leaving a room, removal of the card turns off all the lights, making it impossible to leave a lit room. Every fifth car in Stockholm drives at least partially on alternative fuel.

Some Danish wind turbines have blades almost 300 feet wide. In January 2007, a stormy month, Denmark harvested 36 percent of its electricity from wind, almost double the average (in the United States a comparable figure is one-quarter of 1 percent).

Wind in Denmark can serve up too much of a good thing—in this case, fierce gales that break the super-size blades. To replace the ailing blades has cost Vestas, the owner of a wind farm at Horns Reef, 38 million euros—about $50 million at current exchange. The turbines were ruined not only by strong winds but also by a dunking in salt water.

In 2007, a center-right coalition that took power in Denmark reduced subsidies for wind-power development, and its growth slowed. Suddenly, the tax environment for new wind power development was better in Texas than in Denmark. The Danes, however, had a long head start.

In Iceland, 85 percent of the country's 290,000 people use geothermal energy to heat their homes. Iceland's government, working with Shell and Daimler-Chrysler, in 2003 began to convert Reykjavik's city buses from internal combustion to fuel-cell engines, using hydroelectricity to electrolyze water and produce hydrogen. The next stage is to convert the country's automobiles, then its fishing fleet. These conversions are part of a systemic plan to divorce Iceland's economy from fossil fuels.

Sweden's government includes a Ministry of Sustainable Development (Brazil and Quebec have similar bodies), which during 2006, also included a Commission on Oil Independence. Swedish long-term initiatives include home-grown biofuels, solar, wind, wave power, heat pumps, research into new sources, and technological improvements sponsored by the government.

Swedes are being encouraged to work via teleconferencing and to avoid commuting in gasoline-powered vehicles by setting up home offices (via the Internet). An emphasis also is being placed on public transport, hybrid vehicles, and biodiesel powered cars.

Auto Industry Opposes Oil Dependence

Ulf Perbo, who heads BIL Sweden, the national association for the automobile industry, said that even automakers there want to end oil dependency.

Many people have asked [why BIL] is not against the Oil Commission, but it is not in our interest to be dependent on oil, with regard to the production and sales of cars. Oil is not what interests us; cars are. And oil is going to be a limitation [to the production and sales of cars] in the future. (Johansen 2007, 23)

The Swedish government buys environmentally friendly cars, and some cities offer their drivers free parking. Swedish paper and pulp industries use bark that formerly was wasted to produce energy for their manufacturing processes, as sawmills incinerate wood chips and sawdust to generate power.

Sweden's nine million people have had one of the world's most impressive records on environmental protection for many years. In 2007, according to the government, Sweden's energy use was roughly 35 percent dependent on oil products. The proportion of oil-heated homes in Sweden was down to 8 percent by 2006, as many neighborhoods used hot water from central plants that burn biofuels, often wood-based pellets. In December 2005, all Swedish gasoline stations were required by an act of Parliament to offer at least one alternative fuel. Since the beginning of 2006, householders have been paid to replace oil-burning boilers with environmentally friendly heating systems. Such financial incentives already were available to libraries, aquatic facilities, and hospitals that wanted to switch to more efficient renewable energy.

Sweden has experienced some political bumps in the road on this issue, however. The oil independence panel has its critics. For example, in a country that produces more private cars per capita than any other, the proceedings of the commission have completely omitted the word

"bicycle." Stockholm, however, is laced with well-used bicycle paths.

During the 1970s, energy interests in Sweden pitched nuclear power as an antidote to oil. Seven new nuclear reactors came into operation between 1972 and 1980. Because of environmental considerations, high production costs, and low world market prices, Sweden's substantial uranium reserves (250,000 to 300,000 tons, or about 20 percent of known world reserves) have not been exploited. By 2006, 45 percent of Sweden's electricity still was being generated by nuclear power, and 8 percent from fossil fuels.

Swedish nuclear power sustained a new blow during 2006 because of a near-meltdown at one of the country's reactors. An electricity failure on July 25, 2006, led to the immediate shutdown of the Forsmark 1 reactor after two of four backup generators, which supply power to the reactor's cooling system, malfunctioned for about 20 minutes.

This near-accident revived memories of the Ukraine's Chernobyl, which showered radioactive fallout over much of Sweden during April 1986. By 2010, all 12 of Sweden's nuclear plants will be shut down, to be replaced by natural gas, a fossil fuel, until other sources are available.

Sweden's Legal Measures

Sweden's Carbon Tax, enacted in 1991, supplied an early example of the idea's use on a national scale. Finland, Norway, and Holland later enacted similar taxes. Sweden also levies an energy tax against all fossil fuels. Other taxes are used as incentives to control pollution, including a nitrogen-oxides charge, a sulfur tax, and a tax on nuclear-energy production. In many cases, energy taxes have replaced about 50 percent of the income-tax burden.

The Swedish energy tax varies widely among different products and types of fossil fuels, with by far the highest rates are placed on gasoline. Responding to the new tax system, the use of biofuels (mainly wood-based) for "district" heating soared about 350 percent during the 1990s. Prices of wood-based fuels also fell dramatically as technology improved. Sweden's carbon tax has been credited with reducing fossil-fuel emissions about 20 percent.

Another conservation measure, "congestion chargers," levies tolls to drive a car in downtown Stockholm. These charges became a controversial issue in the Swedish general election during the late summer of 2006. The congestion charge of up to $7.00 a day was narrowly approved by 52 percent in a referendum on September 17. Stockholm already had tested the idea.

The congestion charges reduced auto traffic 20 to 25 percent, while use of trains, buses, and Stockholm's extensive subway system increased. Emissions of carbon dioxide declined 10 to 14 percent in the inner city and 2 percent to 3 percent in Stockholm County. The project also increased the number of environmentally friendly cars, which were exempt from congestion taxes. The Stockholm congestion tax will become permanent in August 2007.

Per Bolund, one of 19 Green Party members in Sweden's 349-member Riksdag, has been watching Green Party initiatives work their way into the political mainstream for many years. The Green Party favored a congestion charge for Stockholm for decades, only to watch conservative forces block it. Ironically, the congestion charge will be imposed under the conservative-right coalition.

The commission on oil dependency also was a long-time Green Party initiative that is now embraced across the Swedish political spectrum. "The Social Democrats stole it from us," Bolund said. He then grinned, and said that Greens must be prepared to be mimicked by the political mainstream to succeed (Johansen 2007, 23).

The Swedish government recently adopted another Green Party idea: a vehicle tax based on carbon-dioxide emissions rather than weight. Bolund pointed out that some cars, such as hybrids, are heavy, but relatively low in emissions.

Bolund said that the present conservative government (in 2007) was emphasizing biofuels too much, and underplaying energy efficiency, which is often not profitable, but *is* the best way to reduce greenhouse-gas emissions. Sweden's metal and forest-products-based heavy industries have profited from cheap hydropower, he said. Many homes still are heated by inefficient baseboard electricity, a holdover from the days of cheap hydropower. Plentiful hydropower has led to waste, Bolund said.

Bolund said that some of Europe's heavy industries, such as automaking, consume three times the electricity per unit of production as similar producers in Central Europe. Saab and Volvo are not energy efficient, he said. When Swedish officials brag that they have reduced the used of oil in home heating to almost zero, said

Bolund, they are ignoring the fact that half that total is from nuclear plants, and the other half from often-inefficient hydropower. "Any time you see those statistics you have to say—yes, but," he said, don't stereotype Sweden as a green heaven. They face the same conflicts as the rest of the world, he emphasized. "We have a long way to go" (Johansen 2007, 23).

FURTHER READING

Johansen, Bruce E. "Scandinavia Gets Serious About Global Warming." *The Progressive*, July 2007, 22–25.

Scotland, Climate Changes in

More than 12,000 forts, castles, and other archaeological sites that stand along Scotland's shorelines are being eroded into the sea, "as rising tides and relentless storms, brought about by global warming, east away at the coastline" (Grant 2001, 5). Among the sites most at risk of eroding into the sea is Scrabster Castle, built in the twelfth century by the Bishop of Caithness, near Thurso. This structure is referenced in the Vikings' Orkney Inga Saga, which dates to 1196.

Dunbar Castle, where Mary Queen of Scots sought refuge after the murder of her second husband, also has been threatened by the sea. Tom Dawson, a researcher at St. Andrews University's School of History, said, "Parts of the coastline are receding up to one meter a year" (Grant 2001, 5). A third-century fort on Orkney, the Borch of Borwick, also has been crumbling into the sea "with pottery and human skeletons falling from ruins into the water" (Grant 2001, 5).

In another sign that Scotland's climate has warmed, hops, which give beer its bitter flavor, have invaded the Scottish countryside. Hops, heretofore confined to the southern British Isles and the southern English Midlands, have been sighted in large patches in at least a dozen locations along roadways near Pathhead, Midlothian. Botanics specialist Brian Moffat was quoted in the *Glasgow Herald* as saying, "Finding one hop plant here is usually a little bit of a freak event; finding so many is really extraordinary" ("Changes in Climate" 2000, 13).

Elsewhere in Scotland, two of the area's five ski resorts were put up for sale early in 2004 after they suffered large financial losses. The owners of Glencoe and Glenshee put the resorts on the market after deciding that they could no longer afford to keep them open. Mild winters and lack of snow have left the winter sports industry in Scotland in poor financial shape. With the pace of global warming accelerating, some in Scotland expected that its skiing industry would cease to exist within 20 years.

Fred Last, president of the Royal Caledonian Horticultural Society, who has recorded changes in plants for a quarter century, said he had observed significant changes in Scotland. Based on his continued observations of 800 species, he said, "There is good evidence that some plants are flowering earlier than before but the patterns of flowering have changed. Some flowering plants are showing about three weeks earlier, gardeners will need to think about their herbaceous borders to fill in any possible gaps that may occur" (Stewart 2002, 8). "If all the predictions are right, 2050 could turn out to be an excellent year for Scottish wine," quipped the *Glasgow Herald* (Crilly 2002, 11).

FURTHER READING

"Changes in Climate Bring Hops Northward." *Glasgow Herald*, September 29, 2000, 13.
Crilly, Rob. "2050 to Be Good Year for Scottish Wine: Global Warming Will Bring Grapes North." *Glasgow Herald*, November 20, 2002, 11.
Grant, Christine. "Swelling Seas Eating Away at Country's Monuments." *The Scotsman*, December 24, 2001, 5.

Sea Birds Starve as Waters Warm

The survival of seabirds on New Zealand's sub-Antarctic islands is related to supplies of krill that are declining as waters warm. Populations of rockhopper penguins, a species with brilliant yellow eyebrows that nests on the rocks of wind-blown Campbell Island, have declined by more than 95 percent since the 1940s. A breed of albatross called gray-headed mollymawks has declined by 84 percent. Muttonbirds are down by a third and elephant seals by about half, according to a report in the *New Zealand Herald* (Collins 2002).

A study by New Zealand's National Institute of Water and Atmospheric Research (NIWAR) said that the collapse of the rockhopper penguin population is due to the birds' inability to find enough food in a less-productive marine ecosystem. "The cause is reduced productivity of the ocean where they are feeding," said NIWAR scientist Paul Sagar. "They eat mainly krill but they do also eat small fish and squid" (Collins 2002).

This research measured the amount of food available to the penguins in the sea during the preceding 120 years by analyzing feathers from rockhoppers alive today compared with those of 45 museum specimens from the Antipodes and Campbell Island. Results indicated declining numbers of the phytoplankton in the rockhoppers' diet (Collins 2002).

Sagar said that further research was required to discern why phytoplankton populations had declined; some scientists asserted that the declining populations result from water that has become too warm for a species that is adapted to the sub-Antarctic. "It could be a long-run natural cycle, with the possibility that polar water coming up from the south may not be coming as far north as it used to," Sagar said. "It could be that the position of the currents has changed so the productive areas have moved further south or north, away from where the penguins are feeding. If it's happening over such a long time, I wouldn't put it down specifically to global warming" (Collins 2002). He said that the study used the rockhoppers as an "indicator species," because it was likely that the same changes in climate and phytoplankton populations also were causing a decline in other birds' populations.

Another study, at New Zealand's Otago University, indicated that muttonbird numbers have dropped by a third on one of their major breeding islands. Conservation Department scientist Peter Moore, who studied the rockhoppers on Campbell Island in 1996, said that their numbers declined from 1.6 million breeding pairs in 1942 to 103,000 pairs in 1985 and had continued to fall at a similar rate since (Collins 2002). Sagar said there was evidence that some of the adult birds were feeding better-quality food to their chicks, while their own diet declined, but Moore said the orphaned chicks often died as well. During 1990 many surviving chicks from a yellow-eyed penguin colony were fed and reared by humans after their parents died. But after the chicks were released, 99 percent of them disappeared (Collins 2002).

Sea-bird population declines have spanned the world, for similar reasons. In Scotland, for example, guillemots, Arctic terns, kittiwakes, and other seabird colonies on the Shetland and Orkney islands in 2004 experienced one of their worst breeding seasons in memory. The Royal Society for the Protection of Birds (RSPB) observed few chicks on the breeding cliff ledges. Sandeels, the small fish on which the birds feed, have migrated northward, probably because of warming waters, placing the birds' traditional food supply largely out of reach, so they have had difficulty reproducing and feeding their young.

FURTHER READING

Collins, Simon. "Birds Starve in Warmer Seas." *New Zealand Herald*, November 14, 2002, n.p. (LEXIS)

Sea-Level Rise, Worldwide Survey

On a practical level, rising seas provoked by melting ice and thermal expansion of seawater will be the most notable challenge related to global warming (ranging from inconvenience to disaster) for many millions of people around the world. Human beings have an affinity for the open sea. Thus, many major population centers have been built within a mere meter or two of mean sea level. From Mumbai (Bombay) to London to New York City, many millions of people will find warming seawater lapping at their heels during coming years. Sea levels have been rising very slowly for a century or more, and the pace will increase in coming years.

The last time that temperatures were 3° C higher than today was the Middle Pliocene, three million years ago. The seas at that time were 50 to 115 feet higher than today. The level of carbon dioxide in the atmosphere at that time peaked at about 425 parts per million (ppm). The level in 2008 had broached 385 ppm and had been increasing 2–3 ppm per year. Thus, by the end of this century, if not before, we probably will have enough warming "in the pipeline" to raise sea levels as much as 80 feet. One billion people today live within 25 meters (80 feet) of sea level.

During the twenty-first century, the phrase "environmental refugee" may become more familiar around the world. Lowland residents who could be forced out of their homes by a three-foot rise in sea levels during the next century include 26 million people in Bangladesh, 70 to 100 million in China, 20 million in India, and 12 million in the Nile Delta of Egypt (Gelbspan 1997b, 162). In Egypt, a one-meter sea-level rise

could cost 15 percent of the country's gross national product, including much of its agricultural base (Edgerton and Natural Resources Defense Council 1991, 72–73) A 14-inch rise in sea levels could flood 40 percent of the mudflats that ring Puget Sound, obliterating a significant habitat for shellfish and waterfowl (Gough 1999, 48). These locations are only a few of a great many examples, because the Earth's junctures of seas and rivers have been important crossroads for human trade (as well as fertile farming areas) throughout human history. Many river deltas are densely populated and vulnerable to even a small amount of sea-level rise.

Anticipated global warming will raise worldwide sea level not only because of melting ice, but also because the molecular structure of liquid water expands as it is heated. The same amount of liquid water will occupy more volume as it warms.

Sea-Level Rise Estimates, 2006

During late March 2006, reports were appearing in the journal *Science* that melting ice, principally from Greenland and the West Antarctic Ice Sheet, could contribute to a rise in sea levels of several meters within a century—a startling upward revision of previous estimates. Jonathan T. Overpeck and colleagues said in *Science*,

> Sea-level rise from melting of polar ice sheets is one of the largest potential threats of future climate change. Polar warming by the year 2100 may reach levels similar to those of 130,000 to 127,000 years ago that were associated with sea levels several meters above modern levels; both the Greenland Ice Sheet and portions of the Antarctic Ice Sheet may be vulnerable. The record of past ice-sheet melting indicates that the rate of future melting and related sea-level rise could be faster than widely thought. (Overpeck et al. 2006, 1747)

"The question is: Can we predict sea level? And the answer is no," said David Holland, who directs New York University's Center for Atmosphere Ocean Science. Holland, an oceanographer, added that this could mean researchers will just have to watch the oceans to see what happens: "We may observe the change much more than we ever predict it" (Eilperin 2007b, A-6). Thus, the wide range in estimates for sea-level rise by the end of the century—7.8 inches to two feet from the Intergovernmental Panel on Climate Change (IPCC) 2007 assessment. How quickly will temperature rise convert into melting ice and sea-level rise? No one really knows. A three-foot increase in sea level could turn at least 60 million people into refugees, the World Bank estimated (Eilperin 2007b, A-6).

Michael Oppenheimer, a Princeton University professor of geosciences and international affairs, said that if either the Greenland (which would raise the world ocean 23 feet) or the West Antarctic Ice Sheet (17 feet) collapses (one would not likely melt without the other), "[i]t would destroy coastal civilization as we know it" (Eilperin 2007b, A-6). The topography of land under the thick sheets of ice (which affects its flow) is largely unknown, and few measurements are taken of the remote oceans in polar regions, adding to uncertainty.

Rapid Sea-Level Rise

Sea levels have risen quickly in the past, even without human provocation. During the last deglaciation, for example, a large and rapid injection of glacial meltwater from the Laurentide Ice Sheet to the North Atlantic (called meltwater pulse 1a or mwp-1a) raised sea level by about 20 meters over only about two centuries.

"Even a foot rise [of world sea levels] is a pretty horrible scenario," said Stephen P. Leatherman, director of the Laboratory for Coastal Research at Florida International University in Miami. On low-lying and gently sloping land like coastal river deltas, a sea-level rise of only one foot could send water thousands of feet inland. Hundreds of millions of people worldwide make their homes in such deltas; virtually all of coastal Bangladesh lies in the delta of the Ganges River. Over the long term, much larger sea-level rises would render the world's coastlines unrecognizable, creating a whole new series of islands (Rudolf 2007). "Here in Miami," Dr. Leatherman said, "we're going to have an ocean on both sides of us" (Rudolf 2007).

Among the important results of warming seas will be coastal erosion, shoreline inundation because of higher tide levels, higher storm surges, and saltwater intrusion into coastal estuaries and groundwater supplies. A report by the World Wildlife Fund (WWF) said: "Scientific evidence strongly suggests that global climate change already is affecting a broad spectrum of marine species and ecosystems, from tropical coral reefs to polar ice-edge communities" (Mathews-Amos and Berntson 1999).

The net effect of sea-level rise in any particular place must be adjusted for the gradual rise or fall of the land itself. Some areas that were covered with ice during the last glacial maximum are rising—for example, Stockholm, Sweden (Silver and DeFries 1990, 94). Parts of Canada and Scotland also are slowly rising following the melting of glaciers several thousand years ago, while much of the U.S. Atlantic coast has subsided about a foot during the twentieth century. Along a relatively level, sandy shoreline, such a change may cause a beach to lose as much as 100 feet (Edgerton and Natural Resources Defense Council 1991, 25). Many Atlantic beaches have been receding as much as three feet per year; in some areas of the Gulf of Mexico coast, the rate averages five feet per year (Edgerton and Natural Resources Defense Council 1991, 78; Silver and DeFries 1990, 92). By 2007, in the town of Surfside Beach, Texas, 65 miles from Houston, the shoreline had retreated 200 feet in 25 years, leaving several homes that were once well landward of the ocean, and which have since been raised on stilts, awash in the tides (Hudson 2007, B-1).

By 1990, coastal Louisiana was subsiding 0.4 inch per year, an unusually rapid rate, due to the removal of oil, gas, and groundwater (Silver and DeFries 1990, 96). Because it is subsiding so quickly, the Mississippi River Delta in Louisiana is probably the area of the United States that is most vulnerable to sea-level rises caused by global warming (Silver and DeFries 1990, 97). Land loss in Louisiana due to subsidence and rising waters has been estimated at about one million acres during the twentieth century; by 1990, roughly 50 square miles a year were being lost. The highest point in the city of New Orleans is only 13 feet above mean sea level, as the ground under the city sinks three feet per century.

Barrier islands and sand-spits of land along of the Atlantic and Gulf Coasts are among the most vulnerable areas to sea-level rises. Most coastal barrier islands are long, narrow spits of sand with ocean on one side and a bay on the other. Typically, the oceanfront shore of any given island usually ranges from two to four meters above high tide, while the bay side is less than a meter above high water. Thus, even a one-meter rise in sea level would threaten much of these lands (and their ranks of hotels, businesses, and homes) with saltwater inundation. Erosion, moreover, threatens the high parts of these islands and is generally viewed as a more immediate problem than the inundation of barrier islands' bay sides.

Coastal erosion is an additional problem in the Houston-Galveston area. At Sargent Beach, the shoreline eroded about 1,000 feet between 1956 and the early 1990s. Galveston built a seawall to protect against oceanic flooding during the early 1900s, following a disastrous hurricane and storm surge there in 1900. When the seawall was constructed, it faced a beach that averaged about 300 feet wide. During ensuing years, the beach has eroded; by about 1940, more rocks were required to protect the original seawall. Most of these rocks sank during the following few years, requiring even more rocks (North, Schmandt, and Clarkson 1995, 172).

Additional Sea-Level Projections

Coastlines worldwide could be devastated by a sea-level rise of more than five feet by the end of the century, according to a study published on December 16, 2007, in *Nature Geoscience*. The team led by Eelco Rohling of Britain's National Oceanography Centre asserted that the sea-level rise during this century could match the rise 124,000 years ago, when the Earth's climate warmed because its orbit changed. In Great Britain, for example, an increase of that magnitude could flood London and low-lying land such as the Fens and the marshes of Essex and North Kent. Almost every beach in Britain would be under water. Several million people would lose their homes because of coastal flooding ("Britain's Sea Levels" 2007). The New York City area, home to 20 million people, has 2,400 kilometers of coastline, more than 2,000 bridges and tunnels, most with entrances less than three meters above sea level (Lynas 2007, 157–158).

E. J. Rohling and colleagues wrote that "[t]he last interglacial period, Marine Isotope Stage (MIS) 5e, was characterized by global mean surface temperatures that were at least 2° C warmer than present. Mean sea level stood 4 to 6 meters higher than the modern sea level." Much of the sea-level rise was a result of the Greenland ice sheet's melting. The most astounding finding of this group was that short-term fluctuations in sea level may have varied as much as 10 meters, although "so far it has not been possible to constrain the duration and rates of change of these shorter-term variations" (Rohling et al. 2007). The average rate of sea-level rise was 1.6 meters per century, with temperature rises "comparable

to projections for future climate change under the influence of anthropogenic greenhouse-gas emissions" (Rohling et al. 2007).

Rohling added that in the interglacial period, 124,000 to 119,000 years ago, Greenland was 5.4° to 9° F (3° to 5° C) warmer than now, which is similar to the warming period expected in the next 50 to 100 years unless drastic measures are taken to control pollution ("Britain's Sea Levels" 2007).

Experts Hedging Their Estimates

By 2005, many oceanographers were hedging their estimates of sea-level rise during the twenty-first century upward. Richard B. Alley and several other climate-change specialists who specialize in the future of ice and oceans wrote that while sea level may rise about half a meter during the century due to warming climate around the world, "Recent observations of startling changes at the margins of the Greenland and Antarctic ice sheets indicate that dynamical responses to warming may play a much greater role in the future mass balance of the ice sheets than previously considered [and] sea-level projections may need to be revised upward" (Alley et al. 2005, 456).

Ice shelves near the edges of Greenland and Antarctica are shrinking most quickly, even as ice in areas of East Antarctic increases due to heavier snowfalls, also caused at least in part by warming temperatures. Parts of inland Greenland also have been experiencing heavier snowfalls, causing the ice sheet there to grow by six to seven centimeters a year between 1992 and 2003. Alley and colleagues commented that "[i]ce shelves are susceptible to attack by warming-induced increases of melt-water ponding in crevasses that cause hydrologically driven fracturing and by warmer sub-shelf waters that increase basal melting" (Alley et al. 2005, 458).

The IPCC 2007 assessment was severely criticized for underestimating potential sea-level rise during the twenty-first century. The report reduced its worst-case sea-level rise from 88 centimeters to 59, but excluded real-world evidence that had not been included in models. Bob Correll, chairman of the Arctic Climate Impact Assessment, said that any prediction of less than one meter would "not be a fair reflection of what we know" (Pearce 2007a, 8). The IPCC also failed to include warnings by the British Antarctic Survey that the Antarctic peninsula is warming more quickly than almost any place on Earth and that the West Antarctic Ice Sheet is "unstable and contributing significantly to sea-level rise" (Pearce 2007a, 8).

The level of the world's seas and oceans has been rising slowly for much of the twentieth century, a millimeter or two a year, enough to produce noticeable erosion on 70 percent of the world's sandy beaches, including 90 percent of sandy beaches in the United States (Edgerton and Natural Resources Defense Council 1991, 18). In some areas, such as the U.S. Gulf Coast between New Orleans and Houston, the withdrawal of underground water and oil is causing some areas to sink as the oceans rise.

Sea-Level Rise May Accelerate

Writing in the March 2004 edition of *Scientific American*, James Hansen, director of the NASA Goddard Institute for Space Studies in New York City, warned that catastrophic sea-level increases could arrive much sooner than anticipated by the IPCC (Holly 2004). The IPCC estimated sea-level increases of roughly half a meter over the next century if global warming reaches several degrees Celsius above temperatures seen in the late 1800s. Hansen warned that if recent growth rates of carbon-dioxide emissions and other greenhouse gases continue during the next 50 years, the resulting temperature increases could provoke large increases in sea levels with potentially catastrophic effects.

Hansen warned that because so many people live on coastlines within a few meters of sea level, a relatively small rise could endanger trillions of dollars worth of infrastructure. Additional warming already "in the pipeline" could take us halfway to paleoclimatic levels that raised the oceans five to six meters above present levels during the Eemian period, about 120,000 to 130,000 years ago (Hansen 2004a, 73). Past interglacials have been initiated with enough ice melt to raise sea levels roughly a meter every 20 years, "which was maintained for several centuries" (Hansen 2004a, 73).

An important issue in global warming, wrote Hansen, is sea-level change, as related to "the question of how fast ice sheets can disintegrate" (Hansen 2004a, 73). "In the real world," wrote Hansen, "[i]ce-sheet disintegration is driven by highly non-linear processes and feedbacks." For example, higher sea levels can physically lift

marine ice shelves that prevent land ice sheets from sliding into the ocean. This effect accelerates the breakup of the land ice (Holly 2004). In addition, melting glacier water flows downward through holes in the ice to the bottom of the ice mass, where it serves as a lubricant that further accelerates the disintegration of the land ice and its flow into the sea.

Although buildup of glaciers is gradual, "once an ice sheet begins to collapse, its demise can be spectacularly rapid" (Hansen 2004a, 74). The darkening of ice by black-carbon aerosols (soot), pollution associated with the burning of fossil fuels, also accelerates melting. While the timing of melting is uncertain, wrote Hansen, "global warming beyond some limit will make a large sea-level change inevitable for future generations" (Hansen 2004a, 75). Hansen estimated that such a limit could be crossed with about 1° C of additional worldwide warming. This amount is below even the most conservative estimates of the IPCC for the next 50 years. Hansen recommended restriction of methane and soot emissions to balance slow growth in carbon dioxide. That, plus improved energy efficiency and increased use of renewable energy sources may buy time. He added that new technologies may be developed "that we have not imagined." The question, he concludes, is "Will we act soon enough?" (Hansen 2004a, 77)

Hansen described a world in which great ice sheets accumulate slowly and disintegrate quickly: "The great ice sheets on Greenland and Antarctica require millennia to grow, because their rate of growth depends on the snowfall rate in a cold, relatively dry, place. Ice sheet disintegration, on the other hand, is a wet process that can proceed rapidly" (Hansen 2006b, 19).

An "Unfolding Planetary Disaster of Monstrous Proportions"

With a temperature increase of 2° to 3° C on average (more in polar regions), Hansen sees human contributions to the Earth's greenhouse-gas overload provoking massive sea-level increases of about 25 meters within a few centuries.

If additional human-made global warming (above that in 2000) is so large, say 2–3° C, that the expected equilibrium (long-term) sea level rise is of the order of 25 meters, there would be the potential for a continually unfolding planetary disaster of monstrous proportions. If additional warming is kept less than 1° C there may still be the possibility of initiating ice sheet response that begins to run out of our control. However, the long term change that the system would be aiming for would be "only" several meters, at most, and, because the energy imbalance would be much less, the time required to reach a given sea level change would be longer, thus yielding a situation with better opportunities for both adaptation and mitigation. (Hansen 2006b, 21)

Hansen developed his case:

If humanity follows a business-as-usual course with global warming of at least 2–3° C, we should anticipate the likelihood of an eventual sea level rise of 25 meters ± 10 meters. It is not possible to say just how long it would take for sea level to change, as ice sheet disintegration begins slowly until feedbacks are strong enough to evoke a highly non-linear cataclysmic response. Global warming of 2–3° C would cause larger polar warmings, leaving both Greenland and West Antarctica dripping in summer melt-water. It is my opinion that 2–3°C global warming would likely cause a sea level rise of at least ~6 m within a century. Although ice sheet inertia may prohibit large change for a few decades, it is plausible that rapid change would begin this century under the Business-as-Usual climate forcing scenario. The Earth's history reveals numerous cases in which sea level increased several meters per century. Although the paleoclimate cases may have involved disintegration of ice sheets at slightly lower latitudes, the driving forces were far weaker than the presumed anthropogenic forcings later this century. With global warming of 2–3° C, the Greenland and West Antarctic ice sheets would be at least as vulnerable as the paleoclimate ice sheets. (Hansen 2006b, 21–22)

In particular, according to Hansen:

The East Coast of the United States, including many major cities, is particularly vulnerable, and most of Florida would be under water with a 25 meter sea level rise. Most of Bangladesh and large areas in China and India also would be under water.... The population displaced by a 25 meter sea level rise, for the population distribution in 2000, would be about 40 million people on the East Coast of the United States and 6 million on the West Coast. More than 200 million people in China occupy the area that would be under water with a sea level rise of 25 m. In India it would be about 150 million and in Bangladesh more than 100 million. (Hansen 2006b, 22–23)

Effects of sea-level rise would be felt most acutely in times of weather emergencies, according to Hansen:

The effects of a rising sea level would not occur gradually, but rather they would be felt mainly at the time of storms. Thus, for practical purposes, sea level rise being spread over one or two centuries would be difficult to deal with. It would imply the likelihood of a need to continually rebuild above a transient coastline. (Hansen 2006b, 23)

Sea-Level Rise: "We're Losing the Battle"

At least 70 percent of sandy beaches around the world were receding; in the United States, roughly 86 percent of East Coast barrier beaches (excluding evolving spit areas) have experienced erosion during the last century (Zhang, Douglas, and Leatherman 2004, 41). "We're losing the battle," said Stanley Riggs, a geologist at East Carolina University in Greenville, North Carolina (Boyd 2001, A-3). The Outer Banks of North Carolina have been eroding rapidly. "Highway 12 is falling into the ocean. What was once the third row of houses [on the beach] is now the first row," Riggs told a National Academy of Sciences conference on coastal disasters (Boyd 2001, A-3). A French and American space satellite named "Jason" was launched in 2001 to monitor the upward creep of the seas, In orbit, 830 miles above the Earth, Jason's radar altimeter will be able to calculate the sea level within an accuracy of one inch, according to Ghasser Asrar, NASA's associate administrator for Earth science (Boyd 2001, A-3).

Sea-level rise provoked partly by global warming poses an acute threat to barrier islands and nearby low-level wetlands in the U.S. Middle Atlantic states, according to a report by the Environmental Protection Agency, U.S. Geological Survey, and other agencies issued early in 2009. Not only is the coast eroding quickly (a foot per century, and forecast to accelerate), but many parts of the Middle Atlantic states' coasts are densely populated.

The report (available at http://www.climatescience.gov) said that "it is virtually certain" that coastal headlands, spits, and barrier islands will erode faster than they have in the past (Dean 2009). With a sea-level rise of two feet in a century (forecast by the Intergovernmental Panel on Climate Change), "some barrier islands in this region will cross a threshold" and begin to break up, the report said. The islands along the Outer Banks of North Carolina are the most vulnerable (Dean 2009). Such a rise also will swamp many coastal wetlands.

Since 1900, for example, mean sea levels have risen 12.3 inches in New York City; 8.3 inches in Baltimore; 9.9 inches in Philadelphia; 7.3 inches in Key West, Florida; 22.6 inches in Galveston, Texas; and 6 inches in San Francisco (Boyd 2001, A-3). The rate of sea-level rise has been accelerating over time. At the port of Baltimore, at the head of Chesapeake Bay, for example, the water level crept up at only about one-tenth of an inch per year for much of the twentieth century. After 1989, however, the level rose by half an inch per year, according to Court Stevenson, a researcher at the University of Maryland's Center for Environmental Science (Boyd 2001, A-3). Sea levels have risen 12 to 20 inches on the Maine coast, and as much as two feet along Nova Scotia in 250 years, according to an international team of researchers. Global warming is the main factor, said Roland Gehrels, of England's University of Plymouth. He said that the rate of sea-level rise accelerated during the twentieth century, "as industrialization swept the globe" ("Global Warming Blamed" 2001, 20-A).

Islands Disappear in the Ganges Delta

Shyamal Mandal lives at the edge of ruin on Ghoramara Island, which, at two square miles, had shrunk by half in less than 40 years. As described by Soimini Sengupta in the *New York Times*:

> In front of his small mud house lies the wreckage of what was once his village on this fragile delta island near the Bay of Bengal. Half of it has sunk into the river. Only a handful of families still hang on so close to the water, and those that do are surrounded by reminders of inexorable destruction: an abandoned half-broken canoe, a coconut palm teetering on a cliff, the gouged-out remnants of a family's fish pond. All that stands between Mr. Mandal's home and the water is a rudimentary mud embankment, and there is no telling, he confessed, when it, too, may fall away. "What will happen next, we don't know," he said, summing up his only certainty. (Sengupta 2007a)

The rising sea is only one reason why the island is disappearing. Rivers from the Himalayas, emptying into the Bay of Bengal, have swollen with glacial icemelt in recent years,

changing the shape and size of islands in the Ganges River Delta. In 30 years, about 31 square miles of the islands have gone under water, according to a study by Sugata Hazra, an oceanographer at Jadavpur University in Calcutta, which shares the same delta. More than 600 families have lost their homes. Sheikh Suleman, now nearing 60, recalled at time when his harvest of coconuts was so plentiful that his wife would give them freely to their neighbors. Now, he said, she has to beg for a coconut (Sengupta 2007a).

More than 600 families were displaced by 2007, as fields were submerged. Two islands vanished entirely. Ironically, rising levels of atmospheric carbon dioxide are playing a major role in the drowning of these islands, people who live there use nearly no fossil fuels. According to an account in the *New York Times*, several hundred families have moved to a displaced people's camp on Sagar, a nearby island (Sengupta 2007a). The islands also are more exposed to intensifying cyclones from the Bay of Bengal in much the same way that the sinking of coastal wetlands of the Mississippi River exposed New Orleans to the storm surge of Hurricane Katrina during 2005. "Nature didn't create this place for humans to cut the forests and chase out the tigers and wildlife," said Tushar Kanjilal, founder of the Tagore Society for Rural Development. "We are killing the Sundarbans. Our government, the people themselves, we are all together killing it. After 50 years, will they exist?" (Sengupta 2007a).

Sea Levels Rise from Southern Florida to India, China, and the Nile Delta

On Cape Sable, on the far southwestern edge of Florida, boaters, sportsmen, and scientists have watched as a rising sea level has swallowed a freshwater marsh. A field of saw grass is now a saltwater mangrove swamp. The Cape Sable seaside sparrow, an endangered bird, has fled northward. With old photographs and tidal gauge records, University of Miami professor Harold Wanless, chairman of the university's geology department, found that since the 1930s, with sinking land and rising seas, the relative water level has risen nine inches, he said. "Freshwater marshes on Cape Sable are now evolving into more or less open marine waters," he said. "We're not talking about global warming as something that will happen in the future. It's happening right now. All the king's horses and all the king's men won't be able to put Cape Sable together again" (Harden and Eilperin 2006, A-3).

Alarm over rising sea levels and subsidence in Shanghai, China's largest city (population 16 million), has prompted officials to consider building a dam across its main river, the Huangpu. "Its main function is to prevent the downtown areas from being inundated with floods," Shen Guoping, an urban planning official, told the *China Daily* ("Shanghai Mulls" 2004). Rising water levels of the Huangpu, provoked by rising sea levels due to global warming as well as subsidence has resulted in construction of floodwalls hundreds of kilometers in length. At the same time, subsidence caused by pumping of groundwater and rapid construction of skyscrapers has averaged more than 10 millimeters a year.

A half to one-meter rise in sea levels could submerge three of India's biggest cities (Bombay, Calcutta, and Madras) by 2020, according to Rajiv Nigam, a scientist with India's Geological Oceanography Division. Nigam said that a one-meter rise in sea level could cause five trillion rupees (US$108 billion) worth of damage to property in India's Goa province alone. "If this is the quantum of damage in a small state like Goa that has only two districts, imagine the extent of property loss in metros like Bombay," Nigam added at a workshop in the National College in Dirudhy, Tamil Nadu ("Warming Could Submerge" 2003).

By the year 2000, rising sea levels were nibbling up to 150 meters a year from the low-lying, densely populated Nile River Delta. At Rosetta, Egypt, a seawall two stories high has slowed the march of the sea, which is compounded by land subsidence in the delta, but "sea walls cannot stop the rising [salinity of] the palm groves and fields adjoining the shore" (Bunting 2000, 1). A one-meter rise in sea level could drown most of the Nile Delta, 12 percent of Egypt's arable land, home, in 2004, to seven million people.

Threats to Water Supplies

As sea levels rise, cities along the U.S. East and Gulf Coasts (and other coastal locales around the world) may find saline water seeping into their drinking-water supplies. Cities most at risk in the United States may be New York City and Philadelphia. In the Philadelphia area, a sea-level rise of a third of a meter would require a 12 percent rise in reservoir capacity to prevent saltwater intrusion into system intakes on the Delaware River (Cline 1992, 127). Other cities around the world that lie near seacoasts may face similar problems: Amsterdam, Rotterdam, Liverpool, Istanbul, Venice, Barcelona, Gothenburg, St. Petersburg, Calcutta, among others.

New York City, Philadelphia, and much of California's Central Valley get their water from areas that are just upstream from areas in which the water is salty during droughts. Farmers in central New Jersey as well as the city of Camden rely on the Potomac-Raritan-Magothy aquifer, which could become salty if sea level rises. The South Florida Water Management District already spends millions of dollars per year to prevent Miami's Biscayne aquifer from becoming salty (Miller 1989). A rise in sea level also would enable saltwater to penetrate farther inland, as well as upstream into rivers, bays, wetlands, and aquifers, which would be harmful to some aquatic plants and animals, and would threaten human uses of water. Increased salinity probably will threaten oyster harvests in the Delaware and Chesapeake Bays (Gunter 1974).

Along the U.S. Eastern Seaboard, oceanfront property, some of the most valuable real estate in the country, may be vulnerable to sea-level rises caused by global warming. "In Massachusetts," wrote author and activist Bill McKibben, "[b]etween three thousand and ten thousand acres of ocean-front land worth between $3 billion and $10 billion might disappear by 2025, and that figure does not include land lost to growing ponds and bogs [created] as the rising sea lifts the water table" (McKibben 1989, 112–113).

Chesapeake Bay "is subject to all the SLR [sea-level rise] impacts: erosion, inundation of low-lying lands and wetlands loss, saltwater intrusion into aquifers and surface waters, higher water tables, and increased flooding and storm damage" (Leatherman 1992, 17). Crab populations that rely on aquatic vegetation for protection during their early life stages are declining in the Chesapeake Bay. In the same area, many small islands, which have been important rookeries for several species of birds, are eroding. The Cape Hatteras lighthouse, first built in 1803 (then, as now, the tallest lighthouse in the United States), is being moved inland. It was 1,600 feet from shore when built, but 120 feet from open water when it was moved this year ("Cape Hatteras" 1999, A-16).

Texas may face special hazards from sea-level rise along the Gulf Coast. In the modulated tones of science, a report outlining possible effects of the greenhouse effect on Texas sketched severe problems with land subsidence and sea-level rise along parts of the state's coast:

The Houston-Galveston urban region, with its high groundwater table, subsidence, and long history of severe flooding along its coastline and bayous, will be particularly vulnerable, and could experience permanent loss of urban land in sensitive areas, necessitating extensive relocation programs. (North, Schmandt, and Clarkson 1995, 170)

In addition, a rise in sea level may imperil drinking water supplies in the coastal regions of Texas, the state's fastest-growing region in terms of human population and development. The Houston-Galveston area is additionally vulnerable because the land is relatively flat for several miles inland and has been experiencing subsidence (sinking) as groundwater and oil have been removed for human consumption. The cities of Beaumont, Port Arthur, Orange, Freeport, and Corpus Christi subsided as much as a foot between 1906 and 1974 (North 1995, 171). Some small areas have sunk as much as 10 feet during that period. Subsidence rates could increase as more groundwater is withdrawn.

FURTHER READING

Alley, Richard B., Peter U. Clark, Philippe Huybrechts, and Ian Joughin. "Ice-Sheet and Sea-Level Changes." *Science* 310 (October 21, 2005):456–460.

Boyd, Robert S. "Rising Tides Raises Questions: Satellites Will Provide Exact Measurements." Knight-Ridder Newspapers in *Pittsburgh Post-Gazette*, December 9, 2001, A-3.

"Britain's Sea Levels 'Will Rise 5 Feet This Century.'" *London Daily Mail*, December 17, 2007. (LEXIS)

Bunting, Madeleine. "Confronting the Perils of Global Warming in a Vanishing Landscape: As Vital Talks Begin at the Hague, Millions Are Already Suffering the Consequences of Climate Change." *London Guardian*, November 14, 2000, 1.

"Cape Hatteras, N.C. Lighthouse Lights Up Sky from New Perch." *Omaha World-Herald*, November 14, 1999, A-16.

Cline, William R. *The Economics of Global Warming.* Washington, D.C.: Institute for International Economics, 1992.

Dean, Cornelia. "Study Warns of Threat to Coasts from Rising Sea Levels." *New York Times*, January 17, 2009. http://www.nytimes.com/2009/01/17/science/earth/17sea.html.

Edgerton, Lynne T., and the Natural Resources Defense Council. *The Rising Tide: Global Warming and World Sea Levels.* Washington, D.C.: Island Press, 1991.

Eilperin, Juliet. "Clues to Rising Seas Are Hidden in Polar Ice." *Washington Post*, July 16, 2007b, A-6. http://www.washingtonpost.com/wp-dyn/content/article/2007/07/15/AR2007071500882_pf.html.

Gelbspan, Ross. *The Heat Is On: The High Stakes Battle over Earth's Threatened Climate.* Reading, Mass.: Addison-Wesley Publishing Co., 1997b.

"Global Warming Blamed for Rising Sea Levels." Associated Press in *Omaha World-Herald*, November 25, 2001, 20-A.

Gough, Robert. "Stress on Stress: Global Warming and Aquatic Resource Depletion." *Native Americas* 16, no. 3/4 (Fall/Winter 1999):46–48. http://nativeamericas.aip.cornell.edu.

Gunter, G. "An Example of Oyster Production Decline with a Change in the Salinity Characteristics of an Estuary—Delaware Bay, 1800-1973." *Proceedings of the National Shellfish Association* 65 (1974):3–13.

Hansen, James. "Defusing the Global Warming Time Bomb." *Scientific American* 290, no. 3 (March 2004a):68–77.

Hansen, James E. "Declaration of James E. Hansen." *Green Mountain Chrysler-Plymouth-Dodge-Jeep et al. Plaintiffs v. Thomas W. Torti, Secretary of the Vermont Agency of Natural Resources et al. Defendants.* Case Nos. 2:05-CV-302 and 2:05-CV-304, Consolidated. United States District Court for the District of Vermont. August 14, 2006b. http://www.giss.nasa.gov/~dcain/recent_papers_proofs/vermont_14aug20061_textwfigs.pdf.

Harden, Blaine, and Juliet Eilperin. "On the Move to Outrun Climate Change: Self-Preservation Forcing Wild Species, Businesses, Planning Officials to Act." *Washington Post*, November 25, 2006, A-3.

Holly, Chris. "Sea-level Rise Seen As Key Global Warming Threat." *The Energy Daily* 32, no. 36 (February 25, 2004), n.p. (LEXIS)

Hudson, Kris. "Whose Beach Is This, Anyway?" *Wall Street Journal*, December 12, 2007, B-1, B-8.

Leatherman, Stephen P. "Coastal Land Loss in the Chesapeake Bay Region: An Historical Analog Approach to Global Change Analysis." In *The Regions and Global Warming: Impacts and Response Strategies*, ed. Jurgan Schmandt and Judith Clarkson, 17–27. New York: Oxford University Press, 1992.

Lynas, Mark. *Six Degrees: Our Future on a Hotter Planet.* London: Fourth Estate (HarperCollins), 2007.

Mathews-Amos, Amy, and Ewann A. Berntson. "Turning up the Heat: How Global Warming Threatens Life in the Sea." World Wildlife Fund and Marine Conservation Biology Institute, 1999. http://www.worldwildlife.org/news/pubs/wwf_ocean.htm.

McKibben, Bill. *The End of Nature.* New York: Random House, 1989.

Miller, T., J. C. Walker, G. T. Kingsley, and W. A. Hyman. "Impact of Global Climate Change on Urban Infrastructure." In *Potential Effects of Global Climate Change on the United States: Appendix H, Infrastructure,* ed. J. B. Smith and D. A. Tirpak. Washington, D.C.: U.S. Environmental Protection Agency, 1989.

North, Gerald R., Jurgen Schmandt, and Judith Clarkson. *The Impact of Global Warming on Texas.* Austin: University of Texas Press, 1995.

Overpeck, Jonathan T., Bette L. Otto-Bliesner, Gifford H. Miller, Daniel R. Muhs, Richard B. Alley, and Jeffrey T. Kiehl. "Paleoclimatic Evidence for Future Ice-Sheet Instability and Rapid Sea-Level Rise." *Science* 311 (March 24, 2006):1747–1750.

Pearce, Fred. "But Here's What They Didn't Tell Us." *New Scientist,* February 10–16, 2007a, 6–9.

Rahmstorf, Stefan. "A Semi-Empirical Approach to Projecting Future Sea-Level Rise." *Science* 215 (January 19, 2007): 368–370.

Rohling, E. J., K. Grant, Ch. Hemleben, M. Siddall, B. A. A. Hoogakker, M. Bolshaw, and M. Kucera. "High Rates of Sea-level Rise during the Last Interglacial Period." *Nature Geoscience* (December 16, 2007), doi: 10.1038/ngeo.2007.28.

Rudolf, John Collins. "The Warming of Greenland." *New York Times,* January 16, 2007. http://www.nytimes.com/2007/01/16/science/earth/16gree.html.

Sengupta, Somini. "Sea's Rise in India Buries Islands and a Way of Life." *New York Times,* April 11, 2007a. http://www.nytimes.com/2007/04/11/world/asia/11india.html.

"Shanghai Mulls Building Dam to Ward off Rising Sea Levels." Agence France Presse, February 9, 2004. (LEXIS)

Silver, Cheryl Simon, and Ruth S. DeFries. *One Earth, One Future: Our Changing Global Environment.* Washington, D.C.: National Academy Press, 1990.

"Warming Could Submerge Three of India's Largest Cities: Scientist." Agence France Presse, December 6, 2003. (LEXIS)

Zhang, Keqi, Bruce C. Douglas, and Stephen P. Leatherman. "Global Warming and Coastal Erosion." *Climatic Change* 64, nos. 1 and 2 (May 2004):41–58.

Seawater Cooling Grid in Tokyo

Tokyo city planners during 2002 announced plans to build the world's largest cooling system and fill it with seawater to reduce temperatures by 2.6° C—or about the amount that average temperatures in the city increased during the twentieth century. General warming is being enhanced in Tokyo by an intensifying urban heat-island effect as population density increases

and more air conditioning (with its waste heat) is being used to cool tall buildings.

This huge construction project, estimated to cost $300 billion, would require

> a lattice of pipes under 250 hectares (about 600 acres) of the city, drawing in cold water from Tokyo Bay. Heat would be forced into the network through a second set of pipes connected to air conditioning systems in the buildings above. The heat would be absorbed by the seawater, which would then flow back into the bay. (Ryall 2002, 20)

Skeptics of the project fear that dumping warmed seawater in Tokyo Bay could cause environmental damage.

Tokyo's average temperature rise of 2.9° C during the twentieth century was five times the average worldwide rate of warming. The number of days when the city experiences temperatures of 30° C or more has doubled during the last two decades (Ryall 2002, 20). "This is a radical approach but that's what the situation demands," said Tadafumi Maejima, director of the Japan District Heating and Cooling Association, which was commissioned to draw up the plan. "Nothing like this has been attempted anywhere in the world, but we could be ready to start work in as little as two years" (Ryall 2002, 20).

According to a report in the *London Times,*

> The pipes would initially spread beneath 123 hectares of the Marunouchi district, the city's financial heart. Later they would extend through Kasumigaseki, where the Diet (parliament), and the ministries are based. A further phase, covering an area north of Tokyo station, would use water from a sewage treatment plant. (Ryall 2002, 20)

According to the association's report, "The amount of heat emitted by buildings needs to be cut, especially in the heart of the city where demand for air conditioners is concentrated, to break the vicious cycle of summer warming" (Ryall 2002, 20).

FURTHER READING

Ryall, Julian. "Tokyo Plans City Coolers to Beat Heat." *London Times,* August 11, 2002, 20.

Sequestration of Carbon Dioxide

On a society-wide scale, James E. Hansen, director of NASA's Goddard Institute for Space

Studies, believes that the single most important solution to global warming is a moratorium on construction of new coal-fired power plants until technology for carbon-dioxide capture and sequestration is available. About a quarter of power plants' carbon-dioxide emissions will remain in the air "forever" (i.e., more than 500 years), long after new technology is refined and deployed. As a result, Hansen expects that all power plants without adequate sequestration will be obsolete and slated for closure (or at least retrofitting) before mid-century (Hansen 2007b).

Hansen believes that:

Coal will determine whether we continue to increase climate change or slow the human impact. Increased fossil fuel CO_2 in the air today, compared to the pre-industrial atmosphere, is due 50% to coal, 35% to oil and 15% to gas. As oil resources peak, coal will determine future CO_2 levels. Recently, after giving a high school commencement talk in my hometown, Denison, Iowa, I drove from Denison to Dunlap, where my parents are buried. For most of 20 miles there were trains parked, engine to caboose, half of the cars being filled with coal. If we cannot stop the building of more coal-fired power plants, those coal trains will be death trains—no less gruesome than if they were boxcars headed to crematoria, loaded with uncountable irreplaceable species. (Hansen 2007c)

According to an account by Andrew Revkin in the *New York Times*, "The components for capturing carbon dioxide and disposing of it underground are already in use, particularly in oil fields, where pressurized carbon dioxide is used to drive the last dregs of oil from the ground" (Revkin 2006k). David Keith, an energy expert at the University of Calgary, said "We just need to build the damn things on a billion-dollar scale" (Revkin 2006k). A major project in the United States is a proposed 285-megawatt Futuregen power plant under construction with funding from the federal Energy Department, along with private and international partners, that was scheduled to be built in Illinois. It was canceled in 2008, however, because of cost overruns and lack of progress.

A typical new coal-fired power plant operates for several decades after its construction. Even in 2007, with widespread concern about global warming, roughly one large coal-burning plant was being commissioned per week, most of them in China. "We've got a $12 trillion capital investment in the world energy economy and a turnover time of 30 to 40 years," said John P. Holdren, a physicist and climate expert at Harvard University and president of the American Association for the Advancement of Science. "If you want it to look different in 30 or 40 years, you'd better start now" (Revkin 2006k).

The Makings of a Moratorium

Environmental groups tightened their focus on proposed coal-fired power plants during 2007. Environmental Defense and the Natural Resources Defense Council assembled "strike forces" to mobilize opposition to new plants state by state. These strike forces played a role in obtaining cancellation of plants in Florida and Texas. In New Mexico, for example, the groups intervened into a dispute over whether to constrict a new power plant on the Navajo Nation, where the state government, which is opposed to the project, has no direct power to prevent it. The plant's carbon footprint would equal 1.5 million average automobiles. Coal-fired electricity contributes more than half of the 57 million tons of annual carbon-dioxide emissions in New Mexico. Together, two existing plants in that state emit 29 million tons (Barringer 2007b).

By mid-2008, several new coal-powered generating plants were being canceled or postponed across the United States. By that time, 645 coal-fired plants were producing about half the country's electricity; as recently as May 2007 more than 150 new ones had been planned to meet electricity demand that was rising at a 2.7 percent annualized rate. A private equity deal worth $32 billion involving TXU Corporation trimmed eight of 11 planned coal plants, as similar plants were scuttled in Florida, North Carolina, Oregon, and other states.

About two dozen coal plants were cancelled during 2006 and 2007, according to the National Energy Technology Laboratory in Pittsburg, an agency of the U.S. Department of Energy. Citibank downgraded the stocks of coal-mining companies in mid-July, saying "prophesies of a new wave of coal-fired generation have vaporized" (Smith 2007c, A-1).

Climate-change concerns have been cited when coal plants were cancelled, especially in Florida, where rising sea levels from melting ice in the Arctic, Antarctic, and mountain glaciers already have been eroding coastlines. Florida's Public Service Commission is now legally required to give preference to alternative energy projects over new fossil-fuel generation of electricity. The states of Washington and California have been moving toward similar requirements. Xcel Energy and Public Service of Colorado were allowed to go ahead with a 750-megawatt coal-fired power plant only after it agreed to obtain 775 megawatts of wind power.

Coal Plants' Emissions Tallied

In mid-July 2007, the Environmental Integrity Project released a list of the United States' 378 largest coal-fired power-plant emission sources of carbon dioxide. Nevada Power's Reid Gardner coal-fired power plant was listed as the number-one source by emission rating ("Dirtiest Power Plants" 2007). Environmental Integrity Project attorney Ilan Levin said,

While Congress is poised to seriously consider legislation to limit the greenhouse gases that made 2006 the hottest year on record, the electric power industry is racing to build a new fleet of coal-fired power plants that rely on conventional combustion technologies that would only accelerate global warming. ("Dirtiest Power Plants" 2007)

Emissions from 2,100 coal-fired power plants on the planet emit a third of humankind's carbon dioxide. In the United States, roughly 600 coal-powered plants produce 30 percent of the country's seven million metric tons of greenhouse gases, as much as all the cars and all other industries combined. In 2006, China generated 80 percent of its power with coal and was adding one new major generating plant every two weeks (Kintisch 2007b, 184).

The best way to make additional coal-fired plants unnecessary would be to decrease or eliminate electrical demand. Every kilowatt used must be examined. A great deal of outdoor lighting, for example, could be dimmed or eliminated. Calgary, Alberta, for example, reduced its municipal electrical bill $2 million a year by dimming or eliminating street lights. The California Department of Transportation gives preference on highways to reflectors and other passive guides over streetlights. The safety value of floodlighting is often overrated. Full light merely gives burglars and vandals illumination in which to do their work. Motion-sensor lights are much better for capturing activity. The San Antonio public schools switched off lights in its buildings and parking lots at night and found that vandalism fell sharply (Owen 2007, 30).

Problems with the Disposal of Carbon Dioxide in the Oceans

Disposal of carbon dioxide in the deep oceans has been proposed as one method of mitigating global warming. However, many proposals to inject human-generated carbon dioxide into the oceans ignore the possible effects of such sequestration on life at these levels. Brad A. Seibel and Patrick J. Walsh examined these effects, finding that increased deep-water carbon-dioxide levels result in decreases of seawater pH (increasing acidity), which can be harmful to sea creatures, "as has been demonstrated for the effects of acid rain on freshwater fish" (Seibel and Walsh 2001, 319). They find that "a drop in arterial pH by just 0.2 would reduce bound oxygen in the deep-sea crustacean *Glyphocrangon vicaria* by 25 percent" (Seibel and Walsh 2001, 320). The same drop in arterial pH would reduce bound oxygen in the mid-water shrimp *Gnathophausia ingens* by 50 percent.

According to Seibel and Walsh, "Deep-sea fish hemoglobins are even more sensitive to pH" (Seibel and Walsh 2001, 320). Small increases in carbon-dioxide levels and resulting decreases of pH levels "may trigger metabolic suppression in a variety of organisms," they explained, as "low pH has been shown to inhibit protein synthesis in trout living in lakes rendered acidic through anthropogenic effects" (Seibel and Walsh 2001, 320). Seibel and Walsh cited research by R. L. Haedrich that "any change that takes place too quickly to allow for a compensating adaptive change within the genetic potential of finely adapted deep-water organisms is likely to be harmful" (Seibel and Walsh, 2001, 320). Seibel and Walsh concluded that "[a]vailable data indicate that deep-sea organisms are highly sensitive to even modest pH changes.... Small perturbations in CO_2 or pH may thus have important consequences for the ecology of the deep sea" (Seibel and Walsh 2001, 320).

"Through various feedback mechanisms, the ocean circulation could change and affect the retention time of carbon dioxide injected into the deep ocean, thereby indirectly altering oceanic carbon storage and atmospheric carbon dioxide concentration," said Atul Jain, a professor of atmospheric sciences at the University of Illinois at Urbana-Champaign. "Where you inject the carbon dioxide turns out to be a very important issue" ("Global Warming Could Hamper" 2002). To investigate the possible effects of feedbacks between global climate change, the ocean carbon cycle, and oceanic carbon sequestration, Jain and graduate student Long Cao developed an atmosphere-ocean, climate-carbon cycle model of intermediate complexity. The researchers then used the model to study the effectiveness of oceanic carbon sequestration by the direct injection of carbon dioxide at different locations and ocean depths ("Global Warming Could Hamper" 2002).

Jain and Cao found that climate change has an important impact on the oceans' ability to store carbon dioxide. The effect was most pronounced in the Atlantic Ocean. "When we ran the model without the climate feedback mechanisms, the Pacific Ocean held more carbon dioxide for a longer period of time," Cao said. "But when we added the feedback mechanisms, the retention time in the Atlantic Ocean proved far superior. Based on our initial results, injecting carbon dioxide into the Atlantic Ocean would be more effective than injecting it at the same depth in either the Pacific Ocean or the Indian Ocean" ("Global Warming Could Hamper" 2002). *See also:* Carbon Capture and Sequestration

FURTHER READING

Barringer, Felicity. "Navajos and Environmentalists Split on Power Plant." *New York Times*, July 27, 2007b. http://www.nytimes.com/2007/07/27/us/27navajo.html.

"Dirtiest Power Plants in the USA Named." Environment News Service, July 26, 2007. http://www.ens-newswire.com/ens/jul2007/2007-07-26-05.asp.

"Global Warming Could Hamper Ocean Sequestration." Environment News Service, December 4, 2002. http://ens-news.com/ens/dec2002/2002-12-04-09.asp.

Hansen, James E. "Political Interference with Government Climate Change Science." Testimony of James E. Hansen to Committee on Oversight and Government Reform. United States House of Representatives, March 19, 2007b.

Hansen, James E. "Coal Trains of Death." James Hansen's E-mail List, July 23, 2007c.

Kintisch, Eli. "Making Dirty Coal Plants Cleaner." *Science* 317 (July 13, 2007b):184–186.

Owen, David. "The Dark Side: Making War on Light Pollution." *The New Yorker*, August 20, 2007, 28–33.

Revkin, Andrew C. "Budgets Falling in Race to Fight Global Warming." *New York Times*, October 30, 2006k. http://www.nytimes.com/2006/10/30/business/worldbusiness/30energy.html.

Seibel, Brad A., and Patrick J. Walsh. "Potential Impacts of C02 Injection on Deep-sea Biota." *Science* 294 (October 12, 2001):319–320.

Smith, Rebecca. "New Plants Fueled by Coal Are Put on Hold." *Wall Street Journal*, July 25, 2007c, A-1, A-10.

Shipping: Sails Reborn

Some cargo ships that have been propelled solely by fossil fuels are using sails to reduce fuel consumption by as much as one third. One example is the MV Beluga SkySails (Kleiner 2007, 272). The sails, produced in Bremen, Germany, by Beluga Shipping, can cover as much as 5,000 square feet and can be manipulated to take advantage of wind direction, acting like parafoils that generate lift as well as propulsion—as much as a 5,000 kilowatt engine.

International shipping (like airlines) has long been exempt from greenhouse-gas limits, but some owners expect them. They are reacting to rapidly rising prices of oil and other fossil fuels. By weight, cargo ships carry 400 times as much as airlines, four times that of trucks, and six times railroads' loads. The average ship carries a ton of cargo using one-fourth to half as much energy, even without fuel-efficiency restrictions (Kleiner 2007, 272).

Shipping can become more energy efficient by reducing ships' speeds. Doubling speed increases fuel consumption eight times. Attention is being paid to more efficient hull and propeller designs, as well as a redesign of ships' engines. Improved hull designs alone may improve efficiency by as much as 15 percent (Kleiner 2007, 272).

On October 5, 2007, California Attorney General Edmund G. Brown Jr. joined three environmental groups (Oceana, Friends of the Earth, and the Center for Biological Diversity) in a request that the U.S. Environmental Protection Agency strictly regulate greenhouse-gas emissions for cargo, cruise, and other commercial ships that use the oceans. The petition said that ocean-going vessels together emit more carbon dioxide than any single nation except the United States, the Russian Federation, China, Japan, India, and Germany, said Brown ("EPA Petitioned" 2007).

FURTHER READING

"EPA Petitioned to Limit Greenhouse Gases From Ships." Environment News Service, October 5, 2007. http://www.ens-newswire.com/ens/oct2007/2007-10-05-094.asp.

Kleiner, Kurt. "The Shipping Forecast." *Nature* 449 (September 20, 2007):272–273.

Skiing Industry and Ice Melt

Around the world, ski resorts have found themselves threatened by warming temperatures that turn snow to slush and shorten their seasons. The largest ski resorts in the United States have found their average season shortened from an average of 158 days a year in 1997 to 146 days in 2005–2006. Even artificial snow does not work if temperatures are above freezing. As a result, many ski resorts are offering activities that do not require snow, such as golf and mountain biking. In Europe, some resorts have turned into aquatic parks and shopping malls, planning for a long-term warming future.

Higher-than-usual European temperatures interfered with the Alpine Skiing World Cup in 2006, raising fundamental questions about the future of a sport that depends on snow and ice. "It will very quickly be a big crisis for us if we continue canceling races in December," said Atle Skaardal, who supervised the women's portion of the tour for the International Ski Federation. "I think it's very critical, not only for racing but for public skiing, which also has a big impact on racing. We all have to hope for colder temperatures and snowfall in Europe" (Vinton 2006). The event took place on a pristine strip of artificial snow, on a racecourse otherwise strewn with vegetation and rocks. Some ski resorts in the Alps have wrapped glaciers and snowfields with foam insulation in an attempt (usually futile) to shelter glaciers in the summer.

The Alps are the warmest they have been in at least 1,250 years. A few more degrees of warmth will melt nearly all the snow that Alpine resorts require. Bruno Abegg, a researcher at the University of Zurich, said low-lying resorts faced an insuperable problem. "Let's put it this way," he said, "I wouldn't invest in Kitzbühel." One of the resorts, sits in a low Tyrolean valley, at an altitude of only 2,624 feet (Landler 2006). Val d'Isère, in France, and St. Moritz, in Switzerland—which are twice as high—were forced to cancel World Cup races during 2006, lacking snow (Landler 2006). The Alps are warming twice as quickly as the global average. In 1980, 75 percent of Alpine glaciers were advancing; now, 90 percent are retreating (Landler 2006).

St. Moritz, Switzerland, canceled World Cup races scheduled for December 9 and 10, 2006, saying that temperatures were too high even for artificial snow (Vinton 2006). A month earlier, a race in Sölden, Austria, had been canceled because rain melted snow off the Rettenbach Glacier. "Of course we're all very worried about the future of our sport," said Anja Paerson of Sweden, the gold medalist in slalom at the Olympic Winter Games in Turin, Italy. "Every year we have more trouble finding places to train," said Paerson (Vinton 2006). Indoor skiing venues have opened in countries such as Germany, Japan, and the United Arab Emirates. Some Scottish Highlands ski areas are being redone as mountain-biking destinations. In the United States, resorts in the Pacific Northwest got a harbinger last season when a warm winter led to a 78 percent drop in skier visits (Shaw 2006).

At the bottom of the Hahnenkamm, a downhill course in the Austrian Alps, the slope ended in a grassy field during mid-December 2006. Artificial snow was abandoned because temperatures hovered persistently above freezing. "Of course I'm nervous about the snow, but what am I supposed to do?" said Signe Kramheller-Reisch, as she walked in a bare field near a hotel her family owns. "We have classic winters and we have non-classic winters" (Landler 2006).

Ski Slopes to Swimming Pools

As snow becomes scarcer in Europe's Alps, resorts are transforming ski resorts into spas and shopping centers, advertising them as escapes from sweltering low-land cities. In Davos, Switzerland, the InterContinental Resort will by 2010 become a huge spa with 186 luxury hotel rooms, residential apartments, shops, and conference rooms. "A lot of people are telling us: You guys are doing fine because you're far above the critical height line where ski areas will have a problem," said Armin Egger, former director of Davos Tourism. "But we know if about 40 percent of skiing areas in the European Alps will be gone in 50, 100 years, then we will have a problem as well" (Williams 2007). Davos earns money hosting United Nations conferences on climate change.

Atop Little Matterhorn, the highest point in the Alps that tourists can reach on cable cars at 13,120 feet above sea level, the featured attraction has become indoor swimming pools. At the new Aqua Dome, near the Austrian ski resort of

Solden, according to an account in the *New York Times*,

It's hard to miss the three enormous concrete bowls that resemble outdoor birdbaths. Each contains a different soaking experience: one is a super-size whirlpool tub, the second has a battery of massage jets, and the third is filled with saltwater and has piped-in underwater music. Inside, a dome-topped spa has two more pools and a waterfall. (Williams 2007)

A Green-Skiing Movement

A "green-skiing movement" has arisen with its center in Aspen, Colorado. The Aspen Skiing Co. during 1997 hired Auden Schendler as an environmental-affairs director. A year later, radical environmentalists torched ski operations at Vail to emphasize the industry's role in environmental destruction, an event that Schendler called a "wake-up call." He told Daniel Shaw of *Grist Magazine* that "[c]limate change should be driving everything we all do" (Shaw 2006). His company has closely examined its operations to conserve energy and water and reduce greenhouse-gas emissions, operating a small hydroelectric plant, as well as a slope groomer powered by biodiesel fuel and wind power. The National Ski Areas Association now operates a "sustainable slopes program" that has enrolled about 180 of 492 ski resorts in the United States (Shaw 2006).

Mammoth Mountain, on the eastern slope of California's Sierra Nevada since 2000, has reduced electricity use 9 percent and cut propane use by 70,000 gallons per year. "All the people I talk to in this industry, they're scared," Isaacs said. "I'm scared. Global warming trumps everything. If it continues, we won't even be able to make snow" (Shaw 2006). *See also:* Contrarians

FURTHER READING

Elliott, Christopher. "Skiing Goes Downhill." *National Geographic*, December 2007, 12.

Landler, Mark. "Global Warming Poses Threat to Ski Resorts in the Alps." *New York Times*, December 16, 2006. http://www.nytimes.com/2006/12/16/world/europe/16austria.html?pagewanted=print.

Shaw, Daniel. "Global Warming Pushes Ski Industry Downhill." *Grist Magazine*, June 19, 2006. http://www.alternet.org/story/31991/.

Vinton, Nathaniel. "Changing Climate Is Forcing World Cup Organizers to Adapt." *New York Times*, November 27, 2006. http://www.nytimes.com/2006/11/27/sports/othersports/27ski.html.

Williams, Gisela. "Resorts Prepare for a Future Without Skis." *New York Times*, December 2, 2007. http://travel.nytimes.com/2007/12/02/travel/02ski-global.htm.

Snow Pack: Sierra Nevada and Northern California Cascades

Global warming is severely shrinking the Sierra Nevada's snow pack, which is crucial to sustaining California's economy and population. The California water system, one of the most highly engineered networks of dams and aqueducts in the world, supports the state's 35 million people as well as some of the world's richest irrigated farmland. California reservoirs are used for flood control as well as water supply. In the winter, they help mitigate floods; in the summer, they release water during seasonal droughts.

Snowmelt provides roughly 70 percent of the western United States' water flow. During recent decades, the snowmelt season has advanced from early spring to late winter. Spring temperatures in the Sierra Nevada have increased 2° to 3° F since 1950, bringing peak snowmelt two to three weeks earlier and prompting trees and flowers to bud one to three weeks sooner. Western rivers are seeing their peak runoff five to 10 days sooner than 50 years ago.

Floods and Drought

In California, even slightly higher temperatures could lead to greater flooding, an overall decline in water supplies, a decrease in summer hydroelectric-power production, and a widespread disruption of flora and fauna in the 400-mile-long Sierra Nevada. Similar effects are expected in the rivers that spring from the snowfields of the Northern California Cascades (Vogel 2001, A-1). Within a human lifetime, Californians will begin to see a shift in precipitation that will bring less snow and more rain to the mountains, according to scientists at the Scripps Institution of Oceanography in La Jolla, who calculated the effect of a 2° to 4° C temperature increase over 60 years, a pace of global warming that the Intergovernmental Panel on Climate Change (IPCC) considers likely.

"This whole state clings to the Sierra Nevada," said Jeffrey Mount, chairman of the geology department at the University of California at Davis. "The health of the Sierra Nevada, the hydrology of the Sierra Nevada, is everything to California. I don't think you can overstate that" (Vogel 2001, A-1). "The snow pack acts like a big natural reservoir," said Noah Knowles, the Scripps researcher who used information from mountain weather stations and the state's annual snow pack surveys to track the effect of a warming trend. "If you lose that, management-wise, it's like losing reservoir storage space" (Vogel 2001, A-1).

The Scripps report detailed expected effects of a warmer, drier climate on the mountains at various elevations. Sierra mountaintops are so cold that they will not thaw with slight warming. In the middle and lower elevations, however, a warmer atmosphere is provoking more rain and less snow. Rain runs off immediately, which would heighten the risk of floods in the Sacramento and San Joaquin valleys. Major rivers that begin in the Sierra would also run higher in the winter, making them more dangerous (Vogel 2001, A-1).

The Scripps research on the Sierra Nevada estimated the expected change in water content of the snow pack for each four-square-kilometer section of the mountain range. According to this analysis the snow pack, especially at lower elevations, could shrink 80 percent or more. The Nancy Vogel of the *Los Angeles Times* wrote that,

> Under the regimen described by the Scripps researchers, dam operators would have to keep reservoirs lower in the winter to protect against overflow during heavy rains. And there would be less chance of the reservoir refilling in the spring, because so much of the season's precipitation would have already fallen as rain, not snow." (Vogel 2001, A-1)

Researchers at the University of California at Santa Cruz came to similar conclusions. Lisa Sloan, an associate professor of earth sciences at the University of California, led a study that was published online June 7, 2002, by *Geophysical Research Letters*. "Everybody has guessed at the effects on water resources, but now we have numbers and locations. It's a lot different from the standard arm-waving," Sloan said. "Our hope is that this kind of study will give state and regional officials a more reliable basis for planning how to cope with climate change" ("Global Warming Threatens" 2002).

This model indicated increased rainfall in northern California and unchanged precipitation in the south. Snow accumulation in the mountains decreases in this scenario. In March, for example, the scientists' model showed an additional eight inches of rain falling in the central Sierra Nevada, while the depth of the snow pack at the end of March dropped by 13 feet ("Global Warming Threatens" 2002). "With less precipitation falling as snow and more as rain, plus higher temperatures creating increased demand for water, the impacts on our water storage system will be enormous," Sloan said ("Global Warming Threatens" 2002).

Heat-Related Deaths

Elsewhere in California, a study published during the late summer of 2004 by the *Proceedings of the National Academy of Sciences* projected that by the end of the twenty-first century rising temperatures could contribute to a sevenfold increase in heat-related deaths in Los Angeles. Given a scenario in which fossil-fuel use continues at its present pace, the study forecast that summertime high temperatures could increase by 15° F in some inland cities, giving them a climate much like that of Death Valley in 2000. The same scenario also anticipated a reduction of 73 to 90 percent in the snow pack in the Sierra Nevada, resulting in disrupted water supplies from the San Francisco Bay Area to the Central Valley (Hayhoe et al. 2004, 12,422). In Los Angeles, the study's worst-case scenario forecast that the number of days of extreme heat could increase by four to eight times. It projected that heat-related deaths in Los Angeles, which it said averaged 165 annually during the 1990's, "could double or triple under the moderate scenario and grow as much as seven times under the harsher one" (Murphy 2004).

California Snow Pack in 2007

California mountain snow pack on May 1, 2007, was only 29 percent of the average, in the midst of a record drought for the entire area that may be tied to global warming's influence on steering currents in the upper atmosphere. Similar problems are afflicting Australia in the Southern Hemisphere ("California Snowpack" 2007). Snow melt is important in California not only for urban use, but also for hydroelectric power and agricultural irrigation.

California Governor Arnold Schwarzenegger said that

> We … need additional storage because scientists say that global warming will eliminate 25 percent of our snow pack by the half of this century, which will mean less snow stored in the mountains, which will mean more flooding in the winter and less drinking water in the summer. ("California Snowpack" 2007)

The governor continued:

> That is why it is important that we don't let that much water run off into the ocean before capturing it. We must plan ahead and build more above the ground water storage to make sure that that does not happen. It is my responsibility as the governor to not only think one year ahead, or two years ahead, but to think 10, 20, 30, 40 years ahead. I want to make sure that by 2050 we have enough water, and we protect our people from huge floods. ("California Snowpack" 2007)

Some communities in California began voluntary water rationing. The San Francisco Public Utilities Commission asked its 2.4 million customers to reduce water use 10 percent or face mandatory water rationing by summer 2007. The East Bay Municipal Utilities District asked its customers to repair plumbing leaks and to water lawns no more often than three days a week.

"With more precipitation falling as rain we face more flooding and less water in the snowpack to flow to our cities and fields in the summer," said California Department of Water Resources Director Lester Snow. "Obviously, this increases the need for having more sources to draw from, to ensure that our economy and communities have enough water," said Snow ("Low Snowpack" 2007).

FURTHER READING

"California Snowpack 71 Percent below Normal." Environment News Service, May 4, 2007. http://www.ens-newswire.com/ens/may2007/2007-05-04-09.asp#anchor5.

"Global Warming Threatens California Water Supplies." Environment News Service, June 4, 2002. http://ens-news.com/ens/jun2002/2002-06-04-09.asp#anchor3.

Hayhoe, Katherine, Daniel Cayan, Christopher B. Field, Peter C. Frumhoff, Edwin P. Maurer, Norman L. Miller, Susanne C. Moser, Stephen H. Schneider, Kimberly Nicholas Cahill, Elsa E. Cleland, Larry Dale, Ray Drapek, R. Michael Hanemann, Laurence S. Kalkstein, James Lenihan, Claire K. Lunch, Ronald P. Neilson, Scott C. Sheridan, and Julia H. Verville. "Emissions Pathways, Climate Change, and Impacts on California." *Proceedings of the National Academy of Sciences* 101, no. 34 (August 24, 2004):12422–12427.

"Low Snowpack Shows Critical Need for Water Storage and Conservation." California Department of Water Resources. Advisory, May 1, 2007. http://www.publicaffairs.water.ca.gov/newsreleases/2007/050107snow.cfm.

Murphy, Dean. "Study Finds Climate Shift Threatens California." *New York Times*, August 17, 2004. http://www.nytimes.com/2004/08/17/national/17heat.html.

Vogel, Nancy. "Less Snowfall Could Spell Big Problems for State." *Los Angeles Times*, June 11, 2001, A-1.

Solar Influences on Climate

During the Little Ice Age, roughly 1600 to 1800, New Yorkers occasionally walked from Manhattan to Staten Island across the frozen harbor. Changes in the sun's activity may have had a role in this centuries-long cold episode. The cold spell that doomed Viking settlements in Greenland from 1420 to 1570 also occurred during a decline of the sun's magnetic activity, measured by sunspot activity. In 2006, astronomers said that the sun was about to enter another quiet period. Others asked whether this cycle might stall some of Earth's warming influences from greenhouse gases. A few climate contrarians even heralded a new ice age, regardless of rapidly rising greenhouse-gas levels in the atmosphere (Haigh 2001, 2109).

How Powerful Is the Sun's Influence on Climate?

Humankind's use of fossil fuels is only one influence on climate, although it does grow relatively more powerful as the atmosphere's overload of greenhouse gases grows. Another important shaper of climate has been cycles initiated by the sun's generation of the energy that sustains all life on Earth.

A team writing in *Nature* found that the level of sunspot activity during the last two-thirds of the twentieth century was "exceptional," the highest in roughly 8,000 years (Solanki et al. 2004, 1084–1086). While the sun's cycles have had long-term effects on climate, these authors asserted that "solar variability is unlikely to have been the dominant cause of the strong warming during the past three decades" (Solanki et al. 2004, 1084).

The level of greenhouse gases in the atmosphere influences surface temperature, but not in a simple one-on-one relationship. As Michael E. Mann, Raymond S. Bradley, and Michael K. Hughes point out in *Nature*, the level of greenhouse gases in the atmosphere is one part in a cacophony of influences on a given climate at any particular time and place. In addition to the possibility of warming caused by greenhouse gases during the past century, there is evidence that both solar irradiance and explosive volcanism have played an important role in forcing climate variations over the past several centuries (Mann, Bradley, and Hughes 1998, 779).

James E. Hansen, director of the Goddard Institute for Space Studies, described a role for the sun in climate change that is overshadowed by Earthbound forcings such as greenhouse-gas levels (in the long term) and La Niña/El Niño (ENSO) cycles, which produce pronounced short-term changes:

This cyclic solar variability yields a climate forcing change of about 0.3 W/m2 [watts per square meter] between solar maxima and solar minima. [Although solar irradiance of an area perpendicular to the solar beam is about 1366 W/m2, the absorption of solar energy averaged over day and night and the Earth's surface is about 240 W/m2.] Several analyses have extracted empirical global temperature variations of amplitude about 0.1° C associated with the 10–11 year solar cycle, a magnitude consistent with climate model simulations, but this signal is difficult to disentangle from other causes of global temperature change including unforced chaotic fluctuations. The solar minimum forcing is thus about 0.15 W/m2 relative to the mean solar forcing. For comparison, the human-made GHG [greenhouse gas] climate forcing is now increasing at a rate of about 0.3 W/m2 per decade....

If the sun were to remain "stuck" in its present minimum for several decades, as has been suggested in analogy to the solar Maunder Minimum of the seventeenth century, that negative forcing would be balanced by a 5-year increase of greenhouse gases. Thus, in the current era of rapidly increasing greenhouse gases, such solar variations cannot have a substantial impact on long-term global warming trends is about to get underway. ("GISS 2007 Temperature Analysis" 2008)

In a paper published online by *Science* (www.scienceexpress.org), paleoceanographer Gerard Bond of the Lamont-Doherty Earth Observatory in Palisades, New York, and colleagues reported that the climate of the northern North Atlantic has warmed and cooled nine times in the past 12,000 years in step with the waxing and waning of the sun. "It really looks like the sun has mattered to climate," said Richard Alley. "The Bond, et al. data are sufficiently convincing that [solar variability] is now the leading hypothesis," said Alley, to explain the roughly 1,500-year oscillation of climate seen since the last ice age, including the Little Ice Age of the seventeenth century (Kerr 2001, 1431).

This cycle is now in a rising mode, and "could also add to the greenhouse warming of the next few centuries," according to a report by Richard Kerr in *Science* (Kerr 2001, 1431). According to Alley (cited by Kerr), solar variations can "gain leverage on the atmosphere" by changing circulation patterns in the stratosphere, which then effects the lower atmosphere and, finally, ocean circulation, where these variations affect such climate drivers as the rate at which "deep water" forms in polar regions (Kerr 2001, 1432).

Clues from Lake Sediments

Utilizing lake sediments from southwestern Alaska, Feng Sheng Hu and colleagues have found indications that "small variations in solar irradiance induced cyclic changes in northern high-latitude environments" (Hu et al. 2003, 1890). These changes, on century-long time scales, were detected in Holocene climate, "similar between the sub-polar regions of the North Atlantic and North Pacific, possibly because of sun-ocean-climate linkages" (Hu et al. 2003, 1890).

Solar activity measured by the abundance of sunspots was slowing during the twenty-first century after a hundred years of greater-than-usual intensity, with a rising number of solar flares, sunspots, and geomagnetic storms (Radowitz 2003). This change may partially counteract the effects of greenhouse forcing during the twenty-first century. During the twentieth century, the sun is thought to have added between 4 percent and 20 percent of observed temperature increases (Radowitz 2003).

British Atlantic Survey research suggests that in years to come the sun will exert less of an impact on climate. Mark Clilverd, who led this study, said:

This work is speculative and relies on the idea that the Sun shows regular cycles of activity on timescales of 10 to 10,000 years, and that its heat output and activity are related.... We believe the work is well-grounded and the effect of solar activity on Earth's environmental system will not increase in the way it has during the [twentieth] century. (Radowitz 2003)

The reduction in solar activity may be temporary, however. Clilverd said he expected solar activity to return to twentieth-century levels by 2200.

A study by Swiss and German scientists also has suggested that increasing radiation from the sun may be responsible for a part of recent warming. Sami Solanki, director of the Max Planck Institute for Solar System Research in Gottingen, Germany, who led the research, said:

The Sun has been at its strongest over the past 60 years and may now be affecting global temperatures. The Sun is in a changed state. It is brighter than it was a few hundred years ago and this brightening started relatively recently—in the last 100 to 150 years. (Leidig and Nikkhah 2004, 5)

Solanki said that the brighter sun and higher levels of greenhouse gases have contributed to recent sharp rises in the Earth's temperature. Determining the exact proportion of the warming caused by changes in solar activity is impossible with present knowledge, he said.

Solanki's research team measured several hundred years' records of sunspots, which are believed to have intensified the sun's energy output. Sunspot activity has been increasing for the past 1,000 years, they found, in a trend that roughly conforms to temperature records. Along with other research (cited above), these researchers found that the increase in sunspot activity was especially notable during the twentieth century. Gareth Jones, an English climate researcher, said that Solanki's findings were inconclusive because the study had not incorporated other potential climate-change factors. "The Sun's radiance may well have an impact on climate change but it needs to be looked at in conjunction with other factors such as greenhouse gases, sulphate aerosols and volcano activity," he said (Leidig and Nikkhah 2004, 5).

FURTHER READING

"GISS 2007 Temperature Analysis." Goddard Institute for Space Studies. January 15, 2008. http://www.columbia.edu/~jeh1/mailings/20080114_GISTEMP.pdf.

Haigh, Joanna D. "Climate Variability and the Influence of the Sun." *Science* 294 (December 7, 2001):2109–2111.

Hu, Feng Sheng, Darrell Kaufman, Sumiko Yoneji, David Nelson, Aldo Shemesh, Yongsong Huang, Jian Tian, Gerard Bond, Benjamin Clegg, and Thomas Brown. "Cyclic Variation and Solar Forcing of Holocene Climate in the Alaskan Subarctic." *Science* 301 (September 26, 2003):1890–1893.

Kerr, Richard A. "A Variable Sun Paces Millennial Climate." *Science* 294 (November 16, 2001):1431–1432.

Lean, J., J. Beer, and R. Bradley. "Reconstruction of Solar Irradiance since 1610: Implications for Climate Change." *Geophysical Research Letters* 22 (1995):3195–3198.

Leidig, Michael and Roya Nikkhah. "The Truth about Global Warming—It's the Sun That's to Blame: The Output of Solar Energy Is Higher Now Than for 1,000 Years, Scientists Calculate." *London Sunday Telegraph*, July 18, 2004, 5.

Mann, Michael E., Raymond S. Bradley, and Michael K. Hughes. "Global-scale Temperature Patterns and Climate Forcing over the Past Six Centuries." *Nature* 392 (April 23, 1998):779–787.

Radowitz, John von. "Calmer Sun Could Counteract Global Warming." Press Association, October 5, 2003. (LEXIS)

Solanki, S. K., I. G. Usoskin, B. Kromer, M. Schussler, and J. Beer. "Unusual Activity of the Sun during Recent Decades Compared to the Previous 11,000 Years." *Nature* 431 (October 28, 2004):1084–1086.

Solar Power

Solar power has advanced significantly since the days of inefficient photovoltaics. In California, solar power is being built into roof tiles, and talk is that nanotechnology will some day make any surface the sun hits a source of power—windows, for example. Experiments have been undertaken with a new form of solar energy—Concentrating Solar Power (CSP).

CSP uses

a mirror in the shape of a parabola to focus light onto a black pipe with a heat-transfer fluid inside. The fluid is used to boil water into steam, which turns a generator that can produce 64 megawatts. The newest solar-thermal technology involves building a "power tower," a tall structure flanked by thousands of mirrors, each of which pivots to focus light on the tower, heating fluid. That design can work even in places with weaker sunlight than a desert. (Wald 2008d)

Rows of solar-power panels. (Courtesy of Shutterstock)

In our lifetimes, as private homes using alternative energy sources feed power into the electrical grid, meters may run backward, paying householders for contributed power.

With solar power still three to five times as expensive as coal, per kilowatt hour, inventors are working on new technologies and mass production to reduce that gap. A measure of the interest in solar in California (spurred by state law requiring that 20 percent of the state's electricity come from renewable sources by 2010), the U.S. Bureau of Land Management by mid-2008 had received applications for projects that could cover 78,490 acres adjacent to the Joshua Tree National Park—125 projects that, if they are built, could replace 70 industrial-size coal-fired power plants (Maloney 2008, H-2).

A breakthrough in solar power has been sought since the days of Thomas Edison. In a conversation with Henry Ford and the tire tycoon Harvey Firestone in 1931, shortly before Edison died, he said, "I'd put my money on the sun and solar energy. What a source of power! I hope we don't have to wait until oil and coal run out before we tackle that" (Revkin 2006k).

"Solar power has captured the public imagination," wrote Andrew C. Revkin and Matthew L. Wald in the *New York Times,*

> Panels that convert sunlight to electricity are winning supporters around the world—from Europe, where gleaming arrays cloak skyscrapers and farmers' fields, to Wall Street, where stock offerings for panel makers have had a great ride, to California, where Gov. Arnold Schwarzenegger's "Million Solar Roofs" initiative is promoted as building a homegrown industry and fighting global warming. (Revkin and Wald 2007b)

For all the excitement, however, solar power in 2006 contributed only 0.01 percent of the U.S. electricity supply. At the same time, worldwide, one coal-fired electric plant is being built every *week.* "Most of the environmental stuff out there now is toys compared to the scale we need to really solve the planet's problems," said Vinod Khosla, a prominent Silicon Valley entrepreneur who focuses on energy (Revkin and Wald 2007b).

Solar-photovoltaic projects grew 48 percent in the United States in 2007; by mid-2008, 11 concentrating solar projects were in operation and

20 were in various stages of planning, according to the Solar Energy Industries Association (Frosch 2008, A-13). As of early 2008, 10 solar-power plants were in advanced planning in California, Arizona, and Nevada. On a sunny early afternoon, plants will produce as much power as three standard-size nuclear reactors, at a fraction of the cost. They can, for example, be built in two years, while a nuclear plant takes a decade or longer. At the same time, eight plants were being built in Spain, Algeria, and Morocco. Other projects were planned in Israel, Mexico, China, South Africa, and Egypt, according to a count kept by Frederick H. Morse, formerly in charge of solar energy at the Energy Department and now a consultant (Wald 2008a).

Solar Power's Scale Grows

The scale of solar power is growing to industrial scale. In 2008, photovoltaic solar power installations were being planned in San Luis Obispo County, California, that will produce 800 megawatts of power under bright sun by 2013 (according to present plans), enough to equal the output of a large coal-fired (or small nuclear) plant. The power will be sold to Pacific Gas and Electric. The solar array, in two locations, will cover 12.5 square miles. This is many times the scale of any previous solar array. The largest previous photovoltaic solar array in the United States, at Nellis Air Force Base, Nevada, produces 14 megawatts. A CSP array in Nevada (NevadaOne) produced 64 megawatts as of 2008.

By 2008, the solar power industry was adding several thousand jobs in production and installation per year in the United States. SunPower, which manufactures silicon-based cells, reported 2007 revenue of more than $775 million, more than triple the same figure for 2006. Three-quarters of U.S. demand for solar power came from California. During 2007, about 100 megawatts of solar capacity was installed in California, a 50 percent increase in a year, according to the Solar Energy Industries Association.

As a source of energy, solar has a lot of room to grow, however. In 2007, it represented less than one-tenth of 1 percent of the $3 trillion global energy market.

Efficiency Increases

Until recently, photovoltaic cells have been able to convert only about 20 percent of the solar energy they receive into energy. By 2007, however, a team of researchers at the University of Delaware had converted 42.8 percent in an experimental project (a prototype has yet to be built). They are aiming at 50 percent. The work was being done under a $12 million Department of Defense contract to develop battlefield electronics. The group uses a novel light-splitting technique that markedly increases efficiency. While the concept is viable in the laboratory, cost has been a major problem. During 2007, however, the DuPont company put $100 million on the line to make the concept commercially viable (Kintisch 2007c, 583–584).

The cost of solar power has been declining sharply, from $22 per watt in 1980, to $6 per watt in 1990, and to $2.70 in 2005. Economies of scale, as well as improvements in efficiency and less expensive construction materials may bring solar energy down to cost that competes with fossil-fuel generation by about 2015 (Service 2008a, 718).

The silicon solar panels that dominate the industry today may be replaced by new technologies that combine several light-absorbing materials to capture different portions of the solar spectrum, or solar cells manufactured in rolls of thin copper-indium gallium selenide film atop a metal foil. Nanotechnology plays a role in some designs for future solar-generating technology that is been theorized, but not yet commercialized. While today's silicon cells convert about 15 to 20 percent of sunlight to electricity in the field (up to 24 percent under perfect laboratory conditions), new technologies that have broached the realm of theory (and some in design, but not commercialization) raise that figure to 40, 60, even 80 percent (Service 2008a, 718-720). Photovoltaics made of plastic may dramatically reduce manufacturing costs.

Sometimes solar power leaps the fossil-fuel age. Consider Tecnosol, a small company in Nicaragua, where more than half of rural people have no access to electricity. In some areas, most notably in the remote eastern provinces, firewood is still the main source of fuel for cooking. Many people suffer respiratory diseases from wood smoke or spend scarce money on kerosene. Tecnosol in 2007 was installing 25,000 solar units, cutting carbon-dioxide emissions by 150,000 tons over the life of the equipment (Mallaby 2007, A-17).

Concentrating Solar Power

CSP is much more powerful than photovoltaic cells. A rooftop photovoltaic complex might power a small office building, while a CSP complex near Seville, Spain, can generate 11 megawatts, enough electricity for a small town (Abboud 2006, A-4). A similar-size development was under way by 2008 in the Mojave Desert of California designed to serve 400,000 homes. The CSP mirrors track the sun and concentrate its power on single points, generating steam that runs a turbine. Some of the heat also is stored in oil or molten salt to run the turbine after sunset or when clouds obscure the sun. Such new technologies may increase the potential of solar power and bring down its cost, now $0.12 to $0.15 per kilowatt hour, compared with an average of $0.04 for coal-fired energy.

According to the consulting firm Emerging Energy Research, 45 CSP projects were in planning stages around the world by 2007, including some in the United States. The Spanish government has set a goal of 500 megawatts of solar power by the year 2010. Spain is presently subsidizing CSP development, requiring utility companies to buy their power at above-market rates. Abengoa Bioenergy plans to eventually build enough CSP capacity to supply all of Seville, about 180,000 homes.

A city the size of Seville is just the beginning for solar power. Several large arrays of solar thermal power plants across the Sahara Desert could supply most if not all of the power needs for Europe, the Middle East, and North Africa, where plans were being made by 2007 to establish a renewable energy "supergrid." Thermal solar stores solar energy in a heat-retaining fluid and uses it to drive turbines. One plan calls for roughly 1,000 100-megawatt power plants worth of solar and other renewable energies (Feresin 2007, 595). The technology for this system is now available, says Gerhard Knies, a retired physicist based in Hamburg, Germany. The grid "could offer unlimited, cheap, and carbon-dioxide free energy to Europe," said Knies (Feresin 2007, 595). Knies has joined with other advocates of the idea.

While the energy will be cheap once infrastructure has been built, getting to that point could cost at least 400 billion euros (almost US$600 billion). Thus, cheap coal is not endangered just yet. Europeans may not be willing to place their energy fates in the hands of North African states. The idea of the Sahara as solar-energy bank actually goes back as far as Frank Shuman, an inventor who, working in Philadelphia during 1913, built the first prototype thermal solar plant in Egypt. Today's advocates propose that solar power could be the major component of an energy system that also will use wind and biomass fuels.

In 2007, in the desert north of Tucson, Arizona Public Service was experimenting with CSP, using an array of mirrors there focuses sunlight and heats mineral oil up to 550° F; the heat then evaporates a liquid hydrocarbon, which runs a generator to make electricity. The array includes six rows of mirrors, each nearly a quarter-mile long, or about 100,000 square feet of reflective space. This demonstration project produces one megawatt of power, enough to power a large shopping center. Acciona Solar Power, the company that installed the CSP experimental array, also in 2007 was planning a 350-acre plant near Boulder City, Nevada, with a capacity of 64 megawatts, also using CSP. Arizona Public Service and several other utilities were considering a joint project to build a 250-megawatt CSP plant (Wald 2007a).

Zero Energy Homes and Retail Stores

Some new housing developments already have been designed to stop the electric meter completely, with enough solar power from their roofs to supply a home's every necessary kilowatt. Premier Gardens in the Sacramento area touted "zero energy homes" in a subdivision that produced 300 megawatts of electricity a year, using photovoltaic roof tiles with integrated solar technology developed and sold by General Electric (Rogers 2006, 2-RE).

In England, Susan Roaf's solar roof fuels her electric car. Roaf, an architect at England's Oxford Brookes University, has designed a solar house fitted with photovoltaic cells that harvest the sun's energy and convert it into electricity. According to a report in the *London Times*, "Her system is so efficient that she uses it to charge her electric car's batteries and makes a profit by selling 57 percent back to the national network" (O'Connell 2002). Using similar technology, cities could someday become self-sufficient power generators without the use of fossil fuels.

In some northern Chinese farming villages, homes that a few years ago relied solely on low-grade coal for heat and cooking now utilize solar-heated rooftop water tanks (Landauer 2002, F-4). About 85 percent of Israeli households have solar water heaters, which the government estimates lighten the country's overall energy burden by 3 percent. Solar heaters have been used in Israel since the 1950s. The global energy crisis of 1979 reminded Israelis of their reliance on foreign sources for oil and coal. Solar water heaters have been required in new homes by law in Israel since 1980. About 20 companies, employing 4,000 people, manufacture the systems and sell them for $300 to $1,000 each. A mid-range system can pay for itself in energy savings in three or four years and, if maintained, can last more than 10 years (Kaplow 2001, 1-P).

Eric Doub, owner of Ecofutures Building of Boulder, Colorado, has built a five-bedroom family home with his wife Catherine Childs and two children that produces more energy than it consumes. It even has a hot tub. To do it, he spent $1.38 million (the house itself cost $987,000) and cut his firm's employment by 10 people, 40 percent of its workforce. By 2008, the four-year-old project had become a show place, with 2,600 visitors. The house now has a name: Solar Harvest, and it was built mainly with salvaged materials. It includes 12 used solar panels that heat water in a 6,000-gallon underground hot-water tank purchased from a dairy farm (Feder 2008).

During 2008, retail chains such as Safeway, Wal-Mart Stores, Kohl's, and Whole Foods Market installed solar arrays on the roofs of their stores to generate large amounts of electricity to save money in the long run, reduce their use of coal-fired energy, and beat a December 31, 2008, deadline for U.S. federal tax breaks. Many states also offer tax incentives, with California, New Jersey, and Connecticut at the forefront. In many urban areas, retail stores are major electricity consumers, and solar panels can supply about 10 to 40 percent of what a store uses.

"It's very clear that green energy is now front and center in the minds of the business sector," said Daniel M. Kammen, an energy expert at the University of California–Berkeley. "Not only will you see panels on the roofs of your local stores, but I suspect very soon retailers will have stickers in their windows saying, 'This is a green energy store'" (Rosenbloom 2008). While solar power still costs as much as $0.25 to $0.30 per kilowatt hour, many retailers believe that economies of scale will being the price down over time, as fossil-fuel prices (now as low as 0.04 to 0.06 cents

per kilowatt hour for coal-fired power) will rise. The trend follows a pattern in Europe, where solar tax incentives in countries such as Germany are permanent.

Appropriate-Scale Technology

Solar power creates opportunities for decentralized "appropriate-scale" technology, especially in countries with large rural populations. One example is India, which averages 210 days a year of nearly direct sunlight, a large rural population, and a tradition of local, basic, small-scale problem-solving, which stems from Mahatma Gandhi. Gandhi turned homespun cloth from a small spinning machine into a powerful political symbol regarding the centralized weaving industry that was controlled by the British.

By 1995, six thousand villages in India that had no access to conventional electric power grids were drawing electricity from banks of photovoltaic solar cells. Using the same model of small-scale, locally controlled technology, photovoltaic modules and solar cooking stoves are being used increasingly in India's villages. Many villagers also use biogas digesters that convert the dung of cows and other animals to energy. The resulting methane is burned as energy before it bubbles into the atmosphere as a greenhouse gas. New technology also allows dung to be turned to an energy-rich sludge without smoke and fire. A million digesters were operating in India by 1990, despite the fact that one of them costs about $50 (with half the amount paid by a government subsidy), or almost one-fifth of the average rural Indian's annual cash income. (Oppenheimer and Boyle 1990, 137, 139) The digester-financing program is administered by the Indian federal government's Department of Nonconventional Energy.

These programs should not leave an impression that India, as a whole, is reducing its greenhouse-gas emissions. In India's cities, a growing middle class is creating demands for more energy, most of it generated from fossil fuels, especially coal. India has only 1 percent of the world's coal reserves, but it is fourth among the world's nations in coal combustion. During the late 1980s, India had only 55,000 megawatts of electrical-generating capacity, twice the capacity of New York State (Oppenheimer and Boyle 1990, 137) but serving a population of more than 800 million. These figures suggest that electricity generation via fossil fuels is still in its infancy on the Indian subcontinent, even with recent solar-power development on a small scale.

Solar Power in Cloudy Germany

Even Germany, land of fog and weeping skies, was finding a place for solar cells so sensitive that a hot shower on a cloudy day was no longer a problem. By 2006, half the world's new solar capacity was being installed in Germany, despite the fact that Germany has only half as many sunny days as Portugal, a more obvious solar success story (Whitlock 2007). The German solar panels work on drizzly days, although they generate only a quarter to half the electricity as on a sunny day.

Coal mining was Espenhain's largest employer under the East German regime that collapsed during 1989, providing 8,000 jobs. After German reunification, the mining jobs vanished. "This region was known as the dirtiest in all of Europe," said Juergen Frisch, mayor of Espenhain. "The solar plant came at a very good time for Espenhain. It's helped to change our image," Frisch added. Unlike the coal mines, the solar plant makes almost no noise, save for the low thrum of a few outdoor air-conditioning units that cool the electrical transformers. The plant, with 33,500 solar panels, sits on a 37-acre site in a field off a rural road and requires scarcely any maintenance (Whitlock 2007).

German law, enacted in 2000, requires the country's utility companies to subsidize new solar installations by buying their electricity and using it on their grids at marked-up rates that allow small solar enterprises to earn profits. Wind and biofuels have preferred status under German law. As the world's sixth-biggest producer of carbon-dioxide emissions, Germany is trying to slash its output of greenhouse gases and wants renewable sources to supply a quarter of its energy needs by 2020 (Whitlock 2007). Germany also has decided to phase out all nuclear power plants by 2020.

During 2006, German exports amounted to 15 percent of worldwide sales of solar panels and other photovoltaic equipment, according to industry officials. German companies hope to double their share of the global market, which amounted to $9.5 billion in 2006 and is growing by about 20 percent annually, said Carsten Koernig, managing director of the German Solar Industry Association (Whitlock 2007). "It's been very important to create the necessary market in Germany," Koernig said. "We not only want to master the German market, but to conquer the world market as well." For now, the technology remains expensive and barely registers as a fraction of total energy production—less than 0.5 percent. The government hopes to increase that figure to 3 percent by 2020 (Whitlock 2007).

In northwestern Germany, Freiburg has been working to become the world's first "solar city," with a "solar-powered train station, energy-efficient row houses, innumerable rooftop photovoltaic systems and, high on a hill overlooking the vineyards, the world-famous Heliotrop, a high-tech cylindrical house that rotates to follow the sun" (Roberts 2004, 192).

Freiberg is home to the Fraunhofer Institute for Solar Energy Systems, where scientists have been seeking breakthrough research that will reduce the cost of photovoltaic solar energy to competitive levels with fossil fuels and nuclear power, including a "multilayer" photovoltaic cell with an efficiency of 40 percent, twice present levels. Germany and Japan have implemented tax subsidies for rooftop solar energy systems. Given trends in research and cost reduction, solar energy in regions with abundant sunshine (such as the Middle East, the Mediterranean, and the U.S. Southwest) may be cost-effective before the year 2010.

Solar Power a Priority in Portugal

In Portugal by 2007, a new solar array spread across 150 hilly acres near Serpa, about 120 miles southeast of Lisbon. This project was old-fashioned photovoltaic, constructed by General Electric Energy Financial Services and PowerLight Corp. of the United States in partnership with the Portuguese company Catavento. "This is the most productive [largest capacity] solar plant in the world," its backers asserted. "It will produce 40 percent more energy than the second-largest one, Gut Erlasse in Germany," said Howard Wenger, speaking for Power Light Corp. ("Largest-producing Solar" 2007).

Unlike frequently misty Germany, southern Portugal is among the sunniest places in Europe, with as much as 3,300 hours of sunlight annually, nine hours during an average day. This solar array, when finished, produced enough power to supply 8,000 homes. It replaced 30,000 tons worth of annual greenhouse-gas emissions ("Portugal Celebrates" 2007). This plant's photovoltaic system uses silicon solar cell technology to convert sunlight directly into electricity, producing 20 gigawatts per year. "This project is successful because Portugal's sunshine is plentiful, the solar power technology is proven, government policies are supportive, and we are investing … to help our customers meet their environmental challenges," said Kevin Walsh, managing director and leader of renewable energy at GE Energy Financial Services ("Portugal Celebrates" 2007).

Portugal, which depends almost entirely on imported energy, has been developing wave and solar power projects, as well as wind farms, with plans to supply about 750,000 homes before 2010. Portugal has been exploring new hydropower projects and plans to invest 8 billion euros (US$10.8 billion) in renewable energy projects within five years. Socialist Prime Minister Jose Socrates anticipated during January 2007 that 45 percent of Portugal's will come from renewable sources by 2010 ("Portugal Celebrates" 2007).

Businesses and Public Facilities Go Solar

Corporations and retail stores seized on solar power in 2007 and 2008 as a money-saving strategy. Sharp, the television maker, has equipped its Kameyama (Japan) plant, with solar panels and window film capable of producing 5.2 megawatts. Google began building a rooftop solar-powered generation system at its Mountain View, California, headquarters generating 1.6 megawatts, enough to power about 1,000 California homes.

By 2008, Google was using solar power for almost a third of the electricity consumed by office workers at its headquarters, excluding power consumed by data centers that power many of Google's Web services worldwide. Data centers usually consume about 10 times more electricity than buildings that house office workers. "We are going to be producing roughly 30 percent of the power that we use," David Radcliffe, vice president of real estate at Google, told Reuters in an interview. "This is for our corporate-office people center," he said (Auchard and Anderson 2006).

In 2008 the Fresno Yosemite International Airport, began to install the largest solar electric project at any airport in the United States, said Fresno's Mayor Alan Autry. At about the same time, the Hawaii Department of Transportation Airports Division solicited proposals from private companies to develop photovoltaic solar power systems within two years.

FURTHER READING

Abboud, Leila. "Sun Reigns on Spain's Plains: Madrid Leads a Global Push to Capitalize on New Solar-power Technologies." *Wall Street Journal*, December 5, 2006, A-4.

Auchard, Eric, and Leonard Anderson. "Google Plans Largest U.S. Solar-Powered Office." Reuters in *Washington Post*, October 16, 2006. http://www.washingtonpost.com/wp-dyn/content/article/2006/10/16/AR2006101601100_pf.html.

Feder, Barnaby J. "The Showhouse That Sustainability Built." *New York Times*, March 26, 2008. http://www.nytimes.com/2008/03/26/business/businessspecial2/26boulder.html.

Feresin, Emiliano. "Europe Looks to Draw Power from Africa." *Nature* 450 (November 29, 2007):595.

Frosch, Dan. "Citing Need for Assessments, U.S. Freezes Solar Energy Projects." *New York Times*, June 27, 2008, A-13.

Kaplow, Larry. "Solar Water Heaters: Israel Sets Standard for Energy." *Atlanta Journal-Constitution*, August 5, 2001, 1-P.

Kintisch, Eli. "Light-Splitting Trick Squeezes More Electricity Out of Sun's Rays." *Science* 317 (August 3, 2007c):583–584.

Kirchner, Stephanie. "Be Aggressive about Passive [Solar Power]." *TIME*, April 9, 2007. http://www.time.com/time/printout/0,8816,1603747,00.html.

Landauer, Robert. "Big Changes in Our China Suburb." *Sunday Oregonian*, October 20, 2002, F-4.

"Largest-producing Solar Power Plant Inaugurated." *Kuwait Times*, March 29, 2007. http://www.kuwait-times.net/read_news.php?newsid=MTU0ODkxMzk3Mw.

Mallaby, Sebastian. "Carbon Policy That Works: Avoiding the Pitfalls of Kyoto Cap-and-Trade." *Washington Post*, July 23, 2007, A-17. http://www.washingtonpost.com/wpdyn/content/article/2007/07/22/AR2007072200884_pf.html.

Maloney, Peter. "Solar Projects Draw New Opposition." *New York Times* (Business of Green), September 24, 2008, H-2.

O'Connell, Sanjida. "Power to the People." *London Times*, May 20, 2002, n.p. (LEXIS)

Oppenheimer, Michael, and Robert H. Boyle: *Dead Heat: The Race against the Greenhouse Effect.* New York: Basic Books, 1990.

"Portugal Celebrates Massive Solar Plant" Associated Press in *New York Times*, March 28, 2007. http://www.nytimes.com/aponline/technology/AP-Portugal-Solar-Power-Plant.html.

Revkin, Andrew C. "Budgets Falling in Race to Fight Global Warming." *New York Times*, October 30,

2006k. http://www.nytimes.com/2006/10/30/business/worldbusiness/30energy.html.

Revkin, Andrew C., and Matthew L. Wald. "Solar Power Captures Imagination, Not Money." *New York Times*, July 16, 2007b. http://www.nytimes.com/2007/07/16/business/16solar.html.

Roberts, Paul. *The End of Oil: The Edge of a Perilous New World*. Boston: Houghton-Mifflin, 2004.

Rogers, Paul. "Solar Energy Heats Up." *Omaha World-Herald*, October 15, 2006, 1-RE, 2-RE.

Rosenbloom, Stephanie. "Giant Retailers Look to Sun for Energy Savings." *New York Times*, August 11, 2008. http://www.nytimes.com/2008/08/11/business/11solar.html.

Service, Robert F. "Solar Power: Can the Upstarts Top Silicon?" *Science* 319 (February 8, 2008a):718–720.

Smith, O. Glenn. "Harvest the Sun—from Space." *New York Times*, July 23, 2008. http://www.nytimes.com/2008/07/23/opinion/23smith.html.

Wald, Matthew. "What's So Bad about Big?" *New York Times*, March 7, 2007a. http://www.nytimes.com/2007/03/07/business/businessspecial2/07big.html.

Wald, Matthew L. "Turning Glare into Watts." *New York Times*, March 6, 2008a. http://www.nytimes.com/2008/03/06/business/06solar.html.

Wald, Matthew L. "Two Large Solar Plants Planned in California." *New York Times*, August 15, 2008d. http://www.nytimes.com/2008/08/15/business/15solar.html.

Whitlock, Craig. "Cloudy Germany a Powerhouse in Solar Energy." *Washington Post*, May 5, 2007, A-1. http://www.washingtonpost.com/wp-dyn/content/article/2007/05/04/AR2007050402466_pf.html.

Soot: A "Wild Card" in Global Warming

Atmospheric soot is more plentiful worldwide than most scientists had previously thought. It also is contributing to the rapid heating of the Earth's atmosphere, according to an increasing body of research. Microscopic carbon particles in air pollution have long been linked to respiratory ailments. Scientists for several years have been improving their understanding of how smoke in the air interacts with sunlight and chemicals to influence global warming (Polakovic 2003, A-30).

"Soot is the wild card," said Stanford University climatologist Stephen Schneider. "And this stuff can get pretty uncertain. It's very complex. It could have any number of effects." Among the outcomes, soot particles might help form clouds that reflect sunlight, or concentrate in clouds, heat them up and evaporate them to contribute to global warming. Also, the effect depends on how high soot accumulates in the atmosphere and the landscape below it, Schneider said (Polakovic 2003, A-30).

Dirty snow containing even small amounts of soot (measured in parts per billion) may be responsible for as much as a quarter of recent temperature rises in polar regions, according to NASA research. James Hansen and Larissa Nazarenko, climate specialists at the Goddard Institute for Space Studies in New York City, said that even small amounts of soot contained in fossil-fuel effluvia (most notably diesel exhausts) absorb more sunlight and inhibit the reflection of light and its attendant heat back into space.

Soot causes snow to melt more quickly, eventually contributing to rising sea levels, Hansen and Nazarenko said in the *Proceedings of the National Academy of Sciences* (Hansen and Nazarenko 2004, 423–428). Before this study, soot usually had not been factored into climate models projecting global warming's speed and scope.

By 2008, scientists were quantifying the role of black carbon in global warming, and finding that the Intergovernmental Panel on Climate Change (IPCC) assessments had underestimated it. While the IPCC's fourth assessment (2007) had estimated that soot warms the atmosphere by 0.2 to 0.4 watts per square meter, work by Veerabhadran Ramanathan of the Scripps Institution of Oceanography and Gregory Carmichael of the University of Iowa put that value at 0.9 watts per square meter, compared with 1.66 watts per square meter for carbon dioxide. This would make soot the second most important contributor to greenhouse forcing, ahead of methane and other trace gases.

Human-created black carbon may be injected into the air from burning of grasslands in Africa and rainforests in Brazil, diesel emissions from trucks, cooking fires using coal (mainly in China), and the use of cow dung as an energy source in India. These sources are so prevalent that they have cast a brown cloud of pollution across much of southern and eastern India. Unlike carbon dioxide, however, a given quantity of soot remains resident in the atmosphere for only a few days before settling to Earth. Carbon dioxide can remain in the air for a century or much longer. Thus, once point sources are reduced, the contribution of soot to warming could be reduced relatively rapidly (Service 2008b, 1745).

The "Dirty Snow" Effect

Hansen and colleagues asserted that soot from diesel engines and burned wood should be reduced before it reaches the Earth's large snow packs. The "dirty snow effect," as Hansen calls it, could add a new wrinkle to the debate over global warming and its causes. "I think that this is an important climate [forcing] that has been overlooked," Hansen said. "I searched through the several thousand pages of past IPCC documents. It is never mentioned" (Nesmith 2003b, 3-A). "In the developing world, it is more a matter of replacing biofuels with cleaner fuels or burning them more cleanly," he said. "One of the most effective ways for them to clean up the air is electrification, so that fuels are burned at a power plant where it is easier to eliminate emissions." Hansen added that "the bottom line is that the technologies exist for both developed and developing countries to greatly reduce emissions, but there needs to be a concerted effort to do that" (Nesmith 2003b, 3-A).

Hansen and Nazarenko commented that

> For a given forcing it [soot] is twice as effective as CO_2 in altering global surface air temperature. This indirect soot forcing may have contributed to global warming of the past century, including the trends toward early springs in the Northern Hemisphere, thinning arctic sea ice, and melting land ice and permafrost.... Reducing soot emissions, thus restoring snow albedos to pristine high levels, would have the double benefit of reducing global warming and raising the global temperature at which dangerous anthropogenic interference occurs. (Hansen and Nazarenko 2004, 423)

According to an account by Gary Polakovic in the *Los Angeles Times*,

> Understanding airborne soot is critical to determining how Earth's climate has been altered since the Industrial Revolution and how it might change in the future. About 1 million tons of so-called black carbon is floating above the Earth. About 40 percent comes from fossil-fuel power plants and factories and diesel-powered vehicles in industrialized countries. (Polakovic 2003, A-30)

The rest of the atmosphere's load of soot comes from smoke from forest burning or stoves using wood or dung to cook and heat homes in developing nations.

Additionally, an increasing amount of airborne soot over southern Asia alters monsoon precipitation patterns and influences climate around the world, including "large warming over northern Africa and cooling over the southern United States, all superimposed upon a more general global-scale warming" (Chameides and Bergin 2002, 2214).

Soot's Effects May Diminish

Theodore Anderson and colleagues wrote in *Science* that while aerosols presently act as a damper on global warming, their effect will diminish in the future:

> Although even the sign of the current total forcing is in question, the sign of the forcing by the middle of the 21st century will certainly be positive. The reason is that GHGs [greenhouse gases] accumulate in the atmosphere, whereas aerosols do not. Even if the most negative value of aerosol forcing shown in the figures turns out to be correct, the current range of plausible emissions scenarios indicates that GHG forcing will exceed aerosol forcing somewhere between 2030 and 2050. Thus, despite current uncertainties, forward calculations lead to the unambiguous conclusion that anthropogenic activity will inevitably result in a strong, positive forcing of Earth's climate system. (Anderson et al. 2003, 1104)

Surabi Menon of NASA and Columbia University joined with James Hansen and colleagues, writing in *Science* that their climate models associate soot from diesel engines—and incomplete burning of such fuels as wood, crop residue, cow dung, and other sources, mainly in Asia—with several local climatic trends in China and India. These trends include

> increased summer flooding in south China, increased drought in north China, and moderate cooling in China and India while the rest of the world has been warming [on average].... We found precipitation and temperatures changes in the model that were comparable to those observed if the aerosols included a large proportion of absorbing black carbon (soot), similar to observed amounts. Absorbing aerosols heat the air, alter regional atmospheric stability and vertical motions, and affect the large-scale circulation and hydrological cycle with significant regional climate effects. (Menon et al. 2002, 2250)

Unlike carbon-dioxide emissions, which add to global warming by trapping heat in the atmosphere, soot emissions may contribute to global warming and climate change by absorbing sunlight, heating the air and making the atmosphere more unstable, according to this study (Heilprin 2002).

The effects of soot in China also may be intensified by overfarming, overgrazing, and destruction of forests. In northern China, where dust storms have become frequent in Beijing and nearby areas, "with adhered toxic contaminants ... [that are] cause for public-health concern" (Menon et al. 2002, 2250). Some residual dust from these storms has been detected as far away as North America. In addition, according to sources cited in this research, soot aerosols "are carcinogenic and ... a major cause of deaths associated with particulate air pollution" (Menon et al. 2002, 2250).

Reducing the amount of black carbon or soot may help diminish the intensity of floods in the south and droughts in the northern areas of China, in addition to having human health benefits. The research, based on data from Chinese ground stations provided by Yunfeng Luo of the Chinese Academy of Sciences, may determine whether a similar pattern exists in India (Heilprin 2002). Menon and Hansen used a climate computer model developed by NASA and pollution data from 46 monitoring stations in China. They used computer models to run four simulations estimating the effects of soot on precipitation cycles over China. In addition to its effects in China, the computer simulation also showed "a pattern of larger warming over northern Africa and cooling over the southern United States from the soot, against a background of overall increases in global temperatures" (Bowman 2002). *See also:* Aerosols and Climate Change

FURTHER READING

Anderson, Theodore L., Robert J. Charlson, Stephen E. Schwartz, Reto Knutti, Olivier Boucher, Henning Rodhe, and Jost Heintzenberg. "Climate Forcing by Aerosols—a Hazy Picture." *Science* 300 (May 16, 2003):1103–1104.

Andreae, Meinrat O. "The Dark Side of Aerosols." *Nature* 409 (February 8, 2001):671–672.

Bowman, Lee. "Soot Could Be Causing a Lot of Bad Weather." Scripps Howard News Service, September 26, 2002. (LEXIS)

Chameides, William L., and Michael Bergin. "Climate Change: Soot Takes Center Stage." *Science* 297 (September 27, 2002):2214–2215.

Hansen, James, and Larissa Nazarenko. "Soot Climate Forcing via Snow and Ice Albedos." *Proceedings of*

the National Academy of Sciences 101, no. 2 (January 13, 2004):423–428.

Heilprin, John. "Study Says Black Carbon Emissions in China and India Have Climate Change Effects." Associated Press, September 26, 2002. (LEXIS)

Jacobson, Mark. "Strong Radiative Heating Due to the Mixing State of Black Carbon in Atmospheric Aerosols." *Nature* 409 (February 8, 2001):695–697.

Menon, Surabi, James Hansen, Larissa Nazarenko, and Yunfeng Luo. "Climate Effects of Black Carbon Aerosols in China and India." *Science* 297 (September 27, 2002):2250–2253.

Nesmith, Jeff. "Dirty Snow Spurs Global Warming: Study Says Soot Blocks Reflection, Hurries Melting." *Atlanta Journal-Constitution*, December 23, 2003b, 3-A.

Polakovic, Gary. "Airborne Soot Is Significant Factor in Global Warming, Study Says." *Los Angeles Times*, May 15, 2003, A-30.

Sato, Makiko, James Hansen, Dorothy Koch, Andrew Lacis, Reto Ruedy, Oleg Dubovik, Brent Holben, Mian Chin, and Tica Novakov. "Global Atmospheric Black Carbon Inferred from AERONET." *Proceedings of the National Academy of Sciences* 100, no. 11 (May 27, 2003):6319–6324.

Service, Robert F. "Study Fingers Soot as Major Player in Global Warming." *Science* 319 (March 28, 2008b):1745.

Spruce Bark Beetles in Alaska

On Alaska's Kenai Peninsula, a forest nearly twice the size of Yellowstone National Park has been dying. "Century-old spruce trees stand silvered and cinnamon-colored as they bleed sap," from spruce bark beetle infestations spurred by rising temperatures, wrote reporter Tim Egan of the *New York Times* (Egan 2002a, A-1). Temperatures hit the middle 80s there by mid-May 2002, as park rangers worried that tinder-dry, warm conditions could provoke major wildfires. During 15 years (1988–2003), 40 million spruce trees on the Kenai Peninsula have died. The beetle infestations have nearly reached Anchorage, where "[v]isitors flying into the city's airport cross islands covered with the bristling, white skeletons of dead trees that are easily visible through the plane windows" (Lynas 2004a, 60).

Alaskan author Charles Wohlforth described a plague of bark beetles:

On certain spring days in the mid-1990s, clouds of spruce bark beetles took flight among the big spruce trees around Kachemak Bay, 120 miles south of Anchorage. They could be seen from miles away, rolling down the Anchor River Valley. People who witnessed the arrival sometimes felt like they were in a horror film, the air thick with beetles landing in their eyes and catching in their hair, and knew when it happened that their trees were destined to turn red and die. (Wohlforth 2004, 238)

The six-legged spruce beetle, which is about a quarter-inch long, takes to the air in the spring, looking for trees on which to feed. When beetles find a vulnerable group of trees, they will signal to other beetles "a chemical message," Holsten said. They then burrow under the bark, feeding on woody capillary tissue that the tree uses to transport nutrients. Healthy spruce trees produce chemicals (terpenes) that usually repel beetles. The chemicals cannot overwhelm a mass infestation of the type that has been taking place, however (Egan 2002b, F-1). As a spruce dies, green needles turn red, and then silver or gray. According to Egan's account, "Ghostly stands of dead, silver-colored spruce—looking like black and white photographs of a forest—can be seen throughout south-central Alaska, particularly on the Kenai. Scientists estimate that 38 million spruce trees have died in Alaska in the current outbreak" (Egan 2002b, F-1).

More than four million acres of white spruce trees on the Kenai Peninsula were dead or dying by 2004 from an infestation of beetles, the worst devastation by insects of any forest in North America. Beetles have been gnawing at spruce trees in Alaska for many thousands of years, but with rapid warming since the 1980s their populations have exploded (Egan 2002b, F-1). A series of warmer-than-average years in Alaska allowed the spruce bark beetles to reproduce at twice their historic rate. "Hungry for the sweet lining beneath the bark," wrote Egan, "[t]he beetles have swarmed over the stands of spruce, overwhelming the trees' normal defense mechanisms.... The dead spruce forest of Alaska may well be one of the world's most visible monuments to climate change." By 2002, nearly 95

percent of the spruce on the Kenai Peninsula had been destroyed by the beetles, leaving a tinderdry forest ripe for major wildfires that will further devastate the habitats of resident moose, bear, salmon, and other creatures (Egan 2002b, F-1).

"The chief reason why the beetle outbreak has been the largest and the longest is that we have had an unprecedented run of warm summers," said Edward Berg, 62, a long-time student of the Kenai Peninsula ecosystem (Egan 2002b, F-1). Berg has been piecing together the causes of the forest's demise. "His [Berg's] work is very convincing; I would even say unimpeachable," said Glenn Juday, a forest ecologist at the University of Alaska. "For the first time, I now think beetle infestation is related to climate change" (Egan 2002b, F-1). Ed Holsten, who studies insects for the Forest Service in Alaska, said he thinks that climate change is only one reason for the beetle outbreak. The trees on the Kenai are old and ripe for beetle outbreaks, he said. If they had been logged, or burned in fire, it might have kept the bugs down, Holsten said (Egan 2002b, F-1).

"It's very hard to live among the dead spruce; it's been a real kick in the teeth," said Berg. "We all love this beautiful forest" (Egan 2002b, F-1). *See also:* Bark Beetles Spread across Western North America; Pine Beetles in Canada

FURTHER READING

Egan, Timothy. "Alaska, No Longer so Frigid, Starts to Crack, Burn, and Sag." *New York Times*, June 16, 2002a, A-1.

Egan, Timothy. "On Hot Trail of Tiny Killer in Alaska." *New York Times*, June 25, 2002b, F-1.

Lynas, Mark. *High Tide: The Truth about Our Climate Crisis.* New York: Picador/St. Martins, 2004a.

Wohlforth, Charles. *The Whale and the Supercomputer: On the Northern Front of Climate Change.* New York: North Point Press/Farrar, Strauss and Giroux, 2004.

T

Tanganyika, Lake: Warming Waters Choke Life

Two independent teams of scientists studying central Africa's Lake Tanganyika, Africa's second-largest body of fresh water, have found a microcosm of crisis regarding global warming. The scientists have found that warming at the lake's surface has impaired mixing of nutrients, reducing its population of fish. These reductions have affected the local economy as fishing yields fell by a third or more during 30 years, with more declines anticipated. Heretofore, Lake Tanganyika's fish had supplied 25 to 40 percent of the protein consumed by neighboring peoples in parts of Burundi, Tanzania, Zambia, and the Democratic Republic of Congo.

Lake Tanganyika is a tropical body of water that experiences relatively high temperatures year-round, so the scientists were rather surprised to discover that further warming affected its nutrient balance so significantly. Like other deep-water lakes, however, Tanganyika's waters utilize temperature differences at various depths to mix water and nutrients. Such mixing is critical in tropical lakes with sharp temperature gradients that stratify layers of water, with warm, less-dense layers on top of nutrient-rich waters below.

"Climate warming is diminishing productivity in Lake Tanganyika," Catherine M. O'Reilly and colleagues wrote. "In parallel with regional warming patterns since the beginning of the twentieth century," they continued, "a rise in surface-water temperature has increased the stability of the water column" (O'Reilly et al. 2003, 766).

A regional decrease in average wind speed over the lake has contributed to reduced mixing of the 1,470-meter-deep lake, "decreasing deep-water nutrient upwelling and entrainment into surface waters" (O'Reilly et al. 2003, 766). Fish yields have declined roughly 30 percent, the scientists wrote, in an example "that the impact of regional effects of global climate change on aquatic ecosystem functions can be larger than that of local anthropogenic activity or overfishing" (O'Reilly et al. 2003, 766).

Lake Tanganyika is especially vulnerable because year-round tropical temperatures accelerate biological processes, "and new nutrient inputs from the atmosphere or rock weathering cannot keep up with the high rates of algal photosynthesis and decomposition" (Verschuren 2003, 733).

Lake Tanganyika is the second-deepest lake in the world and the second richest in terms of biological diversity; it has at least 350 species of fish, with new ones being discovered regularly. Nutrient mixing has been vital for its biodiversity (Connor 2003c). Piet Verburg, of the University of Waterloo, in Canada, and O'Reilly of the University of Arizona, who led the studies, found that warmer temperatures and less windy weather in the region has been starving the lake's life of essential salts that contain nitrogen and sulfur (Verburg, Hecky, and Kling 2003, 505–507). Verburg and colleagues utilized profiles of temperature changes in the lake between 110 and 800 meters, and found that that degree of temperature stratification has tripled since 1913.

O'Reilly and colleagues, writing in *Nature*, suggested that the lake's productivity, measured by the amount of photosynthesis, has diminished by 20 percent, which could easily account for the 30 percent decline in fish yields. The scientists said that climate change, rather than overfishing, was mainly responsible for the collapse in Tanganyika's fish stocks. With additional warming, fish populations in the lake are expected by the scientists to decline further (Connor 2003c).

"The human implications of such subtle, but progressive, environmental changes are potentially dire in this densely populated region of the world, where large lakes are essential natural resources for regional economies," the scientists said. Dirk Verschuren, a freshwater biologist at Ghent University in Belgium, said that both studies could explain why sardine catches in Lake Tanganyika have declined between 30 and 50 percent since the late 1970s (Verschuren 2003, 733). "Since overexploitation is at most a local problem on some fishing grounds, the principal cause of this decline has remained unknown," Verschuren said.

> Taken together ... the data in the two papers provide strong evidence that the effect of global climate change on regional temperature has had a greater impact on Lake Tanganyika than have local human activities. Their combined evidence covers all the important links in the chain of cause and effect between climate warming and the declining fishery. (Connor 2003c)

FURTHER READING

Connor, Steve. "Global Warming Is Choking the Life out of Lake Tanganyika." *London Independent*, August 14, 2003c, n.p. (LEXIS)

O'Reilly, Catherine M., Simone R. Alin, Pierre-Denis Plisnier, Andrew S. Cohen, and Brent A. McKee. "Climate Change Decreases Aquatic Ecosystem Productivity of Lake Tanganyika, Africa." *Nature* 424 (August 14, 2003):766–768.

Verburg, Piet, Robert E. Hecky, and Hedy Kling. "Ecological Consequences of a Century of Warming in Lake Tanganyika." *Science* 301 (July 25, 2003):505–507.

Verschuren, Dirk. "Global Change: The Heat on Lake Tanganyika." *Nature* 424 (August 14, 2003):731–732.

Tar Sands

As petroleum becomes scarcer, and more expensive, fossil fuels that are even higher in greenhouse-gas emissions are coming to the fore—including tar sands (principally in Alberta, Canada), liquid fuels distilled from coal, and oil shale. The mining of tar sands and oil shales disfigure the Earth; they are a form of strip mining that is nearly impossible to restore. Coal emits roughly twice the greenhouse gases of oil once it is manufactured into a usable fuel—more than that, if one counts the energy necessary to transport and market it.

By 2012, the mining of tar sands in Canada may consume as much natural gas as the country

uses for home heating (Kolbert 2007c, 49). The gas is used to produce synthetic oil and products from it, such as gasoline. Tar sands require about 15 to 40 percent more energy in manufacture compared with conventional crude oil; oil shales require about twice as much. Converting tar sands into oil costs about as much as oil at $30 a barrel, however. With oil pushing $100 a barrel in late 2007, tar sands were becoming profitable for many fossil-fuel companies (Kolbert 2007c, 49, 50).

Under present accounting practices, greenhouse-gas emissions and damage to the land are regarded as "externalities." They do not count.

FURTHER READING

Kolbert, Elizabeth. "Unconventional Crude: Canada's Synthetic-fuels Boom." *The New Yorker*, November 12, 2007c, 46–51.

Temperatures, Cold Spells

Early in 2000, an Associated Press photograph in the *Omaha World-Herald* showed a group of Muslim worshippers bowing toward Mecca from the steps of a mosque in Jerusalem that had just been cleared of a very rare 15-inch snowstorm ("Mideast Snow" 2000, 4). Eight years later, nearly to the day, snow again lay atop mosques and temples in Jerusalem, as the *New York Times* carried a front-page photograph of Arabs in full traditional dress throwing snowballs at each other in Jordan ("A Snowy Day" 2008, A-1). Unusual snowstorms in Shanghai and other cities in southern China paralyzed that country's rail and airline systems for several days during January 2008. Climate contrarians raved at reports the previous winter of unusual (if short-term) snowfalls in Buenos Aires, Argentina, and Johannesburg, South Africa, as evidence that global warming had been staunched. Perhaps a new ice age was on the way, some asserted.

Weather is still variable, and even as many reports indicate a warming Earth on average, specific episodes of record cold and snowfall still occur. While it snowed in Jordan early in 2008, however, New York City was reporting its first January in recorded history with no snow, and Stockholm, Sweden, was experiencing a rainy winter.

In point of meteorological fact, having experienced its warmest January through October on

Two Orthodox Jews wrapped in prayer shawls stand in freshly-fallen snow as they pray at the Western Wall, Judaism's holiest site, in Jerusalem's Old City January 28, 2000, after a rare snowstorm that dumped 40 centimeters (15 inches) of snow in the area. (AP/Wide World Photos)

record in the year 2000, November and December in the United States were the *coldest* months on record. By the first day of winter (December 21), Omaha had its snowiest December, at 20.4 inches, surpassing the old record, set in 1969, of 19.9 inches. Chicago got twice as much the same

month. In Omaha, the temperature for the month averaged 9.5° F below average.

Throughout the winter, this paradoxical season continued to build its resume, reminding everyone that, although climate may be the plot, weather is the story line. By February 1, 2001,

Omaha had its average annual snowfall of 32.5 inches, to which was added another eight inches in a blizzard eight days later—another day, another snow storm and cold wave, and yet another emphatic reminder that the Earth is not traveling a straight line back to the climatic days of the dinosaurs—not just yet.

The Cold Spells of 2007–2008

Marc Morano, the communications director for the Republican minority on the Senate Environment and Public Works Committee, an assistant to Oklahoma Senator (and long-time ardent contrarian) James Inhofe, went so far as to brag on his blog that global warming was now over: "Earth's 'Fever' Breaks: Global COOLING Currently Under Way" (Revkin 2008b).

Most climatologists said this was regular weather in a cool, La Niña phase. Many scientists said that these events did not undermine the longer warming trend.

"The current downturn is not very unusual," said Carl Mears, a scientist at Remote Sensing Systems, a private research group in Santa Rosa, California. "Temperatures are very likely to recover after the La Niña event is over," he said (Revkin 2008b). "Climate skeptics typically take a few small pieces of the puzzle to debunk global warming, and ignore the whole picture that the larger science community sees by looking at all the pieces," said Ignatius G. Rigor, a climate scientist at the Polar Science Center of the University of Washington in Seattle (Revkin 2008b).

James E. Hansen examined the temperature record and found the winter of 2007–2008 rather average for a strong La Niña,

> to expose the recent nonsense that has appeared in the blogosphere, to the effect that recent cooling has wiped out global warming of the past century, and the Earth may be headed into an ice age. On the contrary, these misleaders have foolishly (or devilishly) fixated on a natural fluctuation that will soon disappear. (Hansen 2008b)

Hansen commented:

> The past year (2007) witnessed a transition from a weak El Niño to a strong La Niña (the latter is perhaps beginning to moderate already, as the ocean waters near Peru are beginning to warm). January 2007 was the warmest January in the period of instrumental data in the GISS analysis, while, October 2007 was Number 5 warmest, November 2007

was 8th warmest, December 2007 was 8th warmest, and January 2008 was 40th warmest. (Hansen 2008b)

NASA's Goddard Institute for Space Studies, the World Meteorological Organization, the National Oceanic and Atmospheric Administration, and Britain's Hadley Center determined that 2008 was between the 7th and 12th warmest since systematic meteorological record-keeping began in 1880—not as warm as most immediately prior years, but warmer than average for a period during which the Pacific Ocean experienced a strong, cool phase of the Pacific Decadal Oscillation, which usually lasts 5 to 20 years. The nine warmest years in the record, according to GISS, have occurred since 1998. The Arctic continued in a marked warming trend. During most of the 1980s and 1990s, the Pacific experienced the oscillation's warm phase. During December 2008, the convergence of a La Niña and a Pacific Decadal Oscillation, both cooling influences in the Pacific Ocean, helped produce snowfall in some unusual places, including New Orleans, Louisiana, and Las Vegas, Nevada. *See also:* Contrarians (Skeptics)

FURTHER READING

Hansen, James. Cold Weather. Accessed March 3, 2008b. http://www.columbia.edu/~jeh1/mailings/20080303_ColdWeather.pdf.

"Mideast Snow Disrupts Life, Prayers." *Omaha World-Herald*, January 29, 2000, 4.

Revkin, Andrew C. "Skeptics on Human Climate Impact Seize on Cold Spell." *New York Times*, March 2, 2008b. http://www.nytimes.com/2008/03/02/science/02cold.html.

"A Snowy Day in—Jordan?" *New York Times*, February 1, 2008, A-1.

Temperatures, Recent Warmth

While no single temperature reading "proves" that the Earth's atmosphere is undergoing a sustained warming spell caused by human activities, several hundred readings over many years establish evidence that indicates a trend. Averages as well as daily temperature reports indicate that trend. According to an analysis by the Goddard Institute for Space Studies, the year 2007 tied for second warmest in the period of instrumental data, behind the record warmth of 2005. The year 2007 tied 1998, in a remarkably different climatic context. First, temperatures in 1998 leapt 0.2° C above the prior record, a very steep increase, aided by the most intense El Niño conditions in a century. Nine years later, a La Niña condition was in sway, which usually cools world temperatures (2008, also a La Niña year, was slightly cooler worldwide).

The unusual warmth in 2007 also was noteworthy because it occurred as solar irradiance was at a minimum in the recent 10- to 11-year cycle. Thus, the 2007 averages, according to the Goddard Institute for Space Studies (GISS), "continue the strong warming trend of the past thirty years that has been confidently attributed to the effect of increasing human-made greenhouse gases" ("GISS 2007 Temperature Analysis" 2008).

The GISS summary rebutted assertions by contrarians that warming has been stopped in its tracks:

> Because both of these natural effects were in their cool phases in 2007, the unusual warmth of 2007 is all the more notable. It is apparent that there is no letup in the steep global warming trend of the past 30 years. "Global warming stopped in 1998" has become a recent mantra of those who wish to deny the reality of human-caused global warming. The continued rapid increase of the five-year running mean temperature exposes this assertion as nonsense. In reality, global temperature jumped two standard deviations above the trend line in 1998 because the "El Niño of the century" coincided with the calendar year, but there has been no lessening of the underlying warming trend. ("GISS 2007 Temperature Analysis" 2008)

Record Warmth in 2006 and 2007

The winter of 2006–2007 was the warmest on record in the United States; 11 of the last 12 previous years on the world temperature record also had been the warmest on record since 1880. In August 2007, Atlanta, Georgia, saw five of its 10 hottest days on record during one heat wave. That same summer, Phoenix, Arizona, set a record for number of days above 110° F (32 days).

The world experienced its warmest period on record during the 2006–2007 Northern Hemisphere winter, according to the National Oceanic and Atmospheric Administration (NOAA). The NOAA report said the globally averaged combined land and sea-surface temperature for December to February was the highest since records began in 1880. During that period, temperatures worldwide were above average, except in Saudi Arabia, Iraq, and a few areas in the central United States ("World Breaks" 2007). The global average was 0.72° C higher than the previous record in 2004, considered to be a very large amount ("World Breaks" 2007).

In late December 2006, bears at the Moscow Zoo, in Russia, which usually begin to hibernate in early November, finally fell asleep after weeks of insomnia resulting from record mild temperatures. The season was Moscow's warmest on record.

With unusual warmth in Siberia, Canada, northern Asia, and Europe, the world's land areas were 3.4° F warmer than an average during the following January, according to the U.S. National Climatic Data Center in Asheville, North Carolina. That didn't just nudge past the old record set in 2002, but it broke that mark by 0.81° F.

In meteorological terms, such records usually are broken by hundredths of a degree (Borenstein 2007b). "That's pretty unusual for a record to be broken by that much," said the data center's scientific services chief, David Easterling. "I was very surprised." Scientists took an unusual step of running computer climate models "just to make sure that what we're seeing was real," Easterling said. "From one standpoint it is not unusual to have a new record because we've become accustomed to having records broken," said Jay Lawrimore, climate monitoring branch chief. But January, he said, was a bigger jump than the world has seen in about 10 years (Borenstein 2007b). Siberia was 9° F warmer than average, as Eastern Europe' temperatures rose 8° F above average, and Canada averaged more than 5° F degrees higher.

Flowers in January

With temperatures broaching 73° F in Washington, D.C., during the first week of January 2007, some scientists ventured that the year might become the warmest in the instrumental record with the aid of a brief El Niño episode. Joel Achenbach of the *Washington Post* wrote:

> Never has good weather felt so bad. Never have flowers inspired so much fear. Never has the warm caress of a sunbeam seemed so ominous. The weather is sublime, it's glorious, it's *the end of the world*. January is the new March. The daffodils are busting out everywhere. It's porch weather. Put on a T-shirt and shorts, fire up the grill.... Everyone

out for volleyball! The normal high for this time of year is 43 degrees; yesterday's high at Reagan National was a record-breaking 73. And yet it's all a guilty pleasure. Weather is both a physical and a psychological phenomenon. Meteorology, meet eschatology. We've read the articles, we've seen the Gore movie, we've calculated our carbon footprint, and we're just not intellectually capable anymore of fully enjoying warm winter weather.... Greenland melting faster than the Wicked Witch of the West. Just ain't right. Ain't natural. Cherry blossoms during the NFL playoffs? *Run for your lives.* (Achenbach 2007)

Winter was so late in Moscow during the winter of 2006–2007 that even a chance of snow flurries was newsworthy. Peter Finn wrote in the *Washington Post:*

Scattered flurries teased Moscow ... with the promise of a real winter, the birthright of a city whose people take pride in trudging through snow and in ice fishing and cross-country skiing in white countryside beyond the outer beltway. The winter of 2006 has yet to arrive, however, and Muscovites are deeply discombobulated. "I want snow. I want the New Year's feeling," said Viktoria Makhovskaya, a street vendor who sells gloves and mittens. "This is a disgusting winter. I don't like it at all." (Finn 2006, A-1)

The Met Office, Britain's national weather service, and the University of East Anglia released data at about the same time indicating that 2006 was the warmest year in Britain since 1659, when recordkeeping began (Finn 2006, A-1). Swiss trees put out leaves in mid-winter, as meadows around low-altitude ski resorts turned green. "We are currently experiencing the warmest period in the Alpine region in 1,300 years," Reinhard Boehm, a climatologist at Austria's Central Institute for Meteorology and Geodynamics, told the Associated Press in Vienna. Boehm was one of the authors of a European Union–funded climate study that found similar warming periods in the tenth and twelfth centuries. But, he said, it's warmer now, and "it will undoubtedly get warmer in the future" (Finn 2006, A-1). July 2006, was the warmest month in the recorded history of the Netherlands. On June 14, 2006, Denver, Colorado's high rose to 102° F, the earliest the temperature had broken three digits there on the observational record. The previous record, also 102° F, had been reached on June 23, 1954. On July 15, 2006, South Dakota set an all-time record high of 120° F. Pierre,

South Dakota, and Fergus Falls, Minnesota, hit 117° F the same day.

Malcolm Ritter of the Associated Press sketched the national weather scene:

Let's put it this way: People played golf this winter in Maine. In shorts. Buttercups have been blooming in Montana. In Ohio, an ice-free Lake Erie allowed an early start to seasonal ferry service. And the sap started running early in Vermont. While January plunged much of Europe and Russia into the deep freeze, it appeared to be remarkably mild across the United States. Federal scientists haven't calculated yet whether it ranks as the warmest January on record nationwide, but "it's certainly going to be right up there," said Michael Halpert, a meteorologist at the National Oceanic and Atmospheric Administration's Climate Prediction Center. (Ritter 2006)

Minneapolis–St. Paul had its warmest January in 160 years. Ice sculptures at the St. Paul Winter Carnival melted and broke up in puddles of water nearly as quickly as they were carved. Several ice-fishing contests in Minnesota that had been held yearly for as long as memory stretched were canceled or moved because of thin ice. Temperatures in Bismarck, North Dakota, stayed above zero the entire month, a record. The Nonesuch River Golf Club in Scarborough, Maine, hosted 250 players on January 21 and had to turn away 200 more (Ritter 2006).

The National Climatic Data Center reported that July 2006 was the second hottest on record in the United States, at 77.2° F, ranking with the Dust Bowl years of 1936 (77.5° F) and 1934 (77.1° F), Mid-summer 2006 was notable for a coast-to-coast heat wave across the United States at the same time that temperatures reached record levels in Europe. England experienced its highest July temperature on record (97° F) as London reached 96° F, and temperatures in the Underground subway reached 115° F.

Paris, France, roasted at 100° F. Oklahoma City hit 107° F as parts of Interstate 44 were closed because they buckled in the heat. Omaha hit 106° F. On July 23, Phoenix, Arizona's minimum was 97° F, after a record-high maximum of 112° F. Sacramento's low July 23 was 79° F, after a high of 107° F. On July 27, 2006, Sacramento experienced the last of 11 consecutive 100-degree days, a record. Salt Lake City's low July 23 was 77° F, after a high of 100° F on July 22. On July 31 at 4 A.M., the temperature in Sioux City, South Dakota, stood at 85° F. On August 2, the temperatures at 3 A.M. were 86° F in Kansas City and Des Moines. The average night-time temperature in Los Angeles was 7° F warmer than it was a century ago (Harden 2006, A-1).

Record High Minimums

Several, record-high minima across the United States during July 2006 kept the national average temperature on course for a record high during the first seven months of the year. Scorching temperatures in July, particularly strings of hot nights, were almost certainly related in part to the continuing buildup of heat-trapping smokestack and tailpipe gases linked to global warming, said Jay Lawrimore of the National Climatic Data Center. "The long-term trend we're seeing cannot be explained without the influence of greenhouse gases," Lawrimore said (Revkin 2006g). On August 2, 2006, LaGuardia Airport in New York City tied its all-time record for highest overnight low, at 86° F. In two weeks, at least 17 cities in the U.S. West set record high minimums.

"Rising temperatures in Thailand over the past 50 years indicate that the kingdom is heading for hotter, more stressful nights," according to meteorological data researched by Atsamon Limsakul, a scientist at Thailand's Environmental Quality Promotion Department. Thailand's average minimum temperature rose 4.9° C between 1951 to 2003, compared with a 1.9° C jump in the average maximum temperature. "The rise in maximum and minimum temperatures at different rates has caused a progressive narrowing of daily temperature ranges, meaning the night does not become much cooler than the day," Limsakul told *The Nation*. "When the temperature does not drop much at night, bodies can't cool down from the heat. The stress that accumulated from the daytime is more likely to stay with us through the night," warned Limsakul ("Thailand Weather" 2007).

Historic Heat in California

In California, also during the summer of 2006, a heat wave killed more than 140 people, the highest toll in that state since the widespread advent of air conditioning in the 1950s. This heat wave was exceptional for its length (almost two weeks), the record temperatures, and—most notable of all—the fact that temperatures did not cool at night, as is usual in California. Most

of the victims were elderly who lived alone without air conditioning, as well as field workers.

One unidentified man, a fieldworker around 40 years of age, reached a hospital emergency room only after his body temperature reached 109.9° F. Unusually high humidity and night-time high temperatures played a role in the death toll. Roughly 16,500 cows, 1 percent of the state's dairy herd, died of the heat, according to California Dairies, the state's largest milk cooperative. Panting, miserable cows, which lack the benefit of sweat glands, yielded 10 percent to 20 percent less milk than usual (Steinhauer 2006). The walnut harvest was damaged when the intensity of sun and heat burned the nuts inside their shells.

Power consumption records tumbled in California during the scorching summer of 2006. On July 22, the temperature in Woodland Hills, near Los Angeles, reached 119° F. Most startling were the minimum temperatures in many cities of the western United States, where low dew points usually cause temperatures to fall on summer nights, even after warm days. Seattle had a high of 97° F followed by a record-high low of 72° F. Boise reached a low of 77° F on July 24, after a high of 98° F (Tehran, Iran, was 104° F and 84° F that same day). Advertisements for the documentary *An Inconvenient Truth* (*New York Times* on July 21, 2006, B-14) ran reports of record highs, headlining "How Hot Is It in Your City?"

April 2006 was the warmest on the instrumental record in the United States, the National Climatic Data Center reported. The temperatures nationwide were 4.5° F above average in the 48 contiguous states. Record heat continued in 2007, as Phoenix registered 29 days at 110° F or higher.

Average temperatures in California increased almost 2° F, between 1950 and 2000, according to data from NASA and California State University, Los Angeles. Land-use patterns as well as generalized greenhouse warming played a role in the temperature rise. The state's urban areas, most notably Southern California and the San Francisco Bay area, experienced the largest temperature increases. Rural, nonagricultural regions warmed the least. The Central Valley warmed the slowest, while coastal areas warmed faster, and the southeast desert warmed most quickly of all. Only one area was cooler: a small part of the state's rural northeast interior ("California's Temperature" 2007).

Patzert said the increased rate of minimum temperatures has led to narrower daily temperature ranges throughout the state. "California nights are heating up, giving us a jump start on hotter days," said Bill Patzert of NASA's Jet Propulsion Laboratory in Pasadena, one of the report's co-authors. "This is primarily due to increased urbanization, not increases in cloudiness or precipitation. Rainfall and snowfall didn't increase significantly for most California stations during the study period" ("California's Temperature" 2007). "California's complex topography and large latitude range give it some of the most diverse microclimates in North America," said Steve LaDochy of California State University, also a co-author of the report. "Climate change models and assessments often assume global warming's influence here would be uniform. That is not the case," said the authors ("California's Temperature" 2007).

Temperatures at a Critical Level

According to NASA scientists, the world's temperatures in 2006 reached levels not previously seen in thousands of years. James Hansen, director of NASA's Goddard Institute for Space Studies, joined other scientists to conclude that because of a rapid warming trend during the past 30 years, the Earth is now reaching and passing through the warmest levels in the current interglacial period, which has lasted nearly 12,000 years. This evidence implies that we are getting close to dangerous levels of human-made pollution, said Hansen.

"The most important result found by these researchers," according to a NASA statement, "is that the warming in recent decades has brought global temperature to a level within about 1° C (1.8° F) of the maximum temperature of the past million years" ("NASA Study" 2006). According to Hansen,

> That means that further global warming of 1° C defines a critical level. If warming is kept less than that, effects of global warming may be relatively manageable. During the warmest interglacial periods the Earth was reasonably similar to today. But if further global warming reaches 2° or 3° C, we will likely see changes that make Earth a different planet than the one we know. The last time it was that warm was in the middle Pliocene, about three million years ago, when sea level was estimated to have been about 25 meters (80 feet) higher than today. ("NASA Study" 2006)

Temperatures Rise in the Arctic

Several data sources, including those from Russian North Pole measurements (1950–1991), manned drifting camps, arrays of buoys, and satellite retrievals, all indicate an increase in Arctic Ocean surface-air temperatures (SAT). Wintertime SAT anomalies averaged over the Arctic and the midlatitudes indicate two characteristic warming events: the first during the middle 1920s to 1940 and the second after 1980 (Semenov and Bengtsson 2003). The warming in the first part of the twentieth century had its largest amplitude in the Arctic, above 60 degrees north latitude. The warming today is more widespread than that of the 1920s and 1930s.

Josefino Comiso of NASA's Goddard Space Flight Center, Greenbelt, Maryland, compiled surface-temperature measurements from several points in the Arctic between 1981 and 2001, using satellite thermal infrared data. Large warming anomalies were found over sea ice, Eurasia, and North America, but temperatures in Greenland varied from no change to a slight cooling in some areas. Temperature increases were more rapid during the 1990s than the 1980s. Warming lengthened the melt season (when temperatures were above the freezing point) by 10 to 17 days, and the rate of warming between 1981 and 2001 was eight times as rapid as during the previous century (Comiso 2002, 1956; Comiso 2003, 3498).

"There may be a natural part of it, but there's something else being superimposed on top of it," glaciologist Lonnie Thompson said of accelerating Arctic ice melt indicated by Comiso's data. "And it matches so many other lines of evidence of warming. Whether you're talking about borehole temperatures, shrinking Arctic sea ice, or glaciers, they're telling the same story" (Revkin 2001a, A-1).

According to NASA satellite surveys, perennial (year-round) sea ice in the Arctic has been declining at a rate of 9 percent per decade (Stroeve et al. 2005). During 2002, summer sea ice was at record low levels, a trend that persisted in 2003 and 2004 (Stroeve et al. 2005). During the past 35 years, Arctic sea ice also has thinned by more than 40 percent—from an average of nine feet thick to about five feet (Toner 2003d, 1-A). According to research by scientists at University College London and the British Met Office's Hadley Centre for Climate Prediction and Research, Arctic ice thinned from 3.5 meters

(11.5 feet) 30 years ago to less than two meters in 2003. By the middle of the twenty-first century, according to NASA projections, the Arctic could be ice-free during the summer months (Comiso 2003, 3498).

Warmth at Street Level

At street level, on a day-to-day basis, surges of unusual warmth could be dramatic. In Anchorage, Alaska, on 20 days during November 2002, the low temperature was higher than the average high. The city ended the month with no snow on the ground, a most unusual event. During October and November, Anchorage had less than 10 percent of its average snowfall. At the same time, the U.S. Northeast and mid-Atlantic experienced repeated snowfalls. More snow was recorded that winter in Richmond, Virginia, than in Anchorage. A snow drought also afflicted Fairbanks. For the first time in recorded history, Fairbanks during the winter of 2002–2003, did not record a single night-time temperature below −40° F.

November and early December 2001 was a very warm month, especially in the north and eastern sections of the United States. John Kahionhes Fadden, Mohawk culture bearer, said on December 2 that there was no snow at his house high in the Adirondacks of far upstate New York—quite unusual for that date. Buffalo, New York, reported no snow in November for the first time since records had been kept. Marquette, Michigan, reported a record high average, 39.9° F, 10.4° F above the average. (Weather is still variable, of course. Two years later, Fadden reported copious snow and temperature as low as minus 36° F.)

Omaha's November 2001 average temperature was 49.5° F, 10.4° F above average, an all-time record high (dating from 1871), exceeding the previous record set two years earlier by a remarkable 2.4° F. Some trees were putting out buds in late fall. The local newspaper contained no speculation about global warming in its page-one article on the unusual warmth. The headline on the piece bid readers to "enjoy" the warm spell (Gaarder 2001, A-1).

On December 5, 2001, in Omaha, the temperature was 65° F at 6 A.M., a record daily high before the sun came up. The warm, humid early morning was followed by thunder, lightning, and a torrential downpour, portending passage of a summer-style cold front. The December 6, 2001,

edition of the *Omaha World-Herald* published a page-one story headlined "Warm Weather a Mixed Blessing." Instead of climate change, however, the news report under the headline concentrated on flagging sales of snowmobiles in a season that had been, to that date, utterly bereft of snow.

The unusual weather was not restricted to Omaha. Josh Rubin commented in the *Toronto Star*: "A rose by any other name might smell as sweet. But blooming in Toronto, in December?" (Rubin 2001, B-2). Rubin wrote: "Roses are blooming, bulbs are sprouting, golfers are still hitting the links and climatologists are puzzled" (Rubin 2001, B-2). "People are coming in and asking what to do about their roses still blooming, and wondering if it's safe. Some of their trees are starting to bud. It's pretty weird for this time of year," said John Manning, manager of White Rose Nursery in East York (Rubin 2001, B-2). At Stouffville's Rolling Hills Golf Club, manager John Finlayson said the phone had been ringing off the hook with golfers requesting tee times. "It's pretty much been the busiest it's been all year," said Finlayson, who kept the course open well past the usual end of the season. "We'd normally close down some time in November, and it wouldn't usually be so busy late in the season," said Finlayson (Rubin 2001, B-2).

At about the same time, *Boston Globe* columnist Derrick Z. Jackson wrote:

> On Thanksgiving morning here in the cradle of the new North, I went to my backyard to clip a full bunch of cilantro for my guacamole. Earlier that week, I snipped four nice-sized eggplants from the vine, as well as my last jalapeno peppers to smoke into the chipotles that would go into the guacamole. More fresh cilantro is still poking up out of the ground. It is not soaring to reach the summer's sun, but its greens are still vibrant and the taste still a pungent surprise. A garden in New England in December, three weeks before the winter solstice, used to be a non sequitur.... Being this excited about seeing my garden grow, it is hard to see the madness behind the miracle. Behind the heat of my jalapenos, I know there is supposed to be a throb in my head: "glo ... bal ... war ... ming ... glo-bal ... war-ming ... globalwarming." (Jackson 2001, A-23)

Jackson wrote that the New England Regional Assessment Group, an arm of the federal U.S. Global Change Research Program, had forecast climatic warming by 2090 that could give Boston the late twentieth-century climate of Richmond, Virginia, or Atlanta, Georgia.

> That rise could mean a lot more than cilantro and eggplant. It may mean more smog, acid rain, and red tides. The sugar maples of our spring syrup and our fall foliage could be in trouble. And those poor beach-front millionaires.... If Boston becomes Atlanta, what does Atlanta become? The Sahara? The poor beachfront millionaires get Atlantis. (Jackson 2001, A-23)

Jackson described "[t]he vigorous multitudes who ran and biked along the Charles River in tank tops, the smiling families who picked up Christmas trees in T-shirts, and the shoppers who walked in their shorts as if this were Tampa would rather have this than blustery zero-degree wind chill" (Jackson 2001, A-23).

By late 2001, Canada was in the midst of 17 straight warmer-than-usual seasons. Weather is never a one-way street, however. To be climatically fair, the next winter in eastern Canada (as well as the eastern United States) was colder and snowier than usual, following a global-warming tease comprising several seasons of unusual warmth. In 2003–2004, the northeastern quadrant of the United States and adjacent Canada experienced one of its coldest winters on record. Boston (at 20.7° F) and New York City (at 24.8° F) had their coldest months (both January, in this case) in 70 years. At the same time, the rest of the United States was warmer than usual. Early in 2005, the Boston area was hit by a monumental blizzard that blasted easy certainties about global warming with as much as three feet of snow.

Temperatures during 2005

The year 2005 was the second-warmest worldwide on the instrumental record (since 1861), and the warmest in the Northern Hemisphere, according to the World Meteorological Organization and the U.S. National Climatic Data Center. The average temperature worldwide was 58.1° F, one-tenth of a degree shy of the 1998 record of 58.2° F. The 1998 record at the time was much warmer than any previous year and was aided by a strong El Niño event, which usually raises temperatures. The 2005 average, by contrast, occurred with no aid from an El Niño event. Warming was strongest over the Arctic, where loss of sea ice is changing reflectivity to aid future warming.

After a hot summer and a November that was 4.5° F above average, Omaha (and many other areas in the Midwest and Northern Rockies) were afflicted with their coldest first week of December on record. Omaha's temperature averaged 9.3° F for December 1 through 8, compared with the previous record of 13.1° F in 1913. By Christmas, however, temperatures in Omaha had risen to 15° to 20° F *above* average several days in a row. *See also:* Heat Waves

FURTHER READING

Achenbach, Joel. "March in January! Or Is It Mayday? It's Nice Out There, But Global Warming Dampens the Fun." *Washington Post* January 7, 2007, D-1. http://www.washingtonpost.com/wp-dyn/content/article/2007/01/06/AR2007010601215.html.

Borenstein, Seth. "January Weather Hottest by Far." Associated Press, February 15, 2007b. (LEXIS)

"California's Temperature Rising." Environment News Service, April 9, 2007. http://www.ens-newswire.com/ens/apr2007/2007-04-09-09.asp#anchor6.

Comiso, J.C. "A Rapidly Declining Perennial Sea Ice Cover in the Arctic." *Geophysical Research Letters* 29, no. 20 (October 18, 2002):1956–1960.

Comiso, Josefino. "Warming Trends in the Arctic from Clear Sky Satellite Observations." *Journal of Climate* 16, no. 21 (November 1, 2003):3498–3510.

Finn, Peter. "In Balmy Europe, Feverish Choruses of 'Let It Snow.'" *Washington Post*, December 20, 2006, A-1. http://www.washingtonpost.com/wp-dyn/content/article/2006/12/19/AR2006121901681_pf.html.

"First Half of Year was Warmest on Record for U.S." Associated Press, July 14, 2006. (LEXIS)

Gaarder, Nancy. "State Enjoying 'Exceptional' Warmth." *Omaha World-Herald*, December 4, 2001, A-1, A-2.

"GISS 2007 Temperature Analysis." Goddard Institute for Space Studies. January 15, 2008. [http://www.columbia.edu/~jeh1/mailings/20080114_GISTEMP.pdf]

Harden, Blaine. "Tree-Planting Drive Seeks to Bring a New Urban Cool: Lower Energy Costs Touted as Benefit." *Washington Post*, September 4, 2006, A-1. http://www.washingtonpost.com/wp-dyn/content/article/2006/09/03/AR2006090300926_pf.html.

Jackson, Derrick Z. "Sweltering in a Winter Wonderland." *Boston Globe*, December 5, 2001, A-23.

Kerr, Richard A. "Yes, It's Been Getting Warmer in Here Since the CO_2 Began to Rise." *Science* 312 (June 30, 2006b):1854.

Lederer, Edith M. "U.N. Report Says Planet in Peril." Associated Press, August 13, 2002.

McGuire, Bill. *Surviving Armageddon: Solutions for a Threatened Planet.* New York: Oxford University Press, 2005.

"NASA Study Finds World Warmth Edging Ancient Levels." NASA, September 25, 2006. http://www.giss.nasa.gov/research/news/20060925.

Revkin, Andrew C. "A Message in Eroding Glacial Ice: Humans Are Turning Up the Heat," *New York Times*, February 19, 2001a, A-1.

Revkin, Andrew C. "Last 7 Months Were Warmest Stretch on Record." *New York Times*, August 8, 2006g. http://www.nytimes.com/2006/08/08/science/earth/08brfs-006.html.

Ritter, Malcolm. "This Is Winter? Much of Nation Basked in Warm January." Associated Press, February 3, 2006. (LEXIS)

Rubin, Josh. "Toronto's Blooming Warm: Gardens, Golfers Spring to Life as Record High Nears." *Toronto Star*, December 5, 2001, B-2.

Semenov, Vladimir A., and Lennart Bengtsson. *Modes of Wintertime Arctic Temperature Variability*. Hamburg, Germany: Max Planck Institut fur Meteorologie, 2003.

Smith, Craig S. "One Hundred and Fifty Nations Start Groundwork for Global Warming Policies." *New York Times*, January 18, 2001, 7.

Steinhauer, Jennifer. "In California, Heat Is Blamed for 100 Deaths." *New York Times*, July 28, 2006. http://www.nytimes.com/2006/07/28/us/28heat.html.

Stroeve, J. C., M.C. Serreze, F. Fetterer, T. Arbetter, W. Meier, J. Maslanik, and K. Knowles. "Tracking the Arctic's Shrinking Ice Cover: Another Extreme September Minimum in 2004" *Geophysical Research Letters* 32, no. 4 (February 25, 2005), L04501. http://dx.doi.org/10.1029/2004GL021810.

"Thailand Weather: Hot Nights Ahead for Thailand, Bangkok." Deutsche Presse-Agentur." October 29, 2007. (LEXIS)

Toner, Mike. "Arctic Ice Thins Dramatically, NASA Satellite Images Show." *Atlanta Journal-Constitution*, October 24, 2003d, 1-A.

"World Breaks Temperature Records." *Guardian Unlimited*, March 16, 2007. (LEXIS)

Temperatures and Carbon-Dioxide Levels

A key question—some argue it is *the* key issue—in climate change is the determination of how the level of carbon dioxide and other greenhouse gases in the atmosphere relate to a specific rise or fall in temperature after other "forcings" have been factored in. For example, how much will global temperatures rise, on average, if the carbon-dioxide level doubles, all other factors being equal (which they rarely are)? Humankind has been altering the Earth's climate for several thousand years; cultivation of rice, for example, raises methane levels; biomass burning increases the amount of carbon dioxide in the air (White 2004, 1610). In effect, human interference with the carbon cycle began, in a small way, the first time that a human being cooked his dinner over an open fire.

David W. Lea, writing in *Journal of Climate*, said that several models have produced varying answers to this question. To sharpen the debate, Lea turned to paleoclimatic data describing both tropical sea-surface temperatures and Antarctic ice cores. Using proxy records from the Eastern equatorial Pacific Ocean and Vostok (Antarctic) ice cores, Lea arrived at an estimate of "a tropical climate sensitivity of 4.4° to 5.6° C (error estimated at plus or minus 1° C) for a doubling of atmospheric CO_2 concentration" (Lea 2004, 2170). This result, according to Lea, "[s]uggests that the equilibrium response of tropical climate to atmospheric CO_2 changes is likely to be similar to the upper end of available global predictions from global models" (Lea 2004, 2170).

Temperatures Cycle with Carbon-Dioxide Levels—or Not

During most of the Earth's history, temperatures have risen and fallen roughly in tandem with carbon-dioxide levels. To add another mystery to the equation, however, this has not *always* been the case. Jan Veizer and colleagues presented a temperature record using the oxygen isotopic composition of tropical marine fossils describing two periods with mismatches between carbon-dioxide levels and temperatures: the late Ordovician (440 million years ago) and the Jurassic and early Cretaceous (120 to 220 million years ago). Veizer and colleagues' findings indicated that occasionally rising levels of carbon dioxide were accompanied by a *drop* in average temperatures(Veizer, Godderis, and Francois 2000, 698–701).

During the Jurassic, 145 million to 208 million years ago, for example, carbon-dioxide levels may have been as much as 10 times present levels, but temperature averages were cooler. Veizer believes that the sun may be a major driver of climate change, disrupting the relationship of temperature and carbon dioxide (Veizer, Godderis, and Francois 2000, 698). This conclusion is far from unquestioned, however. Some scientists question the climate proxies used in these studies, asserting that they may be producing inaccurate data (Kump 2000, 651; Pearson et al. 2001, 481).

Lee R. Kump has commented in *Nature*: "Either the CO_2 proxies are flawed, or our understanding of the relationship between CO_2 and climate … needs rethinking" (Kump 2000, 651). Kump believes that "[l]ack of close correspondence between climate change and proxy indicators of atmospheric CO_2 may force us to reevaluate the proxies, rather than disallow the notion that substantially increased atmospheric CO_2 will indeed lead to marked warming in the future" (Kump 2000, 652).

Climate modelers have long been perplexed by an apparent discrepancy between "climate models with increased levels of carbon dioxide [which] predict that global warming causes heating in the tropics [compared to] … investigations of ancient climates based on paleodata [that] have generally indicated cool tropical temperatures during supposed greenhouse episodes" (Pearson et al. 2001, 481). During the late Cretaceous (about 970 million years ago) and Eocene (about 50 million years ago), for example, the poles were believed to have been ice free, and tropical sea-surface temperatures have been estimated to have been between 15° and 23° C, based on oxygen isotope paleothermometry of surface-dwelling planktonic forminifer shells, "which provide proxy information on ocean temperatures" (Kump 2001, 470). This data indicate an ice-free Earth in which tropical sea-surface temperatures were cooler than today in most areas—an unusual circumstance by present-day standards. Either the Earth was quite different at that time, or the proxies being used to estimate temperatures are flawed, or both.

Debating Proxies

Writing in *Nature* during 2001, Paul N. Pearson and colleagues questioned the validity of such data "on the grounds of poor preservation and diagenetic alteration" (Pearson et al. 2001, 481). They presented new data "from exceptionally well-preserved foraminifera shells extracted from impermeable clay-rich sediments," which indicated temperatures of at least 28° to 32° C (compared with 25° to 27° C today) in the tropics during these periods. (Foraminifera shells are valuable in paleoclimatic research because they can be dated.)

Such estimates would indicate warming in the tropics consistent with the rest of the Earth during periods of high atmospheric carbon dioxide. These new data indicate that "for those parts of the Late Cretaceous and Eocene epochs that we have sampled, tropical temperatures were at least as warm as today, and probably several degrees warmer.… [allowing] us to dispose of the 'cool tropic paradox' that has bedeviled the study of past warm climates" (Pearson et al. 2001, 486).

This study confirmed the relationship between past episodes of elevated atmospheric carbon dioxide and global warming. These results fit well with existing knowledge that temperature-sensitive organisms, such as corals, often were displaced northward during times when the atmosphere's carbon-dioxide levels were relatively high (Kump 2001, 471). Such episodes are being used as predictors of conditions in the future, when "fossil-fuel burning is likely to drive atmospheric CO_2 to perhaps six times the preindustrial level" (Kump 2001, 470).

How High Will Temperatures Go?

How high will temperatures go if greenhouse-gas levels continue to increase as they have been? Under business-as-usual greenhouse-gas emissions, by 2100, New York City may feel like Las Vegas does today (with additional humidity) and San Francisco will have a climate comparable to that of today's New Orleans. By the same date, Boston will have average temperatures similar to current temperatures in Memphis, Tennessee, according to a report, "The Cost of Climate Change." According to the report, released on May 23, 2008, by researchers at Tufts University (and commissioned by the Natural Resources Defense Council), temperatures will rise 13° F on average in the United States, and 18° F in Alaska.

FURTHER READING

Kump, Lee R. "What Drives Climate?" *Nature* 408 (December 7, 2000):651–652.

Kump, Lee R. "Chill Taken out of the Tropics." *Nature* 413 (October 4, 2001):470–471.

Lea, David W. "The 100,000-year Cycle in Tropical SST [Sea-surface Temperatures], Greenhouse Forcing, and Climate Sensitivity." *Journal of Climate* 17, no. 11 (June 1, 2004):2170–2179.

Pearson, Paul N., Peter W. Ditchfield, Joyce Singano, Katherine G. Harcourt-Brown, Christopher J. Nicholas, Richard K. Olsson, Nicholas J. Shackleton, and Mike A. Hall. "Warm Tropical Sea-surface

Temperatures in the Late Cretaceous and Eocene Epochs." *Nature* 413 (October 4, 2001):481–487.

Veizer, Jan, Yves Godderis, and Louis M. Francois. "Evidence for Decoupling of Atmospheric CO_2 and Global Climate during the Phanerozoic Eon." *Nature* 408 (December 7, 2000):698–701.

White, James C. "Do I Hear a Million?" *Science* 304 (June 11, 2004):1609–1610.

Thermohaline Circulation

Scientists studying past climate epochs have found times when the oceans' circulation lost its vital character, perhaps even shut down. Analyses of ice cores, deep-sea sediment cores, and other geologic evidence have clearly demonstrated that the Conveyor has abruptly slowed or halted many times in Earth's past. That has caused the North Atlantic region to cool significantly and brought long-term drought conditions to other areas of the Northern Hemisphere—over time spans as short as years to decades. According to climate scientist Thomas F. Stocker,

> Evidence from paleoclimatic archives suggests that the ocean atmosphere system has undergone dramatic and abrupt changes with widespread consequences in the past. Climatic changes are most pronounced in the North Atlantic region where annual mean temperatures can change by 10° C and more within a few decades. Climate models are capable of simulating some features of abrupt climate change. These same models also indicate that changes of this type may be triggered by global warming. (Stocker, Knutti, and Plattner 2001, 277)

Meltwater Releases

At the end of the last ice age (between 8,200 and 12,800 years ago), ice-core records from Greenland indicate abrupt temperature declines at about the same time that massive amounts of cold freshwater were released from a huge body of glacial meltwater, which glaciologists today call Lake Agassiz, that reached from today's Canadian prairies eastward to Quebec and southward to Minnesota. Reaching the Atlantic Ocean through the St. Lawrence Valley and Hudson Bay, "such a large amount of low-density fresh water would have reduced the density of the North Atlantic surface water considerably, preventing it from sinking and thus slowing down

(or perhaps even shutting off completely) the Gulf Stream," wrote glaciologist Doug Macdougall in *Frozen Earth: The Once and Future Story of Ice Ages* (2004, 110).

Evidence from the Irish Sea Basin as recently as 19,000 years ago indicates a large reduction in the strength of North Atlantic deep-water formation that provoked cooling of the North Atlantic at the same time that global sea levels were rising rapidly, a response to melting ice. According to Clark and colleagues, "These responses identify mechanisms responsible for the propagation of deglacial climate signals to the Southern Hemisphere and tropics while maintaining a cold climate in the Northern Hemisphere" (Clark et al. 2004, 1141).

Jochen Erbacher studied an anoxic event "in the restricted basins of the western Tethys and North Atlantic" during the mid-Cretaceous period, about 112 million years ago that appears to have been caused by "increased Thermohaline stratification" (Erbacher et al. 2001, 325). The Western Tethys was a sea formed as continents were separating on the site of the present-day North Atlantic Ocean. "Ocean anoxic events were periods of high carbon burial that led to drawdown of atmospheric carbon dioxide, lowering of bottom-water oxygen concentrations and, in many cases, significant biological extinctions," commented Erbacher and colleagues (Erbacher et al. 2001, 325). "We suggest that that the partial tectonic isolation of the various basins in the Tethys and Atlantic, a low sea level, and the initiation of warm global climates may be important factors in setting up oceanic stagnation" during this event (Erbacher et al. 2001, 327).

According to research by Helga Kleiven and colleagues in *Science,*

> An outstanding climate anomaly 8200 years before the present (B.P.) in the North Atlantic is commonly postulated to be the result of weakened overturning circulation triggered by a freshwater outburst. New stable isotopic and sedimentological records from a northwest Atlantic sediment core reveal that the most prominent Holocene anomaly in bottom-water chemistry and flow speed in the deep limb of the Atlantic overturning circulation begins at ~8.38 thousand years B.P., coeval with the catastrophic drainage of Lake Agassiz. The influence of Lower North Atlantic Deep Water was strongly reduced at our site for ~100 years after the outburst, confirming the ocean's sensitivity to freshwater forcing. The similarities between the timing and duration of the pronounced deep circulation

changes and regional climate anomalies support a causal link. (Kleiven et al. 2008, 60)

Ocean "Weather" and "Climate"

Ocean circulation is not unlike surface weather. It varies widely over short as well as long periods. Ocean circulation experiences a kind of "weather," as well as a longer-term "climate." Scientists have studied past epochs, and found that the deep Atlantic Ocean during the height of the last ice age appears to have been quite different from today. One group reviewed observations that implied the

> Atlantic meridional overturning circulation during the Last Glacial Maximum was neither extremely sluggish nor an enhanced version of present-day circulation. The distribution of the decay products of uranium in sediments is consistent with a residence time for deep waters in the Atlantic only slightly greater than today. However, evidence from multiple water-mass tracers supports a different distribution of deep-water properties, including density, which is dynamically linked to circulation. (Lynch-Stieglitz et al. 2007, 66)

North Atlantic Deep Water was warmer during the last interglacial than it is today and probably warmed Antarctic waters, accelerating ice loss and raising sea levels. J. C. Duplessy and colleagues wrote that

> Oxygen isotope analysis of benthic foraminifera in deep sea cores from the Atlantic and Southern Oceans shows that during the last interglacial period, North Atlantic Deep Water (NADW) was 0.4° ± 0.2 C warmer than today, whereas Antarctic Bottom Water temperatures were unchanged. Model simulations show that this distribution of deep water temperatures can be explained as a response of the ocean to forcing by high-latitude insolation. The warming of NADW was transferred to the Circumpolar Deep Water, providing additional heat around Antarctica, which may have been responsible for partial melting of the West Antarctic Ice Sheet. (Duplessy, Roche, and Kageyama 2007, 89)

Another study found that

> An exceptional analogue for the study of the causes and consequences of global warming occurs at the Palaeocene/Eocene Thermal Maximum, 55 million years ago. A rapid rise of global temperatures during this event accompanied turnovers in both marine and terrestrial biota' as well as significant changes in ocean chemistry and circulation. Here

we present evidence for an abrupt shift in deep-ocean circulation using carbon isotope records from fourteen sites. These records indicate that deep-ocean circulation patterns changed from Southern Hemisphere overturning to Northern Hemisphere overturning at the start of the Palaeocene/Eocene Thermal Maximum. This shift in the location of deep-water formation persisted for at least 40,000 years, but eventually recovered to original circulation patterns. These results corroborate climate model inferences that a shift in deep-ocean circulation would deliver relatively warmer waters to the deep sea, thus producing further warming. Greenhouse conditions can thus initiate abrupt deep-ocean circulation changes in less than a few thousand years, but may have lasting effects; in this case taking 100,000 years to revert to background conditions. (Nunes and Norris 2006, 60)

M. Latif and colleagues asked, in the *Journal of Climate* (2006, 4631–4637), "Is the Thermohaline Circulation Changing?" Their answer is that it has changed considerably during the last century, mainly as a result of natural multidecadal climate variability, and that it may change in coming decades because of melting Arctic ice freshening the Nordic Sea. However, these researchers do believe that "such a weakening will not exceed the range of multi-decadal variability [which they take to be 40 percent] for several decades" (Latif et al. 2006, 4635).

The most prominent cold period during the Holocene, 8200 years ago, spanned several hundred years and was caused by an influx of meltwater into the North Atlantic. Evidence from a North Atlantic deep-sea sediment core reveals that the largest climatic perturbation in the present interglacial, the 8,200-year event, is marked by two distinct cooling events in the subpolar North Atlantic at 8,490 and 8,290 years ago. An associated reduction in deep flow speed provides evidence of a significant change to a major down-welling limb of the Atlantic meridional overturning circulation. The existence of a distinct surface freshening signal during these events strongly suggests that the sequenced surface and deep ocean changes were forced by pulsed meltwater outbursts from a multistep final drainage of the proglacial lakes associated with the decaying Laurentide Ice Sheet margin.

An Analogue in the Mid-Cretaceous Period?

Seeking an analogue to a world dominated by severe global warming, Paul A. Wilson and

Richard D. Norris investigated the climate of the Mid-Cretaceous period, "a time of unusually warm polar temperatures, repeated reef-drownings in the tropics, and a series of oceanic anoxic events.... with maximum sea-surface temperatures 3° to 5° C warmer than today" (Wilson and Norris 2001, 425). This was period with considerable greenhouse forcing not unlike the end of the twenty-first century may resemble, except that the forcing was natural, not anthropogenic. Writing in *Geophysical Research Letters*, Orsi and colleagues found no evidence of a recent reduction in generation of Antarctic Bottom Water (Chin 2001, 575). The researchers did not rule out the possibility, however, that global warming could affect formation rates of North Atlantic Deep Water or Antarctic Bottom Water in the future.

The possibility that water temperatures rose to 15° C or perhaps higher in the Arctic during the Cretaceous period also has been explored by Hugh C. Jenkyns and colleagues in *Nature* (Jenkyns et al. 2004, 888–892). At the time, atmospheric carbon-dioxide levels may have been three to six times today's levels, a "super greenhouse" caused mainly by volcanic out-gassing. Today's climate models have had a difficult time accommodating such warmth in the Arctic without raising projected temperatures at lower latitudes to levels that would have been intolerable for animals and plants that lived at the time.

Thermohaline Circulation: Present-Day Debate

A study of North Atlantic ocean circulation published in late 2005 (Bryden et al. 2005, 655) reported a 30 percent reduction in the meridional overturning (thermohaline) circulation at 26.5 degrees north latitude, based on readings taken during 1957, 1981, 1992, and 1998. Harry Bryden at the Southampton Oceanography Centre in the United Kingdom, strung buoys in a line across the Atlantic from the Canary Islands to the Bahamas, "and found that the flow of water north from the Gulf Stream into the North Atlantic had faltered by 30 percent since the 1990s." Less ocean water was going north on the surface and less was coming south along the bottom (Pearce 2007b, 147). Bryden said at the time that he was not sure whether the change he described was temporary or part of a long-term trend. This analysis was, however, "the first observational evidence that such a decrease of

the oceanic overturning circulation is well underway" (Quadfasel 2005, 565).

By 2007, scientific consensus was leaning away from Bryden's analysis, as changes earlier taken to be long-term erosion were being regarded, instead, as part of cyclical variation. Stuart Cunningham of the National Oceanography Centre at Southampton, United Kingdom, and colleagues found that the thermohaline circulation varied by 25 percent in one year (Schiermeier 2007b, 844–845).

The idea that changes in thermohaline circulation might cool Europe substantially as most of the world warms has been losing traction among scientists, even as popular entertainment promotes it. Movies such as *The Day after Tomorrow* (2004) presented the idea in cinematic fashion. Reality intruded, however, as temperatures rose in Europe, and climate models forecast more of the same. Richard Kerr wrote in *Science* that the ocean-circulation system is "prone to natural slowdowns and speedups. Furthermore, researchers are finding that even if global warming were slowing the conveyor and reducing the supply of warmth to high latitudes, it would be decades before the change would be noticeable above the noise" (Kerr 2006d, 1064). These findings came from the Rapid Climate Change (RAPID) Program, whose researchers moored 19 light-weight cables with instruments along 26.5 degrees north latitude from West Africa to the Bahamas to measure ocean flows.

"The concern had previously been that we were close to a threshold where the Atlantic circulation system would stop," said Susan Solomon, a senior scientist at the National Oceanic and Atmospheric Administration. "We now believe we are much farther from that threshold, thanks to improved modeling and ocean measurements. The Gulf Stream and the North Atlantic Current are more stable than previously thought" (Gibbs 2007).

The United Nations Intergovernmental Panel on Climate Change said in its 2007 assessment on the basis of 23 climate models that substantial cooling of Europe is not occurring and that a marked disruption of thermohaline circulation is unlikely during the twenty-first century. The IPCC assessment did say that the circulation probably would weaken by about 25 percent through the year 2100, given increased melting of the Greenland Ice Cap and more rainfall in the Arctic that will cause more freshwater to flow southward. However, any resulting cooling of

Europe will probably be overwhelmed by a general worldwide warming trend.

"The bottom line is that the atmosphere is warming up so much that a slowdown of the North Atlantic Current will never be able to cool Europe," said Helge Drange, a professor at the Nansen Environmental and Remote Sensing Center in Bergen, Norway (Gibbs 2007).

Thermohaline circulation is only one contributor to Europe's mild climate. Prevailing winds play a part. In addition, the Greenland Ice Cap would have to melt quickly to stop the ocean conveyor, according to present theories, "The ocean circulation is a robust feature, and you really need to hit it hard to make it stop," said Eystein Jansen, a paleoclimatologist who directs the Bjerknes Center for Climate Research, also in Bergen. "The Greenland ice sheet would not only have to melt, but to dynamically disintegrate on a huge scale across the entire sheet" (Gibbs 2007). Any collapse would probably require centuries. *See also:* Ocean Circulation, Worldwide

FURTHER READING

Bryden, Harry L., Hannah R. Longworth, and Stuart A. Cunningham. "Slowing of the Atlantic Meridional Overturning Circulation at 25° N." *Nature* 438 (December 1, 2005):655–657.

Chin, Gilbert. "No Deepwater Slowdown?" Editor's Choice. *Science* 293 (July 27, 2001):575, citing Orsi et al., *Geophysical Research Letters* 28 (2001):2923.

Clark, Peter U., A. Marshall McCabe, Alan C. Mix, and Andrew J. Weaver. "Rapid Rise of Sea Level 19,000 Years Ago and Its Global Implications." *Science* 304 (May 21, 2004):1,141–1,144.

Duplessy, J. C., D. M. Roche, and M. Kageyama. "The Deep Ocean During the Last Interglacial Period." *Science* 316 (April 6, 2007):89–91.

Ellison, Christopher R. W., Mark R. Chapman, and Ian R. Hall. "Surface and Deep Ocean Interactions during the Cold Climate Event 8200 Years Ago." *Science* 312 (June 30, 2006):1929–1932.

Erbacher, Jochen, Brian T. Huber, Richard D. Norris, and Molly Markey. "Increased Thermohaline Stratification as a Possible Cause for an Ocean Anoxic Event in the Cretaceous Period." *Nature* 409 (January 18, 2001):325–327.

Gibbs, Walter. "Scientists Back Off Theory of a Colder Europe in a Warming World." *New York Times,* May 15, 2007. http://www.nytimes.com/2007/05/15/science/earth/15cold.html.

Jenkyns, Hugh C., Astrid Forster, Stefan Schouten, and Jaap S. Sinninghe Damste. "High Temperatures in the Late Cretaceous Arctic Ocean." *Nature* 432 (December 16, 2004):888–892.

Kerr, Richard A. "Atlantic Mud Shows How Melting Ice Triggered an Ancient Chill." *Science* 312 (June 30, 2006c):1,860.

Kerr, Richard A. "False Alarm: Atlantic Conveyor Belt Hasn't Slowed Down After All." *Science* 314 (November 17, 2006d):1,064.

Kleiven, Helga (Kikki) Flesche, Catherine Kissel, Carlo Laj, Ulysses S. Ninnemann, Thomas O. Richter, and Elsa Cortijo. "Reduced North Atlantic Deep Water Coeval with the Glacial Lake Agassiz Freshwater Outburst." *Science* 3129 (January 4, 2008):60–64.

Latif, M., C. Boning, J. Willebrand, A. Biastoch, J. Dengg, N. Keenlyside, U. Schweckendiek, and G. Madec. "Is the Thermohaline Circulation Changing?" *Journal of Climate* 19, no. 18 (September 15, 2006):4631–4637.

Lynch-Stieglitz, Jean, Jess F. Adkins, William B. Curry, Trond Dokken, Ian R. Hall, Juan Carlos Herguera, Joîl J.-M. Hirschi, Elena V. Ivanova, Catherine Kissel, Olivier Marchal, Thomas M. Marchitto, I. Nicholas McCave, Jerry F. McManus, Stefan Mulitza, Ulysses Ninnemann, Frank Peeters, Ein-Fen Yu, and Rainer Zahn. "Atlantic Meridional Overturning Circulation during the Last Glacial Maximum." *Science* 316 (April 6, 2007): 66–69.

Macdougall, Doug. *Frozen Earth: The Once and Future Story of Ice Ages.* Berkeley: University of California Press, 2004.

Nunes, Flavia, and Richard D. Norris. "Abrupt Reversal in Ocean Overturning during the Palaeocene/Eocene Warm Period." *Nature* 439 (January 5, 2006):60–63.

Pearce, Fred. *With Speech and Violence: Why Scientists Fear Tipping Points in Climate Change.* Boston: Beacon Press, 2007b.

Quadfasel, Detlef. "Oceanography: The Atlantic Heat Conveyor Slows." *Nature* 438 (December 1, 2005):565–566.

Schiermeier, Quirin. "Ocean Circulation Noisy, Not Stalling." *Nature* 448 (August 23, 2007b):844–845.

Stocker, Thomas F., Reto Knutti, and Gian-Kasper Plattner. "The Future of the Thermohaline Circulation—a Perspective." In *The Oceans and Rapid Climate Change: Past, Present, and Future,* ed. Dan Seidov, Bernd J. Haupt, and Mark Maslin, 277–293. Washington, D.C.: American Geophysical Union, 2001.

Wilson, Paul A., and Richard D. Norris. "Warm Tropical Ocean Surface and Global Anoxia during the Mid-Cretaceous Period." *Nature* 412 (July 26, 2001):425–429.

Thermohaline Circulation: Debating Points

The idea that failure of the thermohaline circulation could cause colder temperatures in landmasses around the North Atlantic Ocean is highly debatable, as well as politically charged. The debate was given extra force when one of the idea's major proponents backed away from it. Wallace S. Broecker, of Columbia University, heretofore an advocate of a belief that the "'business-as-usual' fossil-fuel track [would] run the risk, late in the twenty-first century, of triggering an abrupt reorganization of the Earth's Thermohaline circulation" has maintained that doubling atmospheric levels of carbon dioxide could "cripple the ocean's conveyor circulation" (Broecker 2001, 83). In 1997 Broecker suggested that if a collapse of the Gulf Stream blocked its warming influence, temperatures in the British Isles could fall by an average of 11° C, plunging Liverpool or Berwick to the same temperatures as Spitsbergen, inside the Arctic Circle. Any dramatic drop in temperature could have devastating implications for agriculture and for Europe's ability to feed itself (Radford 2001b, 3).

In a book published by the American Association of Petroleum Geologists, Broecker basically repudiated his earlier position:

The recent discovery by Gerard Bond that the 1,500-year cycle that paced these glacial disruptions continued in a muted form during times of interglaciation casts a new light on this situation (Bond et al. 1997, 1257). It leads me to suspect that the large and rapid atmospheric changes of glacial time were driven by a sea-ice amplifier. If so, then, because little sea ice will remain at the time of a greenhouse-induced Thermohaline reorganization, perhaps the threat will be far smaller than I had previously envisioned. (Broecker 2001, 83)

Broecker said, "I apologize for my previous sins," of overemphasizing the Gulf Stream's role (Kerr 2002c, 2202).

As is so often the case regarding global warming's putative effects on the Earth's ecosystem, the premises under examination here are open to dispute. Is the Gulf Stream *really* the main climate driver warming Europe's winters? In October 2002, a team of scientists writing in the *Quarterly Journal of the Royal Meteorological Society* asserted that this popular assumption is incorrect. Rather, they maintained, Europe is warmed "by atmospheric circulation tweaked by the Rocky Mountains [and] … summer's warmth lingering in the North Atlantic" (Kerr 2002c, 2202).

Richard Seager of Columbia University's Lamont-Doherty Earth Observatory in Palisades, New York, and David Battisti of the University of Washington headed this study, which sought to determine the relative influences of various influences on European climate. They noted that winds carry five times as much heat out of the tropics to the midlatitudes than oceanic currents. They also estimated that roughly "80 percent of the heat that cross-Atlantic winds picked up was summer heat briefly stored in the ocean rather than heat carried by the Gulf Stream" (Kerr 2002c, 2202). Seager and colleagues relegated the Gulf Stream to the role of a minor player in Europe's wintertime climate. They asserted, however, that the Gulf Stream *does* play a major role in warming Scandinavia and keeping the far northern Atlantic free of ice (Seager et al. 2002, 2563).

FURTHER READING

Broecker, Wallace S. "Are We Headed for a Thermohaline Catastrophe?" In *Geological Perspectives of Global Climate Change*, ed. Lee C. Gerhard, William E. Harrison, and Bernold M. Hanson, 83–95. AAPG [American Association of Petroleum Geologists] Studies in Geology No. 17. Tulsa, OK: AAPG, 2001.

Kerr, Richard A. "European Climate: Mild Winters Mostly Hot Air, Not Gulf Stream." *Science* 297 (September 27, 2002c):2202.

Radford, Tim. "As the World Gets Hotter, Will Britain Get Colder? Plunging Temperatures Feared after Scientists Find Gulf Stream Changes." *London Guardian*, June 21, 2001b, 3.

Seager, R., D. S. Battisti, J. Yin, N. Gordon, N. Naik, A.C. Clement, and M. A. Kane. "Is the Gulf Stream Responsible for Europe's Mild Winters?" *Quarterly Journal of the Royal Meteorological Society* 128 (2002):2,563–2,586.

Thermohaline Circulation: Present-Day Evidence of Breakdown

Inhibition of global ocean circulation by a warming atmosphere is not a mere theory or a paleoclimatic curiosity. Evidence is accumulating that ocean circulation already is breaking down, although researchers are unsure whether this is evidence of a natural cycle, a provocation of warming seawaters, or both (Häkkinen and Rhines 2004, 559). Analysis has been restricted by the scanty nature of data prior to 1978. "These observations of rapid climatic changes over one decade [the 1990s] may merit some concern," according to informed observers (Häkkinen and Rhines 2004, 559).

Evidence suggests that the North Atlantic has cooled while the rest of the world has been warming—a probable signature of thermohaline disruption. During the last half of the twentieth century, research reports indicate a "dramatic" increase in freshwater released into the North Atlantic by melting ice. This "freshening" is well under way (Speth 2004, 61). According to scientists at the Woods Hole Oceanographic Institution, this is "the largest and most dramatic oceanic change ever measured in the era of modern instruments" (Gagosian 2003). By 2002, the amount of freshwater entering the Arctic Ocean was 7 percent more than during the 1930s (Speth 2004, 61).

Meridional overturning circulation in the subpolar North Atlantic Ocean is not a story with a single plot line, however. It can be erratic. Its flow weakened to shallow or nonexistent after 2000, a possible consequence of a warmer climate. During the winter of 2007–2008, however, deep convection in the subpolar gyre in both the Labrador and Irminger seas returned. Scientists analyzing a variety of in situ, satellite, and reanalysis data in *Nature Geoscience* showed that

> contrary to expectations the transition to a convective state took place abruptly, without going through a phase of preconditioning.... [C]hanges in hemispheric air temperature, storm tracks, the flux of fresh water to the Labrador Sea and the distribution of pack ice all contributed to an enhanced flux of heat from the sea to the air, making the surface water sufficiently cold and dense to initiate deep convection. Given this complexity, we conclude that it will be difficult to predict when deep mixing may occur again. (Våge et al. 2009)

Saltier Water

According to oceanographer Ruth Curry, sea-surface waters in tropical regions have become dramatically saltier during the past 50 years, while surface waters at high latitudes in Arctic regions have become much fresher. These changes in salinity accelerated during the 1990s as global temperatures warmed. "This is the signature of increasing evaporation and precipitation" because of warming, Curry said, "and a sign of melting ice at the poles. These are consequences of global warming, either natural, human-caused or, more likely, both" (Cooke 2003, A-2).

Richard A. Kerr, writing in *Science*, said that "To Curry and her colleagues, it's looking as if

something has accelerated the world's cycle of evaporation and precipitation by 5 percent to 10 percent, and that something may well be global warming" (Kerr 2004, 35). These results indicate that freshwater has been lost from the low latitudes and added at high latitudes, at a pace exceeding the ocean circulation's ability to compensate, the authors said ("Study Reports" 2003).

Curry led a team that examined salinity "on a long transect (50 degrees S. to 60 degrees North, or between Iceland in the north and the tip of South America) through the western basins of the Atlantic Ocean between the 1950s and the 1990s" (Curry, Dickson, and Yashayaev 2003, 826). They found "systematic freshening at both poleward ends contrasted with large increases of salinity pervading the upper water columns at lower latitudes." The authors asserted that their data extends "a growing body of evidence indicating that shifts in the oceanic distribution of fresh and saline waters are occurring worldwide in ways that suggest links to global warming and possible changes in the hydrologic cycle of the Earth" (Curry, Dickson, and Yashayaev 2003, 826).

Evaporation and Precipitation

Curry's study suggests that relatively rapid oceanic changes and recent climate changes, including warming global temperatures, may be altering the fundamental planetary system that regulates evaporation and precipitation and cycles freshwater around the globe. An acceleration of Earth's global hydrological cycle could affect global precipitation patterns that govern the distribution, severity, and frequency of droughts, floods, and storms. The same pattern could exacerbate global warming by rapidly adding more water vapor—itself a potent, heat-trapping greenhouse gas—to the atmosphere. It could also continue to freshen northern North Atlantic Ocean waters to a point that could disrupt ocean circulation and trigger further climate changes ("Study Reports" 2003).

Japanese and Canadian scientists reported in *Nature* that the deepest waters of the North Pacific Ocean have warmed significantly across the entire width of the ocean basin (Fukasawa et al. 2004, 825). Masao Fukasawa of the Japan Marine Science and Technology Center in Yokosuka, Japan, and five other researchers measured temperature changes by going to sea on three research vessels and measuring deep-water temperatures across the North Pacific. Then they compared those temperature measurements to measurements made by researchers in 1985.

The Fukasawa team's findings indicated a warming in the deep North Pacific over a "shorter time scale and larger spatial scale than [has] ever been believed," Fukasawa said. "As [far] as I know, our result is the first which shows such a large-scale temperature change in the global Thermohaline circulation," that is, in global heat and salt circulation via ocean currents (Davidson 2004a, A-4). It is much too soon to blame global warming for the deep-sea warming, Reid and other experts cautioned. "To go from this (observation) to say, 'This is global warming,' is just a guess," Reid said. In any case, "this (Fukasawa result) is something that should be watched very carefully" (Davidson 2004a, A-4).

M. J. McPhaden and D. Zhang, writing in *Nature*, also made a case that Thermohaline circulation is slowing:

> Some theories ascribe a central role [in global climate change] to the wind-driven meridional overturning circulation between the tropical and subtropical oceans. Here we show, from observations over the last 50 years, that this overturning circulation has been slowing down since the 1970s, causing a decrease in upwelling of about 25 percent in an equatorial strip between 9 degrees north and 9 degrees south. This reduction in equatorial upwelling of relatively cool water … is associated with a rise in equatorial sea-surface temperatures of about 0.8° C. (McPhaden and Zhang 2002, 603)

Speed of Changes in the Past

Paleoclimatic records indicate that once climate change is "tripped," change may be quite rapid. For example, various proxy records from such sources as ice cores indicate temperature changes of up to 16° C in Greenland in a decade (Stocker et al. 2001, 289). Could such a rapid change, triggered by global warming, shut down the oceans' circulation system within a matter of decades?

Terrence Joyce, chairman of the physical-oceanography department at Woods Hole Oceanographic Institution in Massachusetts, has been trying to raise awareness about this possibility. He said he is "not predicting an imminent climate change—only that once it started (and it is getting more likely) it could occur within 10 years" (Cowen 2002a, 14). Woods Hole director Robert Gagosian feels an urgency to settle the question. He sees enough disturbing information in the North Atlantic data that oceanographers from Woods Hole and other institutions have gathered to elicit "strong evidence that we may be approaching a dangerous threshold" (Cowen 2002a, 14).

Stocker and colleagues wrote: "Although the magnitude of the change is highly uncertain, the models agree that the Thermohaline circulation in the Atlantic will reduce due to the gain of buoyancy associated with the warming and a stronger hydrological cycle" (Stocker et al. 2001, 290). R. B. Thorpe and colleagues considered a number of model variations indicating that various degrees of global warming may decrease thermohaline circulation in the Atlantic between 20 and 60 percent because of temperature change (greater warming at high latitudes) and 40 percent because of salinity change (Thorpe et al. 2001, 3102).

B. Dickson and colleagues wrote in *Nature* that,

> From observations it has not been possible to detect whether the ocean's overturning circulation is changing, but recent evidence suggests that the transport over the sills [in the ocean] may be slackening. Here we show, through the analysis of hydrographic records, that the system of overflow and entrainment that ventilates the deep Atlantic has steadily changed over the past four decades. We find that these changes have already led to sustained freshening of the deep ocean. (Dickson et al. 2002, 832)

Dickson and colleagues' ideas have received support from other studies. An international team of hydrologists and oceanographers reported in *Science* that the flow of freshwater from Arctic rivers into the Arctic Ocean has increased significantly over recent decades. If the trend continues, some scientists believe it could affect the global climate, perhaps leading to cooling in Northern Europe ("Study Reveals" 2002).

Discharge Data from the Six Largest Eurasian Rivers

Bruce J. Peterson of the Marine Biological Laboratory's Ecosystems Center led a research team of scientists from the United States, the

Russian Federation, and Germany whose members analyzed discharge data from the six largest Eurasian rivers that drain into the Arctic Ocean. These rivers, all with headwaters in Russia, account for more than 40 percent of total freshwater inflows from rivers into the Arctic Ocean. Peterson and colleagues found that combined annual discharge from the Russian rivers increased by 7 percent from 1936 to 1999. They contend that this measured increase in runoff is an observed confirmation of what climatologists have been saying for years—that freshwater flow to the Arctic Ocean and North Atlantic will increase with global warming ("Study Reveals" 2002).

"If the observed positive relationship between global temperature and river discharge continues into the future, Arctic river discharge may increase to levels that impact Atlantic Ocean circulation and climate within the 21st century," said Peterson. Annual discharge of the rivers into the Arctic increased about 128 cubic kilometers per year during the 63-year study period. The authors of the study warned that increasing river discharge, coupled with ice melt from Greenland, could have important effects on thermohaline circulation during the twenty-first century (Peterson 2002, 2171–2172; "Study Reveals" 2002).

During mid-2001, other scientists disclosed evidence that the thermohaline circulation is, indeed, breaking down. Researchers in Denmark's Faroe Islands, halfway between Iceland and the northern tip of Britain, found evidence of a 20 percent drop since 1950 in the volume of deep cold water flowing south from the Arctic region through one of several channels into the North Atlantic. Most of the decrease has occurred during the past 30 years, and the rate of decline had accelerated in the past five years. This study, reported in *Nature*, combined waterflow readings from the ocean floor after 1995 with 50 years of temperature and salinity records from weather ships in the area.

Rising Volume of Freshwater in the Norwegian Sea

More evidence of thermohaline breakdown has come from a rising volume of freshwater in the Norwegian Sea from melting sea ice (Connor 2001a, 14). "The seas are warmer and there is more freshwater not just from melting sea ice but from the Siberian rivers. The water has to be

salty and dense or it just won't sink," researcher Bogi Hansen said (Connor 2001a, 14). "Estimating the volume flux conservatively, we find a decrease by at least 20 percent relative to 1950. If this reduction in deep flow from the Nordic seas is not compensated by increased flow from other sources," wrote Hansen and colleagues (Hansen, Turrell, and Sterhus 2001, 927). "It implies a weakened global Thermohaline circulation and reduced inflow of Atlantic water to the Nordic seas," they added.

"This is yet another very important brick in a very solid wall of evidence about the reality of global climate change," said Andrew Weaver, who holds the Canada Research Chair in Atmospheric Studies at the University of Victoria (Calamai 2001, A-18).

The findings most likely indicate an equal drop in the amounts in warm surface water flowing north through the same passage, said Hansen. "The motor driving the ocean conveyer belt appears to be slowing down," Hansen said (Calamai 2001, A-18; Hansen, Turrell, and Sterhus 2001, 927). According to a report by Steve Connor in the *London Independent*,

> The research centered on measurements taken of water movements in a deep-sea channel that separates the Faeroes from northern Scotland. The current in the Faeroe Bank channel pushes about 2 million cubic meters of water every second into the Atlantic, an amount equivalent to about twice the total flow of all the rivers of the world combined. (Connor 2001a, 14)

Hansen explained, "If this reduction we have seen in the Faeroe Bank channel is also seen in the Denmark Strait then we can be sure that the North Atlantic flow has been reduced" (Connor 2001a, 14).

FURTHER READING

Calamai, Peter. "Atlantic Water Changing: Scientists." *Toronto Star*, June 21, 2001, A-18.

Connor, Steve. "Britain Could Become as Cold as Moscow." *London Independent*, June 21, 2001a, 14.

Cooke, Robert. "Waters Reflect Weather Trend: Study Finds Warming Effects." *Newsday*, December 18, 2003, A-2.

Cowen, Robert C. "Into the Cold? Slowing Ocean Circulation Could Presage Dramatic—and Chilly—Climate Change." *Christian Science Monitor*, September 26, 2002a, 14.

Curry, Ruth, Bob Dickson, and Igor Yashayaev. "A Change in the Freshwater Balance of the Atlantic

Ocean over the Past Four Decades." *Nature* 426 (December 18, 2003):826–829.

Davidson, Keay. "Going to Depths for Evidence of Global Warming: Heating Trend in North Pacific Baffles Researchers." *San Francisco Chronicle*, March 1, 2004a, A-4.

Dickson, B., I. Yashayaev, J. Meincke, B. Turrell, S. Dye, and J. Holfort. "Rapid Freshening of the Deep North Atlantic Ocean over the Past Four Decades." *Nature* 416 (April 25, 2002):832–836.

Fukasawa, Masao, Howard Freeland, Ron Perkin, Tomowo Watanabe, Hiroshi Uchida, and Ayako Nishina. "Bottom Water Warming in the North Pacific Ocean." *Nature* 427 (February 26, 2004):825–827.

Häkkinen, Sirpa, and Peter B. Rhines. "Decline of Subpolar North Atlantic Circulation during the 1990s." *Science* 304 (April 23, 2004):555–559.

Hansen, Bogi, William R. Turrell, and Svein Sterhus. "Decreasing Overflow from the Nordic Seas into the Atlantic Ocean through the Faroe Bank Channel since 1950." *Nature* 411 (June 21, 2001):927–930.

Kerr, Richard A. "Climate Change: Sea Change in the Atlantic." *Science* 303 (January 2, 2004):35.

McPhaden, M. J., and D. Zhang. "Slowdown of the Meridional Overturning Circulation in the Upper Pacific Ocean." *Nature* 415 (February 6, 2002):603–607.

Peterson, Bruce J., Robert M. Holmes, James W. McClelland, Charles J. Vorosmarty, Richard B. Lammers, Alexander I. Shiklomanov, Igor A. Shiklomanov, and Stefan Rahmstorf. "Increasing River Discharge to the Arctic Ocean." *Science* 298 (December 13, 2002):2171–2173.

Speth, James Gustave. *Red Sky at Morning: America and the Crisis of the Global Environment.* New Haven, CT: Yale University Press, 2004.

Stocker, Thomas F., Reto Knutti, and Gian-Kasper Plattner. "The Future of the Thermohaline Circulation—A Perspective." In *The Oceans and Rapid Climate Change: Past, Present, and Future*, ed. Dan Seidov, Bernd J. Haupt, and Mark Maslin, 277–293. Washington, D.C.: American Geophysical Union, 2001.

"Study Reports Large-scale Salinity Changes in Oceans: Saltier Tropical Oceans, Fresher Ocean Waters Near Poles Are Further Signs of Global Warming's Impacts on Planet." *Ascribe Newsletter*, December 17, 2003. (LEXIS)

"Study Reveals Increased River Discharge to Arctic Ocean: Finding Could Mean Big Changes to Global Climate." *Ascribe Newsletter*, December 12, 2002. (LEXIS)

Thorpe, R. B., J. M. Gregory, T. C. Johns, R. A. Wood, and J. F. B. Mitchell. "Mechanisms Determining the Atlantic Thermohaline Circulation Response to Greenhouse Gas Forcing in a Non-Flux-Adjusted Coupled Climate Model." *Journal of Climate* 14 (July 15, 2001):3102–3116.

Våge, Kjetil, Robert S. Pickart, Virginie Thierry, Gilles Reverdin, Craig M. Lee, Brian Petrie, Tom A. Agnew, Amy Wong, and Mads H. Ribergaard. "Surprising Return of Deep Convection to the Subpolar North Atlantic Ocean in Winter 2007–2008." *Nature Geoscience* 2 (January 2009). http://www.nature.com/ngeo/journal/v2/n1/abs/ngeo382.html.

Thunderstorms and Tornadoes, Frequency and Severity of

Rising temperatures across parts of the United States by the end of the twenty-first century may double the number of days with conditions likely to produce severe thunderstorms, according to a study by Robert J. Trapp and colleagues in the *Proceedings of the National Academy of Sciences* (2007, 19,719–19,723). The Atlantic and Gulf coastal regions—roughly New York City to Atlanta to New Orleans—stand the greatest chance of increasingly severe storms that include tornadoes, hail, strong winds, and downpours. A warmer atmosphere pumps more water vapor into the air and holds a larger amount before it releases as precipitation.

Street-level evidence supports the study. To cite one of several examples, out-of-season tornadoes killed three people and injured several others from Central Wisconsin (Wheatland) to southwest Missouri on January 7 and 8, 2008, as they strafed several homes. The same outbreak also affected Arkansas, Illinois, Missouri, and Oklahoma.

On February 5 and 6, 2008, an outbreak of more than 70 tornadoes that would have been severe in May killed 50 people in a wide swath from Arkansas to Tennessee. Police in several towns and suburbs of cities (including Memphis) described devastation resembling war zones. Many witnesses expressed awe at the ferocity of the storms, which ripped some houses entirely off their foundations.

Such storms are rare, but they have occurred before. During February 1971, severe tornadoes killed 134 people in Mississippi and Louisiana. The 2008 outbreak appeared to have been the country's second worst for February. On January 21 and 22, 1999, 134 tornadoes scorched the South, killing nine people. Still, the storms in 2008 were notable for their violence and size. One, in Arkansas, was described as six miles wide, a black wall of howling wind and debris.

Unusually severe weather also was being reported in upstate New York. Arthur Einhorn, a retired anthropologist and long-time area resident, described a storm that was unusual in its ferocity:

> We all have heard of "sticker shock" … there's another form I call "wake-up shock." This morning I went out at 7 A.M. to feed the wild creatures at the feeding stations.… a daily routine I generally conduct in bathrobe, slippers (regardless of temps or snow depth), and most often very much still half asleep. This morning it was the "calm after the storm" … very quiet, moderate temp at 39° F, a bit of sun-peek, all the snow gone, and despite many blow-down dead branches here and there, I felt my property had come through the storm reasonably well.… considering we had gusts yesterday that hit 80 mph at times, and just north of us in both New York and adjacent Canada, gusts hit at 100 mph, several people were killed, roofs ripped off, trees smashed cars to pancakes, trucks were tipped over, power went out in many places.… general devastation. (Einhorn 2007)

High-altitude clouds in Earth's tropics associated with severe storms and heavy rainfall have been increasing as the atmosphere warms, according to work by scientists at NASA's Jet Propulsion Laboratory in Pasadena, California. Senior research scientist Hartmut Aumann summarized a study linking severe storms, torrential rain, and hail with seasonal variations in average sea-surface temperature of the tropical oceans. For every 1° C increase in average ocean surface temperature, this research observed a 45 percent rise in frequency of high-level clouds. At the present rate of global warming, 0.13° C per decade, this study inferred that the frequency of such storms may increase 6 percent per decade ("NASA Study" 2008).

FURTHER READING

Einhorn, Arthur. "Sticker Shock" (personal communication). October 10, 2007.
"NASA Study Links Severe Storm Increases, Global Warming." NASA Earth Observatory, December 19, 2008. http://earthobservatory.nasa.gov/Newsroom/view.php?id=36309.
Trapp, Robert J., Noah S. Diffenbaugh, Harold E. Brooks, Michael E. Baldwin, Eric D. Robinson, and Jeremy S. Pal. "Changes in Severe Thunderstorm Environment Frequency During the 21st Century Caused by Anthropogenically Enhanced Global Radiative Forcing." *Proceedings of the National Academy of Sciences* 104 (December 11, 2007):19,719–19,723.

Tipping Points

Among scientists who keep tabs on the pace of global warming, anxiety has been rising that the Earth is reaching an ominous threshold, a point of no return ("tipping point" in some of the scientific literature). Within a decade or two, various feedbacks will accelerate the pace of greenhouse warming past any human ability to contain or reverse it.

Carbon-dioxide levels in the atmosphere are rising rapidly, fed, among other provocations, by increasing fossil-fuel use in the United States, melting permafrost, slash-and-burn agriculture in Indonesia and Brazil, increasing wildfires, and rapid industrialization using dirty coal in China and India. Sir John Houghton, one of the world's leading experts on global warming, told the *London Independent*, "We are getting almost to the point of irreversible meltdown, and will pass it soon if we are not careful" (Lean 2004b, 8).

Global-warming's evidentiary trail is being written in melting ice (Arctic, Antarctic, and mountain glaciers), as well as in warming seas, and their effects on plant and animal life in the oceans that cover two-thirds of the Earth's surface. James E. Hansen, director of NASA's Goddard Institute for Space Studies commented:

> In my opinion there is no significant doubt (probability more than 99 percent) that such additional global warming of 2° C would push the Earth beyond the tipping point and cause dramatic climate impacts including eventual sea level rise of at least several meters, extermination of a substantial fraction of the animal and plant species on the planet, and major regional climate disruptions. Much remains to be learned before we can define these effects in detail, but these consequences are no longer speculative climate model results. Our best estimates for expected climate impacts are based on evidence from prior climate changes in the Earth's history and on recent observed climate trends. (Hansen 2006b, 30)

James Lovelock, 88 years of age in 2006, who works independently in a laboratory in an old barn behind his farmhouse in Devon, St. Giles-on-the-Heath, England, said, "Our global furnace is out of control. By 2020 [to] 2025, you will be able to sail a sailboat to the North Pole. The Amazon will become a desert, and the forests of Siberia will burn and release more methane and plagues will return" (Powell 2006, C-1). "I'm an optimist," Lovelock said. "I think that after the warming sets in and the survivors have settled in near the Arctic, they will find a way to adjust. It will be a tough life enlivened by excitement and fear" (Powell 2006, C-1). *See also:* Feedback Loops

FURTHER READING

Hansen, James E. "Declaration of James E. Hansen." *Green Mountain Chrysler-Plymouth-Dodge-Jeep et al. Plaintiffs v. Thomas W. Torti, Secretary of the Vermont Agency of Natural Resources et al. Defendants.* Case Nos. 2:05-CV-302 and 2:05-CV-304, Consolidated. United States District Court for the District of Vermont. August 14, 2006b. http://www.giss.nasa.gov/~dcain/recent_papers_proofs/vermont_14aug20061_textwfigs.pdf.

Lean, Geoffrey. "Global Warming Will Redraw Map of the World." *London Independent*, November 7, 2004b, 8.

Powell, Michael. "The End of Eden: James Lovelock Says This Time We've Pushed the Earth Too Far. *Washington Post*, September 2, 2006, C-1. http://www.washingtonpost.com/wp-dyn/content/article/2006/09/01/AR2006090101800_pf.html.

Tropical Fish and Sharks off Maine

Following warmer waters, a few sharks were sighted off the coast of Maine during the summer of 2002. According to an account in the *Boston Herald,* "Whether it was blue or mako sharks that invaded the crowded Maine beach, banishing swimmers to the sand, the unusual occurrence is the result of either fish chasing food close to shore or warmer waters luring a more tropical variety to New England waters, marine experts said." The Wells, Maine, shark sightings are "a very unusual circumstance," said Greg Skomol, a shark specialist for the state Division of Marine Fisheries. "We wouldn't expect to see that in Maine or in New England, in general." Skomol added (Richardson 2002, 3).

The sharks swam near the beach three days in a row in a popular resort called Vacationland as the tide was receding, probably hoping to catch whatever baitfish were being washed out of nearby crevices in a rocky ledge. The sharks' usual food supply (mackerel and some herring) may have been lured to shallow waters by unusually warm air temperatures in the 90s (Richardson 2002, 3). "We had hot weather and sometimes that will change the distribution of fish and bait fish that sharks feed on," Skomol

said. "The high temperatures may have increased the mackerel and herring, therefore bringing in the sharks. But with one event, it's hard to predict anything, though it's worth watching" (Richardson 2002, 3).

Bruce Joule, a marine biologist in Boothbay Harbor who advised officials to close the beach when sharks appeared, said he believes water temperature had nothing to do with the sharks. "I don't think the temperature has really anything to do with it," Joule said. "The sharks came in following bait, whether it was natural migration [of bait fish] or the smaller fish were chasing their own food. It's nature" (Richardson 2002, 3). Following Joule's advice, Wells Fire Chief Marc Bellefeuille closed the beach as a precautionary measure, allowing sunbathers into the water only up to their ankles. "This is the first time anyone can remember shark sightings like this," Bellefeuille said (Richardson 2002, 3).

At about the same time, unusually warm water temperatures also brought tropical fish to New England shores, including yellow fin, dolphin fish, white marlin, and sometimes skipjack tuna, Skomol said. "How close they come depends on the water. They're not aggressive. We've been seeing them the last three years" (Richardson 2002, 3). President George W. Bush was photographed during the summer of 2002 landing a large striped bass that his daughter Jenna had caught on a fishing holiday off the coast of Maine. Striped bass is a warmer-water species, which only a generation ago would never have been seen so far north along America's Atlantic coast (Connor 2002c, 12). Bush displayed no clue that he realized his catch was unusual or influenced by climate change.

FURTHER READING

Connor, Steve. "Strangers in the Seas: Exotic Marine Species Are Turning up Unexpectedly in the Cold Waters of the North Atlantic." *London Independent*, August 5, 2002c, 12–13.

Richardson, Franci. "Sharks Take the Bait: Experts: Sightings in Maine an 'Unusual Circumstance.'" *Boston Herald*, August 11, 2002, 3.

Tropical Fish and Warm-Climate Birds Migrate

As cold-water fish abandon waters near Britain, species that usually live in southerly waters have taken their places. Sightings of warm-water fish have been plentiful since about 1990 in British coastal waters and have received considerable publicity in local newspapers. Warm-water species have been turning up regularly off the coasts of Devon and Cornwall for more than a decade; some scientists believe they are clear indicators of global warming. In the late 1980s, southern species such as sunfish and torpedo rays began to appear; by the late 1990s, such visitors were no longer regarded as oddities.

A series of small cold-water marine animals such as copepods were replaced by their warm-water cousins (McCarthy 2002, 13). "As fish are very dependent on the temperature of the water, it is sensible to link these changes with changes in water temperature. They would be consistent with predictions of climate change," said Douglas Herdson of the National Marine Aquarium (McCarthy 2002, 13).

Number and Frequency of Sea Creatures' Migrations

Such sightings are not unprecedented. What has changed is their number and frequency. With migrations of sea creatures have come a number of birds that feed on them. Bob Swann, secretary of the Seabird Group, an international conservation and research organization, said:

> There is much evidence of species previously more associated with more southerly regions of the Atlantic appearing around the British Isles. Climate change can influence oceanic currents and the availability of food—a prime reason for the presence of these birds. Certainly a lot of our breeding seabirds are currently doing very well because they are finding plenty to eat. (Unwin 2001, 7)

A team of British marine biologists analyzed records dating back 40 years and, in 2002, announced a "strong link" between the northward migration of fish and rising sea temperatures. These scientists related the arrival of tropical and semitropical fish off the coast of Cornwall, the southern-most tip of the British mainland, to increases in the average temperature of the North Atlantic Ocean (Connor 2002c, 12).

Surveying records to 1960, the scientists found that "more exotic species of fish are being caught or washed ashore now than ever before and that the sightings can be directly linked to a corresponding rise in sea temperatures." The link is said to be a "significant correlation" and could explain why Cornwall in particular has seen so many exotic species of marine wildlife from warmer regions of the world, according to Tony Stebbing, formerly a biologist with the Plymouth Marine Laboratory, whose work was funded by the Natural Environment Research Council (Connor 2002c, 12).

Mantis Shrimp off England's South Coast

During May 2004, mantis shrimp (warm-water creatures usually found in tropical waters) were caught in trawler nets in Weymouth Bay, Dorset, on England's south coast. Fisherman took the orange-colored shrimp to the nearby SeaLife Centre, where experts identified them. The mantis shrimp (also known as "toe splitter" shrimp) average three inches long. They can strike at up to 100 miles per hour with hammer-like claws, packing a force as powerful as a 0.22-caliber bullet. Two years earlier, another colony of mantis shrimps was found at the north end of Cardigan Bay, Wales.

During mid-summer, 2004, a shoal of two-foot-long gray triggerfish were observed off the British coast. The triggerfish, whose usual habitat is the tropical Atlantic and the Mediterranean, were discovered two miles off the Isle of Purbeck, Dorset. During June 2002, a group of bright purple jellyfish, called by-the-wind sailor fish, were found washed up at Kimmeridge Bay. Their usual habitat is the deep waters of the Mediterranean. Another visitor from southern seas was the four-inch Weaver fish, which buries itself in sand close to shore and releases stinging venom from its dorsal spines when stepped on.

According to an account in the *London Independent*: "This summer's biggest seabird sensation was a red-billed tropic bird (*Phaethon aethereus*), which flew around a yacht about 20 miles south of the Isles of Scilly. This bird previously had not been recorded in northern European waters. The nearest colonies are on the Cape Verde Islands and islets off West Africa. Breeding also occurs on Ascension Island, St. Helena, the West Indies, and on the Red Sea, Persian Gulf, and Arabian Sea islands. Cape Verde and other islands off the coast Africa, including off Scilly and the coasts of Devon and Cumbria, are the source of a series of summer sightings of rare Fea's petrels (*Pterodroma feae*) (Unwin 2001, 7).

Whales, Dolphins, and Jellyfish in British Waters

Whales and dolphins are being observed frequently in British waters as the area has warmed. During one ferry's round-trip between Portsmouth and Bilbao in 2001, wildlife enthusiasts logged sightings of 71 fin whales, four Cuvier's beaked whales, 37 pilot whales, 25 common dolphins, 352 striped dolphins, 120 unidentified dolphins, and seven unidentified large whales. In addition, two sperm whales were spotted in the English Channel from a ferry sailing between Portsmouth and Bilbao (in northern Spain). Newquay's Blue Reef Aquarium took custody of two loggerhead turtles which probably drifted to Britain after damaging their flippers in the open ocean (Unwin 2001, 7).

Stella Turk of the Cornwall Wildlife Trust was quoted in the *London Independent* as saying that a rise in sea temperatures may be a cause of northward migration of tropical sea creatures and birds. "It appears the sea is becoming generally warmer and it could be that such creatures are staying close to Britain throughout the year. An unusually wide range has already been reported this summer and it's still only mid-July—there could be many more surprises over the coming weeks" (Unwin 2001, 7).

In addition to fish and birds, British observers in 2002 sighted hundreds of root-mouthed jellyfish, as well as Rhizo-stoma octopus. Large, disc-shaped Sunfish (*Mola mola*) also have been spotted more frequently in recent summers. Turk said that sightings in 2002 began earlier than before, in May. Another surprise is a report of a six-foot bluefin tuna, *Thunnus thynnus*. On July 14, 2002, about 40 basking sharks were spotted a mile off Perranporth, North Cornwall. A pilot whale was observed at the same time. A spokesman for Falmouth Coast Guard said: "The sharks are huge—they go up to 30 foot [long]. When the water warms up we do get basking sharks here, but it is unusual to get so many" (Unwin 2002, 7). The flying gurnard, unknown in British waters before 1980, also has been sighted more frequently. The gurnards, with their elongated pectoral fins that enable them to move quickly through the water, were first caught in the nets of Cornish fishermen. The sharp-nosed

shark was first netted in 1984 (Connor 2002c, 12).

More Tropical Migrants

Global warming has brought an unexpected benefit to British West Country fishermen who have been struggling to make a living as their traditional catches dwindle. A slight rise in sea temperature has meant that valuable shellfish that once were unable to thrive north of the Channel Islands can now be farmed for export. Disc-like molluscs Haliotis (sea ear), which grow 20 centimeters (seven inches) long have fetched high prices on Japanese markets (Brown and Sutton 2002, 8). This valuable delicacy is known as abalone in most of the world but as "ormers" in Britain. Abalone is best known by tourists as a source of iridescent mother-of-pearl jewelry that is used as an inlay. In Japan, the gonads of the abalone eaten raw are regarded as a particular delicacy, while in California the abalone is consumed in steaks. The people of Guernsey, in England, have traditionally eaten ormer stew (Brown and Sutton 2002, 8).

One notable warm-water visitor to English waters, a single slipper lobster, was caught near the southwest tip of England and later displayed at the Plymouth National Marine Aquarium. The five-inch-long slipper lobster (*Scyllarus arctus*) is usually found near the coasts of the Mediterranean Sea. Only about a dozen have been recorded in United Kingdom waters during the last 250 years. The most recent specimen, caught by fisherman Barry Bennett, was the fifth to be caught in British waters since 1999. The increasing number of slipper lobsters in British waters is one of many indications that warm-water marine species have been moving northward because ocean temperatures are rising (McCarthy 2002, 13).

Other warm-water fish have been found in English waters from two fish families, the breams and the jacks. A Guinea amberjack was first recorded off Guernsey. The previous year, there were five sightings of the almoco jack in the West Country (McCarthy 2002, 13). In 2002, a tropical zebra sea bream, usually resident off the West African coast, was caught for the first time in British waters. The 14-inch fish was accidentally netted by Ross White, 29, a commercial fisherman, near Portland, Dorset, far from its native waters off Senegal and Mauritania. A seahorse also was sighted in the Thames River—a rare, but not unprecedented, event. Another seahorse was sighted in 1976.

During November 2001, Britain's first reported barracuda was caught about 40 miles from the site of the slipper-lobster catch. England is not alone in observing tropical fish; two years after the English barracuda catch, another barracuda was caught on Seattle's waterfront. As they have near England, cold-water cod (of the Pacific variety) are becoming increasingly rare off Washington State's coastline (Stiffler and McClure 2003, A-8). During the summer of 2004, a giant squid, a species that usually ranges no farther north than Mexico's coastal waters, was caught near Maple Bay, in southern British Columbia ("Jumbo Squid" 2004).

Charles Clover of the London *Daily Telegraph* reported that Red Mullet, which were restricted to waters south of the English Channel before 1990, by 2002 were being caught in commercial quantities on both coasts of Scotland. "The largest geographical movement recorded over the past decade, the warmest on record," wrote Clover, "has been made by species of zooplankton, tiny shrimp-like creatures which form the base of the marine food chain. Warm-water species of copepod, as these crustaceans are known, have moved 600 miles northwards up the Bay of Biscay over the past decade, bringing warm-water fish species with them" (Clover 2002b, 14). At the same time, the cold-water copepod, *Calanus finmarchicus*, the main food of the cod and of the sand eel on which the cod also feed has moved north from the North Sea. Martin Edwards, of the Sir Alastair Hardy Laboratory in Plymouth, said the North Sea was in a "transitional state" with the consequences for fish stocks, already endangered as a result of overfishing, hard to predict (Clover 2002b, 14).

Warm-Water Species off Ireland

Warm-water fish have been sighted in Irish waters as well. Among these have been various species of sharks, poisonous puffer fish, loggerhead turtles, and triggerfish. According to a report by Lynne Kelleher in London's *Sunday Mirror*, coastal waters that previously reached a summer maximum of 15° C now commonly reach 20° C, drawing the warm-water species. According to Kelleher,

Fish never seen before are appearing in greater numbers and some are beginning to breed as they become acclimatized. A great white shark was spotted off the coast of Cornwall in recent years and exotic fish such as moray eels, mako sharks and anchovies are swimming off Irish shores. (Kelleher 2002, 15)

Kevin Flannery, a marine biologist with the Department of Marine and Natural Resources, said the fish are arriving because of warmer water temperatures. He said, "There is a definite correlation between the rise in temperatures and the arrival of new species of fish in Irish waters." Flannery continued:

The temperatures used to go below 5° C in the winter and go up to 14° or 15° C in the summer. Now they are between 5° and 7° C in the winter and can go between 17° and 20° C in the summer. A moray eel, which is found in Australia and Canada, was caught by a vessel from Waterford off the coast of Cork last year. We have found three puffer fish. One was found off Fenit in Kerry last winter. A number of loggerhead turtles have been found washed up thousands of miles from where they came. (Kelleher 2002, 15)

Large numbers of anchovies, usually found off the coast of Portugal, have been found swimming in waters from Shannon to Kinsale, in Irish waters. Puffer fish, which are poisonous, also have been sighted, probably for the first time (Kelleher 2002, 15). Tropical triggerfish, which have been found in the largest numbers, are small in size but become aggressive to divers and swimmers if they are disturbed during the breeding season. Different types of bream and dory fish also have been sighted in Irish waters.

Flannery said:

The number of tropical species has increased dramatically. There are between 10 to 15 rare species found in recent years. One [fishing] vessel found about 80 triggerfish last year. They can do a lot of damage here to crab, lobster and crayfish that they feed on. They have teeth like a rat and can kill a lobster. Fishermen are finding them inside the lobster pots. Their natural habitat is the tropical waters of Spain and Africa. They won't survive in waters less than 14° C. In 1995, the waters reached temperatures of about 20° C that was one of the highest temperatures. The triggerfish are staying around. They are not just coming in and out to feed. We are finding a number of pregnant triggerfish. This suggests they are acclimatizing. They wouldn't breed unless they were staying around. (Kelleher 2002, 15)

Flannery said that as tropical fish migrate to Ireland, traditional cold-water species such as cod have been leaving its coastal waters. He said: "The rise in temperature could also mean the demise of cod which need cold temperatures. If temperatures go over 17 degrees they will die. There has also been the demise of the Arctic char in our lakes. It is a combination of the pollution and the temperature" (Kelleher 2002, 15).

FURTHER READING

Brown, Paul, and Tony Sutton. "Global Warming Brings New Cash Crop to West Country as Rising Water Temperatures Allow Valuable Shellfish to Thrive." *London Guardian*, December 10, 2002, 8.

Clover, Charles. "Global Warming Is Driving Fish North.'" *London Daily Telegraph*, May 31, 2002b, 14.

Connor, Steve. "Strangers in the Seas: Exotic Marine Species Are Turning up Unexpectedly in the Cold Waters of the North Atlantic." *London Independent*, August 5, 2002c, 12–13.

"Jumbo Squid Has a Message for Us: Changing Global Patterns Are Going to Bring Different Species into Our Waters." *Victoria Times-Colonist* (British Columbia), October 8, 2004, A-10.

Kelleher, Lynne. "Look Who's Here: Tropical Fish Warming to Waters Around Ireland." *London Sunday Mirror*, October 20, 2002, 15.

McCarthy, Michael. "Climate Change Provides Exotic Sea Life with a Warm Welcome to Britain." *London Independent*, January 24, 2002, 13.

Stiffler, Linda, and Robert McClure. "Effects Could Be Profound." *Seattle Post-Intelligencer*, November 13, 2003, A-8.

Unwin, Brian. "Tropical Birds and Exotic Sea Creatures Warm to Britain's Welcoming Waters." *London Independent*, August 20, 2001, 7.

Tropical Glacial Ice Melt

Between 1991 and 2005, the Qori Kalis, the largest outlet glacier of the Quelccaya Ice Cap in the Andes, retreated about 10 times faster than between 1963 and 1978, or 60 meters a year, compared with six meters a year (Thompson et al. 2006, 10536). A 60-meter deep lake has formed in an area that had been frozen only a few decades earlier. Mount Kilimanjaro lost 82 percent of its areal extent between 1912 and 2005 (Thompson et al. 2006, 10536).

Lonnie G. Thompson, who for decades has been taking core samples from the ancient ice of tropical glaciers around the world, told Doug Struck of the *Washington Post* that Earth's climate is undergoing an abrupt change, ending a cooler period that began with a swift "cold snap" in the tropics 5,200 years ago. This shift coincided with the establishment of the first cities and the oldest calendars. Warming that is melting Earth's tropical glaciers indicates to Thompson that the "climate system has exceeded a critical threshold," which has sent tropical-zone glaciers in full retreat and will melt them completely in the near future (Struck 2006d, A-3).

"The current warming at high elevations in the low- to midlatitudes is unprecedented for at least the last two millennia," wrote Thompson and colleagues (2006, 10536). Additionally the researchers noted, "The continuing retreat of glaciers, many having persisted for thousands of years, signals a recent and abrupt change in the Earth's climate system." Paleoclimatic evidence indicates that present warming and glacial retreat may be unprecedented for at least the last 5,200 years (Thompson et al. 2006, 10536). Paleoclimatic records indicate that the first half of the Holocene was warmer than the second half; temperatures in the tropical Pacific Ocean may have been 1.5° to 2° C warmer 8,000 to 11,000 years ago (Thompson et al. 2006, 10541).

Thompson, writing with eight other researchers in an article published in the *Proceedings of the National Academy of Sciences,* said, "the ice samples show that the climate can and did cool quickly, and that a similarly abrupt warming change started about 50 years ago. Humans may not have the luxury of adapting to slow changes" (Struck 2006d, A-3).

This study ends with a cautionary message, under the heading "Implications."

> The recent, rapid, and accelerating retreat of glaciers on a near-global scale suggests that the current increase in the Earth's globally averaged temperature may now have prematurely interrupted the natural progression of cooling in the late Holocene. These observations suggest that within a century human activities may have nudged global-scale climate conditions closer to those that prevailed before 5,000 years ago, during the early Holocene. If this is the case, then Earth's currently retreating glaciers may signal that the climate system has exceeded a critical threshold and that most low-latitude, high-altitude glaciers are likely to disappear in the near future. (Thompson et al. 2006, 10542)

There are thresholds in the system," Thompson said. When they are crossed, "There is the risk of changing the world as we know it to some form in which a lot of people on the planet will be put at risk." He explained, "I think the temperature will continue to rise, the glaciers will continue to melt. Sea levels will continue to rise. I think there is a good indication now that the magnitude of severe storms will rise" (Struck 2006d, A-3).

Thompson's work summarized evidence from around the world and ice-core sampling from seven locations in the South American Andes and the Asian Himalayas. Thompson, whose research has focused on glaciers in the high mountains of the tropics, wrote that the warming there "is unprecedented for at least two millennia." He joined his wife, Ellen Mosley-Thompson, an expert in polar ice sampling, and concluded that the glacial retreat "signals a recent and abrupt change in the Earth's climate system" (Struck 2006d, A-3).

FURTHER READING

Struck, Doug. "Earth's Climate Warming Abruptly, Scientist Says. Tropical-Zone Glaciers May Be at Risk of Melting." *Washington Post*, June 27, 2006d, A-3. http://www.washingtonpost.com/wp-dyn/content/article/2006/06/26/AR2006062601237_pf.html.

Thompson, Lonnie G., Ellen Mosley-Thompson, Henry Brecher, Mary Davis, Blanca León, Don Les, Ping-Nan Lin, Tracy Mashiotta, and Keith Mountain. "Abrupt Tropical Climate Change: Past and Present." *Proceedings of the National Academy of Sciences* 103, no. 28 (July 11, 2006):10536–10543.

Tropical Zones, Expansion of

Warming temperatures in the Earth's atmosphere seem to be affecting circulation on both sides of the Equator, expanding tropical zones and forcing northward desert zones. If one looks at a globe, desert zones appear most frequently about 30 degrees north and south latitude—in the U.S. Southwest, the Sahara, Middle East, Kalahari, and Australia.

These areas often suffer precipitation shortages because of circulation patterns called "Hadley Cells," which cause air to fall (generally), creating high pressure, stable weather, and a lack of rainfall. In areas where air ascends (often nearer the Equator), weather patterns usually favor more frequent rainfall. Research indicates that the zones of tropical weather have expanded 140 to 330 miles north and southward on either side of the Equator, pushing drier areas further into the temperate zones.

A study published on December 2, 2007, in the scientific journal *Nature Geoscience* (an affiliate of London-based *Nature*) indicates that the tropics have been expanding more quickly than climate models had anticipated. Independent teams using four different meteorological measurements found that the tropical atmospheric belt has grown by anywhere between 2 and 4.8 degrees latitude since 1979. That translates to a total north and south expansion of 140 to 330 miles. Dian Seidel, a research meteorologist with the National Oceanic and Atmospheric Administration in Silver Spring, Maryland, lead author of the study, said, "They are big changes.... It's a little puzzling" (Borenstein 2007d). Seidel said that global warming may be coupling with stratospheric ozone depletion and more frequent El Niño conditions to produce expansion of tropical zones.

"Several lines of evidence show that over the past few decades the tropical belt has expanded," Seidel and colleagues wrote.

This expansion has potentially important implications for subtropical societies and may lead to profound changes in the global climate system. Most important, poleward movement of large-scale atmospheric circulation systems, such as jet streams and storm tracks, could result in shifts in precipitation patterns affecting natural ecosystems, agriculture, and water resources.... The observed recent rate of expansion is greater than climate model projections of expansion over the twenty-first century, which suggests that there is still much to be learned about this aspect of global climate change. (Seidel et al. 2007)

Climate scientists Andrew Weaver of the University of Victoria, British Columbia, and Richard Somerville of the Scripps Institution of Oceanography said that computer models often have been underestimating the effects of global warming. "Every time you look at what the world is doing it's always far more dramatic than what climate models predict," Weaver said (Borenstein 2007d).

Dry areas on the edge of the tropics (where weather is governed by descending air in Hadley Cells), such as the U.S. Southwest, parts of the Mediterranean, and southern Australia, could become drier as this effect intensifies. "You're not expanding the tropical jungles, what you're expanding is the area of desertification," Weaver said (Borenstein 2007d). *See also:* Hadley Cells

FURTHER READING

Borenstein, Seth. "Earth's Tropics Belt Expands, May Mean Drier Weather for U.S. Southwest, Mediterranean." Associated Press, December 2, 2007d. (LEXIS)

Seidel, Dian, Qiang Fu, William J. Randel, and Thomas J. Reichler. "Widening of the Tropical Belt in a Changing Climate." *Nature Geoscience* (December 2, 2007), doi: 10.1038/ngeo.2007.38. http://www.nature.com.leo.lib.unomaha.edu/ngeo.

$\mathcal{U}$

"Urban Heat-Island Effect" and Contrarians

Greenhouse skeptics often become particular about sources of human-induced warming, shying away from potential greenhouse-effect warming from various other sources. Such is the case with the urban heat-island effect. Many skeptics go to great lengths to blame rising temperatures on temperature readings taken in cities because they have risen with increasing urbanization. They also seem to forget that a majority of humankind lives in urban areas, where billions of people experience the weather. The urban heat-island effect has been known since at least 1850, when people commented on the urban heat island around the town of St. Louis, Missouri. As urban areas expand, so do their heat-island effects.

Not only do cities generate more heat than the surrounding countryside, but their structures also retain accumulated heat longer than natural surroundings. Some cities record temperatures as much as 10° C (18° F) warmer than nearby countryside on calm, clear nights. The difference decreases with the speed of the wind; with a wind speed of more than 20 miles an hour, the air usually mixes enough to spread the city's heat downwind as quickly as it is created.

Urban areas have many buildings with vertical walls. The walls act as light traps or cavities that capture and store heat during the day and then reemit that heat at night. The skeptics extend the same argument to weather stations at airports, which create heat islands of their own as large, paved runways soak up sunshine, raising temperatures. The burning of jet fuel also contributes heat at airports. Today, a large number of surface-temperature observations are taken at airports, so skeptics object that they overstate the rate of warming.

Many greenhouse skeptics make a clear (and largely imaginary) distinction between global warming caused by greenhouses gases and "urban heat islands," as if they were separate phenomena. Actually, urban heat islands are aggregates of all the ways in which humankind is raising the temperature of the Earth, of which rising greenhouse-gas "forcing" is one important part.

Not only do cities provide emissions centers for all manner of greenhouse gases, but they also replace open fields and forests with concrete and asphalt, which often intensifies heat absorption. Cities contain interior spaces that sometimes kill occupants who have no air conditioning during heat waves. Urban heat islands are centers of industrial, office, and residential waste heat created by machine culture. The cars and trucks that ply urban areas also add both greenhouse gases and waste heat to the atmosphere.

Expanding urban areas worldwide are contributing increasing amounts of human-induced heat that adds momentum to infrared forcing. Early in the twentieth century, for example, Mexico City was a relatively small metropolis covering 86 square kilometers bordering on a large lake with a surface area of 120 square kilometers. At that time, according to a 2000 study in *Climate Change* by Aron Jazcilevich and colleagues, Mexico City's heat-island effect averaged about 1.5° C (Jazcilevich et al. 2000, 515). Mexico City at the end of the twentieth century sprawled across 1,200 square kilometers, while the lake had shrunk to about 10 square kilometers. The heat-island effect at that time (combining the effects of human heat generation with loss of the lake's cooling effect) had reached 8° to 10° C (Gillon 2000, 555).

Dale Quattrochi of NASA's Marshall Space Flight Center has studied the urban heat-island phenomenon by flying NASA aircraft over cities and measuring temperatures using equipment developed for the space program. He found that heat "domes" form over cities, triggering thunderstorms, increasing the production of polluting ozone, and raising local temperatures by as much as 10° F (5.5° C). "Over Atlanta, the heat island is causing the city to create its own weather," he said. "At the end of July and the beginning of August, we have seen a series of thunderstorms generated in the early hours of the morning—when no thunderstorm would normally occur—as a result of heat rising from the city" (Hawkes 2000). Between 1980 and 2000, Atlanta was one of the fastest-growing cities in North America, as it lost 380,000 acres of tree cover and gained 370,000 acres of single-family housing.

FURTHER READING

Gillon, Jim. "The Water Cooler." *Nature* 404 (April 6, 2000):555.

Hawkes, Nigel. "Giant Cities Are Creating Their Own Weather." *The Times of London*, February 23, 2000, n.p. (LEXIS)

Jazcilevich, Arón, Vicente Fuentes, Ernesto Jauregui, and Esteban Luna. "Simulated Urban Climate Response to Historical Land Use Modification in the Basin of Mexico." *Climatic Change* 44 (March 2000):515–536.

U.S. Climate Action Partnership (USCAP)

By early 2007, green credentials were all the rage in corporate America. Previously, when the official line in the Republican-controlled U.S. Senate had been that global warming was a hoax, and "sound science" at the White House meant inviting science-fiction writer Michael Crichton over for lunch, one never would have seen this: the U.S. Climate Action Partnership (USCAP), a coalition of chief executives leading several major corporations, including DuPont, General Electric, Xerox, Rio Tinto (a multinational mining conglomerate), and Duke Energy, arm in arm with leaders of four national environmental groups, all declaring their fidelity to mandatory controls that would reduce greenhouse-gas emissions by 60 to 80 percent in 50 years (Pegg 2007c). By late 2007, USCAP had enrolled 27 of the world's largest companies with combined revenues of $2 trillion and 2.7 million employees ("More CEOs" 2007, 9).

"We share a view that climate change is the most pressing environmental issue of our time and we agree that as the world's largest source of global warming emissions our country has an obligation to lead," said Peter Darbee, chairman and chief executive officer of Pacific Gas and Electric Corporation, California's largest gas and electric utility (Pegg 2007c). The U.S. economy "is the world's locomotive," Darbee told the panel, adding that the members of USCAP "believe it is critical to get the engine pulling in the right direction on climate change" (Pegg 2007c). Senator Barbara Boxer, a California Democrat and chair of the environment committee, said the USCAP recommendations marked a "turning point" in the U.S. debate over controlling greenhouse-gas emissions. "The companies and groups before us today also make clear that by acting now, we can help, not hurt our economy," she said (Pegg 2007c).

Plan for a Partnership

USCAP partners support six recommendations:

- Account for the global dimensions of climate change—U.S. leadership is essential for establishing an equitable and effective international policy framework for robust action on climate;
- Recognize the importance of technology—the cost-effective deployment of existing energy efficient technologies should be a priority;
- Be environmentally effective—mandatory requirements and incentives must be stringent enough to achieve necessary emissions reductions;
- Create economic opportunity and advantage—a climate protection program must use the power of the market to establish clear targets and timeframes;
- Be fair—solutions must account for the disproportionate impact of both global warming and emissions reductions on some economic sectors, geographic regions and income groups;
- Encourage early action—prior to the effective date of mandatory pollution limits, every reasonable effort should be made to reduce emissions. ("Mandatory U.S." 2007)

Senator James Inhofe, a Republican from the oil-producing state of Oklahoma, who had chaired the environment committee (and called global warming a "hoax") had not changed his tune after 2006. He just had no gavel in 2007. As the new committee chair, Barbara Boxer of California, brought down the gavel and said, "Senator Inhofe. You used to do this. Now I do it. Please sit down." Inhofe was still sore, as he called the members of USCAP "climate profiteers," who were motivated by self-interest.

"More and more companies that wish to profit on backs of consumers are coming out of the woodwork to endorse climate proposals in hopes of forcing customers to buy their products or to penalize their competitors," Inhofe said (Pegg 2007c). Oil companies' profits—Exxon had just reported the largest profits in the history of capitalism—did not bother Inhofe at all, as he complained that the companies involved in USCAP, such as DuPont and BP, had invested heavily in renewable energy technologies. He sounded a lot like a typewriter salesman at a computer convention.

Oil Companies Reduce Carbon Footprints

Even as Inhofe blustered, oil companies that once agreed with his position made plans to reduce their carbon footprints. Royal Dutch Shell Chief Executive Jeroen van der Veer displayed that consensus, in an introduction to the company's *Sustainability Report 2006*,

> I have said repeatedly that, for us, the debate about CO_2's impact on the climate is over. I am pleased at how our people are responding to my call to find ways to mitigate CO_2 impacts from fossil fuels. Our focus is on what we can do to reduce CO_2 emissions. We are determined to find better, lower-cost

ways to capture and store CO_2." ("Mandatory U.S." 2007)

Another oil company on USCAP's roster was ConocoPhillips. "We recognize that human activity, including the burning of fossil fuels, is contributing to increased concentrations of greenhouse gases in the atmosphere that can lead to adverse changes in global climate," said chairman and chief executive officer Jim Mulva. "While we believe no one entity can alone address the environmental, economic and technological issues inherent in any solution, ConocoPhillips will show leadership in finding pragmatic and sustainable solutions," Mulva said. ConocoPhillips began to factor the long-term cost of carbon dioxide into its capital spending plans for each of its major projects as it found ways to improve energy efficiency in its facilities, including a specific commitment to a 10 percent improvement in energy efficiency at its U.S. refineries by 2012 ("Mandatory U.S." 2007).

FURTHER READING

"Mandatory U.S. Greenhouse Gas Cap Wins New Corporate Supporters." Environment News Service, May 8, 2007. http://www.ens-newswire.com/ens/may2007/2007-05-08-01.asp.

"More CEOs Call for Climate Action." *Solutions* 38, no. 5 (November 2007):9. Environmental Defense. http://www.environmentaldefense.org.

Pegg, J. R. "U.S. Lawmakers Hear Stern Warnings on Climate Change." Environment News Service, February 14, 2007c. http://www.ens-newswire.com/ens/feb2007/2007-02-14-10.asp.

$\mathcal{V}$

Venice, Sea-Level Rise

Floods have plagued Venice for most of its history, but subsidence and slowly rising seas due to global warming have worsened flooding during the late twentieth and early twenty-first centuries. Venice, which sits atop several million wooden pillars pounded into marshy ground, has sunk by about 7.5 centimeters per century for the past 1,000 years. The rate is now accelerating.

Increased flooding has provoked plans for movable barriers across the entrance to Venice's lagoon. In Venice, the water level rose to 125 centimeters above sea level in June 2002, a record for the month. At the beginning of the twentieth century, St. Mark's Square, the center of the city, was flooded an average of nine times a year. During 2001, it flooded almost 100 times. Venice flooded 111 times during 2003, more than any other year in its lengthy history. In another century, it may flood on a permanent basis ("Heavy Rains" 2002).

Venice has lost two-thirds of its population since 1950; 60,000 people remain in the city, which hosts 12 million tourists a year, who make their way over planks into buildings with foundations rotted by perennial flooding. At the Danieli, one of Venice's most luxurious hotels, tourists arrive on wooden planks raised two feet above the marble floors amidst the suffocating stench from the high water (Poggioli 2002).

"Acqua Alta"

Venice residents and visitors have become accustomed to drills for "acqua alta," high water. A system of sirens much like the ones that convey tornado warnings in the U.S. Midwest sounds when the water surges. Restaurants have stocked Wellington boots and moved their dining rooms upstairs. Venetian gondoliers ask their passengers to shift fore and aft—and watch their heads—as they pass under bridges during episodes of high water (Rubin 2003). Some of the gondoliers have hacked off their boats' distinctive tailfins to clear the bridges brought closer by rising waters.

Faced with rising waters, Venice has proposed construction of massive retractable dikes in an attempt to hold the water at bay, amid considerable controversy. After 17 years of heated debate, the Venice's MOSE (*Modulo Sperimentale Elettromeccanico*) project will cost about US$1 billion.

Some environmentalists assert that the barriers will destroy the tidal movement required to keep local lagoon waters free of pollution and thereby damage marine life. Water quality near Venice is already precarious because pollution has leached into the lagoon from industry, homes, and motor traffic. The Italian Green Party favors shaping the lagoon's entrances to reduce the effects of tides, along with raising pavements as much as a meter inside Venice.

As proposed, the barriers will be constructed at the three entrances to Venice's natural lagoon from the Adriatic Sea. Each barrier is planned to house 79 "flippers" that can be adjusted like the flaps of an aircraft. Installed below the water line, they will be raised when the sea level rises by more than one meter, which at the turn of the millennium was occurring about a dozen times a year (Watson 2001, 18).

During normal tides, according to an account in *Scotland on Sunday*, "The hollow barriers will sit within especially constructed trenches in the bed of the channels connecting the lagoon to the open sea. When a dangerously high tide is forecast, compressed air will be forced into the flippers which will have the effect of squeezing seawater out. As they rise, more water will trickle out to be replaced by air" (Watson 2001, 18). The project is expected to provide as many as 10,000 jobs during 10 years of construction.

MOSE May Run Full Time

By mid-century, the MOSE flood-control system may be running almost all the time, severing the city from the ocean, transforming its neighboring lagoon, according to one observer, "into a stagnant pond with devastating effects on marine life and health" (Poggioli 2002). Many in Venice say that Project MOSE will not help much because it will be constructed to operate only when water rises at least 43 inches.

Environmentalists have argued that the MOSE flood-control system is a construction boondoggle that will turn Venice into a toxic bathtub in which the city's canals will be laced with sludge from surrounding heavy industry, as well as the urban area's human waste. Environmentalists have focused attention on bacteria from animal and human waste in the waters surrounding the city (Petrillo 2003). "Venice has no sewage system; they just dump the stuff right out into the canals. It's not pretty," said Rick Gersberg, a microbiologist. "Normally, the tides come in and flush everything out. But when you cut off the tide, it just sits there" (Petrillo 2003).

Venice's deputy mayor, Gianfranco Bettin, has called MOSE "Expensive, hazardous, and probably useless" (Nosengo 2003, 608).

FURTHER READING

"Heavy Rains Threaten Flood-Prone Venus." *The Straits Times* (Singapore), June 8, 2002, n.p. (LEXIS)

Nosengo, Niccola. "Venice Floods: Save Our City!" *Nature* 424 (August 7, 2003):608–609.

Petrillo, Lisa. "Turning the Tide in Venice." Copley News Service, April 28, 2003. (LEXIS)

Poggioli, Sylvia. "Venice Struggling with Increased Flooding." National Public Radio Morning Edition, November 29, 2002. (LEXIS)

Rubin, Daniel. "Venice Sinks as Adriatic Rises." Knight-Ridder News Service, July 1, 2003, n.p. (LEXIS)

Watson, Jeremy. "Plan to Hold Back Tides of Venice Runs into Flood of Opposition from Greens." *Scotland on Sunday*, December 30, 2001, 18.

Venus

The planet Venus is known today as one of the most hellish environments anywhere, an example of what can happen to a planet in the grip of a runaway greenhouse effect. A scientific case has been made that not long ago, by the standards of geologic time, that planet's atmosphere much more closely resembled Earth's. The geophysical career of Venus long has been a subject of inquiry among leading scientists of global warming, such as James Hansen, who began his professional life as a student of that planet. The climatic transect of Venus may be a cautionary tale for our Earth.

Planetary ecosystems are forever evolving, a fact that may bring cold comfort to students of the greenhouse effect who cast their eyes upon Venus, where catastrophic greenhouse-effect warming has raised surface temperatures to a searing 850° F (464° C), hot enough to melt lead.

Some contemporary theories argue that Venus may have once experienced a climate much more like that of today's Earth, "complete with giant rivers, deep oceans and teeming with life" (Leake 2002, 11). Two British scientists believe that they have found some evidence that rivers the size of the Amazon once flowed for thousands of miles across the Venusian landscape, emptying into liquid-water seas. According to a report by Jonathan Leake in the *London Sunday Times*, these scientists used radar images from a NASA probe to trace the river systems, deltas, and other features they say could have been created only by moving water (Leake 2002, 11).

No one on Earth knew much about Venusian topography or climate until 1990, when NASA's Magellan probe used radar to penetrate the clouds and map the surface. Magellan's images showed that the surface had been carved by large river-like channels that scientists at the time thought were caused by volcanic lava flows. Jones and colleagues reanalyzed the same images using the latest computer technology and found they were too long to have been created by lava (Leake 2002, 11).

Adrian Jones, a planetary scientist at University College of London, who carried out the research, said the findings suggested life on Venus could have evolved on a parallel with Earth's. "If the climate and temperature were right for water to flow, then they would have been right for life, too. It suggests life could once have existed there," said Jones (Leake 2002, 11). Studies compiled by David Grinspoon of the Southwest Research Institute at Boulder, Colorado, suggest that Venus may have been habitable for as long as 2 billion years, before an accelerated greenhouse effect dried its oceans.

The surface of Venus near the Maat Mons, showing lava flows. (NASA)

Roughly 20 space-exploration missions to Venus have returned enough data to construct an image of Venus today as a hellishly hot place, "its skies dominated by clouds of sulfuric acid, poisoned further by hundreds of huge volcanoes that belch lava and gases into an atmosphere lashed by constant hurricane winds" (Leake 2002, 11). Venus differs from Earth in one important respect: it has no tectonic plates that permit stresses to express themselves piecemeal, via earthquakes and volcanic eruptions. The scientists' research suggests that as recently as 500,000 years ago something (perhaps a surge of volcanic eruptions, long contained by the lack of tectonics) triggered runaway global warming that destroyed the Venusian climate and eventually boiled away the oceans (Leake 2002, 11). According to this research, warming on Venus may have been accelerated by heat released into its atmosphere by billions of tons of carbon dioxide from rocks and, possibly, vegetation. Today, Venus' atmosphere is about 95 percent carbon dioxide.

FURTHER READING

Leake, Jonathan. "Fiery Venus Used to be Our Green Twin." *London Sunday Times*, December 15, 2002, 11.

$\mathcal{W}$

Walruses and Melting Ice

Melting Arctic sea ice is separating walrus young from their mothers, and probably killing them. A Coast Guard icebreaker during two months in 2004 found nine lone walrus calves in deep waters flowing from the Bering Sea to the shallower continental shelf of the Chukchi Sea. Researchers reported in the journal *Aquatic Mammals* that the pups, living with their mothers, probably fell into the sea following collapse of an ice shelf that broke up in water where the temperature had increased 6° F in two years. "We were on a station for 24 hours, and the calves would be swimming around us crying," said Carin Ashjian, a biologist at Woods Hole Oceanographic Institution in Massachusetts and a member of the research team. "We couldn't rescue them" (Kaufman 2006). Walrus pups usually live with their mothers for two years.

The breakup of sea ice close to the shore also deprives the adult walrus of a platform from which to dive for food. Remaining sea ice off shore is in waters that are too deep for diving. The researchers, who were funded by the National Science Foundation and the Office of Naval Research, said it was possible that the walrus young were separated from their mothers by a severe storm or that the mothers were killed by hunters from nearby native communities. But they said local hunting ethics frown on killing walrus mothers with young, and it was far more likely that retreating sea ice—rather than a storm—caused the young to become separated (Kaufman 2006).

Shrinking Arctic ice is taking a toll on walrus numbers, as well as those of polar bears. The toll is unknown. "The ice is melting three weeks earlier in the spring than it did 20 years ago, and it's re-forming a month later in the fall," said Carleton Ray of the University of Virginia, who has studied walruses since the 1950s (Angier 2008).

The walrus are imperiled by expanding human industry as well as eroding sea ice. In February 2008, for example, the U.S. Interior Department gave Royal Dutch Shell oil-drilling rights in the Chukchi Sea near the northwestern shore of Alaska, the first exploitation of this prime walrus fishing ground. Walrus find their food in shallow water, and haul out to rest on ice or land. Females and their young use the ice year-round, if it is available. Ice in the Chukchi Sea has been retreating from the shore into waters too deep for walrus to feed.

"As a result," according to an account in the *New York Times*, "females and calves have been forced to abandon the ice in midsummer and follow the males to land. The voyage leaves them emaciated and easily panicked. With the slightest disturbance, the herd desperately heads back into the water, often trampling one another to death as they flee. "The ones that take the brunt of it are the calves," said Chad Jay of the walrus research program at the Alaska Science Center in Anchorage. "Our Russian colleagues have observed thousands of calves killed in episodes of beachside mayhem" (Angier 2008).

FURTHER READING

Angier, Natalie. "Ice Dwellers Are Finding Less Ice to Dwell On." *New York Times*, May 20, 2008. http://www.nytimes.com/2008/05/20/science/20count.html.

Kaufman, Marc. "Warming Arctic Is Taking a Toll: Peril to Walrus Young Seen as Result of Melting Ice Shelf." *Washington Post*, April 15, 2006, A-7. http://www.washingtonpost.com/wp-dyn/content/article/2006/04/14/AR2006041401368_pf.html.

War, Carbon Footprint

Modern machine warfare waged at long distances—such as the present-day United States invasion of Iraq—is hugely carbon-dioxide intensive. Its carbon footprint is huge—and no one seems to be asking exactly how large. The mechanized nature of war has intensified with nearly every conflict as the technology of the industrialized state has been applied to its art and science. Carbon dioxide has become the signature of modern warfare. Witness our age's version of scorched Earth, Iraq's torching of oil wells in and near Kuwait in 1991, trailing plumes of petroleum-laced black smoke.

Global preparations for war (excluding war's actual conduct) has been estimated to produce as much as 10 percent of worldwide carbon-dioxide emissions (Bidlack 1996; Biswas 2000; Majeed 2004). Analysis of paleoclimatic data correlated with historical records of warfare around the world between 1400 and 1900 suggests substantial association of temperature change and frequency of war (Zhang et al. 2007). The increasing mechanization of war since 1900 probably has made this correlation more pronounced.

Military planners now consider climate change a "threat multiplier" that affects national security and postwar rehabilitation of ecosystems as critical as the restoration of peace (CNA Corporation Military Advisory Board and Study Team 2007; Machlis and Hanson 2008, 729). The U.S. Defense Department also now includes climate change in its assessments of global security. Likewise, the Intergovernmental Panel on Climate Change (IPCC) anticipates that effects of global warming will contribute to declining water resources, reduced food security, and increasing migrations of "environmental refugees," all of which are sources of conflict (IPCC 2007).

Horses Succumb to Tanks

Less than a hundred years ago, at the beginning of World War I in Europe, the main motive force that moved men and supplies was the horse and shoe leather. World War I quickly witnessed a dramatic escalation in war's carbon-dioxide production with the advent of aerial bombardment. War is often a powerful technological motor; World War II began with quarter-century-old bi-planes and ended with jet-propelled fighters.

The air war spread with remarkable speed. In 1914, a German zeppelin dropped a bomb on Paris, killing one person. In June 1916, French fliers bombed a circus tent in Karlsruhe, killing 154 German children. By 1917 and 1918, the Germans rained 300 tons of bombs on Britain, killing 1,413 people. The air war escalated during World War II, with routine, large-scale bombing of civilian areas commonplace on both sides (the five-square-mile bombing of Hamburg during July 1945 is one example of many).

Mechanized, carbon-intensive war reached a new level of size and intensity in World War II on the ground, as well as in the air. At Kurst, for example, 7,000 tanks engaged in battle over an area half the size of England. The Soviets amassed 6.7 million men in their final offensive against the Germans, from the Baltic to the Adriatic seas.

The mechanization of the military provided many more opportunities to ramp up carbon-dioxide production during the world wars of the early twentieth century—World War II's Sherman tank got 0.8 miles per gallon. Fighter jets' typical fuel consumption were on the order of hundreds of gallons per hour, maybe 300 or 400 gallons per hour at full thrust, or 100 gallons per hour at cruising speed, during training and missions. During the 1950s and 1960s, U.S. B-52s were in the air at all times to keep the Soviet Union from obliterating the entire U.S. nuclear-armed fleet on the ground. Each B-52 burned hundreds of gallons of fuel per hour while aloft.

The Iraq War's Carbon Footprint

How much carbon dioxide is being injected into the atmosphere by the Iraq war?

- First, add in all the energy used to produce the weapons, transport, and other provisions that are consumed in the war. Begin with transporting 160,000 U.S. troops and 130,000 contractors from North America to Iraq, often by air, with their equipment and provisions.
- Add the cost of running their armed personnel carriers, lodgings, and so forth, as well as the greenhouse gasses added by the conduct of combat itself.
- Add the carbon and other greenhouse gases added to the atmosphere by fires initiated by bombings and other explosions. In Iraq, pay special attention to intentional sabotage of oil pipelines and suicide bombings, as well as improvised explosive devices.
- Add the carbon cost of tending the wounded. In this war, Iraq's emergency room spans nearly

half the world, from airborne surgery to the Landstuhl Regional Medical Center in Germany and hospitals in the United States.

No figures exist that total the carbon footprint of this or any other war.

According to the February 2007 *Energy Bulletin,* the Pentagon is the single largest consumer of oil in the world. Only 35 countries consume more oil. Yet, the official figure of 320,000 barrels of oil per day used only includes vehicle transport and facility maintenance. Does this figure include the jet fuel used to get troops, contractors, war material, and provisions to and from the Iraq war zone?

According to Don Fitz, writing in Z-Net,

That figure does not include energy for manufacture of vehicles, energy for building and dismantling military facilities, energy for construction of roads, and energy consumed while rebuilding whatever the military blows up. Nor does it factor in energy required by the military's partners, NASA and the nuclear industry. Additionally, whenever war or construction razes trees, it eliminates their ability to remove CO_2 from the atmosphere. (Fitz 2007)

The carbon footprint of the Iraq war—and modern combat in general—has yet to be tallied.

FURTHER READING

Bidlack, H. W. "Swords as Plowshares: The Military's Environmental Role." Ph.D. dissertation, University of Michigan, Ann Arbor, 1996.

Biswas, A. K. "Scientific Assessment of the Long-term Consequences of War." In *The Environmental Consequences of War,* ed. J. E. Autsin and C. E. Bruch, 303–316. Cambridge: Cambridge University Press, 2000.

CNA Corporation Military Advisory Board and Study Team. "National Security and the Threat of Climate Change." Alexandria, VA: CNA Corp., 2007.

Fitz, Don. "What's Possible in the Military Sector?" Z-Net, April 30, 2007. http://www.zmag.org/content/showarticle.cfm?ItemID=12705.

IPCC (Intergovernmental Panel on Climate Change). *Climate Change 2007: Synthesis Report.* Geneva, Switzerland: IPCC, 2007.

Machlis, Gary E., and Thor Hanson. "War Ecology." *BioScience* 58, no. 8 (September 2008):729–736.

Majeed, A. "The Impact of Militarism on the Environment: An Overview of Direct and Indirect Effects." Ottawa, Ontario, Canada: Physicians for Global Survival, 2004.

Zhang, D. D., P. Brecke, H. F. Lee, Y.-Q. He, and J. Zhang. "Global Climate Change, War, and Population Decline in Recent Human History." *Proceedings of the Academy of the National Academy of Sciences* 104 (2007):19,214–19,219.

Water Supplies in Western North America

The impact of warming on mountains of California reflects similar changes in other areas of North America that rely on snow pack for water and power. A temperature rise of 2° F could have dramatic impacts on water resources across western North America, according to scientific teams that have warned of reduced snow packs and more intense flooding as temperatures rise. This research was the first time that global climate modelers have worked with teams running detailed regional models of snowfall, rain, and stream flows to predict what warming will do to the area.

The researchers were surprised by the size of the effects generated by a small rise in temperature ("Warmer Climate" 2001). In a warmer world, warmer winters would raise the average snow level, reducing mountain snow packs, the researchers told the American Geophysical Union in San Francisco during 2001 ("Warmer Climate" 2001).

Snow Packs Erode on Sierra Nevada

According to the scientists' models, "Huge areas of the snow pack in the Sierra [Nevada] went down to 15 percent of today's values," said Michael Dettinger, a research hydrologist at the Scripps Institution of Oceanography in La Jolla, California. "That caught everyone's attention" ("Warmer Climate" 2001). The researchers also anticipated that by the middle of the twenty-first century melting snow may cause streams to reach their annual peak flow up as much as a month earlier than at present. With rains melting snow or drenching already saturated ground, the risk of extreme late-winter and early-spring floods will rise, even as the diminishing snow pack's ability to provide water later in the summer declines. Thus, water consumers may face a frequent paradox: spring floods followed by summer drought ("Warmer Climate" 2001).

Because reservoirs cannot be filled until the risk of flooding is past, the models anticipate that within a half-century they will trap only 70 to 85 percent as much runoff as today. This is a particular problem for California, where agriculture, industry, growing population, and environmental needs already compete for limited water supplies ("Warmer Climate" 2001). Observations support the models. Iris Stewart, a climate researcher at the University of California, San Diego, has found that during the last 50 years runoff in the western United States and Canada have been peaking progressively earlier because of a regionwide trend toward warmer winters and springs ("Warmer Climate" 2001).

Snow and Rain Forecasts

Water supplies in the U.S. West could decline by as much as 30 percent by 2050, by one estimate. "This is just one study where we didn't find anything good: It's a train wreck," said marine physicist Tim Barnett of the Scripps Institution of Oceanography in San Diego (Vergano 2002, 9-D). The Accelerated Climate Prediction Initiative (ACPI) pilot study late in 2002 released snow and rain forecasts for specific regions during the next five decades. Funded largely by the Energy Department, the projections said that—

- Reduced rainfall and mountain snow runoff may reduce water released by the Colorado River to cities such as Phoenix and Los Angeles by 17 percent and cut hydroelectric power from dams along the river by 40 percent.
- Along the Columbia River system in the Pacific Northwest, water levels may drop so low that simultaneous use for irrigation and power generation will not permit any salmon spawning. Snow packs that supply the river may drop 30 percent, moving the peak runoff time forward one month.
- In California's Central Valley, "It will be impossible to meet current water system performance levels," which could hurt water supplies, reduce hydropower generation, and cause dramatic increases in saltiness in the Sacramento Delta and San Francisco Bay (Vergano 2002, 9-D).

"The physics are very simple: Higher temperatures mean there is more rain than snow, and the spring melt comes earlier," said Barnett, who headed the two-year project (Vergano 2002, 9-D). Scientists at Scripps, the Pacific Northwest National Laboratory, the National Center for Atmospheric Research, and the University of Washington contributed to the study. Effects of climate change on water resources in the western United States were explored in a special issue of *Climatic Change* on that subject, published early in 2004 (Pennell and Barnett 2004).

Water Supplies and Snow Pack Erosion in the Pacific Northwest

What harm could global warming do in Seattle, a place where a popular joke is that when summer comes, everyone hopes it arrives on a Saturday? When Mark Twain visited, he said that the mildest winter he ever spent was a summer on Puget Sound. Seattle residents should not hold out hope that global warming will improve their chilly, soggy climate, because its most obvious evidence probably will not arrive in the summer. Residents of the Pacific Northwest probably will feel the brunt of climate change in the winter, when snow levels will rise and wash away the next summer's irrigation and power-generating snow pack.

To forecast the severity of snow pack loss from global warming in the Washington Cascades, scientists first took a step back in time. They examined a half-century of temperature and snowfall data at weather stations from Arizona to British Columbia. "The results were striking; I was shocked by the magnitude of the (snow-moisture) declines," said University of Washington climate scientist Phil Mote, who conducted much of the survey. "In some places, particularly in Oregon, we saw declines of 100 percent. It had gotten so warm there was no snow left in April at all" (Welch 2004). Mote studied snow pack records for April 1 in four Northwest states and British Columbia at 145 sites from 1950 through 1992. He found that the amount of snow between 3,000 and 9,000 feet elevation had decreased by an average of 20 percent or more. "I was surprised by the result," said Mote, who works with a group of scientists called the Climate Impacts Group. "There's already a clearer regional signal of warming in the mountains than we expected" (Gordon 2003).

"If you think the water fights we have now are intense … you ain't seen nothing yet," said Ed Miles, University of Washington professor, during the 2004 annual meeting of the American Association for the Advancement of Science in Seattle (Welch 2004). Miles presented evidence that moisture in snow that nourishes the West's network of rivers (and, thus, its farms and cities) has been steadily declining since at least World War II.

The hardest-hit region has been the Cascades, where battles to provide enough water for fish, agriculture, and power have been intensifying for years (Welch 2004). Miles' team found that a small rise in global temperatures since 1950 already had reduced mountain snows across 75 percent of the West. In coastal regions, such as the Cascades and parts of northern California, where winter temperatures are milder, warming during the same period reduced moisture in spring snows by more than 30 percent.

Having surveyed the records, the scientists forecast future snow-pack loses using various global-warming models. Their results indicated that water content of Cascade snows would drop by 59 percent in the coming half-century, according to their most conservative models. The likelihood of precipitation coming as rainfall rather than snow is higher, and with storage reservoirs full, that would mean more early winter flooding. Meanwhile, less snow accumulating in the mountains would mean that spring runoff probably will arrive a month to six weeks earlier than at present (Welch 2004). "The parts of the West best situated to handle water storage are the least vulnerable to temperature changes," Mote said. "Those with the least capacity to store water—the Cascades and Northern California—are most vulnerable. So it's sort of a double-whammy" (Welch 2004).

A report by 46 scientists from several Pacific Northwest institutes and universities released in 2004 (http://inr.Oregonstate.edu) summarized existing information on future warming's probable effects for the region. The report said that temperatures had risen 1° to 3° F in the area over a century, and could rise another 3° F by 2030 and 5.5° F by 2050. The economic infrastructure of the region will be affected significantly by rising snow levels that will decrease the amount of snowmelt available during the summer dry season, as sea levels rise an average of one-half to one inch per year.

FURTHER READING

Gordon, Susan. "U.S. Pacific Northwest Gets Reduced Supply of Snow, Climate Study Says." *Tacoma News-Tribune*, February 7, 2003, n.p. (LEXIS)

Pennell, William, and Tim Barnett, eds. "Special Issue: The Effects of Climate Change on Water Resources in the West." *Climatic Change* 62, no. 1–3 (January and February 2004).

Vergano, Dan. "Global Warming May Leave West in the Dust By 2050, Water Supplies Could Plummet 30 Percent, Climate Scientists Warn." *USA Today*, November 21, 2002, 9-D.

"Warmer Climate Could Disrupt Water Supplies." Environment News Service, December 20, 2001. http://ens-news.com/ens/dec2001/2001L-12-20-09.html.

Welch, Craig. "Global Warming Hitting Northwest Hard, Researchers Warn." Seattle *Times*, February 14, 2004. http://seattletimes.nwsource.com/html/localnews/2001857961_warming14m.html.

Water Vapor, Stratospheric

Tropical wildfires and slash-and-burn agriculture have helped double the moisture content in the stratosphere over the last 50 years, a Yale University researcher concluded after examining satellite weather data ("Biomass Burning" 2002). "In the stratosphere, there has been a cooling trend that is now believed to be contributing to milder winters in parts of the northern hemisphere," said Steven Sherwood, assistant professor of geology and geophysics. "The cooling is caused as much by the increased humidity as by carbon dioxide" ("Biomass Burning" 2002).

Water-vapor assessment by ground-based, balloon, aircraft and satellite measurements shows a global stratospheric water vapor increase of as much as two parts per million by volume (ppm) during the last 45 years, a 75 percent rise. Modeling studies by the University of Reading in England indicate that since 1980 the stratospheric water vapor increase has produced a surface temperature rise that is about half of that attributable to increased carbon dioxide alone.

A Scientific and Industrial Research Organization (CSIRO) computer model anticipated that reductions in methane emissions may worsen conditions in the stratosphere and could drive the ozone levels in 2100 down to 9 percent below 1980 levels. Methane is a greenhouse gas, but it also provides protection against ultraviolet radiation because it produces ozone as it breaks down chemically. According to one observer, "Current global climate change strategy focuses on pushing down methane levels while letting nitrous oxide levels soar. The result is a further drop in ozone levels" (Calamai 2002a, A-8).

"Higher humidity also helps catalyze the destruction of the ozone layer," added Sherwood

("Biomass Burning" 2002). Cooling in the stratosphere causes changes to the jet stream that produce milder winters in North America and Europe. By contrast, harsher winters may result in the high latitudes of the Arctic. Sherwood said that about half of the increased humidity in the stratosphere has been attributed to methane oxidation. No one seems to know, however, what has caused the rest of the additional moisture.

"More aerosols lead to smaller ice crystals and more water vapor entering the stratosphere," Sherwood explained. "Aerosols are smoke from burning. They fluctuate seasonally and geographically. Over decades there have been increases linked to population growth" ("Biomass Burning" 2002). Ozone experts in Canada and the United States said the Australian findings are a serious warning about the risks in trying to manipulate parts of the atmosphere in isolation. "You have to look at all these chemicals and see how they interact and evolve over time," said Tom McElroy, an ozone specialist with the Meteorological Service of Canada (Calamai 2002a, A-8).

Other causes not directly related to human activity also may be increasing stratospheric moisture levels, according to Philip Mote, a University of Washington research scientist. "Half the increase [of water vapor] in the stratosphere can be traced to human-induced increases in methane, which turns into water vapor at high altitudes, but the other half is a mystery," said Mote. "Part of the increase must have occurred as a result of changes in the tropical tropopause, a region about 10 miles above the equator, that acts as a valve that allows air into the stratosphere," added Mote ("Most Serious" 2001). "A wetter and colder stratosphere means more polar stratospheric clouds, which contribute to the seasonal appearance of the ozone hole," said James Holton, University of Washington atmospheric sciences chairman and expert on stratospheric water vapor. "These trends, if they continue, would extend the period when we have to be concerned about rapid ozone depletion," Holton added ("Most Serious" 2001).

FURTHER READING

Betsill, Michelle M. "Impacts of Stratospheric Ozone Depletion." In *Handbook of Weather, Climate, and Water: Atmospheric Chemistry, Hydrology, and Societal Impacts*, ed. Thomas D. Potter and Bradley R. Colman, 913–923. Hoboken, NJ: Wiley Interscience, 2003.

"Biomass Burning Boosts Stratospheric Moisture." *Environment News Service*, February 20, 2002. http://ens-news.com/ens/feb2002/2002L-02-20-09.html.

Calamai, Peter. "Alert over Shrinking Ozone Layer." *Toronto Star*, March 18, 2002a, A-8.

"Most Serious Greenhouse Gas is Increasing, International Study Finds." *Science Daily*, April 27, 2001. http://www.sciencedaily.com/releases/2001/04/010427071254.htm.

Watt-Cloutier, Sheila (1953–)

As international chair of the Inuit Circumpolar Conference from 2002 to 2006, Sheila Watt-Cloutier brought Inuit struggles against persistent organic pollutants and global warming to the world. As president of the Inuit Circumpolar Conference (ICC) in Canada and vice president of the ICC before 2002, Watt-Cloutier was a key figure in negotiating the Stockholm Convention (2001), which banned polychlorinated biphenyls (PCBs), dioxins, and other synthetic pollutants from lower latitudes that have been threatening the lives of all living things in the Arctic, including her people. Forms of dioxin and PCBs had become so prevalent in the Arctic that they biomagnified up the food chain to a point at which Inuit mothers could not breast-feed their babies without endangering their health.

Formed in 1977, the ICC represents the common interests of roughly 155,000 Inuit in northern Canada, Greenland, Alaska, and the Russian Federation. As head of the ICC, Watt-Cloutier traveled the world on jets trying to protect her people, many of whom are only generation off the land, from the effluvia of the industrial world to—as she told a world conference on climate change in Milan, Italy, on December 10, 2003—"bring a human face to these proceedings." She travels the world injecting "life," as she phrases it, into international negotiations with the delicacy of a diplomat and the verve of a social activist (Watt-Cloutier 2003b).

Personal Life

Watt-Cloutier was born in Kuujjuaq, Northern Quebec, December 2, 1953, of Daisy Watt (1921–2002), who was a well-regarded native elder, as well as a musician, interpreter, and healer. Charlie Watt, a brother of Sheila's, served in the

Sheila Watt-Cloutier (1953–). (Inuit Circumpolar Conference)

Canadian federal Senate during the 1980s. After a youth of being raised traditionally, often traveling by dog sled, Watt-Cloutier was taken from her home to attend boarding schools from age 10 in Churchill, Manitoba, and Nova Scotia. Later, she attended Montreal's McGill University during the middle 1970s, specializing in education, human development, and counseling. She also displayed her mother's gift for interpretation in Inuktitut at Ungava Hospital, the beginnings of a 15-year career in Inuit health care.

In 1992, Watt-Cloutier critiqued Northern Quebec's educational system in a report, *Siatunirmut—the Pathway to Wisdom* for the Nunavik Educational Task Force. She also worked in land claims negotiations that led, in part, to the establishment of Nunavut as a semisovereign Inuit province of Canada. Her new career as an Inuit political leader and environmental advocate began during 1995 after she was elected to chair Canada's division of the ICC, to which she was reelected in 1998. She also contributed to academic and ecological studies on the Arctic.

Watt-Cloutier's own life illustrates how the Inuit struggle to maintain some semblance of tradition in a swiftly changing, melting, and now frequently polluted ice world. Watt-Cloutier watches global warming manipulate the seasons from her home overlooking Frobisher Bay in Iqaluit, the capital of the semisovereign Canadian Inuit province of Nunavut, on Baffin Island. The name means "fishing place," an indication of the importance of sustenance from the sea in the traditional Inuit diet. Nunavut means "our home" in the Inuktitut language. Nunavut, a territory four times the size of France, has a population of roughly 25,000, 85 percent of whom are Inuit.

Watt-Cloutier has been active in promoting sustainable development, retention of traditional ecological knowledge, and other forms of education. In 2007, she was nominated for the Nobel Peace prize for her ability to extend what were largely defined as environmental issues to threats to Inuit culture, health, and traditional way of life. She also played a leading role in suing the

United States for negligence regarding global warming in international legal tribunals, including the Organization of American States.

Watt-Cloutier became well known around the world as an adroit diplomat and charismatic speaker. Sometimes, she brought diplomats to tears as she described the effects of persistent organic pollutants (POPs) and global warming on Inuit peoples. She also extended cooperation among Inuit around the Arctic, from Canada to Greenland, Alaska, and Russia.

By 2007, Watt-Cloutier had two grown children and a grandson. Displaying her grandmother's musical talents, Watt-Cloutier's daughter is a traditional Inuit throat-singer and dancer; her son is a pilot who became the youngest pilot hired by Inuit Air (Kafarowski 2007).

"We Feel Like an Endangered Species"

The toxicological and climatic bills for modern industry at the lower latitudes are being left on the Inuits' table in Nunavut. Native people whose diets consist largely of sea animals (whales, polar bears, fish, and seals) have been consuming a concentrated toxic chemical cocktail. Abnormally high levels of dioxins and other industrial chemicals are being detected in Inuit mothers' breast milk.

"As we put our babies to our breasts we are feeding them a noxious, toxic cocktail," said Watt-Cloutier. "When women have to think twice about breast-feeding their babies, surely that must be a wake-up call to the world" (Johansen 2000, 27). More than mother's milk is at stake here. "Greenland has no trees, no grass, no fertile soil," wrote Marla Cone in *Silent Snow* (2005), "which means no cows, no pigs, no chickens, no grains, no vegetables, no fruit. In fact, there is little need for the word "green" in Greenland. The ocean is its food basket" (Cone 2004).

To a tourist with no interest in environmental toxicology, the Inuits' Arctic homeland may seem as pristine as ever during its long, snow-swept winters. Many Inuit still guide sleds onto the pack ice surrounding their Arctic-island homelands to hunt polar bears and seals. Such a scene may seem pristine, until one realizes that the polar bears' and seals' body fats are laced with dioxins and PCBs. Geographically, the Arctic could not be in a worse position for toxic pollution, as a ring of industry in Russia,

Europe, and North America pours pollutants northward on prevailing winds. Arctic ecosystems and indigenous communities are particularly at risk because POPs biomagnify—that is, they increase in potency by several orders of magnitude—with each step up the food chain.

"Imagine for a moment," said Watt-Cloutier,

the shock of the Inuit as they discovered that what has nourished them for generations, physically and spiritually, is now poisoning them. Some Inuit now question whether they should eat locally gathered food; others ask whether it is safe to breast-feed their infants. What sort of public outcry and government action would there be if the same levels of POPs found among Inuit were found in women in Toronto, Montreal, or Vancouver as a result of eating poultry or beef? (Watt-Cloutier and Fenge 2000)

POPs have been linked to cancer, birth defects, and other neurological, reproductive, and immune system damage in people and animals. At high levels, these chemicals also damage the central nervous systems of human beings and other animals. Many of them act as endocrine disrupters, causing deformities in sex organs as well as long-term dysfunction of reproductive systems. These chemicals can interfere with the function of the brain and endocrine system by penetrating the placental barrier between mother and unborn child, scrambling the instructions of the naturally produced chemical messengers. The latter tell a fetus how to develop in the womb and postnatally through puberty; should interference occur, immune, nervous, and reproductive systems may not develop according to the contents of the genes inherited by the embryo.

"At times," said Cloutier, "We feel like an endangered species. Our resilience and Inuit spirit and of course the wisdom of this great land that we work so hard to protect gives us back the energy to keep going" (Watt-Cloutier 2001).

The Inuit Confront Climate Change

In addition to some of the planet's highest levels of toxicity, the Inuit have been forced to confront some of Earth's most rapid rates of global warming. Climate change arrived swiftly and dramatically in the Arctic during the 1990s, including such surprises as warm summer days, winter freezing rain, thunderstorms, and invasions of Inuit villages by heretofore unknown insects and birds. Yellow-jacket wasps have been sighted for the first time, for example, on northern Baffin Island.

The ICC has assembled a human-rights case against the United States (specifically the George W. Bush administration) because global warming is threatening the Inuit way of life. The ICC has invited the Washington-based Inter-American Commission on Human Rights (a body of the Organization of American States) to visit the Arctic to witness the human effects of rapid warming. Introducing the legal action, Watt-Cloutier, said: "We want to show that we are not powerless victims. These are drastic times for our people and require drastic measures" (Brown 2003c, 14). The Inuit say that by repudiating the Kyoto Protocol and refusing to cut carbon-dioxide emissions in the United States, which make up 25 percent of the world total, the White House under George W. Bush imperiled their way of life (Brown 2003c). A hearing was held on the ICC petition on March 1, 2007.

Addressing a U.S. Senate Commerce Committee hearing on global warming August 15, 2004, Watt-Cloutier described the disorienting effect of rapid change on the Inuit: "The Earth is literally melting. If we can reverse the emissions of greenhouse gases in time to save the Arctic, then we can spare untold suffering." She continued: "Protect the Arctic and you will save the planet. Use us as your early-warning system. Use the Inuit story as a vehicle to reconnect us all so that we can understand the people and the planet are one" (Pegg 2004). The Inuits' ancient connection to their hunting culture may disappear within her grandson's lifetime, Watt-Cloutier said. "My Arctic homeland is now the health barometer for the planet" (Pegg 2004).

How quickly is Arctic climate changing? On February 27, 2006, Watt-Cloutier described a heretofore unheard of event in Iqaluit: rain, lightning, thunder, and mud in *February*. The temperature had risen to 6° to 8° C, reaching the average for June, winds had reached 55 miles an hour, and the town was paralyzed in a sea of dirty slush. "Much of the snow has melted on the back of my house and all the roads are already slushy and messy. All planes coming up from the south were cancelled because the runways were icy from the rain," she wrote. "Unfortunately the predictions of the Arctic Climate Impact Assessment are unfolding before my very eyes" (Watt-Cloutier 2006a). Iqaluit hunters expressed concern for the caribou, which would go hungry once freezing temperatures

returned and encased the lichen, their food source, in a crust of ice. Instead of waiting until spring, the usual hunting season, Inuit were taking caribou in early March, believing that they would be too skinny later to be useful as food.

"Snow Age" to *"Space Age"*

"We have," Watt-Cloutier remarked on one occasion,

> gone from the 'snow age' to the 'space age' in one generation. I was born on the land in Nunavik, northern Quebec and we were still traveling by dog team when I was sent to school in southern Canada at the age of ten. Notwithstanding significant economic and social changes, Inuit remain connected with the land and in tune with its rhythms and cycles. (Watt-Cloutier 2004b)

"Inuit live in two worlds—the traditional and the modern," she said. "To this day our hunting based culture remains viable and important. We depend on seals, whale, walrus and other marine mammals to sustain our nutritional, cultural and spiritual well being" (Watt-Cloutier 2004b). *See also:* Arctic Warming and Native Peoples; Legal Liability and Global Warming

FURTHER READING

Brown, Paul. "Inuit Blame Bush for Impending Extinction." *London Guardian*, December 13, 2003c. http://www.smh.com.au/articles/2003/12/12/1071125653832.html.

Cone, Marla. "Dozens of Words for Snow, None for Pollution." *Mother Jones*, January–February 2004. http://www.hartford-hwp.com/archives/27b/059.html.

Cone, Marla. *Silent Snow: The Slow Poisoning of the Arctic.* New York: Grove Press, 2005.

Johansen, Bruce E. "Pristine No More: The Arctic, Where Mother's Milk Is Toxic." *The Progressive*, December 2000, 27–29.

Johansen, Bruce E. "Arctic Heat Wave." *The Progressive*, October 2001, 18–20.

Kafarowski, Joanna. "Cloutier, Sheila-Watt." In *Encyclopedia of the Arctic*. London: Routledge, 2007. Sample Proofs. http://www.routledge-ny.com/ref/arctic/watt.html.

Pegg, J. R. "The Earth is Melting, Arctic Native Leader Warns." Environmental News Service, September 16, 2004.

Watt-Cloutier, Sheila, "Honouring Our Past, Creating Our Future: Education in Northern and Remote Communities." In *Aboriginal Education: Fulfilling the Promise*, ed. Lynne Davis, Louise Lahache, and

Marlene Castellano. Vancouver: University of British Columbia Press, 2000.

Watt-Cloutier, Sheila. Personal communication, March 28, 2001.

Watt-Cloutier, Sheila. "Speech Notes for Sheila Watt-Cloutier, Chair, Inuit Circumpolar Conference. Conference of Parties to the United Nations Framework Convention on Climate Change. Milan, Italy, December 10, 2003b. http://www.inuitcircumpolar. com/index.php?ID=242&Lang=En.

Watt-Cloutier, Sheila. "Canada and Inuit: Addressing Global Environmental Challenges, Remarks by Sheila Watt-Cloutier, Chair, Inuit Circumpolar Conference at the Inaugural Environmental Protection Service Inuit Speaker Series." Ottawa, Ontario, January 16, 2004b. http://www.inuitcircumpolar. com/index.php?ID=250&Lang=En.

Watt-Cloutier, Sheila. Personal communication, February 27, 2006a.

Watt-Cloutier, Sheila, and Terry Fenge. "Poisoned by Progress: Will Next Week's Negotiations in Bonn Succeed in Banning the Chemicals that Inuit are Eating?" Press Release, Inuit Circumpolar Conference, March 14, 2000. http://www.inuitcircumpolar. com/index.php?ID=250&Lang=En.

Wave Power and Shipping

A natural energy-saving marriage—shipping and wave power—was tested during 2008 by the three-ton *Suntory Mermaid II* on a voyage of 3,780 nautical miles from the Hawaii Yacht Club in Honolulu to the Kii Channel near the East Coat of Japan, intended as the longest voyage under wave power. The boat's propulsion system, designed by Yutaka Terao of the Tokai University School of Marine Science and Technology, is mounted in the bow, rather than the stern, and pulls it, rather than pushing (Geoghegan 2008, D-3).

Horizontal fins that rise and fall with wave action propel the craft as they would a dolphin. Wave-propelled boats have been tested since at least 1895, and as early as 1935 one achieved a speed of five knots off the coast of Long Beach, California. It was an 18-inch model, however, not a full-scale sailing vessel.

Weather conditions (which determine wave action) can influence sailing conditions greatly. The *Suntory Mermaid II* was equipped with a sail and a motor, should wave power fail. The trip took two and a half months at an average speed of three knots.

FURTHER READING

Geoghegan, John. "Long Ocean Voyage Set for Vessel that Runs on Wave Power." *New York Times*, March 11, 2007, D-3.

West Antarctic Ice Sheet

During the 1990s, the stability of the West Antarctic Ice Sheet (which comprises about a quarter of the Earth's largest mass of frozen water) became a subject of intense scientific inquiry. A vibrant debate has grown up regarding the future of the ice sheet, with assurances of stability on one side, and speculation of future collapse on the other.

A report issued by the National Academy of Sciences during 1991 asserted that melting of the West Antarctic Ice Sheet is unlikely, "and virtually impossible before the end of the next century" (National Academy of Sciences 1991, 23) According to climate models used in this report, several centuries of rising temperatures will be required before the ice sheet disintegrates.

The idea that global warming could provoke the disintegration of the West Antarctic Ice Sheet was aired as theory by glaciologists as early as 1979 (Hughes, Fastook, and Denton 1979; Bentley 1980). J. H. Mercer has suggested that the West Antarctic Ice Sheet fell apart during an interglacial period about 125,000 years ago without an added boost from the burning of carbon-based fuels. T. J. Hughes has examined the geophysical mechanisms that may cause the West Antarctic Ice Sheet to collapse, and J. T. Hollin has examined evidence of major ice sheet "surges" in the past, which led to 10 to 30 meter rises of sea level in less than 100 years (Barry 1978).

An Unusual Warm Spell

Scientists in 2007 announced that an unusual warm spell had traversed parts of Antarctica two years earlier, covering an area the size of California. Using its QuikScat satellite, the team measured snowfall accumulation and melting in Antarctica and Greenland from July 1999 through July 2005. Son Nghiem of NASA's Jet Propulsion Laboratory in Pasadena and Konrad Steffen, director of the Cooperative Institute for Research in Environmental Sciences at the University of Colorado-Boulder, led the research.

Antarctica as a whole has shown little warming (even slight cooling in some areas) in the recent past with the exception of the Antarctic Peninsula (which is warming rapidly). By 2007, however, large regions of the continent were showing the first signs of warming's impacts, as interpreted by this satellite analysis, said Steffen. "Increases in snowmelt, such as this in 2005, definitely could have an impact on larger-scale melting of Antarctica's ice sheets if they were severe or sustained over time," he said ("NASA: Vast Areas" 2007).

The observed melting occurred in several regions, some far inland at high latitudes (in one instance within 310 miles of the South Pole) and at elevations to 6,100 feet, where melt had heretofore never been seen. Judging from the satellite's survey of snowfall that had turned to ice from melting and refreezing, temperatures in the affected areas rose above the freezing point and in some cases remained there for about a week.

Maximum air temperatures at the time of the melting were unusually high, reaching more than 5° C (41° F) in one of the affected areas. They remained above melting for about a week.

Larsen A Ice Shelf Disintegrates

In early 1995, the Larsen A Ice Shelf disintegrated during a single storm after years of shrinking gradually. "The speed of the final breakup was unprecedented, and followed several of the warmest summers on record for this portion of the Antarctic," said Ted Scambos of the University of Colorado at Boulder-based National Snow and Ice Data Center ("Satellite Images" 1997). "Ice shelves appear to be good bellwethers for climate change, since they respond to change within decades, rather than the years or centuries sometimes typical of other climate systems," said Scambos ("Satellite Images" 1997).

During 1998 and 1999, several reports described the retreat of ice along the shores of the Antarctic Peninsula. Reports indicated that an additional 1,100 square miles of ice had melted during 1998. At about the same time, David Vaughan, a researcher with British Antarctic Survey, and Scambos reported that the Larsen B and Wilkins shelves on the Antarctic Peninsula were in full retreat. These ice sheets had been retreating slowly for about 50 years, losing about 2,700 square miles during that period. A loss of 1,100 square miles in one year (1998) thus represented a major acceleration of the ice sheets' erosion. During that year as much ice was lost as had melted during the preceding four decades.

"This may be the beginning of the end for the Larsen Ice Shelf," said Scambos as the ice sheet crumbled in 1999. "This is the biggest ice shelf yet to be threatened," Scambos said. "The total size of the Larsen B Ice Shelf is more than all the previous ice that has been lost from Antarctic ice sheets in the past two decades" ("Satellite Images" 1997). Until its dissolution, the Larsen B was the northernmost ice shelf in Antarctica, and therefore "on the front line of the warming trend," said Scambos ("Satellite Images" 1997).

Questions about the Pine Island Glacier

Radar images from satellite observations of the Pine Island Glacier in Antarctica taken during the 1990s indicate that it has been shrinking rapidly. The shrinking of this glacier is important, "because it could lead to a collapse of the West Antarctic Ice Sheet," said Eric Rignot, a computer-radar scientist at the Jet Propulsion Laboratory in California, who led the study. "We are seeing a … glacier melt in the heart of Antarctica," said Rignot ("Melting Antarctic Glacier" 1998). "The continuing retreat of Pine Island glacier could be a symptom of the WAIS [West Antarctic Ice Sheet] disintegration," said Craig Lingle, a glaciologist at the University of Alaska in Fairbanks, who is familiar with the study ("Melting Antarctic Glacier" 1998).

The Pine Island Glacier is important because it is part of a stream of ice that moves more rapidly than the ice cap surrounding it. This glacier is part of an ice stream that runs from the interior of the West Antarctic Ice Sheet into the surrounding ocean waters. If a glacier in this ice stream melts more quickly from the bottom than snow accumulates on its top, the net icemelt goes into the ocean, raising sea levels.

A "disaster scenario," as described by Richard Alley, a glaciologist at Pennsylvania State University, has the Pine Island Glacier retreating enough to "make a hole in the side of the ice sheet.… The remaining ice would drain through that hole" ("Melting Antarctic Glacier" 1998). Once enough ice had drained through the hole, the West Antarctic Ice Sheet might eventually collapse, raising average sea levels around the world 15 to 20 feet in a few years. Such an increase in mean sea level would flood roughly 30 percent of Florida and Louisiana; 15 to 20 percent of the District of Columbia, Maryland, and Delaware; and 8 to 10 percent of the Carolinas and New Jersey. Inundation of coastal areas would have a similar impact around the world (Schneider and Chen 1980). Among the flooded areas would be the centers of some of the world's great urban and commercial centers, from New York City to Bombay, Calcutta, and Manila.

During 1998, Eric Rignot wrote in *Science* that West Antarctica's Pine Island Glacier was retreating at 1.2 kilometers a year (plus or minus 0.3 kilometers) and that its ice was thinning 3.5 (plus or minus 0.9) meters per year. "The fast recession of the Pine Island Glacier, predicted to be a possible trigger for the disintegration of the West Antarctic Ice Sheet, is attributed to enhanced basal melting of the glacier floating tongue by warm ocean waters," Rignot wrote (Rignot 1998, 549). This glacier is widely believed to be "the ice sheet's weak point" (Kerr 1998a, 499). While the accelerated melting of this glacier does not portend an immediate disintegration of the West Antarctic Ice Sheet, Alley wrote that "most models indicate [that the retreat] would speed up if it kept going" (Kerr 1998a, 499). One observer was quoted in this context as stating that a quick collapse of the West Antarctic Ice Sheet would "back up every sewer in New York City" (Kerr 1998a, 500). Rignot speculated that warmer ocean waters were causing Pine Island's rapid bottom melting. "This is one of the most sensitive ice sheets to climatic change. For many, many years we have neglected the importance of bottom melting," Rignot said ("Melting Antarctic Glacier" 1998).

"The sudden appearance of thousands of small icebergs suggests that the shelves are essentially broken up in place and then flushed out by storms or currents afterward," said Scambos (Britt 1999). The Larsen and Wilkins ice shelves have been melting since the 1950s, and scientists had expected them to fall apart, but the disintegration occurred more quickly than anticipated. "We have evidence that the shelves in this area have been in retreat for 50 years, but those losses amounted to only about 7,000 square kilometers," said David Vaughan, a researcher with the Ice and Climate Division of the British Antarctic Survey. "To have retreats totaling 3,000 square kilometers in a single year is clearly an escalation. Within a few years, much of the Wilkins ice shelf will likely be gone" (Britt 1999).

A Century-Long Warming

Speculation regarding the future of the West Antarctic Ice Sheet takes place in a climatic context that includes paleoclimate records and instrument readings indicating that the western Antarctic peninsula has warmed significantly during the last century, with the trend accelerating at the end of the period. Temperatures in this area rose 4° to 5° C during the century, according to the World Wildlife Fund. Temperatures in the area averaged above longer-term averages more than 75 percent of the time during the last third of the twentieth century. The number of annual days above freezing has increased by two to three weeks, mainly during the most recent quarter century. This temperature increase tends to erode coastal ice sheets: "Large areas of ice shatter as the melt water percolates into fractures, and deep cracks are forced open to the base of the ice sheet by the weight of the water.... Once ice sheets weaken to a critical point, they may collapse very suddenly" (Mathews-Amos 1999).

For several decades, scientists at the Palmer Station, astride the northwestern edge of Antarctica's Palmer Peninsula, have watched steady changes in the environment, as well as the flora and fauna it supports. Annual average temperatures at the station have risen 3° to 4° F since the 1940s; winter averages have climbed 7° to 9° F during the same period (Petit 2000, 66). Southern elephant seals, which can weigh as much as 8,800 pounds, usually raise their young near the Falkland Islands to the north, but, in recent years, several hundred of them have been living year-round on the Palmer Peninsula. Fur seals, virtually unknown near the Palmer station before 1950, have colonized the area around the station in the thousands, as areas once covered by ice and snow year-round have been sprouting "low grass, tiny shrubs, and mosses" (Petit 2000, 67).

At the same time, animals that depend on sea ice for food have been migrating southward. Adelie penguins (18-inch birds that look like they are wearing tuxedoes) have become scarcer near the Palmer Station, because they eat krill that arrives with sea ice. Adelie populations within two miles of the Palmer Station have declined by about 50 percent in 25 years, including a 10 percent decline during two years in the late 1990s (Petit 2000, 68). About 1950, four of five winters at Palmer Station produced extensive sea ice; by the end of the century, the average was two winters in five. The Adelie penguins have been moving their range southward toward areas that heretofore had been too cold and barren for them.

Scientists who study this situation have no firm estimates regarding how much warmer the Earth would have to become to provoke a general collapse of the West Antarctic Ice Sheet. Some give it 6° C (10° F) and a century or two. Others believe the West Antarctic Ice Sheet as a whole may not drain into the sea for several centuries, if at all. Uncertainty regarding the future of the West Antarctic Ice Sheet has introduced a great deal of variability into models that estimate how much global warming may cause worldwide sea levels to rise.

How Ice Moves

The speed with which Antarctic ice may melt depends not only on how much temperatures rise, but also on the ways in which ice moves within the ice cap. Jonathan L. Bamber, David G. Vaughan, and Ian Joughin have been studying these "rivers" of subsurface Antarctic ice. "It has been suggested," they wrote, "that as much as 90 percent of the discharge from the Antarctic Ice Sheet is drained through a small number of ... ice streams and outlet glaciers fed by relatively stable and inactive catchment areas." Their research suggests that "each major drainage basin is fed by complex systems of tributaries that penetrate up to 1,000 kilometers from the grounding line to the interior of the ice sheet" (Bamber, Vaughan, and Joughin 2000, 1248). Such "complex flows" are noted throughout the Antarctic ice sheet by these researchers.

Bamber, Vaughan, and Jouglin asserted that "this finding has important consequences for the modeled or estimated dynamic response time of past and present ice sheets to climate forcing" (Bamber, Vaughan, and Joughin 2000, 1248). The researchers also find evidence of similar (although smaller in scale) ice sheet dynamics in Greenland. "This evidence," they wrote, "challenges the view that the Antarctic plateau is a slow-moving and homogenous region" (Bamber, Vaughan, and Joughin 2000, 1250). These researchers also contend that the dynamics of large ice flows are too complex for present-day models to predict, so climate modelers have very little idea how global warming will affect the

largest of the Earth's remaining ice masses. *See also:* Antarctica and Climate Change

FURTHER READING

Bamber, Jonathan L. David G. Vaughan, and Ian Joughin. "Widespread Complex Flow in the Interior of the Antarctic Ice Sheet." *Science* 287 (February 18, 2000):1248–1250.

Barry, R. G. "Cryospheric Responses to a Global Temperature Increase." In *Carbon Dioxide, Climate, and Society: Proceedings of a IIASA Workshop Co-sponsored by WMO, UNEP, and SCOPE, February 21–24, 1978,* ed. Jill Williams, 169–180. Oxford, UK: Pergamon Press, 1978.

Bentley, Charles. "Response of the West Antarctic Ice Sheet to CO_2-Induced Global Warming." In *Environmental and Societal Consequences of a Possible CO_2-Induced Climate Change,* Vol. 2. Washington, D.C.: Department of Energy, 1980.

Hughes, T.J., J. L. Fastook, and G. H. Denton. *Climatic Warming and the Collapse of the West Antarctic Ice Sheet.* Orono, Maine: University of Maine Press, 1979.

Kerr, Richard A. "West Antarctica's Weak Underbelly Giving Way?" *Science* 281 (July 24, 1998a):499–500.

"Melting Antarctic Glacier Could Raise Sea Level." Reuters. July 24, 1998. http://benetton.dkrz.de:3688/homepages/georg/kimo/0254.html.

"NASA: Vast Areas of West Antarctica Melted in 2005." Environment News Service, May 28, 2007. http://www.ens-newswire.com/ens/may2007/2007-05-28-09.asp#anchor1.

National Academy of Sciences. *Policy Implications of Greenhouse Warming.* Washington, D.C.: National Academy Press, 1991.

Petit, Charles W. "Polar Meltdown: Is the Heat Wave on the Antarctic Peninsula a Harbinger of Global Climate Change?" *U.S. News and World Report,* February 28, 2000, 64–74.

Rignot, E. J. "Fast Recession of a West Antarctic Glacier." *Science* 281 (July 24, 1998):549–551.

"Satellite Images Show Chunk of Broken Antarctic Ice Shelf." University of Colorado News and Events for the Media, April 16, 1997. http://cires.colorado.edu/news/press/1997/97-04-16.html.

Schneider, Stephen H., and R. S. Chen. "Carbon Dioxide Warming and Coastline Flooding: Physical Factors and Climatic Impact." *American Review of Energy* 5 (1980):107–140.

West Nile Virus and Warming

Until 1999, West Nile Virus had never even been detected in North America. No one knows exactly how the virus reached the United States. Once it arrived, however, West Nile Virus spread

Centers for Disease Control Director Julie Louise Gerberding, speaking at an August 2003 West Nile Virus press briefing. (Greg Knobloch/Centers for Disease Control and Prevention)

rapidly across the continent; by 2001, it had infected 29 species of mosquitoes, 100 species of birds, and many mammals, including humans. By the summer of 2002, West Nile had reached 36 states, as well as the southern regions of eastern Canada (Grady 2002, F-2). By 2003, most of the continental United States was reporting West Nile Virus. Global warming may be a contributing factor, due to warm winters and pervasive summer droughts that seem to favor the spread of the mosquito-borne virus.

The disease initially brings fever, aches, and profound fatigue, sometimes followed by paralysis and other neurological complications, including meningitis and encephalitis, which can leave a victim physically disabled and brain-damaged. Between 1999 and 2004, West Nile Virus infected more than 16,000 people in the United States, killing more than 600, and afflicting 6,500 others with severe neurological problems (Chase 2004, A-1).

Regarding West Nile Virus, Epstein said that drought helps the mosquito species *Culex pipiens*, which plays a major role in spreading West Nile (Grady 2002, F-2). Epstein added that drought may wipe out darning needles, dragonflies, and amphibians, which destroy mosquitoes.

Drought also may aid the spread of infection by drawing thirsty birds to the pools and puddles where mosquitoes breed. "Hot weather plays a role, too," Epstein said. "Warmth increases the rate at which pathogens mature inside mosquitoes" (Grady 2002, F-2).

FURTHER READING

Chase, Marilyn. "As Virus Spreads, Views of West Nile Grow Even Darker." *Wall Street Journal*, October 14, 2004, A-1, A-10.
Grady, Denise. "Managing Planet Earth: On an Altered Planet, New Diseases Emerge as Old Ones Re-emerge." *New York Times*, August 20, 2002, F-2.

White Christmases: Soon to Be a Memory?

Statistics provided by researchers at the Oak Ridge National Laboratory that examined weather records of 16 cities, mainly in the northern United States after 1960, indicated that the number of white Christmases declined between the 1960s and the 1990s. In Chicago, for example, the number of white Christmases (defined as at least one inch of snow on the ground) dropped from seven in the 1960s to two during the 1990s. In New York, the number declined from five in the 1960s to one during the 1990s. Detroit had just three white Christmases during the 1990s compared with nine in the 1960s ("Are White" 2001). The snowfall analysis was performed by Dale Kaiser, a meteorologist with the Carbon Dioxide Information Analysis Center at the Department of Energy's Oak Ridge National Laboratory, and Kevin Birdwell, a meteorologist in the lab's Computational Science and Engineering Division.

FURTHER READING

"Are White Christmases Just a Memory?" Environment News Service, December 21, 2001. http://ens-news.com/ens/dec2001/2001L-12-21-09.html.

Wildfires

In 1991, 13 percent of the U.S. Forest Service's budget was spent on firefighting; by 2006, however, that figure was 43 percent. Between the 1970s and the year 2000, the western U.S., firefighting season expanded 92 days a year ("Harper's Index" 2006, 17). Increasing heat and drought spur fires that, in turn, add extra carbon dioxide to the atmosphere, increasing future warming, in a feedback loop.

Increasing fires in recent years may be only the beginning. Wildfires in the Western United States probably will increase in the coming decades, according to a study that links episodic fire outbreaks in the past five centuries with periods of warming sea-surface temperatures in the North Atlantic. Warming surface waters in the North Atlantic correspond to episodes of drought and fires in the West that are recorded in tree rings studied by the researchers, according to lead researcher Thomas Kitzberger of the University of Comahue in Argentina ("Western Wildfires" 2007).

The team analyzed nearly 34,000 individual fire scar dates from tree rings, primarily ponderosa pine and Douglas fir, at 241 sites, the largest record of tree rings linked to past wildfires ever assembled. Wildfire frequency in Washington, Oregon, California, Colorado, New Mexico, Arizona, and South Dakota was found to be affected. "This study underscores the value of building large networks of high-resolution fire history data to better understand how climate may affect fire regimes over large areas of the globe," said Kitzberger ("Western Wildfires" 2007).

Fires and Atmospheric Conditions

Warmer temperatures and changing atmospheric circulation patterns provoking drought in the western United States have dramatically increased both the number of large forest wildfires and the length of the season during which they occur during the past 35 years. Specifically, according to one study published in 2006, the number of fires from 1986 to 2005 increased fourfold, and the area burned increased sixfold, compared with 1979 through 1985 (Westerling et al. 2006, 940–943). The same study found that the average fire season (the period of time when major fires were taking place) had increased by 78 days, and the average length of major fires had risen from 7.5 days to 37.1 days (Running 2006, 927).

The authors wrote that increases in spring and summer temperatures by 0.9° C and mountain snow packs' melting one to four weeks earlier had contributed to the increased number and duration of wildfires. These fires also add carbon dioxide to the atmosphere—worldwide, according to one calculation, roughly 40 percent of human-generated emissions. Compounding all of this, according to A. L. Westerling and colleagues, warmer, longer growing seasons at high elevations reduce the amount of carbon removed from the atmosphere by living trees, especially during droughts (Westerling et al. 2006, 943). Adding it all up, the forests of the western United States are likely to become a net source of carbon to the atmosphere, rather than a sink.

Westerling and colleagues compiled a comprehensive time series of large forest wildfires in the western United States for the period from 1970 to 2003, and compared those data with corresponding observations of climate, hydrology, and land surface conditions (Westerling et al. 2006, 940–943). Wildfire activity increased suddenly in the mid-1980s. Hydroclimate and fires are closely related, and climate variation has been the primary cause of the increase in fires during the period of their study, although land-use changes can also be important. Longer springs and summers that could result as the world warms will continue to lengthen the fire season and continue to cause more large wildfires.

Smoke from fires in Southern California, October 2007. (NASA)

Westerling and colleagues wrote:

Western United States forest wildfire activity is widely thought to have increased in recent decades, yet neither the extent of recent changes nor the degree to which climate may be driving regional changes in wildfire has been systematically documented. Much of the public and scientific discussion of changes in western United States wildfire has focused instead on the effects of 19th- and 20th-century land-use history. We compiled a comprehensive database of large wildfires in western United States forests since 1970 and compared it with hydroclimatic and land-surface data. Here, we show that large wildfire activity increased suddenly and markedly in the mid-1980s, with higher large-wildfire frequency, longer wildfire durations, and longer wildfire seasons. The greatest increases occurred in mid-elevation, Northern Rockies forests, where land-use histories have relatively little effect on fire risks and are strongly associated with increased spring and summer temperatures and an earlier spring snowmelt. (Westerling et al. 2006, 940)

Fires in California

Reports from fires in Southern California during October 2007 were apocalyptic—more than a million people routed from their homes, many huddled in stadiums and fairgrounds; highways choked with the fleeing multitude; hundreds of homes burned to the ground; fire fighters completely overwhelmed as a hurricane-force Santa Ana winds, heating as they descended from the mountains, drove flames through brush dried by a record drought, burning 500,000 acres within a week in 100-degree afternoon temperatures and air nearly devoid of humidity. The phrase "environmental refugees" is being used more often, and the fires and late-October highs in the 80s along the east coast are being tied into global warming on the *NBC Evening News*. In some cases, the fires moved so quickly that firefighters found themselves trapped as they wrapped themselves in fireproof aluminum fabric to survive.

As is so often the case, climatic extremes have combined with other factors to produce disaster on an epic scale. An extremely strong Santa Ana fanned flames into suburbs that had been expanded onto land prone to wildfires. Winds gusted as strong as 111 miles an hour. An inventory by University of Wisconsin researchers found that about two-thirds of the new buildings in Southern California over the past decade were on land susceptible to wildfires, said Mike Davis, a historian at the University of California at Irvine and author of environmental and social histories of the region that have anticipated the toll of wildfires. "It gives you some parameters for understanding the current situation," Davis said.

Another way to look at it is you simply drive out the San Gorgonio Pass, where the winds blow over 50 mph over a hundred days a year and you have new houses standing next to 50-year-old chaparral.... You might as well be building next to leaking gasoline cans. (Vick and Geis 2007, A-1)

The Fires Next Time

Even before the fires of 2007, scientists had warned that they were coming. Fires that charred nearly three-quarters of a million acres in California during the fall of 2003 could presage increasingly severe fire danger as global warming weakens more forests through disease and drought. Warmer, windier weather and longer, drier summers could result in higher firefighting costs and greater loss of lives and property, according to researchers at the Lawrence Berkeley National Laboratory and the U.S. Forest Service. Both the number of out-of-control fires and the acreage burned are likely to increase, more than doubling losses in some regions, according to a study published in the scientific journal *Climatic Change*. While the study examined Northern California, "the concern for Southern California would be much higher," because that region is drier for longer periods, said researcher Evan Mills of the Lawrence Berkeley lab (Thompson 2003).

According to this study, a doubling of atmospheric carbon dioxide will provoke fires that "burn more intensely and spread faster in most locations." According to models developed by Jeremy S. Fried, Margaret S. Torn, and Evan Mills, the number of "escapes" (fires that exceed initial containment efforts) doubles present frequencies, according to their models. Contained fires also burn 50 percent more land under the warmer and windier conditions anticipated with a doubling of carbon-dioxide concentration in the atmosphere. The researchers stressed that their projections

represent a *minimum* expected change, or best-case forecast. In addition to the increased suppression costs and economic damages, changes in fire severity of this magnitude would have widespread impacts on vegetation distribution, forest condition, and carbon storage, and greatly increase the risk to property, natural resources, and human life. (Fried, Torn, and Mills 2004, 169)

The researchers projected at least a 50 percent increase in out-of-control fires in the southern San Francisco Bay area and a 125 percent increase in the Sierra Nevada foothills, with a more than 40 percent increase in the area burned. The state's northern coast saw no significant change under the computer model and conditions used in the study (Thompson 2003).

The study's projections use conservative forecasts that do not take into account increased lightning strikes and the spread of volatile grasslands into areas now dominated by less flammable vegetation. Even potentially wetter winters simply mean more growth, providing additional fuel for summer fires, according to the study. "Fires may be hotter, move faster, and be more difficult to contain under future climate conditions," said Robert Wilkinson of the University of California, Santa Barbara, School of Environmental Science and Management, in a federal report on the impact of climate change on California. "Extreme temperatures compound the fire risk when other conditions, such as dry fuel and wind, are present" (Thompson 2003). Damage may be aggravated by constriction of homes in brushlands that are vulnerable to fires.

By mid-summer 2004, Western North America was experiencing its worst wildfire season on record, as the usual season was only beginning, fueled by the worst drought in at least 500 years and rising temperatures. Nearly four million acres of forest from Alaska to southern California had burned by the end of July. Visitors to Yosemite National Park were advised not to overexert themselves because of the "very unhealthy quality" of the air, caused by pollution from fires. Usually, western wildfires reach their peak in October. In 2004, however, they started earlier than any other season on record. By early May, there had been 77 fires in the state of Washington alone, compared with 22 in 2003 (Lean 2004a, 20).

Wildfires and Boreal (Northern) Forests

A report by Greenpeace International suggests that between 50 and 90 percent of the Earth's existing boreal forests are likely to disappear if atmospheric levels of carbon dioxide and other greenhouse gases double. These forests comprise a third of the Earth's remaining tree cover, about 15 million square kilometers, across Russia (where they are called "taiga"), Canada, the United States, Scandinavia, parts of the Korean Peninsula, China, Mongolia, and Japan. Large forests also clothe many mountain ranges outside of these zones. In total, boreal forests cover about 10 percent of the world's land area.

The Greenpeace report indicates that global warming's toll on the boreal forests had begun by the early 1990s. The report warns that decaying forests may provide an extra boost to rising carbon-dioxide levels, causing warming to feed upon itself. The decline of boreal forests also endangers more than one million indigenous people who live in them, including the Dene and Cree of Canada, the Sami (Lapplanders) of Norway, Sweden and Finland, the Ainu of northern Japan, and the Nenets, Yakut, Udege, and Altaisk of Siberia (Jardine 1994). Some of the forests' larger animals, such as the Siberian tiger, already are near extinction. The Greenpeace report concludes that rapid logging of the boreal forests is intensifying pressure on animal life, and accelerating the release of even more carbon dioxide and other greenhouse gases into the atmosphere.

If boreal forests continue to decline, their burning and rotting could contribute to the release of as much as 225 billion tons of extra carbon dioxide into the atmosphere, raising current levels by a third, accelerating the pace of warming. Trees could have difficulty colonizing the thawing tundra to the north of their present ranges because they simply cannot migrate quickly enough, and because the treeless tundra cannot evolve quickly enough to sustain them.

Pests that may invade boreal forests under warming conditions include the western spruce budworm, the Douglas-fir tussock moth, and the mountain pine beetle (Kurz et al. 1995, 127). Kurz and colleagues concluded that "biospheric feedbacks from temperate and boreal forest ecosystems will be positive feedbacks that further enhance the carbon content of the global atmosphere" (Kurz et al. 1995, 129) By 1998, spruce budworms had devoured 50 million acres (20 million hectares) of Alaskan woodland.

Kevin Jardine sketched the role of insect outbreaks in the anticipated destruction of boreal forests:

> In moderation, the forest usually benefits from insect outbreaks because they reduce the likelihood of catastrophic fire by helping to eliminate older stands and diseased trees. However, populations of insects such as bark beetles, the Siberian silkworm and the spruce budworm can explode and devastate millions of acres of forest. The life cycles of the spruce budworm (*Choristoneura fumiferana*) and the spruce bark beetle (*Ips typographus*) are strongly influenced by climate, with both species likely to increase in numbers during the kind of warmer, drier weather predicted in a global warming world.... A Canadian government study released in 1987 showed that conifers were growing on average up to 65 percent slower than in the 1940s and 1950s, because of spruce budworm outbreaks, and possibly acid rain. (Jardine 1994)

See also: Forest Fires as Feedback Mechanism

FURTHER READING

Fried, Jeremy S., Margaret S. Torn, and Evan Mills. "The Impact of Climate Change on Wildfire Severity." *Climatic Change* 64 (May 2004):169–191.

"Harper's Index." *Harper's Magazine*, September 2006, 17.

Jardine, Kevin. "The Carbon Bomb: Climate Change and the Fate of the Northern Boreal Forests." Ontario, Canada: Greenpeace International, 1994. http://dieoff.org/page129.htm.

Kurz, Werner A., Michael J. Apps, Brian J. Stocks, and Jan A. Volney. "Global Climate Change: Disturbance Regimes and Biospheric Feedbacks of Temperate and Boreal Forests." In *Biotic Feedbacks in the Global Climate System: Will the Warming Feed the Warming?* ed. George M. Woodwell and Fred T. MacKenzie, 119–133. New York: Oxford University Press, 1995.

Lean, Geoffrey. "Worst U.S. Drought in 500 Years Fuels Raging California Wildfires." *London Independent*, July 25, 2004a, 20.

Running, Steven W. "Is Global Warming Causing More, Larger Wildfires? *Science* 313 (August 18, 2006):927–928.

Thompson, Dan. "Experts Say California Wildfires Could Worsen with Global Warming." Associated Press, November 12, 2003. (LEXIS)

Vick, Karl, and Sonya Geis. "California Fires Continue to Rage Evacuation May Be Largest, Officials Say." *Washington Post*, October 24, 2007, A-1. http://www.washingtonpost.com/wp-dyn/content/article/2007/10/23/AR2007102300347_pf.html.

Westerling, A. L., H. G. Hidalgo, D. R. Cayan, and T. W. Swetnam. "Warming and Earlier Spring Increase Western U.S. Forest Wildfire Activity." *Science* 313 (August 18, 2006):940–943.

"Western Wildfires Linked to Atlantic Ocean Temperatures." Environmental News Service, January 3, 2007. http://www.ens-newswire.com/ens/jan2007/2007-01-03-09.asp#anchor2.

Wildlife, Arctic: Musk Oxen, Reindeer, and Caribou

Retreating ice may imperil walrus and caribou populations in the Arctic as well as polar bears. Dwane Wilkin, a reporter for the *Nunatsiaq News*, reporting from Iqaluit, summarized the consequences of global warming for that area: "The good news is that sailing through the Northwest Passage will finally be a cinch. The bad news? Well, global warming will probably drive musk oxen, polar bears and Peary caribou into extinction. And other species—including humans—will face declining food sources"

(Wilkin 1997). These combined stresses could lead to "complete reproductive failure" of the caribou, in a worst-case scenario (George 2000).

Arctic precipitation may increase as much as 25 percent with the advent of rapidly rising temperatures during a century. When precipitation falls as heavy, wet snow and ice, caribou and other grazing animals have a harder time finding food during lean winter months. Heavier snow cover may lead to smaller, thinner animals that will be forced to go further afield for food in winter. The same animals will be plagued by increasing numbers of insects during warmer summers. Climate change is already making it tougher for browsing animals in the Arctic, such as caribou and reindeer, to survive the seven months of each year when they must find and eat large amounts of lichen and moss buried beneath snow and ice (Calamai 2002b, A-23).

Temperatures near the freezing point are more likely to cause precipitation to fall as freezing rain rather than snow, forming a barrier several centimeters thick. Given warmer weather in the Arctic, chances of ice replacing snow are rising, according to U.S. climate expert Jaakko Putkonen, an earth-sciences professor at the University of Washington. Based on two decades of work at the Spitsbergen archipelago in the Scandinavian Arctic, Putkonen calculated that five centimeters of rain falling on snow is enough to form an impenetrable ice barrier (Calamai 2002b, A-23).

"You only need one really big event to have a disastrous effect," said Putkonen. A single heavy rainfall on Russia's snowy Kamchatka peninsula led to the death of 5,000 reindeer, he said (Calamai 2002b, A-23). During the winter of 1996–1997, an estimated 10,000 reindeer died of starvation on Russia's far northeast Chukotka peninsula, after ice formed a thick crust over pastures, precluding reindeer grazing. Three thousand Sami in northern Sweden tend reindeer that have been suffering from climate change that often has changed snow to ice. During the 1990s, unusually rainy autumns caused ice crust on the moss that feed reindeer. According to one account, "Herd sizes ... tumbled, bringing not just financial hardship but also the threat of cultural decay" (Roe 2001).

In addition to separating grazing animals from their food, ice also raises temperatures at ground level by sealing in latent heat, encouraging the spread of fungi and toxic molds that attack the lichen. Signs of this problem were reported among reindeer in the Scandinavian

Arctic; experts say the same threat exists for more than three million caribou in migratory herds that sustain scores of aboriginal communities in northern Canada. Computer modeling suggests that the area vulnerable to such rain-on-snow events could increase by 40 percent worldwide by the 2080s under current global warming projections, with significant expansion into parts of central Canada that are home to both woodland and barren ground caribou (Putkonen and Roe 2003, 1188).

"They're the same animal genetically, only those in Europe and Asia have been domesticated while here they roam wild," said Don Russell, a biologist who studies caribou in the Yukon for the Canadian Wildlife Service. "They need about three kilograms dry weight every day— about a garbage-bag full. And they can spend about half their time just digging though the snow with their feet and feeding," said Russell (Calamai 2002b, A-23).

In the Arctic National Wildlife Refuge, spring has been arriving earlier. Consequently, caribou have had difficulty migrating from wintering areas in time to take advantage of periods of maximum springtime plant growth. Warming during springs since 1990 have been the earliest in nearly 40 years. By the time the animals reached the plain, their principal food plant had gone to seed. Caribou herd populations could decline significantly, should future climate and vegetation patterns prevent proper nourishment of calves. High Arctic Peary Caribou and musk oxen may even become extinct.

FURTHER READING

Calamai, Peter. "Global Warming Threatens Reindeer." *Toronto Star*, December 23, 2002b, A-23.

George, Jane. "Global Warming Threatens Nunavut's National Parks." *Nunatsiaq News*, May 19, 2000. http://www.nunatsiaq.com/archives/nunavut000531/nvt20519_18.html.

Putkonen, J. K., and G. Roe. "Rain-on-snow Events Impact Soil Temperatures and Affect Ungulate Survival." *Geophysical Research Letters* 30, no.4 (2003):1188–1192.

Roe, Nicholas. "Show Me a Home Where the Reindeer Roam." *London Times*, November 10, 2001, n.p. (LEXIS)

Wilkin, Dwayne. "A Team of Glacial Ice Experts Say Mother Nature's Thermostat Has Kept the Eastern Arctic at about the Same Temperature since 1960." *Nunatsiaq News*, May 30, 1997. http://www.nunanet.com/~nunat/week/70530.html#7.

Wildlife, Arctic: Threats to Harp Seals in Canada

According to an account by Colin Nickerson in the *Boston Globe*, early melting of ice in Canada's Gulf of St. Lawrence, "is wreaking havoc on harp seals which give birth on the floes and causing economic hardship for hard-pressed fishermen who depend on the controversial spring hunt" (Nickerson 2002, A-1). Several hundred drowned seal pups washed up on the shores of Newfoundland during 2002 after their mothers, unable to find ice on which to haul-out, gave birth in open water. Seal pups need at least 12 days on the ice before they complete nursing.

Fishermen who kill the seals with clubs or high-powered rifles in a much-criticized hunt for pelts, vitamin-rich oils, and sex organs for the Asian aphrodisiac trade also have found their livelihood threatened by the retreat of ice that is reducing harp-seal populations. Seal oil is rich in omega-3 fatty acids that may be helpful in reducing blood cholesterol levels.

In earlier years, ice usually extended from Quebec's Magdalen Islands southward to Prince Edward Island. By late March, according to Nickerson the floes usually teem with hundreds of thousands of seal mothers and their pups. "In five days of flying over the entire region, we haven't been able to spot a single seal pup," said Rick Smith, a marine biologist and Canadian director of the antisealing group from Prince Edward Island. "Usually, there are 200,000 to 300,000 harp seals born in the Gulf of St. Lawrence" (Nickerson 2002, A-1). "The seals need ice, but whether there has been a real reproductive failure this year remains to be seen," said Ian McLaren, professor of biology at Nova Scotia's Dalhousie University and an authority on seals. "One year's loss of pups is not necessarily a catastrophe," said McLaren (Nickerson 2002, A-1).

FURTHER READING

Nickerson, Colin. "An Early Melting Hurts Seals, Hunters in Canada." *Boston Globe*, April 1, 2002, A-1.

Wind Power

By 2008, wind power, now cost-competitive with most fossil-fuel energy sources, was the fastest-growing source of energy in the United States

Wind turbines. (Courtesy of Getty Images/PhotoDisc)

and the world. Wind power had become so popular that a shortage of parts was causing installations to fall behind demand. A new phrase in the English language, "wind-rich," describes an area with a relatively steady, unimpeded access to turbine-ready breezes. Some surprising pitchmen for wind (one of them being former President George W. Bush) believe that the United States may derive as much as 20 percent of its electrical power from wind energy by roughly the year 2020. Electricity generation accounts for about 40 percent of U.S. greenhouse gases.

Spain's tiny industrial state of Navarre, which generated no wind power in 1996, by 2002 generated 25 percent of its electricity that way. By 2007, 60 percent of its electricity was from renewable sources (mainly wind, with some solar), and plans to raise that proportion to 75 percent by 2010. At the end of 2006, the wind sweepstakes stood at Germany 20,652 megawatts, Spain 11,614, the United States 11,5745, India 6,228, and Denmark 3,101 (Fairless 2007a, 1048).

Wind power has become more adaptable, from huge turbines with blades the length of football fields to household-size units. Turbines may be mounted on land or offshore; engineering is being developed to install floating wind farms in depths of 200 meters (660 feet) or more.

Capacity Increases Rapidly

By 2008, wind-power capacity in the United States was increasing hand over fist, but the electrical grid has not kept up with the additional supply in many areas. A $320 million 200-turbine array at Maple Ridge Wind farm in Upstate New York was forced to shut down for lack of transmission capacity even when the wind blows, for example. In places, the grid is almost a century old, with 200,000 miles of lines and 500 owners. "We need an interstate transmission superhighway system," said Suedeen G. Kelly, a member of the Federal Energy Regulatory Commission. "The windiest sites have not been built, because there is no way to move that electricity from there to the load centers," he said (Wald 2008e). Many of the best wind-power production areas are sparsely populated. Bill Richardson, governor of New Mexico who was federal energy secretary under President Bill Clinton, said, "We still have a third-world grid.… With the federal government not investing, not setting good regulatory mechanisms, and basically taking a back seat on everything except drilling and fossil fuels, the grid has not been modernized, especially for wind energy" (Wald 2008e).

Europe's largest onshore wind farm, able to generate enough power for 320,000 homes, was approved in 2008 by the Scottish government. Announcing the new wind farm approval ahead of the World Renewable Energy Congress in Glasgow, First Minister Alex Salmond said the 152-turbine Clyde wind farm near Abington in South Lanarkshire would make Scotland the green energy capital of Europe.

In 2008, wind turbines in the United States generated about 48 billion kilowatt hours of electricity, enough to power about 4.5 million homes. That is a significant rise from previous years, but still only 1.2 percent of the country's electrical demand (Tomsho 2008, AS-11). Wind power in the United States in 2007 grew by 45 percent, adding 5,244 megawatts of capacity, a third of all the new electrical capacity in the country. In one year, wind energy employment in the United States doubled to about 20,000.

Meanwhile, in mid-July 2008, Texas regulators approved a $4.93 billion wind-power transmission project, including plans to carry enough power to supply 18,500 megawatts, or 3.7 million homes, from wind-rich western parts of the state to major population centers like Dallas, Houston, Austin, and San Antonio.

An Oilman Boosts Wind Power

At about the same time, Texas billionaire T. Boone Pickens, who made his fortune in oil, undertook a massive campaign in favor of alternative energy, most notably wind power, to reduce U.S. demand for petroleum. Pickens told a press conference in New York on July 8, 2008: "I've been an oil man all my life, but this is one emergency we can't drill our way out of. But if we create a new renewable energy network, we can break our addiction to foreign oil" ("Oilman Pickens" 2008).

In 1970, said Pickens, the United States imported 24 percent of its oil. In 2008, that figure was 70 percent and growing. At current oil prices (then about $135 a barrel), the United States could spend $700 billion on oil imports in one year, more than four times the annual cost of the Iraq war. The United States, with 4 percent of the world's population, consumes 25 percent of its oil production. Pickens started a media campaign advancing his plan to cut U.S. oil consumption one-third in 10 years. Much of the plan calls for massive development of wind energy.

Pickens has invested in the $2 billion Pampa Wind Project, in the Texas Panhandle, that probably will be the world's largest wind farm when it opens in 2014. Between 1,700 and 2,000 wind turbines at Pampa are expected to generate more than 4,000 megawatts, enough to power 1.3 million households. Suddenly, late in 2008, with the value of his investments sinking rapidly, Pickens, delayed his huge wind-power project. The price of a barrel of oil had crashed from $140 to $50 at the time, and Pickens said that fossil-fuel prices would have to rise again to make his wind-power megaproject financially feasible.

The 11,000 residents of Hull, Massachusetts, near Boston, embraced wind power in a big way, with two turbines by 2008 and more planned. Hull's power plant previously had no generating capacity of its own and was buying energy from another source at an average of about $0.08 per kilowatt hour. Their wind turbines supply it for about half as much (Tomsho 2008, A-11). Also, during 2008, the government of Sweden authorized tripling the country's wind-power capacity with a wind farm of about 100 turbines at Stora Middelgrund in the Kattegatt, some 30 kilometers west of the Holland coast parallel with Halmstad.

A Chic Wind Turbine Out Back

Wind turbines were popping up in some unexpected places. Parts of the U.S. "rust belt," along the south shores of the Great Lakes, are retooling as wind-energy centers. Witness Lackawanna, New York, a suburb of Buffalo, where, according to a report in the *New York Times*,

on the 2.2-mile shoreline above a labyrinth of pipes, blackened buildings and crumbling coke ovens that was once home to a behemoth Bethlehem Steel plant: eight gleaming white windmills with 153-foot blades slowly turning in the wind off Lake Erie, on a former Superfund site where iron and steel slag and other industrial waste were dumped during 80 years of production. (Staba 2007)

The wind farm is part of the New York State "brownfields" program, which turns former low-level toxic sites to productive uses. To local boosters, the "rust belt" has become the "wind belt." The turbines [called "Steel Winds"] cost $4.5 million each to build. Power lines once used by the steel plant now carry electricity from the turbines, while paved roads, rail lines, and an industrial port built by Bethlehem are now wind-turbine infrastructure.

A wind turbine out back has become a corporate chic status symbol, a brandishing of "green" credentials. They have been popping up all over the country, many of them *ad hoc* corporate efforts. *Outside Magazine* (a subsidiary of *Rolling Stone*) switched 90 percent of its Santa Fe, New Mexico, office's electricity needs to wind power. Burgerville, a chain in Washington and Oregon, was running all its restaurants on wind power.

Wind Power Becomes Cost-Competitive

By the early twenty-first century, wind power was becoming competitive in cost with electricity generated by fossil fuels, as its use surged. After 2004, wind power became a less expensive way to generate electricity than fossil fuels in some situations, as its use exploded by 20 to 25 percent annually. As its use grows, the unit price falls (Flannery 2005, 268–269).

While wind power still was a tiny fraction of energy generated in the United States, some areas of Europe (Denmark, as well as parts of Germany and Spain) were using it as a major source. Germany's northernmost state of Schleswig-Holstein has been using wind for 28 percent of its electricity (Brown 2003, 201). By 2003, 18 percent of Denmark's electricity was being derived from wind; Germany's northernmost state of Schleswig-Holstein has been using wind for 28 percent of its electricity (Brown 2003, 201).

Advances in wind-turbine technology adapted from the aerospace industry have reduced the cost of wind power from $0.38 per kilowatt hour (during the early 1980s) to $0.03 to $0.06. This rate is competitive with costs of power generation from fossil fuels, but costs vary according to site. Major corporations, including Shell International and British Petroleum, have been moving into wind power.

Wind capacity in the Pacific Northwest, where it is often used with hydroelectric, soared from only 25 megawatts in 1998 to a projected 3,800 megawatts by 2009. During 2006, Washington State added 428 megawatts of wind power, trailing only Texas in new installations. According to Randall Swisher, executive director of the American Wind Energy Association, the electrical grid in the Northwest is especially inviting for wind-power developers because of its hydroelectric distribution network, relatively reliable wind, progressive utility companies, and new state laws that establish preferences for renewable energy.

State law now requires utilities to work toward generating 15 to 25 percent of their electricity from sources that do not generate carbon dioxide. Utilities in 2007 planned to add 6,000 megawatts of wind power within a few years, more than doubling capacity. Hydroelectric dams have been built in river gorges and other valleys that are natural conduits for wind. The transmission line connecting the dams with urban areas has been built with considerable surplus capacity—a built-in network for wind power. Farmers in the area have been earning $2,000 to $4,000 per year for each wind turbine on their properties (Harden 2007, A-3).

Canada during 2006 became a serious player in wind energy, said Robert Hornung, president of the Canadian Wind Energy Association. "Canada's is on the cusp of a wind energy boom as provincial governments are now targeting to have a minimum of 10,000 megawatts of installed wind energy capacity in place by 2015," Hornung said ("Global Wind Power" 2007).

Asia experienced the strongest increase in installed capacity outside of Europe, with an addition of 3,679 megawatts, taking the continent more than 10,600 megawatts, about half that of Germany. In 2006, the continent grew by 53 percent and accounted for 24 percent of new installations. China more than doubled its total installed capacity by installing 1,347 megawatts of wind energy in 2006, a 70 percent increase over 2005. This brings China up to 2,604 megawatts of capacity, making it the sixth-largest market worldwide ("Global Wind Power" 2007).

The Chinese market was boosted by the country's new Renewable Energy Law, which entered into force on January 1, 2006. "Thanks to the Renewable Energy law, the Chinese market has grown substantially in 2006, and this growth is expected to continue and speed up," said Li Junfeng of the Chinese Renewable Energy Industry

Association. "According to the list of approved projects and those under construction, more than 1,500 MW will be installed in 2007," said Li. "The goal for wind power in China by the end of 2010 is 5,000 MW, which according to our estimations will already be reached well ahead of time" ("Global Wind Power" 2007).

Denmark's Example

Denmark was dependent on imported oil during the 1970s, and made an enduring commitment to achieve energy independence when supplies were embargoed and its economy devastated. Wind power is an important part of this strategy and, as a result, Denmark has become a world leader in wind-turbine technology. Work on Danish turbines is a major reason why the technology today generates electricity that competes in price with oil, coal, and nuclear power. In the meantime, Denmark has built infrastructure that provides several thousand jobs.

In matters of advanced technology, Denmark dominates the worldwide wind-power industry; Danish companies have supplied more than half the wind turbines now in use worldwide, making wind-energy technology one of the country's largest exports. Some Danish wind turbines now have blades almost 300 feet wide—the length of a football field. During January 2007, a very stormy month, Denmark harvested 36 percent of its electricity from wind, almost double the usual.

In 2007, a center-right coalition that took power in Denmark reduced subsidies for wind-power development, and its growth slowed. Suddenly, the tax environment for new wind-power development was better in Texas (see below) than in Denmark. The Danes, however, had a very long head start.

Danish wind energy has experienced technical setbacks, as Danish wind operators, hoping to bypass local objections and take advantage of stronger, steadier air currents, have tried to build giant turbines at sea (some are now more than 300 feet high and have blade assemblies nearly that wide). In one case, in 2004, turbines at Horns Reef, some 10 miles off the Danish coast, broke down because the equipment had been damaged by storms and salt water. Vestas, a Danish manufacturer, fixed the problem by replacing the equipment at a cost of 38 million euros. But Peter Kruse, the head of investor relations for Vestas, warned that the lesson from Horns Reef was that wind farms at sea would remain far more expensive than those on land. "Offshore wind farms don't destroy your landscape," Mr. Kruse said, but the added installation and maintenance costs are "going to be very disappointing for many politicians across the world" (Kanter 2007).

The Danish wind-power industry also provides 150,000 families with shares of profits from the national electricity grid. Thousands of rural Danish residents have joined wind-power cooperatives, buying turbines and leasing sites to build them, often on members' land. At the same time, power companies in Denmark are now taxed for each ton of carbon dioxide emitted above a low (and, over time, gradually declining) limit.

Wind Turbines' Problems: "Visual Pollution," Bird Kill, and Fickleness

Wind turbines have not been universally welcomed. Some of their neighbors complain that they cause "visual pollution." *Cape Wind: Money, Celebrity, Class, Politics, and the Battle for Our Energy Future on Nantucket Sound*, by Wendy Williams and Robert Whitcomb (2007), is devoted to the battle over this issue on Massachusetts' Cape Cod, with some high-octane names (one of them being Senator Edward Kennedy). The course of their complaints is a wind farm of the same name on the Massachusetts coast's Nantucket Sound. Perhaps a coal strip mine would be a more salubrious sight, but no one receiving coal-fired energy on Cape Cod has to witness the mining of the coal. The Cape Wind project is no garden-variety project. Its plans include 130 turbines producing 3.6 megawatts each. Each 3.6-megawatt machine uses a 175-foot blade. On a windy day, at maximum capacity, all 130 machines would produce as much power as a modest-size electricity plant burning coal or natural gas.

The power capacity of wind turbines may be restricted only by the size of a blade that can be hauled to a site, mounted, and maintained with some degree of structural integrity. A turbine at 250 feet takes advantage of winds that are 20 percent stronger at that elevation than at 150 feet, said Dr. Mark Z. Jacobson, an associate professor at Stanford's department of civil and environmental engineering (Wald 2007a).

The bigger the turbine, the greater the chance it may kill birds—about 40,000 of them a year in the United States by 2006. That is one bird per 30 turbines per year, however, and a small fraction of the millions killed each year by domestic cats (Marris and Fairless 2007, 126).

A major problem for wind energy is that the hottest days (when power demand is highest) are usually the least windy. "As a result, wind turns out to be a good way to save fuel, but not a good way to avoid building plants that burn coal," wrote Matthew Wald in the *New York Times*.

> A wind machine is a bit like a bicycle that a commuter keeps in the garage for sunny days. It saves gasoline, but the commuter has to own a car anyway. Without major advances in ways to store large quantities of electricity or big changes in the way regional power grids are organized, wind may run up against its practical limits sooner than expected. In many places, wind tends to blow best on winter nights, when demand is low. (Wald 2006)

Robert E. Gramlich, a policy director with the American Wind Energy Association, said that wind energy could be integrated into an electrical grid on a large enough area that ebbing wind in one part could be balanced by continuing breezes in another (Wald 2006).

Noise and vibrations from large wind turbines that some people compare to a jet taking off have provoked some of their nearby neighbors to complain of insomnia, headaches, and insomnia. Niña Pierpont, a pediatrician who lives in Malone, New York, has given the condition a name (Wind Turbine Syndrome) and plans to write a book on the topic. A physician, Amanda Harry, who lives in Great Britain, said that the turbines also can cause anxiety, vertigo, and depression. Mariana Alves-Pereira, of Portugal, an acoustical engineer, asserted that wind-turbine noise and vibration may contribute to strokes and epilepsy. The European Union, in 2008, exonerated wind turbines of culpability in any medical condition except occasional loss of sleep (Keen 2008, A-3). Curing the syndrome is usually simple: a matter of distance requiring, in real-estate legal language, proper setbacks of usually 1,000 to 2,500 feet. Some neighbors insist on setbacks of a mile or more to keep home values from declining.

Wind Power in Texas Oil Country

While Texas has a reputation as a conservative state run by oil barons, it has become one of the most progressive environments for wind energy in the United States. George W. Bush, himself once an oilman, helped start the "wind rush" when he was governor by signing a bill requiring the state's power infrastructure to include 3 percent from alternative energy by 2009. By December 2004, state planners recommended that by 2025 10 percent of Texas' power should come from renewable sources.

"That's just money you're hearing," said wind farmer Louis Brooks, who receives $500 a month for each of 78 wind turbines on his land near Sweetwater, Texas, as the turbines spun a low hum (Krauss 2008). By 2008, Texas was obtaining more than 3 percent of its electricity, enough to supply power to one million homes, from wind turbines. Some of the turbines are twice as high as the Statue of Liberty, with blades that span as wide as the wingspan of a jumbo jet.

"Texas has been looking at oil and gas rigs for 100 years, and frankly, wind turbines look a little nicer," said Jerry Patterson, the Texas land commissioner, whose responsibilities include leasing state lands for wind energy development. "We're No. 1 in wind in the United States, and that will never change" (Krauss 2008).

In 2007, the Texas Public Utility Commission approved transmission lines across Texas that will carry up to 25,000 megawatts of wind energy by 2012, a 500 percent increase from 2008. Shell and the TXU Corporation, a utility based in Dallas, are planning a 3,000-megawatt wind farm in the Texas Panhandle. In 2007, Texas had installed wind power of 4,356 megawatts. Land values in some counties with wind farms have nearly doubled.

Cielo Wind Power of Austin has been buying 60-year "wind rights" from ranchers in Texas with plans to build the Noelke Hill Wind Ranch for $130 million. The wind ranch plans to generate 240 megawatts of power, enough to provide electricity to 80,000 homes through TXU. Several companies, among them American Electric power and General Electric, have plans to spend as much as $1 billion on wind energy in the midst of the Permian Basin, which has heretofore been known as oil country. Wind speed in the area averages 16 miles an hour and is consistent enough to bend juniper trees to their sides.

By late 2002, Cielo had built $300 million worth of wind-power turbines in Texas and sold them to various power companies, "jutting more than 200 feet into the air with blades 200 feet in diameter, to capture high-velocity wind patterns"

(Herrick 2002, B-3). Texas by late 2002 was generating enough wind power to supply about 300,000 homes (Herrick 2002, B-3). A standard line used by wind developers with the ranchers in the area is: "You've been getting your hat blown off your head your whole life. It's time to stop cussin' and make some money" (Herrick 2002, B-3). Wind turbines placed roughly one per 25 acres earn the average landowner about $3,200 each (Herrick 2002, B-3).

Wind has become truly big business in Texas. By 2007, Royal Dutch Shell and a wind-development company owned by Goldman Sachs Group Inc. were building wind-power infrastructure in the Texas Hill Country around Silverton. The companies were building some of the largest wind farms in the world, planning hundreds of turbines costing about $2 million each (Ball 2007b). The area, once home to oil rigs, has a relentless wind that blows so steadily that the local flora (sagebrush and mesquite) grow at an angle. The area's canyons tend to funnel the wind.

In the same area, Shell, a company diversifying from its oil roots, was planning a 120-square-mile wind farm, still another that was being touted as "the largest in the world," occupying a land area five times the size of Manhattan Island. Shell's new wind farm will produce electricity equal to a coal-fired power plant (Ball 2007b, A-11).

Shell and other companies are betting that wind power's time has come. Their lobbying contacts in Austin, the state capital, are poised to make a reality of this hunch. Large companies, such as General Electric, which manufactures wind turbines, also have joined the wind lobby. Presently, federal tax rules allow a company engaged in wind-power generation to reduce its tax bill by $0.019 per kilowatt hour. The U.S. Congress has authorized the credit several times to be used for short lengths of time, and then reauthorized it after expiration. Texas law also requires utilities to buy wind power when and where it is available as a proportion of generating capacity. Shell in 2007 was selling wind power from its 30-square-mile Brazos wind farm to TXU, which is under pressure by environmentalists to reduce its reliance on coal-generated power. The Brazos wind farm contains 160 turbines that have a capacity of 160 megawatts, if the wind blows. If it does not, the capacity is zero. Local ranchers can collect as much as $80,000 a year from energy companies for a 640-acre section of land under a wind farm (Ball 2007b, A-11).

Household Wind Turbines

Herschel Carter is one of a growing number of people in the United States who have been installing household-scale wind-power generators, which are available for $10,000 to $15,000. Herschel, 62, a retired insurance agent, watched the amount of electricity he consumes from the grid fall to nearly half (1,752 to 993 kilowatt hours) in the first month his wind turbine operated (March 2007). His electric bill for that month fell $68. Payback will take several years, but Carter regards the turbine as a form of energy security because his wife is on oxygen and needs electricity for it. The unit has a battery that stores energy from windy days for calm ones (McNulty 2007, 7).

The United States invented household-scale wind turbines about 1920 and is a world leader in their technology. More than 90 percent of household wind turbines have been installed there. Growth in the number of household wind turbines has been averaging 14 to 25 percent a year between 2000 and 2006 worldwide. At least a half-acre of land is required for a residential wind turbine, and regulators must permit a tower of at least 35 feet (the higher the tower, the more efficient the turbine).

Smaller wind turbines may be available. A miniature wind turbine developed by the Scottish company Windsave can be mounted on an urban roof. Its four-foot blades can provide part of an average home's electricity (company publicity says one-third) at a cost of about $2,000 (McGuire 2005, 175).

FURTHER READING

Ball, Jeffrey. "The Texas Wind Powers a Big Energy Gamble." *Wall Street Journal*, March 12, 2007b, A-1, A-11.

Brown, Lester R. *Plan B: Rescuing a Planet under Stress and a Civilization in Trouble.* New York: Earth Policy Institute/W.W. Norton, 2003.

Fairless, Daemon. "Renewable Energy: Energy Go-Round." *Nature* 447 (June 28, 2007a):1046–1048.

Flannery, Tim. *The Weather Makers: How Man Is Changing the Climate and What It Means for Life on Earth.* New York: Atlantic Monthly Press, 2005.

"Global Wind Power Generated Record Year in 2006." Environment News Service, February 12, 2007. http://www.ens-newswire.com/ens/feb2007/2007-02-12-04.asp.

Harden, Blaine. "Air, Water Powerful Partners in Northwest: Region's Hydro-Heavy Electric Grid Makes for Wind-Energy Synergy." *Washington Post*, March 21, 2007, A-3. http://www.washingtonpost.com/wp-dyn/content/article/2007/03/20/AR2007032001634_pf.html.

Herrick, Thaddeus. "The New Texas Wind Rush: Oil Patch Turns to Turbines, as Ranchers Sell Wind Rights." *Wall Street Journal*, September 23, 2002, B-1, B-3.

Johnson, Keith. "Renewable Power Might Yield Windfall." *Wall Street Journal*, March 22, 2007, A-8.

Kanter, James. "Across the Atlantic, Slowing Breezes." *New York Times*, March 7, 2007. http://www.nytimes.com/2007/03/07/business/businessspecial2/07europe.html.

Keen, Judy. "Neighbors at Odds over Noise from Wind Turbines." *USA Today*, November 4, 2008, A-3.

Krauss, Clifford. "Move over, Oil, There's Money in Texas Wind." *New York Times*, February 23, 2008. http://www.nytimes.com/2008/02/23/business/23wind.html.

Marris, Emma, and Daemon Fairless. "Wind Farms' Deadly Reputation Hard to Shift." *Nature* 447 (May 10, 2007):126.

McGuire, Bill. *Surviving Armageddon: Solutions for a Threatened Planet.* New York: Oxford University Press, 2005.

McNulty, Sheila. "U.S. Power Generation Answer Is Blowing in the Wind." *London Financial Times*, April 24, 2007, 7.

"Oilman Pickens Promotes the Wind." Environment News Service, July 8, 2008. http://www.ens-newswire.com/ens/jul2008/2008-07-08-091.asp.

Staba, David. "An Old Steel Mill Retools to Produce Clean Energy." *New York Times*, May 22, 2007. http://www.nytimes.com/2007/05/22/nyregion/22wind.html.

Tomsho, Robert. "Wind Shift in Energy Debate." *Wall Street Journal*, June 19, 2008, A-11.

Wald, Matthew. "It's Free, Plentiful and Fickle." *New York Times*, December 28, 2006. http://www.nytimes.com/2006/12/28/business/28wind.html.

Wald, Matthew. "What's So Bad About Big?" *New York Times*, March 7, 2007a. http://www.nytimes.com/2007/03/07/business/businessspecial2/07big.html.

Wald, Matthew. "Wind Energy Bumps into Power Grid's Limits." *New York Times*, August 27, 2008e. http://www.nytimes.com/2008/08/27/business/27grid.html.

Smith, Rebecca. "Wind, Solar Power Gain Users." *Wall Street Journal*, January 18, 2008, A-6.

Wine Grapes and Warming

Significant warming could pose major problems for the U.S. wine-grape industry, reducing the area suitable for growing premium grapes by 50 to 80 percent by the end of the twenty-first century, according to a study published in the *Proceedings of the National Academy of Sciences* (White et al. 2006, 11,217–11,222). The article's authors anticipate that intensifying weather problems for grapes in such areas as California's Napa and Sonoma valleys, including more extremely hot days, according to Noah Diffenbaugh of the department of earth and atmospheric sciences at Purdue University (Schmid 2006a).

Using a climate model from the Intergovernmental Panel on Climate Change, the authors of this study asserted that "potential premium wine-grape production area in the conterminous United States could decline by up to 81 percent by the end of the 21st century" (White et al. 2006, 11,217). An increasing number of days with daily maximum temperatures above 95° F could eliminate wine-grape production in many present growing areas. Grape production probably will be restricted by century's end to a narrow band along the west coast and cooler areas of the Northeast and Northwest, many of which face problems related to excess moisture.

At temperatures above 95° F, many wine grapes cannot maintain photosynthesis, so sugars in the grapes break down, according to Diffenbaugh (Schmid 2006a). "We have very long-term studies of how this biological system (of vineyards) responds to climate," said Diffenbaugh, and that gives the researchers confidence in their projection. Diffenbaugh is a co-author of the paper (Schmid 2006a).

James A. Kennedy, a professor of food science and technology at Oregon State University, said he was shocked by the report on the potential effects on wine grapes. "The lion's share of the industry is in California, so it's a huge concern from a wine quality standpoint," he said. For people in the industry "this paper is going to be a bit of a shocker" (Schmid 2006a).

French Grape Harvest Moves Up 40 Days in 50 Years

Grapes that were harvested in the French Alsace district during 1978 in mid-October ripened in mid-September by 1998. In 2007, the harvest began on August 24, dramatic evidence that warming temperatures gave changed growing seasons. "I noticed the harvest was getting earlier before anybody had a name for it," said

59-year-old Ren Mur, the eleventh generation of his family to produce wine from the clay and limestone slopes of the Vosges Mountains near the German border. "When I was young, we were harvesting in October with snow on the mountaintops. Today we're harvesting in August" (Moore 2007, A-1).

Grapes are sensitive to temperature and other weather conditions, and detailed records have been kept on their responses to various climatic conditions. "The link of wine to global warming is unique because the quality of wine is very dependent on the climate," said Bernard Seguin, an authority on the impact of global warming and viniculture at the French National Agronomy Institute. "For me, it is the ultimate expression of the consequences of climate change (Moore 2007, A-1).

Provence and other southern regions of France have become too dry and hot for grapes. Pests also move north with warmer temperatures, including the leafhopper, which is spreading yellow-leaf disease in Alsace vineyards for the first time. In some Mediterranean vineyards, weather has become so warm that grapes are harvested at night. Vineyards are being relocated to higher elevations. Some French growers are planning to move to England, where the wine-making industry is reemerging. In France, vintners no longer add sugar to the wines to improve their flavors and alcohol content. Warmer summers do that. Temperatures in Alsace have risen 3.5° F in 30 years.

British Columbia Wines

In the Western Hemisphere, by 2007 southern British Columbia was developing a wine-grape industry for the first time. Located near Tappen, adjacent to the Okanagan Valley, the occasional early frost still caused problems at harvest time in the middle of October. The grapes grown there are increasingly showing up in some of the world's most expensive Chardonnay and Cabernet Sauvignon wines. The ancient French names stem from regions where warming temperatures have begun to ruin temperature-sensitive harvests.

Before 1990, Canadian wine was a joke, along the lines of the Nebraska Navy, but with longer, warmer summers, it has become serious business as farmers abandon dairy cows for grapes. Temperatures have risen 5° F over 60 years, and the growing season has extended almost two weeks.

Winter temperatures have risen substantially, enough to enable survival of grape vines (Belkin 2007, A-1, A-20). By 2007, British Columbia hosted 136 wineries, and its wines were winning international competitions.

FURTHER READING

Belkin, Douglas. "Northern Vintage: Canada's Wines Rise with Mercury." *Wall Street Journal*, October 15, 2007, A-1, A-20.

Moore, Molly. "In Northern France, Warming Presses Fall Grape Harvest into Summertime." *Washington Post*, September 2, 2007, A-1. http://www.washingtonpost.com/wp-dyn/content/article/2007/09/01/AR2007090101360_pf.html.

Schmid, Randolph E. "Climate Change Could Devastate U.S. Wineries." Associated Press, July 10, 2006a. (LEXIS)

White, M. A., N. S. Diffenbaugh, G. V. Jones, J. S. Pal, and F. Gioirgi. "Extreme Heat Reduces and Shifts United States Premium Wine Production in the 21st Century." *Proceedings of the National Academy of Sciences* 103, no. 30 (July 25, 2006):11,217–11,222.

Wintertime Warming and Greenhouse-Gas Emissions

Accumulation of greenhouse gases explains why the Northern Hemisphere is warming more quickly during wintertime months than the rest of the world during the last 30 years, according to a computer climate model developed by NASA scientists. They found that greenhouse gases, more than any other factor, increase the strength of polar winds that regulate Northern Hemisphere climate in winter. The polar winds that play a large role in the wintertime climate of the Northern Hemisphere blow in the stratosphere, eventually mixing with air and influencing weather close to the Earth's surface. The findings of Drew Shindell, Gavin Schmidt, and other atmospheric scientists from NASA's Goddard Institute for Space Studies and Columbia University appeared in the April 16, 2001, issue of the *Journal of Geophysical Research* (Shindell et al. 2001, 7193).

Shindell and colleagues asserted that increases in greenhouse gases contribute to persistence of stronger polar winds into the springtime and contribute to a warmer early spring climate in the Northern Hemisphere. A stronger wind circulation around the North Pole increases temperature difference between the pole and the midlatitudes. Shindell said that the Southern Hemisphere is not affected by increasing greenhouse gases the same way, because it is colder and the polar wind circulation over the Antarctic is already very strong (O'Carroll 2001).

"Surface temperatures in the Northern Hemisphere have warmed during winter months as much as to 9° F during the last three decades, over 10 times more than the global annual average 0.7° F," said Shindell. "Warmer winters will also include more wet weather in Europe and western North America, with parts of western Europe the worst hit by storms coming off the Atlantic" (O'Carroll, 2001). Year-to-year changes in the polar winds are quite large, according to Shindell, "But over the past 30 years, we have tended to see stronger winds and warming, indicative of continually increasing greenhouse gases" (O'Carroll 2001).

FURTHER READING

O'Carroll, Cynthia. "NASA Blames Greenhouse Gases for Wintertime Warming." *UniSci*, April 24, 2001. http://unisci.com/stories/20012/0424011.htm.

Shindell, D. T., G. A. Schmidt, R. L. Miller, and D. Rind. "Northern Hemisphere Winter Climate Response to Greenhouse Gas, Ozone, Solar, and Volcanic Forcing." *Journal of Geophysical Research* 106 (2001):7193–7210.

Worldwide Climate Linkages

Rising temperatures in one area may affect vital climate patterns (including rainfall) in others, hundreds or thousands of miles away. Warming of the Indian Ocean, for example, may be at least partially responsible for declining rainfalls in the Sahel region of the southern Sahara that have contributed to devastating droughts. A modeling study by Alessandra Giannini of the International Institute for Climate Prediction in Palisades, New York, has linked rising Indian Ocean sea-surface temperatures with generally declining rainfall across the Sahel between 1930 and 2000. The Indian Ocean has warmed more rapidly than any other oceanic body on Earth (Giannini, Saravanan, and Chang 2003, 1027–1030; Kerr 2003b, 210). Studies by Giannini and colleagues linked variability of rainfall in the Sahel between 1930 and 2000 to "response of the African summer monsoon to oceanic forcing, amplified by land-atmosphere interaction" (Giannini, Saravanan, and Chang 2003, 1027).

Warmer oceanic temperatures tend to weaken convergence that determines the intensity of the monsoon from Senegal to Ethiopia. Another modeling study, by Mojib Latif of the University of Kiel, Germany, produced similar conclusions. "We found," he said, "[t]hat the Indian Ocean is probably the most important agent in driving decadal changes in Sahel rainfall" (Kerr 2003b, 210). Rising ocean temperatures do not explain all of the Sahel's drought problems, however; Giannini and colleagues wrote that temperature change may be responsible for only 25 to 35 percent of the observed change (Zeng 2003, 1000).

FURTHER READING

Giannini, A., R. Saravanan, and P. Chang. "Oceanic Forcing of Sahel Rainfall on Interannual to Interdecadal Time Scales." *Science* 302 (November 7, 2003):1027–1030.

Kerr, Richard A. "Warming Indian Ocean Wringing Moisture From the Sahel." *Science* 302 (October 10, 2003b):210–211.

Zeng, Ning. "Drought in the Sahel." *Science* 302 (November 7, 2003):999–1000.

Bibliography

Abboud, Leila. "Sun Reigns on Spain's Plains: Madrid Leads a Global Push to Capitalize on New Solar-power Technologies." *Wall Street Journal*, December 5, 2006, A-4.

Abboud, Leila. "How Denmark Paved Way to Energy Independence." *Wall Street Journal*, April 16, 2007, A-1, A-13.

Abboud, Leila, and John Biers. "Business Goes on an Energy Diet." *Wall Street Journal*, August 27, 2007, R-1, R-4.

Abrahamson, Dean Edwin. "Global Warming: The Issue, Impacts, Responses." In *The Challenge of Global Warming*, 3–34. Washington, DC: Island Press, 1989.

Abram, Nerilie J., Michael K. Gagan, Malcolm T. McCulloch, John Chappell, and Wahyoe S. Hantoro. "Coral Reef Death during the 1997 Indian Ocean Dipole Linked to Indonesian Wildfires." *Science* 301 (August 15, 2003):952–955.

Abramowitz, Michael. "U.S. Joins G-8 Plan to Halve Emissions 2050 Pledge Marks Shift on Issue for Bush." *Washington Post*, July 9, 2008, A-1. http://www.washingtonpost.com/wp-dyn/content/article/2008/07/08/AR2008070800285_pf.html.

"Abrupt Climate Change during Last Glacial Period Could Be Tied to Dust-Induced Global Warming." Press Release, NOAA 96-78, December 4, 1996. http://www.noaa.gov/public-affairs/pr96/dec96/noaa96-78.html.

Achenbach, Joel. "Global Warming Did It! Well, Maybe Not." *Washington Post*, August 3, 2001, B-1. http://www.washingtonpost.com/wp-dyn/content/article/2008/08/01/AR2008080103014_pf.html.

Achenbach, Joel. "March in January! Or Is It Mayday? It's Nice Out There, but Global Warming Dampens the Fun." *Washington Post*, January 7, 2007, D-1. http://www.washingtonpost.com/wp-dyn/content/article/2007/01/06/AR2007010601215.html.

"Acid Rain Emissions Halve in 11 Years." Hermes Database (Great Britain), May 20, 2003 (LEXIS).

Ackerman, A. S., O. B. Toon, D. E. Stevens, A. J. Heymsfield, V. Ramanathan, and E. J. Welton. "Reduction of Tropical Cloudiness by Soot." *Science* 288 (May 12, 2000): 1042–1047.

Adam, David. "Hatchoooooh! Record Numbers of People Are Complaining of Hay Fever." *London Guardian*, June 18, 2003, 4.

Adam, David. "Meltdown Fear as Arctic Ice Cover Falls to Record Winter Low." *London Guardian*, May 15, 2006. http://www.guardian.co.uk/science/story/0,,1774815,00.html.

Adams, J. Brad, Michael E. Mann, and Casper M. Amman. "Proxy Evidence for an El Niño-like Response to Volcanic Forcing." *Nature* 426 (November 20, 2003):274–278.

Adams, Jonathan, Mark Maslin, and Ellen Thomas. "Sudden Climate Transitions during the Quaternary." *Physical Geography* 23 (1999):1–36.

Adams, Marilyn. "Strapped Insurers Flee Coastal Areas." *USA Today*, April 26, 2006, B-1.

Adger, W. N., T. A. Benjaminsen, K. Brown, and H. Svarstad. "Advancing a Political Ecology of Global Environmental Discourses. *Development and Change* 32 (2001):681–715.

Adger, W. N., J. Paavola, S. Huq, and M. J. Mace, eds. *Fairness in Adaptation to Climate Change*. Cambridge, MA: Massachusetts Institute of Technology Press, 2006.

Agarwal, Anil, and Sunita Narain. *Global Warming in an Unequal World: A Case of Environmental Colonialism*. New Delhi, India: Centre for Science and Environment, 1991.

Ainley, D. G., G. Ballard, S. D. Emslie, W. R. Fraser, P. R. Wilson, E. J. Woehler, J. P. Croxall, P. N. Trathan, and E. J. Murphy. "Adélie Penguins and Environmental Change." Letter to the Editor, *Science* 300 (April 18, 2003):429.

"Aircraft Pollution Linked to Global Warming: Himalayan Glaciers Are Melting, with Possibly Disastrous Consequences." Reuters in *Baltimore Sun*, June 13, 1999, 13-A.

"Air New Zealand Jet Flies on Jatropha Biofuel." Environment News Service, December 30, 2008. http://www.ens-newswire.com/ens/dec2008/2008-12-30-02.asp.

Aitken, Mike. "St. Andrews Stymied by Natural Hazard." *The Scotsman*, April 18, 2001, 20.

"Alaskan Glaciers Retreating." Environment News Service, December 11, 2001. http://ens-news.com/ens/dec2001/2001L-12-11-09.html.

Aldhous, Peter. "Global Warming Could Be Bad News for Arctic Ozone Layer." *Nature* 404 (April 6, 2000):531.

"Alexander Cockburn Attacks Global Warming Hoax from the Left." Monbattery. Accessed November 15, 2007. http://www.moonbattery.com/archives/2007/06/alexander_cockb.html.

Alford, Ross A., Kay S. Bradfield, and Stephen J. Richards. "Ecology: Global Warming and Amphibian Losses." *Nature* 447 (May 31, 2007): E-3, E-4.

"Al Gore Praises Swedish Premier for Environmental Leadership." Associated Press, September 8, 2006 (LEXIS).

Allan, Richard P., and Brian J. Soden. "Atmospheric Warming and the Amplification of Precipitation Extremes." *Science* 321 (September 12, 2008):1481–1484.

Allen, Myles. "Film: Making Heavy Weather." *Nature* 429 (June 7, 2004):347–348.

Alley, Richard B. *The Two-Mile Time Machine Ice Cores, Abrupt Climate Change, and Our Future*. Princeton, NJ: Princeton University Press, 2000a.

Alley, Richard B. "Ice-core Evidence of Abrupt Climate Changes." *Proceedings of the National Academy of Sciences of the United States of America* 97, no. 4 (February 15, 2000b):1331–1334.

Alley, Richard B., ed. *Abrupt Climate Change: Inevitable Surprises*. Committee on Abrupt Climate Change, Ocean Studies Board, Polar Research Board, Board on Atmospheric Sciences and Climate, Division of Earth and Life Sciences, National Research Council. Washington, DC: National Academy Press, 2002a.

Alley, Richard B. "On Thickening Ice?" *Science* 295 (January 18, 2002b):451–452.

Alley, Richard B., Mark Fahnestock, and Ian Joughin. "Understanding Glacier Flow in Changing Times." *Science* 322 (November 14, 2008):1061–1062.

Alley, Richard B., J. Marotzke, W. D. Nordhaus, J. T. Overpeck, D. M. Peteet, R. A. Pielke Jr., R. T. Pierrehumbert, P. B. Rhines, T. F. Stocker, L. D. Talley, and J. M. Wallace. "Abrupt Climate Change." *Science* 299 (March 28, 2003):2005–2010.

Alley, Richard B., Peter U. Clark, Philippe Huybrechts, and Ian Joughin. "Ice-Sheet and Sea-Level Changes." *Science* 310 (October 21, 2005):456–460.

Alleyne, Richard, and Ben Fenton. "Heatwave Britain—When the Trees Turn Toxic." *London Daily Telegraph*, May 10, 2004, 3.

Almendares, J., and M. Sierra. "Critical Conditions: A Profile of Honduras." *Lancet* 342 (December 4, 1993):1400–1403.

Alsos, Inger Greve, Pernille Bronken Eidesen, Dorothee Ehrich, Inger Skrede, Kristine Westergaard, Gro Hilde Jacobsen, Jon Y. Landvik, Pierre Taberlet, and Christian Brochmann. "Frequent Long-Distance Plant Colonization in the Changing Arctic." *Science* 316 (June 15, 2007):1606–1609.

"Amazon Deforestation Causing Global Warming, Brazilian Government Says." British Broadcasting Corporation International reports, December 10, 2004 (LEXIS).

Amor, Adlai. "Report Warns of Growing Destruction of World's Coastal Areas." World Resources Institute, April 17, 2001. http://www.dooleyonline.net/media_preview/index.cfm.

Amos, Jonathan. "Swiss Glaciers 'in Full Retreat.'" BBC News, accessed January 7, 2009. http://news.bbc.co.uk/2/hi/science/nature/7770472.stm.

Amstrup, S. C., I. Stirling, T. S. Smith, C. Perham, and G. W. Thiemann. "Recent Observations of Intraspecific Predation and Cannibalism among Polar Bears in the Southern Beaufort Sea." *Polar Biology* 29, no. 11 (2006):997–1002, doi: 10.1007/s00300-006-0142-5.

"Analysis: Climate Warming at Steep Rate." *Los Angeles Times* in *Omaha World-Herald*, February 23, 2000, 12.

Anderson, David M., Jonathan T. Overpeck, and Anil K. Gupta. "Increase in the Asian Southwest Monsoon during the Past Four Centuries." *Science* 297 (July 26, 2002):596–599.

Anderson, J. G., W. H. Brune, and M. H. Proffitt. "Ozone Destruction by Chlorine Radicals within the Antarctic Vortex: The Spatial and Temporal Evolution of ClO/O3, Anticorrelation Based on In Situ ER-2 Data." *Journal of Geophysical Research* 94 (1989):11,465–11,479.

Anderson, J. W. "The History of Climate Change as a Political Issue." The Weathervane: A Global Forum on Climate Policy Presented by Resources for the Future, August 1999. http://www.weathervane.rff.org/features/feature005.html.

Anderson, John B. "Ice Sheet Stability and Sea-Level Rise." *Science* 315 (March 30, 2007):1803–1804.

Anderson, John Ward. "Paris Embraces Plan to Become City of Bikes." *Washington Post*, March 24, 2007a, A-10. http://www.washingtonpost.com/wp-dyn/content/article/2007/03/23/AR2007032301753_pf.html.

Anderson, John Ward. "Europe's Summer of Wild, Wild Weather: Fires, Droughts, and Floods Leave Wake of Destruction." *Washington Post*, August 2, 2007b, A-11. http://www.washingtonpost.com/wp-dyn/content/article/2007/08/01/AR2007080102347_pf.html.

Anderson, John Ward. "U.N. Climate Talks End in Cloud of Discord Industrialized, Developing Nations Still at Odds Over How and When to Cut Emissions." *Washington Post*, September 1, 2007c, A-20. http://www.washingtonpost.com/wp-dyn/content/article/2007/08/31/AR2007083102052_pf.html.

Anderson, Julie. "UNL Student Helps Shed New Light on El Niño." *Omaha World-Herald*, May 14, 2000, 12-B.

Anderson, Rebecca K., Gifford H. Miller, Jason P. Briner, Nathaniel A. Lifton, and Stephen B. DeVogel. "A Millennial Perspective on Arctic Warming from 14-C in Quartz and Plants Emerging from beneath Ice Caps." *Geophysical Research Letters* 35 (2008), L01502, doi: 10.1029/2007/GL032057.

Anderson, Theodore L., Robert J. Charlson, Stephen E. Schwartz, Reto Knutti, Olivier Boucher, Henning Rodhe, and Jost Heintzenberg. "Climate Forcing by Aerosols—A Hazy Picture." *Science* 300 (May 16, 2003):1103–1104.

Andreae, Meinrat O. "The Dark Side of Aerosols." *Nature* 409 (February 8, 2001):671–672.

Andreae, Meinrat O., Chris D. Jones, and Peter M. Cox. "Strong Present-day Aerosol Cooling Implies a Hot Future." *Nature* 435 (June 30, 2005):1187–1190.

Andreae, Meinrat O., D. Rosenfeld, P. Artaxo, A. A. Casta, G. P. Frank, K. M. Longo, and M. A. F. Silva-Dias. "Smoking Rain Clouds over the Amazon." *Science* 303 (February 27, 2004):1337–1342.

Andrews, Edmund L. "Bush Makes a Pitch for Amber Waves of Homegrown Fuel." *New York Times*, February 23, 2007a. http://www.nytimes.com/2007/02/23/washington/23bush.html.

Andrews, Edmund L. "Senate Adopts an Energy Bill Raising Mileage for Cars." *New York Times*, June 22, 2007b. http://www.nytimes.com/2007/06/22/us/22energy.html.

Andrews, Edmund L., and Felicity Barringer. "Bush Seeks Vast, Mandatory Increase in Alternative Fuels and Greater Vehicle Efficiency." *New York Times*, January 24, 2007. http://www.nytimes.com/2007/01/24/washington/24energy.html.

Angier, Natalie. "Ice Dwellers Are Finding Less Ice to Dwell On." *New York Times*, May 20, 2008. http://www.nytimes.com/2008/05/20/science/20count.html.

Annan, Kofi. "As Climate Changes, Can We?" *Washington Post*, November 8, 2006a, A-27. http://www.washingtonpost.com/wp-dyn/content/article/2006/11/07/AR2006110701229_pf.html.

Annan, Kofi. "Global Warming an All-Encompassing Threat." Address to United Nations Conference on Climate Change, Nairobi, Kenya. Environment News Service, November 15, 2006b. http://www.ens-newswire.com/ens/nov2006/2006-11-15-insann.asp.

"Antarctic Ice Loss Speeds Up, Nearly Matches Greenland Loss." NASA Earth Observatory. January 23, 2008. http://www.jpl.nasa.gov/news/news.cfm?release=2008-010.

"Antarctic Ice Shelf Collapse Tied to Global Warming." Environment News Service, October 16, 2006. http://www.ens-newswire.com.

"Antarctic Ocean Losing Ability to Absorb Carbon Dioxide." Environment News Service, May 18, 2007. http://www.ens-newswire.com/ens/may2007/2007-05-18-04.asp.

"Antarctic Sea Ice Has Increased." Environment News Service, August 23, 2002. http://ens-news.com/ens/aug2002/2002-08-23-09.asp#anchor5.

Appenzeller, Tim. "The Big Thaw." *National Geographic*, June 2007, 56–71.

Applebome, Peter. "Human Behavior, Global Warming, and the Ubiquitous Plastic Bag." *New York Times*, September 30, 2007. http://www.nytimes.com/2007/09/30/nyregion/30towns.html.

Archer, C. L., and K. Caldeira. "Historical Trends in the Jet Streams." *Geophysical Research Letters* 35 (April 18, 2008), L08803, doi: 10.1029/2008GL033614.

Archer, David. *Global Warming: Understanding the Forecast*. Malden, MA: Blackwell Publishing, 2007.

Archer, David. *The Long Thaw: How Humans Are Changing the Next 100,000 Years of Earth's Climate*. Princeton, NJ: Princeton University Press, 2008.

Arctic Climate Impact Assessment—Scientific Report. Cambridge: Cambridge University Press, 2006.

"Arctic Ice Melting, with Help." Associated Press in *Omaha World-Herald*, December 3, 1999, 8.

"Arctic Ice Melting Rapidly, Study Says." Associated Press in *New York Times*, September 14, 2006. http://www.nytimes.com/aponline/us/AP-Warming-Sea-Ice.html.

"Arctic Ice Retreating 30 Years Ahead of Projections." Environment News Service, April 30, 2007. http://www.ens-newswire.com/ens/apr2007/2007-04-30-04.asp.

"Arctic Ocean Ice Thinner by Half in Six Years." Environment News Service, September 14, 2007. http://www.ens-newswire.com/ens/sep2007/2007-09-14-03.asp.

"Arctic Region Quickly Losing Ozone Layer." Knight-Ridder News Service in *Omaha World-Herald*, April 6, 2000, 4.

"Arctic Sea Ice Extent Hits Record Low." Environment News Service, August 20, 2007. http://www.ens-newswire.com/ens/aug2007/2007-08-20-01.asp.

"Arctic Sea Ice Melt Accelerating." Environment News Service, October 4, 2006. http://www.ens-newswire.com/ens/oct2006/2006-10-04-02.asp.

"Arctic Sea Ice Melt May Set Off Climate Change Cascade." Environment News Service, March 19, 2006. http://www.ens-newswire.com/ens/mar2007/2007-03-19-06.asp.

Arendt, Anthony A., Keith A. Echelmeyer, William D. Harrison, Craig S. Lingle, and Virginia B. Valentine. "Rapid Wastage of Alaska Glaciers and Their Contribution to Rising Sea Level." *Science* 297 (July 19, 2002):382–386.

"Are White Christmases Just a Memory?" Environment News Service, December 21, 2001. http://ens-news.com/ens/dec2001/2001L-12-21-09.html.

"Arizona Unveils Climate Strategy." Environment News Service, September 11, 2006. http://www.ens-newswire.com.

"Army Vehicle Could Be Iraq's First Hybrid." Military.com. Popular Mechanics. Brittany Marquis. August 13, 2007. http://www.military.com/features/0,15240,145584,00.html.

Arnold, David. "Global Warming Lends Power to a Jellyfish in Narragansett Bay and Long Island Sound: Non-native Species Are Taking Over." *Boston Globe*, July 2, 2002, C-1.

Aronson, Richard B., William F. Precht, Ian G. MacIntyre, and Thaddeus J. T. Murdoch. "Coral Bleach-out in Belize." *Nature* 405 (May 4, 2000):36.

Arrhenius, Savante. "On the Influence of Carbonic Acid in the Air upon the Temperature of the Ground." *London, Edinburgh, and Dublin Philosophical Magazine and Journal of Science*, 5th series (April 1896):237–276.

Arthur, Charles. "Snows of Kilimanjaro Will Disappear by 2020, Threatening World-wide Drought." *London Independent*, October 18, 2002, 7.

Arthur, Charles. "Super El Niño Could Turn Amazon into Dustbowl: British Association for the Advancement of Science." *London Independent*, September 9, 2003a, 6.

Arthur, Charles. "Temperature Rise Kills 90 Percent of Ocean's Surface Coral." *London Independent*, September 18, 2003b, n.p.(LEXIS).

"As Arctic Sea Ice Melts, Experts Expect New Low." Associated Press in *New York Times*, August 28, 2008. http://www.nytimes.com/2008/08/28/science/earth/28seaice.html.

"Asian Brown Clouds Intensify Global Warming." Environment News Service, August 1, 2007. http://www.ens-newswire.com/ens/aug2007/2007-08-01-02.asp.

Asner, Gregory P., David E. Knapp, Eben N. Broadbent, Paulo J. C. Oliveira, Michaael Keller, and Jose N. Silva. "Selective Logging in the Brazilian Amazon." *Science* 310 (October 21, 2005):480–481.

"A Snowy Day in—Jordan?" *New York Times*, February 1, 2008, A-1.

Assel, Raymond A., Frank H. Quinn, and Cynthia E. Sellinger. "Hyrdoclimatic Factors of the Recent Record Drop in Laurentian Great Lakes Water Levels." *Bulletin of the American Meteorological Society* 85, no. 8 (August 2004):1143–1150.

Associated Press. "Pollution Adds to Global Warming." Via Excite! Data feed, 3:23 A.M. ET October 26, 2000. http://apple.excite.com.

"As the World Warms: A Glacial Archive That Documents a Melting Landscape." *New York Times*, June 13, 2006. http://www.nytimes.com/2006/06/13/science/earth/13norw.html.

Atkinson, Angus, Volker Siegel, Evgeny Pakhomov, and Peter Rothery. "Long-term Decline in Krill Stock and Increase in Salps within the Southern Ocean." *Nature* 432 (November 4, 2004):100–103.

Atkisson, Alan. "Letter from Sweden. Fossil-fuel Free by 2020, Maybe." World Changing, August 18, 2006. http://www.worldchanging.com/archives/004832.html.

Auchard, Eric, and Leonard Anderson. "Google Plans Largest U.S. Solar-Powered Office." Reuters in *Washington Post*, October 16, 2006. http://www.washingtonpost.com/wp-dyn/content/article/2006/10/16/AR2006101601100_pf.html.

Augenbraun, Harvey, Elaine Matthews, and David Sarma. "The Greenhouse Effect, Greenhouse Gases, and Global Warming." Goddard Institute for Space Studies, n.d. http://icp.giss.nasa.gov/research/methane/greenhouse.html.

Austin, J., N. Butchart, K.P. Shine. "Possibility of an Arctic Ozone Hole in a Doubled-CO_2 Climate." *Nature* 360 (November 19, 2001):221–225.

"Australia Assesses Fire Damage in Capital." Associated Press in *Omaha World-Herald*, January 20, 2003, A-4.

"Australian Judge Blocks Coal Mine on Climate Grounds." Environment News Service, November 29, 2006. http://www.ens-newswire.com/ens/nov2006/2006-11-29-03.asp.

"Australian Scientists Announce Good News at Last on Global Warming." Agence France Presse, November 25, 2003, n.p.(LEXIS).

"Australia Screws in Compact Fluorescent Lights Nationwide." Environment News Service, February 21, 2007. http://www.ens-newswire.com/ens/feb2007/2007-02-21-01.asp.

"Automakers Join Call for National Greenhouse Gas Limits." Environment News Service, June 27, 2007. http://www.ens-newswire.com/ens/jun2007/2007-06-27-09.asp#anchor4.

Ayres, Ed. *God's Last Offer: Negotiating a Sustainable Future.* New York: Four Walls Eight Windows, 1999.

Baar, Hein J. W. de, and Michel H. C. Stoll. "Storage of Carbon Dioxide in the Oceans." In *Arctic Ecosystems in a Changing Climate: An Ecophysiological Perspective*, ed. F. Stuart Chapin III, Robert L. Jefferies, James F. Reynolds, Gaius R. Shaver, and Josef Svoboda, 143–177. San Diego: Academic Press, 1992.

Babington, Charles. "Party Shift May Make Warming a Hill Priority." *Washington Post*, November 18, 2006, A-6

Bach, Wilfrid. *Our Threatened Climate: Ways of Averting the CO_2 Problem though Rational Energy Use*, trans. Jill Jager. Dordrecht, Germany: D. Reidel, 1984.

"Baffin Island Ice Caps Shrink by Half in 50 Years." Environment News Service, January 28, 2008. http://www.ens-newswire.com/ens/jan2008/2008-01-28-02.asp.

Baker, Andrew C. "Reef Corals Bleach to Survive Change." *Nature* 411 (June 14, 2001):765–766.

Baker, Andrew C., Craig J. Starger, Tim R. McClanahan, and Peter W. Glynn. "Corals' Adaptive Response to Climate Change." *Nature* 430 (August 12, 2004):741.

Baker, Peter. "In Bush's Final Year, the Agenda Gets Greener." *Washington Post*, December 29, 2007, A-1. http://www.washingtonpost.com/wp-dyn/content/article/2007/12/28/AR2007122803046.html?hpid=topnews.

Bala, G., K. Caldeira, A. Mirin, and M. Wickett. "Multicentury Changes to the Global Climate and Carbon Cycle: Results from a Coupled Climate and Carbon Cycle Model." *Journal of Climate*, November 1, 2005, 4531–4544.

Balany, Joan, Josep M. Oller, Raymond B. Huey, George W. Gilchrist, and Luis Serra. "Global Genetic Change Tracks Global Climate Warming in *Drosophila Subobscura*." Science 313 (September 22, 2006):1773–1775.

Baldwin, Mark P., Martin Dameris, and Theodore G. Shepherd. "How Will the Stratosphere Affect Climate Change?" *Science* 316 (June 15, 2007):1576–1577.

Baldwin, Mark P., David W. J. Thompson, Emily F. Shuckburgh, Warwick A. Norton, and Nathan P. Gillett. "Weather from the Stratosphere?" *Science* 301 (July 18, 2003):317–318.

Ball, Jeffrey. "Climate Change's Cold Economics: Industry Efforts to Fight Global Warming Will Hit Consumers' Pockets." *Wall Street Journal*, February 15, 2007a, A-12.

Ball, Jeffrey. "The Texas Wind Powers a Big Energy Gamble." *Wall Street Journal*, March 12, 2007b, A-1, A-11.

Ball, Jeffrey. "The Carbon Neutral Vacation." *Wall Street Journal*, July 30, 2007c, P-1, P-4-5.

Ball, Philip. "Climate Change Set to Poke Holes in Ozone: Arctic Clouds Could Make Ozone Depletion Three Times Worse Than Predicted." *Nature* Science Update, March 3, 2004. http://info.nature.com/cgi-bin24/DM/y/eOCB0BfHSK0Ch0JVV0AY.

Balling. Robert C. *The Heated Debate: Greenhouse Predictions versus Climate Reality.* San Francisco: Pacific Research Institute, 1992.

Bamber, Jonathan L. David G. Vaughan, and Ian Joughin. "Widespread Complex Flow in the Interior of the Antarctic Ice Sheet." *Science* 287 (February 18, 2000):1248–1250.

Bange, Hermann. "It's Not a Gas." *Nature* 408 (November 16, 2000):301–302.

"Bangkok May Slip Beneath Sea." Associated Press in *Omaha World-Herald*, October 29, 2007, 6-A.

Barbeliuk, Anne. "Warmer Globe Choking Ocean." *The Mercury* (Hobart, Australia), March 16, 2002, n.p.(LEXIS).

Barber, D. C., A. Dyke, C. Hillaire-Marcel, A. E. Jennings, J. T. Andrews, M. W. Kerwin, G. Bilodeau, R. McNeely, J. Southon, M. D. Morehead, and J.-M. Gagnon. "Forcing of the Cold Event of 8,200 Years Ago by Catastrophic Drainage of Laurentide Lakes." *Nature* 400 (July 22, 1999):344–351.

Barber, Timothy R., and William M. Sackett. "Anthropogenic Fossil Carbon Sources of Atmospheric Methane." In *A Global Warming Forum: Scientific, Economic, and Legal Overview*, ed. Richard A. Geyer, 209–223. Boca Raton, FL: CRC Press, 1993.

Barber, Valerie A., Glen Patrick Juday, and Bruce P. Finney. "Reduced Growth of Alaskan White Spruce in the Twentieth Century from Temperature-induced Drought Stress." *Nature* 405 (June 8, 2000):668–673.

Barbier, Edward B., Joanne C. Burgess, and David W. Pearce. "Technological Substitution Options for Controlling Greenhouse-gas Emissions." In *Global Warming: Economic Policy Reponses*, ed. Rutiger Dornbusch and James M. Poterba, 109–161. Cambridge, MA: MIT Press, 1991.

Barbraud, Christopher, and Henri Wirmerskirch. "Emperor Penguins and Climate Change." *Nature* 411 (May 10, 2001):183–186.

Barkham, Patrick. "Going Down: Tuvalu, a Nation of Nine Islands—Specks in the South Pacific—Is in Danger of Vanishing, a Victim of Global Warming. As Their Homeland Is Battered by Ferocious Cyclones and Slowly Submerges under the Encroaching Sea, What Will Become of the Islanders?" *London Guardian*, February 16, 2002, 24.

Barnes, Peter. *Who Owns the Sky? Our Common Assets and the Future of Capitalism.* Washington, DC: Island Press, 2001. http://www.ppionline.org/ppi_ci.cfm?knlgAreaID=116&subsecID=149&contentID=3867.

Barnett, Tim P., David W. Pierce, Krishna M. AchutaRao, Peter J. Gleckler, Benjamin D. Santer, Jonathan M. Gregory, and Warren M. Washington. "Penetration of Human-induced Warming into the World's Oceans." *Science* 309 (July 8, 2005):284–287.

Barnett, Tim P., David W. Pierce, Hugo G. Hidalgo, Celine Bonfils, Benjamin D. Santer, Tapash Das, Govindasamy Bala, Andrew W. Wood, Toru Nozawa, Arthur A. Mirin, Daniel R. Cayan, and Michael D. Dettinger. "Human-Induced Changes in the Hydrology of the Western United States." *Science* 319 (February 22, 2008):1080–1083.

Baron, Ethan. "Beetles Could Chew up 80 Percent of B.C. Pine: Report: 'Worst-case Scenario' by 2020 Blamed on Global Warming." *Ottawa Citizen*, September 12, 2004, A-3.

Barrett, Joe. "Ethanol Reaps a Backlash in Small Midwestern Towns." *Wall Street Journal*, March 23, 2007, A-1, A-8.

Barringer, Felicity. "Savage Storms Wreak Havoc across the Washington Region." *New York Times*, June 27, 2006a. http://www.nytimes.com/2006/06/27/us/27rain.html.

Barringer, Felicity. "Officials Reach California Deal to Cut Emissions." *New York Times*, August 31, 2006b. http://www.nytimes.com/2006/08/31/washington/31warming.html.

Barringer, Felicity. "In Gamble, Calif. Tries to Curb Greenhouse Gases." *New York Times*, September 15, 2006c. http://www.nytimes.com/2006/09/15/us/15energy.html.

Barringer, Felicity. "Environmentalists, Though Winners in the Election, Warn Against Expecting Vast Changes." *New York Times*, November 14, 2006d. http://www.nytimes.com/2006/11/14/us/politics/14enviro.html.

Barringer, Felicity. "A Coalition for Firm Limit on Emissions." *New York Times*, January 19, 2007a. http://www.nytimes.com/2007/01/19/business/19carbon.html.

Barringer, Felicity. "Navajos and Environmentalists Split on Power Plant." *New York Times*, July 27, 2007b. http://www.nytimes.com/2007/07/27/us/27navajo.html.

Barringer, Felicity. "Effort to Get Companies to Disclose Climate Risk." *New York Times*, September 18, 2007c. http://www.nytimes.com/2007/09/18/business/18disclose.html.

Barringer, Felicity. "Precipitation across U.S. Intensifies over 50 Years." *New York Times*, December 5, 2007d. http://www.nytimes.com/2007/12/05/us/05storms.html.

Barringer, Felicity. "Lake Mead Could Be Within a Few Years of Going Dry, Study Finds." *New York Times*, February 13, 2008a. http://www.nytimes.com/2008/02/13/us/13mead.html.

Barringer, Felicity. "Flooded Village Files Suit, Citing Corporate Link to Climate Change." *New York Times,* February 27, 2008b, A-16.

Barringer, Felicity. "Businesses in Bay Area May Pay Fee for Emissions." *New York Times*, April 17, 2008c. http://www.nytimes.com/2008/04/17/us/17fee.html.

Barringer, Felicity. "Urban Areas on West Coast Produce Least Emissions Per Capita, Researchers Find." *New York Times*, May 29, 2008d. http://www.nytimes.com/2008/05/29/us/29pollute.html.

Barringer, Felicity. "Polar Bear Is Made a Protected Species." *New York Times*, May 15, 2008e. http://www.nytimes.com/2008/05/15/us/15polar.html.

Barry, R. G. "Cryospheric Responses to a Global Temperature Increase." In *Carbon Dioxide, Climate, and Society: Proceedings of a IIASA Workshop Co-sponsored by WMO, UNEP, and SCOPE, February 21–24, 1978*, ed. Jill Williams, 169–180. Oxford, UK: Pergamon Press, 1978.

Barta, Patrick. "Crop Prices Soar, Pushing Up Cost of Food Globally." *Wall Street Journal*, April 9, 2007a, A-1, A-9.

Barta, Patrick. "Parched Outback: In Australia, a Drought Spurs a Radical Remedy." *Wall Street Journal*, July 11, 2007b, A-1, A-12.

Barta, Patrick. "Jatropha Plant Gains Steam in Global Race for Biofuels." *Wall Street Journal*, August 24, 2007c, A-1, A-12.

Barth, M. C., and J. G. Titus, eds. *Greenhouse Effect and Sea Level Rise: A Challenge for This Generation*. New York: Van Nostrand Reinhold Company, 1984.

Basu, Janet. "Ecologists' Statement on the Consequences of Rapid Climatic Change." May 20, 1997. http://www.dieoff.com/page104.htm.

Bates, Albert K., and Project Plenty. *Climate in Crisis: The Greenhouse Effect and What We Can Do*. Summertown, TN: The Book Publishing Co., 1990.

Battisti, David. S., and Rosamond L. Naylor. "Historical Warnings of Future Food Insecurity with Unprecedented Seasonal Heat." *Science* 323 (January 9, 2009):240–244.

Bauerlein, Valerie. "Intense Storms Tied to Rising Ocean Temperature." *Wall Street Journal*, March 17, 2006, A-2.

Baxter, James. "Canada's Forests at Risk of Devastation: Global Warming Could Ravage Tourism, Lumber Industry, Commission Warns." *Ottawa Citizen*, March 5, 2002, A-5.

Bazzaz, Fakhri A., and Eric D. Fajer. "Plant Life in a CO_2-rich World." *Scientific American*, January 1992, 68–74.

Beaugrand, Gregory, Philip C. Reid, Frédéric Ibañez, J. Alistair Lindley, and Martin Edwards. "Reorganization of North Atlantic Marine Copepod Biodiversity and Climate." *Science* 296 (May 31, 2002):1692–1694.

Beaugrand, Gregory, Keith M. Brander, J. Alistair Lindley, Sami Souissi, and Philip C. Reid. "Plankton Effect on Cod Recruitment in the North Sea." *Nature* 426 (December 11, 2003):661–664.

Becker, Bernie. "Bicycle-Sharing Program to Be First of Kind in U.S." *New York Times,* April 27, 2008. http://www.nytimes.com/2008/04/27/us/27bikes.html.

Behrenfeld, Michael J., Robert T. O'Malley, David A. Siegel, Charles R. McClain, Jorger L. Sarmiento, Gene C. Feldman, Allen J. Milligan, Paul G. Falkowski, Ricardo M. Letelier, and Emmanuel S. Boss. "Climate-driven Trends in Contemporary Ocean Productivity." *Nature* 444 (December 7, 2006):752–755.

Beinecke, Frances. "A Climate for Change: Next Steps in Solving Global Warming." In *The Last Polar Bear: Facing the Truth of a Warming World*, ed. Steven Kazlowski, 175–185. Seattle: Braided River Books, 2008.

Belkin, Douglas. "Northern Vintage: Canada's Wines Rise with Mercury." *Wall Street Journal*, October 15, 2007, A-1, A-20.

Belluck, Pam. "Warm Winters Upset Rhythms of Maple Sugar." *New York Times*, March 3, 2007. http://www.nytimes.com/2007/03/03/us/03maple.html.

Bellwood, D. R., T. P. Hughes, C. Folke, and M. Nystrom. "Confronting the Coral Reef Crisis." *Nature* 429 (June 24, 2004):827–833.

Bengtsson, L., M. Botzet, and M. Esch. "Will Greenhouse Gas-induced Warming over the Next 50 Years Lead to a Higher Frequency and Greater Intensity of Hurricanes? *Tellus* 48A (1996):57–73.

Benjamin, Craig. "The Machu Picchu Model: Climate Change and Agricultural Diversity." *Native Americas* 16, no. 3/4 (Summer/Fall 1999):76–81.

Bennhold, Katrin. "France Tells U.S. to Sign Climate Pacts or Face Tax." *New York Times*, February 1, 2007. http://www.nytimes.com/2007/02/01/world/europe/01climate.html.

Benson, Simon. "Giant Squid 'Taking over the World.'" *Sydney Daily Telegraph*, July 31, 2002, 4.

Bentley, C. R. *West Antarctic Ice Sheet: Diagnosis and Prognosis*. Washington, DC: U.S. Department of Energy, 1983.

Bentley, Charles. "Response of the West Antarctic Ice Sheet to CO_2-Induced Global Warming." In *Environmental and Societal Consequences of a Possible CO_2-Induced Climate Change*, Vol. 2. Washington, DC: Department of Energy, 1980.

Benton, Michael J. *When Life Nearly Died: The Greatest Mass Extinction of All Time*. London: Thames and Hudson, 2003.

Berger, Andre, and Marie-France Loutre. "Climate: An Exceptionally Long Interglacial Ahead?" *Science* 297 (August 23, 2002):1287–1288.

Bering Sea Task Force. *Status of Alaska's Oceans and Marine Resources: Bering Sea Task Force Report to Governor Tony Knowles*. Juneau: State Government of Alaska, March 1999.

Bernard, Harold W., Jr. *Global Warming: Signs to Watch For*. Bloomington: Indiana University Press, 1993.

Betsill, Michelle M. "Impacts of Stratospheric Ozone Depletion." In *Handbook of Weather, Climate, and Water: Atmospheric Chemistry, Hydrology, and Societal Impacts*, ed. Thomas D. Potter and Bradley R. Colman, 913–923. Hoboken, NJ: Wiley Interscience, 2003.

Betts, Richard A. "Offset of the Potential Carbon Sink from Boreal Forestation by Decreases in Surface Albedo." *Nature* 408 (November 9, 2000):187–190.

Bhattacharya, N. C. "Prospects of Agriculture in a Carbon-Dioxide-Enriched Environment." In *A Global Warming Forum: Scientific, Economic, and Legal Overview*, ed. Richard A. Geyer, 487–505. Boca Raton, FL: CRC Press, 1993.

Bianchi, Giancarlo, and I. Nicholas McCave. "Holocene Periodicity in North Atlantic Climate and Deep-ocean Flow South off Iceland. *Nature* 397 (February 11, 1999):515–517.

Bidlack, H. W. "Swords as Plowshares: The Military's Environmental Role." Ph.D. dissertation, University of Michigan, Ann Arbor, 1996.

"Big City Mayors Strategize to Beat Global Warming." Environment News Service, May 15, 2007. http://www.ens-newswire.com/ens/may2007/2007-05-15-01.asp.

"Big U.S. Corporations Urge Quick Carbon Cap-and-Trade Legislation." Environment News Service, November 18, 2008. http://www.ens-newswire.com/ens/nov2008/2008-11-19-091.asp.

Billings, W. D. "Phytogeographic and Evolutionary Potential of the Arctic Flora and Vegetation in a Changing Climate." In *Arctic Ecosystems in a Changing Climate: An Ecophysiological Perspective*, ed. F. Stuart Chapin III, Robert L. Jefferies, James F. Reynolds, Gaius R. Shaver, and Josef Svoboda, 91–109. San Diego: Academic Press, 1992.

"Billions of People May Suffer Severe Water Shortages as Glaciers Melt: World Wildlife Fund." Agence France Presse, November 27, 2003, n.p.(LEXIS).

Bindschadler, Robert. "Hitting the Ice Sheets Where It Hurts." *Science* 311 (March 24, 2006): 1720–1721.

"Biomass Burning Boosts Stratospheric Moisture." Environment News Service, February 20, 2002. http://ens-news.com/ens/feb2002/2002L-02-20-09.html.

Biswas, A. K. "Scientific Assessment of the Long-term Consequences of War." In *The Environmental Consequences of War,* ed. J. E. Autsin and C. E. Bruch, 303–316. Cambridge: Cambridge University Press, 2000.

"Bitter Pill: The Northward Spread of the Okinawan Goya, Warm-weather." Asahi News Service (Japan), January 29, 2003 (LEXIS).

Black, David E. "The Rains May Be A-comin." *Science* 297 (July 26, 2002):528–529.

Blain, Stéphane, Bernard Quéguiner, Leanne Armand, Sauveur Belviso, Bruno Bombled, Laurent Bopp, Andrew Bowie, Christian Brunet, Corina Brussaard, François Carlotti, et al. "Effect of Natural Iron Fertilization on Carbon Sequestration in the Southern Ocean." *Nature* 446 (April 26, 2007):1070–1074.

Blees, Tom. *Prescription for the Planet: The Painless Remedy for Our Energy and Environmental Crises.* Self-published, 2008. http://www.prescriptionfortheplanet.com.

Bloomfield, Janine, and Sherry Showell. *Global Warming: Our Nation's Capital at Risk.* Environmental Defense Fund, 1997. http://www.edf.org/pubs/Reports/WashingtonGW/index.html.

Blunier, Thomas. "'Frozen' Methane Escapes from the Sea Floor." *Science* 288 (April 7, 2000):68–69.

Bly, Laura. "How Green Is Your Valet and the Rest?" *USA Today,* July 12, 2007, D-1, D-2.

Bodhaine, Barry, Ellsworth Dutton, and Renee Tatusko. "Assessment of Ultraviolet Variability in the Alaskan Arctic." Cooperative Institute for Arctic Research, University of Alaska and NOAA. March 6, 2001. http://www.cifar.uaf.edu/ari00/bodhaine.html.

Bolin, Bert, John T. Houghton, Gylvan Meira Filho, Robert T. Watson, M. C. Zinyowera, James Bruce, Hoesung Lee, Bruce Callander, Richard Moss, Erik Haites, Roberto Acosta Moreno, Tariq Banuri, Zhou Dadi, Bronson Gardner, J. Goldemberg, Jean-Charles Hourcade, Michael Jefferson, Jerry Melillo, Irving Mintzer, Richard Odingo, Martin Parry, Martha Perdomo, Cornelia Quennet-Thielen, Pier Vellinga, and Narasimhan Sundararaman. *Intergovernmental Panel on Climate Change. Second Assessment Synthesis of Scientific-Technical Information Relevant to Interpreting Article 2 of the United Nations Framework Convention on Climate Change.* Approved by the IPCC at its eleventh session, Rome, December 11–15, 1995. http://www.unep.ch/ipcc/pub/sarsyn.htm.

Bonan, Gordon. "Forests and Climate Change: Forcings, Feedbacks, and the Climate Benefits of Forests." *Science* 320 (June 13, 2008):1444–1449.

"Borehole Temperatures Confirm Global Warming." February 17, 2000. http://www.cnn.com/2000/NATURE/02/17/boreholes.enn.

Borenstein, Seth. "Arctic Lost 60 Percent of Ozone Layer: Global Warming Suspected." Knight-Ridder News Service, April 6, 2000 (LEXIS).

Borenstein, Seth. "Scientists Worry about Evidence of Melting Arctic Ice." Knight-Ridder News Service in *Seattle Times,* February 18, 2005, A-6.

Borenstein, Seth. "Scientists Find New Global Warming 'Time Bomb' Methane Bubbling up from Permafrost." Associated Press, September 6, 2006a (LEXIS).

Borenstein, Seth. "Future Forecast, Extreme Weather: Study Outlines a Climate Shift Caused by Global Warming." Associated Press, October 21, 2006b, A-2. http://www.washingtonpost.com/wp-dyn/content/article/2006/10/20/AR2006102001454_pf.html.

Borenstein, Seth. "'Smoking Gun' Said to Be in International Climate Report Next Week: Global Warming Here Now." Associated Press, January 23, 2007a (LEXIS).

Borenstein, Seth. "January Weather Hottest by Far." Associated Press, February 15, 2007b (LEXIS).

Borenstein, Seth. "More Severe U.S. Storms Will Come with Global Warming, NASA Researchers Say." Associated Press, August 30, 2007c (LEXIS).

Borenstein, Seth. "Earth's Tropics Belt Expands, May Mean Drier Weather for U.S. Southwest, Mediterranean." Associated Press, December 2, 2007d (LEXIS).

Borenstein, Seth. "Jump in Atlantic Hurricanes not Global Warming, Says Study That Predicts Fewer Future Storms." Associated Press, May 19, 2008 (LEXIS).

Bornemann, Andrè, Richard D. Norris, Oliver Friedrich, Britta Beckmann, Stefan Schouten, Jaap S. Sinninghe Damstè, Jennifer Vogel, Peter Hofmann, and Thomas Wagner. "Isotopic Evidence for Glaciation during the Cretaceous Supergreenhouse." *Science* 319 (January 11, 2008):189–192.

Boswell, Randy. "Southern Butterfly's Trek North Cited as Proof of Global Warming." *Edmonton Journal,* November 19, 2004, A-7.

Both, Christiaan, Sandra Bouhuis, C. M. Lessells, and Marcel E. Visser. "Climate Change and Population Declines in a Long-Distance Migratory Bird." *Nature* 441 (May 4, 2006):81–83.

Both, Christiaan, and Marcel E. Visser. "Adjustment to Climate Change Is Constrained by Arrival Date in a Long-distance Migrant Bird." *Nature* 411 (May 17, 2001):296–298.

Botkin, Daniel R. "Global Warming Delusions." *Wall Street Journal*, October 17, 2007, A-19.

Boudette, Neal E. "Shifting Gears, GM Now Sees Green." *Wall Street Journal*, May 29, 2007, A-8.

Bousquet, P., P. Ciais, J. B. Miller, E. J. Dlugokencky, D. A. Hauglustaine, C. Prigent, G. R. Van der Werf, P. Peylin, E.-G. Brunke, C. Carouge, R. L. Langenfelds, J. Lathière, F. Papa, M. Ramonet, M. Schmidt, L. P. Steele, S. C. Tyler, and J. White. "Contribution of Anthropogenic and Natural Sources to Atmospheric Methane Variability." *Nature* 443 (September 228, 2006):439–443.

Bowen, Gabriel J., David J. Beerling, Paul L. Koch, James C. Zachos, and Thomas Quattlebaum. "A Humid Climate State during the Palaeocene/Eocene Thermal Maximum." *Nature* 432 (November 25, 2004):495–499.

Bowen, Jerry. "Dramatic Climate Change in Alaska." *CBS Morning News*, August 29, 2002 (LEXIS).

Bowen, Mark. *Thin Ice: Unlocking the Secrets of Climate in the World's Highest Mountains.* New York: Henry Holt, 2005.

Bowen, Mark. *Censoring Science: Inside the Political Attack on Dr. James Hansen and the Truth of Global Warming.* New York: Dutton/Penguin, 2008.

Bowles, Scott. "National Gridlock: Traffic Really *Is* Worse Than Ever. Here's Why." *USA Today*, November 23, 1999, 1-A, 2-A.

Bowman, Lee. "Soot Could Be Causing a Lot of Bad Weather." Scripps Howard News Service, September 26, 2002 (LEXIS).

Boxall, Bettina. "Oakland Switches to 'Green' Power." *Los Angeles Times*, June 29, 2000, 3.

Boxall, Bettina. "Epic Droughts Possible, Study Says: Tree Ring Records Suggest That if Past Is Prologue, Global Warming Could Trigger Much Longer Dry Spells Than the One Now in West, Scientists Say." *Los Angeles Times*, October 8, 2004, A-17.

Boyd, Robert S. "Rising Tides Raises Questions: Satellites Will Provide Exact Measurements." Knight-Ridder Newspapers in *Pittsburgh Post-Gazette*, December 9, 2001, A-3.

Boyd, Robert S. "Earth Warming Could Open up a Northwest Passage." Knight-Ridder Newspapers in *Pittsburgh Post-Gazette*, November 11, 2002, A-1.

Boyd, Robert S. "Carbon Dioxide Levels Surge: Global Emissions Rate Is Triple That of a Decade Ago." McClatchy Newspapers in *Calgary Herald*, May 22, 2007, A-8.

Boyd, Robert S. "Arctic Temperatures Hit Record High." Knight-Ridder Washington Bureau, October 16, 2008 (LEXIS).

Boykoff, M. T., and J. M. Boykoff. "Balance as Bias: Global Warming and the U.S. Prestige Press." *Global Environmental Change* 14 (2004):125–136.

Boyle, Robert H. "Global Warming: You're Getting Warmer." *Audubon*, November–December 1999, 80–87.

Boyle, S. T., W. Fulkerson, R. Klingholz, I. M. Mintzer, G. I. Pearman, G. Oinchera, J. Reilly, F. Staib, R. J. Swart, and C.-J. Winter. "Group Report: What Are the Economic Costs, Benefits, and Technical Feasibility of Various Options Available to Reduce Greenhouse Potential per Unit of Energy Service?" In *Limiting Greenhouse Effects: Controlling Carbondioxide Emissions. Report of the Dahlem Workshop on Limiting the Greenhouse Effect*, ed. G. I. Pearman, 229–260. Berlin, September 9–14, 1990. New York: John Wiley & Sons, 1991.

Braasch, Gary. *Earth Under Fire: How Global Warming Is Changing the World.* Berkeley: University of California Press, 2007.

Bradley, Raymond S., Malcolm K. Hughes, and Henry F. Diaz. "Climate in Medieval Time." *Science* 302 (October 17, 2003):404–405.

Bradley, Raymond S., Mathias Vuille, Henry F. Diaz, and Walter Vergara. "Threats to Water Supplies in the Tropical Andes." *Science* 312 (June 23, 2006):1755–1756.

Bradshaw, William E, and Christina M. Holzapfel. "Genetic Shift in Photoperiodic Response Correlated with Global Warming." *Proceedings of the National Academy of Sciences* 98, no. 25 (December 4, 2001):14,509–14,511.

Bradsher, Keith. "Ford Tries to Burnish Image by Looking to Cut Emissions." *New York Times*, May 4, 2001, C-3.

Bradsher, Keith. "China Prospering but Polluting: Dirty Fuels Power Economic Growth." *New York Times* in *International Herald-Tribune*, October 22, 2003, 1.

Bradsher, Keith. "China to Pass U.S. in 2009 in Emissions. *New York Times*, November 7, 2006. http://www.nytimes.com/2006/11/07/business/worldbusiness/07pollute.html.

Bradsher, Keith. "As Asia Keeps Cool, Scientists Worry About the Ozone Layer." *New York Times*, February 23, 2007a. http://www.nytimes.com/2007/02/23/business/23cool.html.

Bradsher, Keith. "Push to Fix Ozone Layer and Slow Global Warming." *New York Times*, March 15, 2007b. http://www.nytimes.com/2007/03/15/business/worldbusiness/15warming.html.

Bradsher, Keith. "A Drought in Australia, a Global Shortage of Rice." *New York Times*, April 17, 2008a. http://www.nytimes.com/2008/04/17/business/worldbusiness/17warm.html.

Bradsher, Keith. "With First Car, a New Life in China." *New York Times*, April 24, 2008b. http://www.nytimes.com/2008/04/24/business/worldbusiness/24firstcar.html.

Bradsher, Keith, and David Barboza. "Pollution from Chinese Coal Casts a Global Shadow." *New York Times*, June 11, 2006. http://www.nytimes.com/2006/06/11/business/worldbusiness/11chinacoal.html.

Bralower, Timothy J. "Volcanic Cause of Catastrophe." *Nature* 454 (July 17, 2008):285–286.

Brandenburg, John E., and Monica Rix Paxson. *Dead Mars, Dying Earth*. Freedom, CA: The Crossing Press, 1999.

Brasseur, Guy P., Anne K. Smith, Rashid Khosravi, Theresa Huang, Stacy Walters, Simon Chabrillat, and Gaston Kockarts. "Natural and Human-induced Pertubations in the Middle Atmosphere: A Short Tutorial." In *Atmospheric Science Across the Stratopause*, David E. Siskind, et al. Washington, DC: American Geophysical Union, 2000.

"Brazil's Leader Speaks Out." Reuters in *New York Times*, February 7, 2007. http://www.nytimes.com/2007/02/07/world/asia/07china.html.

"Breakaway Bergs Disrupt Antarctic Ecosystem." Environmental News Service, May 9, 2002. http://ens-news.com/ens/may2002/2002L-05-09-01.html.

Breed, Allen G. "New Orleans Evacuation Picking Up Steam, but Help Comes Too Late for Untold Number." Associated Press, September 3, 2005 (LEXIS).

Bremner, Charles, Richard Owen, and Mark Henderson. "Heat Wave Uncovers the Grim Secrets of the Snows." *London Times*, August 26, 2000, n.p.(LEXIS).

Brewer, Peter G., Gernot Friederich, Edward T. Peltzer, and Franklin M. Orr, Jr. "Direct Experiments on the Ocean Disposal of Fossil Fuel CO_2." *Science* 284 (May 7, 1999):943–945.

Bright, Becky. "Global Warming: Who Said What—and When." *Wall Street Journal*, March 24, 2008, R-2.

Brikowski, Tom H., Yair Lotan, and Margaret S. Pearle. "Climate-related Increase in the Prevalence of Urolithiasis in the United States." *Proceedings of the National Academy of Sciences* 105, no. 28 (July 15, 2008):9841–9846.

Brinton, Henry G. "Green, Meet God." *USA Today*, November 10, 2008, 11-A.

"Britain's Sea Levels 'Will Rise 5 Feet This Century.'" *London Daily Mail*, December 17, 2007 (LEXIS).

"Britain Urges U.S. to Get Tough on Global Warming." British Broadcasting Corp. Online, June 11, 1997. http://benetton.dkrz.de:3688/homepages/georg/kimo/0254.html.

"British Columbia to Trim Greenhouse Gases, Go Carbon Neutral." Environment News Service, February 14, 2007. http://www.ens-newswire.com/ens/feb2007/2007-02-14-02.asp.

"British Travel Agents Launch Carbon Offset Scheme." Environment News Service, November 28, 2006. http://www.ens-newswire.com/ens/nov2006/2006-11-28-05.asp.

Britt, Robert Roy. "Antarctic Ice Shelves Falling Apart." Explorezone.com, April 9, 1999. http://www.explorezone.com/archives/99_04/09_antarctic_ice.htm.

Broad, William J. "How to Cool a Planet (Maybe)." *New York Times*, June 27, 2006. http://www.nytimes.com/2006/06/27/science/earth/27cool.html.

Broad, William J. "From a Rapt Audience, a Call to Cool the Hype." *New York Times*, March 13, 2007. http://www.nytimes.com/2007/03/13/science/13gore.html.

Broccoli, A. J., and S. Manabe. "Can Existing Climate Models Be Used to Study Anthropogenic Changes in Tropical Cyclone Intensity?" *Geophysical Research Letters* 17 (1990):1917–1920.

Broder, John M. "Compromise Measure Aims to Limit Global Warming." *New York Times*, July 11, 2007. http://www.nytimes.com/2007/07/11/washington/11climate.htm.

Broecker, W. S. "Fate of Fossil Fuel Carbon Dioxide and the Global Carbon Budget." *Science*, 206 (1979):409–418.

Broecker, Wallace S. "Unpleasant Surprises in the Greenhouse?" *Nature* 328 (1987):123–126.

Broecker, Wallace S. "Thermohaline Circulation: The Achilles Heel of Our Climate System: Will Man-made CO_2 Upset the Current Balance?" *Science* 278 (1997):1582–1588.

Broecker, Wallace S. "Are We Headed for a Thermohaline Catastrophe?" In *Geological Perspectives of Global Climate Change,* ed. Lee C. Gerhard, William E. Harrison, and Bernold M. Hanson, 83–95. AAPG [American Association of Petroleum Geologists] Studies in Geology No. 17. Tulsa, OK: AAPG, 2001.

Broecker, Wallace S., and Robert Kunzig. *Fixing Climate: What Past Climate Changes Reveal about the Current Threat—and How to Counter It.* New York: Hill and Wang (Farrar, Straus, and Giroux), 2008.

Broecker, Wallace S., Stewart Sutherland, and Tsung-Hung Peng. "A Possible 20th-Century Slowdown of Southern Ocean Deep Water Formation." *Science* 286 (November 5, 1999):1132–1135.

Brook, Ed. "Paleoclimate: Windows on the Greenhouse." *Nature* 453 (May 15, 2008), 291–292.

Brook, Edward J. "Tiny Bubbles Tell All." *Science* 310 (November 25, 2005):1285–1287.

Brooke, James. "'Heat Island' Tokyo Is Global Warming's Vanguard." *New York Times*, August 13, 2002, A-3.

Brown, Amanda. "New Figures Show Fall in Greenhouse Gas Emissions." Press Association (United Kingdom), March 27, 2003 (LEXIS).

Brown, DeNeen L. "Waking the Dead, Rousing Taboo: In Northwest Canada, Thawing Permafrost Is Unearthing Ancestral Graves." *Washington Post*, October 17, 2001, A-27.

Brown, DeNeen L. "Greenland's Glaciers Crumble: Global Warming Melts Polar Ice Cap into Deadly Icebergs." *Washington Post*, October 13, 2002, A-30.

Brown, DeNeen L. "Hamlet in Canada's North Slowly Erodes: Arctic Community Blames Global Warming as Permafrost Starts to Melt and Shoreline." *Washington Post*, September 13, 2003, A-14.

Brown, Lester R. *Plan B: Rescuing a Planet under Stress and a Civilization in Trouble.* New York: Earth Policy Institute/W.W. Norton, 2003.

Brown, Lester R. "The Earth Is Shrinking." Environment News Service, November 20, 2006. http://www.ens-newswire.com.

Brown, Paul. "Global Warming: Worse Than We Thought." *World Press Review*, February 1999, 44.

Brown, Paul. "Overfishing and Global Warming Land Cod on Endangered List." *London Guardian*, July 20, 2000a, 3.

Brown, Paul. "Global Warming—It's with Us Now: Six Dead as Storms Bring Chaos Throughout the Country." *London Guardian*, October 31, 2000b, 1.

Brown, Paul. [No headline]. *London Guardian*, November 1, 2000c, 1.

Brown, Paul. "Islands in Peril Plead for Deal." *London Guardian*, November 24, 2000d, 21.

Brown, Paul. "Melting Permafrost Threatens Alps: Communities Face Devastating Landslides from Unstable Mountain Ranges." *London Guardian*, January 4, 2001, 3.

Brown, Paul. "Geographers' Conference: Ice Field Loss Puts Alpine Rivers at Risk: Global Warming Warning to Europe." *London Guardian*, January 5, 2002a, 9.

Brown, Paul. "Scientists Warn of Himalayan Floods: Global Warming Melts Glaciers and Produces Many Unstable Lakes." *London Guardian*, April 17, 2002b, 13.

Brown, Paul. "Analysis: Blair Sets out Far-reaching Vision but Where Are the Practical Policies?" *London Guardian*, February 25, 2003a, 13.

Brown, Paul. "Global Warming Is Killing Us Too, Say Inuit." *London Guardian*, December 11, 2003b, 14.

Brown, Paul. "Inuit Blame Bush for Impending Extinction." *London Guardian*, December 13, 2003c. http://www.smh.com.au/articles/2003/12/12/1071125653832.html.

Brown, Paul. "Global Warming Kills 150,000 a Year: Disease and Malnutrition the Biggest Threats, UN Organisations Warn at Talks on Kyoto." *London Guardian*, December 12, 2003d, 19.

Brown, Paul, and Tony Sutton. "Global Warming Brings New Cash Crop to West Country as Rising Water Temperatures Allow Valuable Shellfish to Thrive." *London Guardian*, December 10, 2002, 8.

Browne, Anthony. "How Climate Change Is Killing off Rare Animals: Conservationists Warn That Nature's 'Crown Jewels' Are Facing Ruin." *London Observer*, February 10, 2002a, 15.

Browne, Anthony. "Canute Was Right! Time to Give up the Coast." *London Times,* October 11, 2002b, 8.

Browne, Anthony, and Paul Simons. "Euro-spiders Invade as Temperature Creeps up." *London Times*, December 24, 2002, 8.

Browne, Malcolm W. "Under Antarctica, Clues to an Icecap's Fate: Radar Uncovers a Network of Ice Streams Larger and Faster Than Expected, and More Ominous." *New York Times*, October 26, 1999, F-1, F-6.

"Brush Fires Collapsing Bear Dens." Canadian Press in *Calgary Sun*, November 2, 2002, 18.

Bruun, P. "Sea-level Rise as a Cause of Shore Erosion." *Journal of Waterways and Harbor Division. American Society of Civil Engineers* 88 (1962):117–130.

Bryant, D., L. Burke, J. McManus and M. Spaulding. *Reefs at Risk: A Map-based Indicator of Threats to the World's Coral Reefs.* Washington, DC: World Resources Institute, 1998.

Bryden, Harry L., Hannah R. Longworth, and Stuart A. Cunningham. "Slowing of the Atlantic Meridional Overturning Circulation at 25° North." *Nature* 438 (December 1, 2005): 655–657.

Bueckert, Dennis. "Climate Change Linked to Ill Health in Children." Canadian Press in *Montreal Gazette*, June 2, 2001, A-2.

Bueckert, Dennis. "Forest Fires Taking Toll on Climate: CO_2 from Increased Burning Could Overtake Fossil Fuels as a Source of Global Warming, Prof Warns." Canadian Press in *Edmonton Journal*, September 19, 2002a, A-3.

Bueckert, Dennis. "Climate Change Could Bring Malaria, Dengue Fever to Southern Ontario, Says Report." Canadian Press in *Ottawa Citizen*, October 23b, 2002. http://www.canada.com/news/story.asp?id={B019135A-4FD8-4536-A2F0-908F13560CB2.

Buencamino, Manuel. "Coming Catastrophe?" *BusinessWorld*, June 14, 2004, 21.

Buesseler, Ken O., and Philip W. Boyd. "Will Ocean Fertilization Work?" *Science* 300 (April 4, 2003):67–68.

"Building Consensus to Keep the Earth Cool after 2012." Environment News Service, August 28, 2007. http://www.ens-newswire.com/ens/aug2007/2007-08-28-01.asp,

"Building Parks Can Help to Climate Proof Cities." Environment News Service, June 12, 2007. http://www.ens-newswire.com/ens/jun2007/2007-06-12-04.asp.

Bulkeley, Harriet, and Michele M. Betsill. *Cities and Climate Change: Urban Sustainability and Global Environmental Governance.* London: Routledge, 2003.

Bunkley, Nick. "Seeking a Car That Gets 100 Miles a Gallon." *New York Times*, April 2, 2007a. http://www.nytimes.com/2007/04/02/business/02xprize.html.

Bunkley, Nick. "Detroit Finds Agreement on the Need to Be Green." *New York Times*, June 1, 2007b. http://www.nytimes.com/2007/06/01/business/01auto.html.

Bunting, Madeleine. "Confronting the Perils of Global Warming in a Vanishing Landscape: As Vital Talks Begin at the Hague, Millions Are Already Suffering the Consequences of Climate Change." *London Guardian*, November 14, 2000, 1.

Burnett, Adam W., Matthew E. Kirby, Henry T. Mullins, and William P. Patterson. "Increasing Great Lake-effect Snowfall during the Twentieth Century: A Regional Response to Global Warming?" *Journal of Climate* 16, no. 21 (November 1, 2003):3535–3542.

Burnham, Michael. "Scientists Link Global Warming with Increasing Marine Diseases." Greenwire, October 7, 2003 (LEXIS).

Burroughs, William James. *The Climate Revealed.* Cambridge: Cambridge University Press, 1999.

Burroughs, William James. *Climate Change: A Multidisciplinary Approach*, 2nd ed. New York: Cambridge University Press, 2007.

"Bush Orders First Federal Regulation of Greenhouse Gases." Environment News Service, May 14, 2007. http://www.ens-newswire.com/ens/may2007/2007-05-14-06.asp.

"Bush Previews Abu Dhabi's Planned Carbon Neutral, Car Free City." Environment News Service. January 14, 2008. http://www.ens-newswire.com/ens/jan2008/2008-01-14-01.asp.

"Bush Says Administration Has Not Changed Stance on Global Warming." *The Frontrunner*, August 27, 2004 (LEXIS).

"Bush Snubs Gore's Global Warming Film: Former VP Shoots back, Offers to Host Viewing at White House." Associated Press, May 23, 2006 (LEXIS).

Butler, Declan. "Super Savers: Meters to Manage the Future." *Nature* 445 (February 8, 2007): 586–588.

Buxton, James. "Suspects in the Mystery of Scotland's Vanishing Salmon: Fish Farms, Seals and Global Warming Are All Blamed for What Some See as a Crisis." *London Financial Times*, June 13, 2000, 11.

Byers, Stephen, and Olympia Snowe, co-chairs. International Climate Change Task Force. *Meeting the Climate Challenge: Recommendations of the International Climate Change Task Force*. London: Institute for Public Policy Research, January 2005.

Calamai, Peter. "Atlantic Water Changing: Scientists." *Toronto Star*, June 21, 2001, A-18.

Calamai, Peter. "Alert over Shrinking Ozone Layer." *Toronto Star*, March 18, 2002a, A-8.

Calamai, Peter. "Global Warming Threatens Reindeer." *Toronto Star*, December 23, 2002b, A-23.

Caldeira, Ken, and Philip B. Duffy. "The Role of the Southern Ocean in the Uptake and Storage of Anthropogenic Carbon Dioxide." *Science* 287 (January 28, 2000):620–622.

Caldeira, Ken, and Michael E. Wickett. "Oceanography: Anthropogenic Carbon and Ocean pH." *Nature* 425 (September 25, 2003):365.

"California Air Board Adds Climate Labels to New Cars." Environment News Service, June 25, 2007. http://www.ens-newswire.com/ens/jun2007/2007-06-25-09.asp#anchor7.

"California Snowpack 71 Percent below Normal. Environment News Service, May 4, 2007. http://www.ens-newswire.com/ens/may2007/2007-05-04-09.asp#anchor5.

"California's Temperature Rising." Environment News Service, April 9, 2007. http://www.ens-newswire.com/ens/apr2007/2007-04-09-09.asp#anchor6.

Calvin, William H. *Global Fever: How to Treat Climate Change*. Chicago: University of Chicago Press, 2008.

Came, Rosemarie E., John M. Eiler, Ján Veizer, Karem Azmy, Uwe Brand, and Christopher R. Weidman. "Coupling of Surface Temperatures and Atmospheric CO_2 Concentrations during the Palaeozoic Era." *Nature* 449 (September 13, 2007):198–201.

Campbell, Duncan. "Greenhouse Melts Alaska's Tribal Ways: As Climate Talks Get under Way in Bonn Today, Some Americans Are Ruing the Warming Their President Chooses to Ignore." *London Guardian*, July 16, 2001, 11.

Campbell, Kurt M., ed. *Climatic Cataclysm: The Foreign Policy and National Security Implications of Climate Change*. Washington, DC: Brookings Institution, 2008.

Canadell, Josep G., Corinne Le Quéréc, Michael R. Raupacha, Christopher B. Fielde, Erik T. Buitenhuisc, Philippe Ciaisf, Thomas J. Conwayg, Nathan P. Gillettc, R. A. Houghtonh, and Gregg Marland. "Contributions to Accelerating Atmospheric CO_2 Growth from Economic Activity, Carbon Intensity, and Efficiency of Natural Sinks." *Proceedings of the National Academy of Sciences*. Published online before print October 25, 2007, doi: 10.1073/pnas.0702737104.

Canadell, Josep G., and Michael R. Raupach. "Managing Forests for Climate Change Mitigation." *Science* 320 (June 13, 2008):1456–1457.

"Canadian Tundra Turning Green." Environment News Service, March 3, 2007. http://www.ens-newswire.com/ens/mar2007/2007-03-06-02.asp.

"Cape Hatteras, N.C. Lighthouse Lights Up Sky from New Perch." *Omaha World-Herald*, November 14, 1999, A-16.

Capella, Peter. "Disasters Will Outstrip Aid Effort as World Heats Up: Rich States Could Be Sued as Voluntary Assistance Falters, Red Cross Says." *London Guardian*, June 29, 2001, 15.

Capella, Peter. "Europe's Alps Crumbling: Glaciers Melting in Heatwave." Agence France Presse, August 7, 2003 (LEXIS).

Capiello, Dina. "Adirondacks Climate Growing Hotter Faster." *Albany Times-Union*, September 21, 2002, n.p.

"Carbon Dioxide in Atmosphere Increasing." Associated Press in *New York Times*, October 22, 2007. http://www.nytimes.com/aponline/us/AP-Carbon-Increase.html.

"Carbon Neutral Chic." Editorial, *Wall Street Journal*, July 9, 2007, A-14.

"Carbon Pollution Wreaking Havoc with Amazonian Forest." Agence France Presse, March 10, 2004 (LEXIS).

"Carbon Sinks Cannot Keep Up with Emissions." Environment News Service, May 16, 2002. http://ens-news.com/ens/may2002/2002L-05-16-09.html#anchor3.

Carey, John, and Sarah R. Shapiro. "Consensus Is Growing among Scientists, Governments, and Business That They Must Act Fast to Combat Climate Change." *Business Week*, August 16, 2004, n.p.(LEXIS).

Carlton, Jim. "Some in Santa Fe Pine for Lost Symbol, but Others Move On." *Wall Street Journal*, July 31, 2006, A-1, A-8.

Carlton, Jim. "Citicorp Tries Banking on the Natural Kind of Green." *Wall Street Journal*, September 5, 2007, B-1, B-8.

Carpenter, Betsy. "Feeling the Sting: Warming Oceans, Depleted Fish Stocks, Dirty Water—They Set the Stage for a Jellyfish Invasion." *U.S. News and World Report*, August 16, 2004, 68–69.

Carpenter, Kent E., Muhammad Abrar, Greta Aeby, Richard B. Aronson, Stuart Banks, Andrew Bruckner, Angel Chiriboga, Jorge Cortès, J. Charles Delbeek, Lyndon DeVantier, et al. "One-Third of Reef-Building Corals Face Elevated Extinction Risk from Climate Change and Local Impacts." *Science* 321 (July 25, 2008):560–563.

Carrel, Chris. "Boeing Joins Fight against Global Warming." *Seattle Weekly*, September 17, 1998. http://climatechangedebate.com/archive/09-18_10-27_1998.txt.

Cars and Climate Change. Paris: International Energy Agency, 1993.

Catholic Bishops, U.S. Conference. "Global Climate Change: A Plea for Dialogue, Prudence, and the Common Good: A Statement of the U.S. Catholic Bishops," ed. William P. Fay. June 15, 2001. http://www.ncrlc.com/climideas.html.

Cazenave, Anny. "How Fast Are the Ice Sheets Melting?" *Science* 314 (November 24, 2006):1250–1252.

"Century of Human Impact Warms Earth's Surface." Environment News Service. January 24, 2002. http://ens-news.com/ens/jan2002/2002L-01-24-09.html.

Cerveny, R. S., and R. C. Balling, Jr. 1998. "Weekly Cycles of Air Pollutants, Precipitation and Tropical Cyclones in the Coastal NW Atlantic Region." *Nature* 394 (1998):561–563.

Chameides, William L., and Michael Bergin. "Climate Change: Soot Takes Center Stage." *Science* 297 (September 27, 2002):2214–2215.

Chan, C. L. "Comment on 'Changes in Tropical Cyclone Number, Duration, and Intensity in a Warming Environment.'" *Science* 311 (March 24, 2006):1713.

Chang, Alicia. "Plants Move Up Mountain as Temps Rise, Study Shows." Associated Press, August 11, 2008 (LEXIS).

Chang, Kenneth. "Ozone Hole Is Now Seen as a Cause for Antarctic Cooling." *New York Times*, May 3, 2002, A-16.

Chang, Kenneth. "Warming Is Blamed for Antarctica's Weight Gain." *New York Times*, May 20, 2005, A-22.

Chang, Kenneth. "Climate Shift May Aid Algae Species." *New York Times*, April 18, 2008a, A-11.

Chang, Kenneth. "Strongest Storms Grow Stronger Yet, Study Says." *New York Times*, October 4, 2008b, A-18.

Chang, Kenneth. "Study Finds New Evidence of Warming in Antarctica." *New York Times*, January 22, 2009. http://www.nytimes.com/2009/01/22/science/earth/22climate.html.

"Changes in Climate Bring Hops Northward." *Glasgow Herald*, September 29, 2000, 13.

"Changing Jet Streams May Alter Paths of Storms and Hurricanes." NASA Earth Observatory, April 16, 2008. http://earthobservatory.nasa.gov/Newsroom/MediaAlerts/2008/2008041626579.html.

Chapin, F. S., III., M. Sturm, M. C. Serreze, J. P. McFadden, J. R. Key, A. H. Lloyd, A. D. McGuire, T. S. Rupp, A. H. Lynch, J. P. Schimel, J. Beringer, W. L. Chapman, H. E. Epstein, E. S. Euskirchen, L. D. Hinzman, G. Jia, C.-L. Ping, K. D. Tape, C. D. C. Thompson, D. A. Walker, and J. M. Welker. "Role of Land-Surface Changes in Arctic Summer Warming." *Science* 310 (October 28, 2005):657–660.

Chapman, James. "Early Spring Misery for 12 Million Hay Fever Sufferers." *London Daily Mail*, February 4, 2003, 23.

Charlson, Robert J., John H. Seinfeld, Athanasios Nenes, Markku Kulmala, Ari Laaksonen, and M. Cristina Facchini. "Reshaping the Theory of Cloud Formation." *Science* 292 (June 15, 2001):2025–2026.

Charmantier, Anne, Robin H. McCleery, Lionel R. Cole, Chris Perrins, Loeske E. B. Krunk, and Ben C. Sheldon. "Adaptive Phenotypic Plasticity in Response to Climate Change in a Wild Bird Population." *Science* 320 (May 9, 2008):800–803.

Chase, Marilyn. "As Virus Spreads, Views of West Nile Grow Even Darker." *Wall Street Journal*, October 14, 2004, A-1, A-10.

Chase, Steven. "Our Water Is at Risk, Climate Study Finds." *Toronto Globe and Mail*, August 13, 2002, A-1. http://www.globeandmail.com/servlet/ArticleNews/PEstory/TGAM/20020813/UENVIN/national/national/national_temp/6/6/23.

Chea, Terence. "Environmental Groups Call for Planes to Reduce Emissions." *USA Today*, December 6, 2007, 6-B.

Chen, J. L., C. R. Wilson, and B. D. Tapley. "Satellite Gravity Measurements Confirm Accelerated Melting of Greenland Ice Sheet." *Science* Online, August 10, 2006, doi: 10.1126/science.1129007.

Chengappa, Raj. "The Monsoon: What's Wrong with the Weather?" *India Today*, August 12, 2002, 40.

"Chevron Stockholder Vote on Global Warming Resolution Surprises Annual Meeting." Environmental Media Services/Ozone Action, April 29, 1999. http://www.corpwatch.org/trac/corner/worldnews/other/355.html.

"Chicago Sets Goals for a Cooler City." Environment News Service, September 19, 2008. http://www.ens-newswire.com/ens/sep2008/2008-09-19-092.asp.

Chin, Gilbert. "No Deepwater Slowdown?" Editor's Choice. *Science* 293 (July 27, 2001):575, citing Orsi et al., *Geophysical Research Letters* 28 (2001):2923.

"China Now Number One in Carbon Emissions: USA Number Two." Environment News Service, June 19, 2007. http://www.ens-newswire.com/ens/jun2007/2007-06-19-04.asp.

Chisholm, Sallie W. "Stirring Times in the Southern Ocean." *Nature* 407 (October 12, 2000):685–686.

Choi, Charles. "Rainforests Might Speed up Global Warming." United Press International, April 21, 2003 (LEXIS).

Christensen, Jens H., and Ole B. Christensen. "Severe Summertime Flooding in Europe." *Nature* 421 (February 20, 2003):805.

Christensen, Torben R., Torbjörn Johansson, H. Jonas Åkerman, Mihail Mastepanov, Nils Malmer, Thomas Friborg, Patrick Crill, and Bo H. Svensson. "Thawing Sub-arctic Permafrost: Effects on Vegetation and Methane Emissions." *Geophysical Research Letters* 31, no. 4 (February 20, 2004), L04501, doi: 10.1029/2003GL018680.

Christianson, Gale E. *Greenhouse: The 200-Year Story of Global Warming.* New York: Walker and Company, 1999.

Christie, Maureen. *The Ozone Layer: A Philosophy of Science Perspective.* Cambridge: Cambridge University Press, 2001.

Christy, John R. "My Nobel Moment." *Wall Street Journal*, November 1, 2007, A-19.

Church, John A. "Oceans: A Change in Circulation?" *Science* 317 (August 17, 2007):908–909.

Chylek, Petr, Jason E. Box, and Glen Lesins. "Global Warming and the Greenland Ice Sheet." *Climatic Change* 63 (2004):201–221.

Ciborowski, Peter. "Sources, Sinks, Trends, and Opportunities." In *The Challenge of Global Warming*, ed. Edwin Abrahamson, 213–230. Washington, DC: Island Press, 1989.

Cifuentes, Luis, Victor H. Borja-Aburto, Nelson Gouveia, George Thurston, and Devra Lee Davis. "Hidden Health Benefits of Greenhouse Gas Mitigation." *Science* 252 (August 17, 2001):1257–1259.

Clark, Andrew. "'Open Skies' Air Treaty Threat." *London Guardian*, February 20, 2006. http://www.guardian.co.uk/frontpage/story/0,,1713677,00.html.

Clark, D. A., S. C. Piper, C. D. Keeling, and D. B. Clark. "Tropical Rain Forest Tree Growth and Atmospheric Carbon Dynamics Linked to Interannual Temperature Variation during 1984–2000." *Proceedings of the National Academy of Sciences* 100, no. 10 (May 13, 2003):5852–5857.

Clark, Jayne. "Tours Bear Witness to Earth's Sentinel Species." *USA Today*, November 2, 2007, 1-D, 2-D.

Clark, Jayne. "Lake Powell on the Rise" *USA Today*, May 9, 2008, 1-D, 2-D.

Clark, P. U., A. Marshall McCabe, Alan C. Mix, and Andrew J. Weaver. "Rapid Rise of Sea Level 19,000 Years Ago and Its Global Implications." *Science* 304 (May 21, 2004):1141–1144.

Clark, P. U., J. X. Mitrovica, G. A. Milne, and M. E. Tamisiea. "Sea-level Fingerprinting as a Direct Test for the Source of Global Meltwater Pulse." *Science* 295 (March 29, 2002):2438–2441.

Clark, P. U., N. G. Pisias, T. F. Stocker, and A. J. Weaver. "The Role of the Thermohaline Circulation in Abrupt Climate Change." *Nature* 415 (February 21, 2002):863–868.

Clark, Stuart. "Saved by the Sun?" *New Scientist,* September 16, 2006, 23–36.

Clarke, Tom. "Boiling Seas Linked to Mass Extinction: Methane Belches May Have Catastrophic Consequences." *Nature* Online, August 22, 2003. http://info.nature.com/cgi-bin24/DM/y/eLod0BfHSK0Ch0DYy0AL.

Clavel, Guy. "Global Warming Makes Polar Bears Sweat." Agence France Presse, November 3, 2002.

Clayton, Chris. "U.S.D.A. Will Offer Incentives for Conserving Carbon in Soil." *Omaha World-Herald,* June 7, 2003, D-1, D-2.

Clift, Peter, and Karen Bice. "Earth Science: Baked Alaska." *Nature* 419 (September 12, 2002):129–130.

"Climate Change Acceleration Will Push Claims Bills Higher." *Insurance Day,* March 4, 2004, 1.

"Climate Change Dislocates Migratory Animals, Birds." Environment News Service, November 17, 2006. http://www.ens-newswire.com.

"Climate Change Linked to Doubling of Atlantic Hurricanes." Environment News Service, July 30, 2007. http://www.ens-newswire.com/ens/jul2007/2007-07-30-01.asp.

"Climate Change Pushing Bird Species to Oblivion." Environment News Service, November 14, 2006. http://www.ens-newswire.com/ens/nov2006/2006-11-14-01.asp.

"Climate Change Threatens Loggerhead Turtles." Environment News Service, February 22, 2007. http://www.ens-newswire.com/ens/feb2007/2007-02-22-02.asp.

"Climate-related Perils Could Bankrupt Insurers." Environment News Service, October 7, 2002. http://ens-news.com/ens/oct2002/2002-10-07-02.asp.

"Climate Warms Twice as Fast." British Broadcasting Company Monitoring Asia-Pacific, January 6, 2003 (LEXIS).

Cline, William R. *The Economics of Global Warming.* Washington, DC: Institute for International Economics, 1992.

Cline, William R. *Global Warming and Agriculture: Impact Estimates by Country.* Washington, DC: Peterson Institute, 2007.

Clinton, Bill. "President Clinton's State of the Union Address." *New York Times,* January 20, 1999. http://geography.rutgers.edu/courses/99spring/370sp99/news01_20_99.html#anchor33828.

"Clouds, but No Silver Lining." *London Guardian,* January 24, 2002, n.p.(LEXIS).

Clover, Charles. "Air Travel Is a Threat to Climate." *London Daily Telegraph,* June 5, 1999.

Clover, Charles. "Thousands of Species 'Threatened by Warming.'" *London Daily Telegraph,* August 31, 2000, 9.

Clover, Charles. "Geographers' Conference: Alps May Crumble as Permafrost Melts." *London Telegraph,* January 4, 2001, 12.

Clover, Charles. "2002 Warmest for 1,000 Years.'" *London Daily Telegraph,* April 26, 2002a, 1.

Clover, Charles. "Global Warming Is Driving Fish North.'" *London Daily Telegraph,* May 31, 2002b, 14.

Clover, Charles, and David Millward. "Future of Cheap Flights in Doubt: Ban New Runways and Raise Fares, Say Pollution Experts." *London Daily Telegraph,* November 30, 2002, 1, 4.

CNA Corporation Military Advisory Board and Study Team. "National Security and the Threat of Climate Change." Alexandria, VA: CNA Corp., 2007.

"Coastal Gulf States Are Sinking." Environment News Service. April 21, 2003. http://ens-news.com/ens/apr2003/2003-04-21-09.asp#anchor4.

Cockburn, Alexander. "Dissidents against Dogma." *The Nation,* June 25, 2007. http://www.the-nation.com/doc/20070625/cockburn.

"Cockburn Trilogy of Climate Skepticism." Reprinted from *The Nation.* Part 1. "From Papal Indulgences to Carbon Credits: Is Global Warming a Sin?" Part 2. "Source and Authorities: Dissidents against Dogma." Part 3. "Explosion of the Fearmongers: The Greenhousers Strike Back and Out." New Zealand Climate Science Coalition: Commonsense About Climate Change, 2007. http://nzclimatescience.net/index.php?option=com_content&task=view&id=41&Itemid=1.

Cody, Edward. "Mild Weather Takes Edge off Chinese Ice Festival: Residents of Tourist City Blame Global Warming." *Washington Post,* February 25, 2007, A-19. http://www.washingtonpost.com/wpdyn/content/article/2007/02/24/AR2007022401421_pf.html.

Cohen, Roger. "Bring on the Right Biofuels." *New York Times*, April 24, 2008. http://www.nytimes.com/2008/04/24/opinion/24cohen.html.

Cole, Julia. "A Slow Dance for El Niño." *Science* 291 (February 23, 2001):1496–1497.

Collins, Simon. "Birds Starve in Warmer Seas." *New Zealand Herald*, November 14, 2002, n.p.(LEXIS).

Colwell, Robert K., Gunnar Brehm, Catherine L. Cardelús, Alex C. Gilman, and John T. Longino. "Global Warming, Elevational Range Shifts, and Lowland Biotic Attrition in the Wet Tropics." *Science* 322 (October 10, 2008):258–261.

Comiso, J. C. "A Rapidly Declining Perennial Sea Ice Cover in the Arctic." *Geophysical Research Letters* 29, no. 20 (October 18, 2002):1956–1960.

Comiso, J. C. "Warming Trends in the Arctic from Clear Sky Satellite Observations." *Journal of Climate* 16, no. 21 (November 1, 2003):3498–3510.

"Comment Period Extended on Polar Bear Extinction Threat." Environment News Service, October 2, 2007. http://www.ens-newswire.com/ens/oct2007/2007-10-02-091.asp.

Commoner, Barry. *Making Peace with the Planet.* New York: Pantheon, 1990.

Cone, Marla. "Dozens of Words for Snow, None for Pollution." *Mother Jones*, January–February 2004. http://www.hartford-hwp.com/archives/27b/059.html.

Cone, Marla. *Silent Snow: The Slow Poisoning of the Arctic.* New York: Grove Press, 2005.

"Conferences Tackle Key Issues in Air Conditioning, Refrigeration." AScribe Newswire, June 23, 2004 (LEXIS).

Connor, Steve. "Ozone Layer over Northern Hemisphere Is Being Destroyed at 'Unprecedented' Rate." *London Independent*, March 5, 2000a, 5.

Connor, Steve. "Global Warming Is Blamed for First Collapse of a Caribbean Coral Reef." *London Independent*, May 4, 2000b, 12.

Connor, Steve. "Britain Could Become as Cold as Moscow." *London Independent*, June 21, 2001a, 14.

Connor, Steve. "Catastrophic Climate Change 90 Percent Certain." *London Independent*, July 20, 2001b, 15.

Connor, Steve. "Malaria Could Become Endemic Disease in U.K." *London Independent*, September 12, 2001c, 14.

Connor, Steve. "El Niños' Rise May Be Linked to Pacific Current Slowdown." *London Independent*, February 7, 2002a, n.p.

Connor, Steve. "World's Wildlife Shows Effects of Global Warming." *London Independent*, March 28, 2002b, 11.

Connor, Steve. "Strangers in the Seas: Exotic Marine Species Are Turning Up Unexpectedly in the Cold Waters of the North Atlantic." *London Independent*, August 5, 2002c, 12–13.

Connor, Steve. "Polar Sea Ice Could Be Gone by the End of the Century." *London Independent*, March 10, 2003a, 5.

Connor, Steve. "Global Warming May Wipe Out a Fifth of Wild Flower Species, Study Warns." *London Independent*, June 17, 2003b, n.p.(LEXIS).

Connor, Steve. "Global Warming Is Choking the Life out of Lake Tanganyika." *London Independent*, August 14, 2003c, n.p.(LEXIS).

Connor, Steve. "Peat Bog Gases Accelerate Global Warming." *London Independent*, July 8, 2004a, 9.

Connor, Steve. "Meltdown: Arctic Wildlife Is on the Brink of Catastrophe—Polar Bears Could Be Decades from Extinction." *London Independent*, November 11, 2004b, n.p.(LEXIS).

Connor, Steve. "Global Warming Is Twice as Bad as Previously Thought." *London Independent*, January 27, 2005, 10.

Conover, Ted. "Capitalist Roaders." *New York Times Sunday Magazine*, July 2, 2006. http://www.nytimes.com/2006/07/02/magazine/02china.html.

Constantineau, Bruce. "Weather Wreaking Havoc on British Columbia Farms: Global Warming Has Increased the Intensity and Frequency of Weather Events, Making It Difficult for Farmers to Compete in the Marketplace." *Vancouver Sun*, March 3, 2003, D-3.

"Contrails Linked to Temperature Changes." Environment News Service, August 8, 2002. http://ens-news.com/ens/aug2002/2002-08-08-09.asp#anchor4.

Cook, A. J., A. J. Fox, D. G. Vaughan, and J. G. Ferrigno. "Retreating Glacier Fronts on the Antarctic Peninsula over the Past Half-Century." *Science* 308 (April 22, 2005):541–544.

Cook, Edward R., Connie A. Woodhouse, C. Mark Eakin, David M. Meko, and David W. Stahle. "Long-Term Aridity Changes in the Western United States." *Science* 306 (November 5, 2004):1015–1018.

Cooke, Robert. "Is Global Warming Making Earth Greener?" *Newsday*, September 11, 2001, C-3.

Cooke, Robert. "Scientists: Pacific Slower at Surface—Data May Help Explain Trend in El Niño Events." *Newsday*, February 7, 2002a, A-46.

Cooke, Robert. "Global Warming Is in the Air." *Newsday*, September 20, 2002b, A-34.

Cooke, Robert. "Waters Reflect Weather Trend: Study Finds Warming Effects." *Newsday*, December 18, 2003, A-2.

Cookson, Clive. "Global Warming Triggers Epidemics in Wildlife." *London Financial Times*, June 21, 2002, 4.

Cookson, Clive, and Victoria Griffith. "Blame for Flooding May Be Misplaced: Climate Change Global Warming May Not Be the Reason for Recent Heavy Rainfall in Europe and Asia." *London Financial Times*, August 15, 2002, 6.

Cookson, Craig, and David Firn. "Breeding Bugs That May Help Save the World: Craig Venter Has Found a Large Project to Follow the Human Genome." *London Financial Times*, September 28, 2002, 11.

Cooperman, Alan. "Evangelical Body Stays Course on Warming: Conservatives Oppose Stance." *Washington Post*, March 11, 2007a, A-5. http://www.washingtonpost.com/wp-dyn/content/article/2007/03/10/AR2007031001175_pf.html.

Cooperman, Alan. "Eco-Kosher Movement Aims to Heed Tradition, Conscience." *Washington Post,* July 7, 2007b, A-1. http://www.washingtonpost.com/wp-dyn/content/article/2007/07/06/AR2007070602092_pf.html.

Copland, L., D. R. Mueller, and L. Weir. "Rapid Loss of the Ayles Ice Shelf, Ellesmere Island, Canada. *Geophysical Research Letters* 34 (2007), L21501.

"Corn-ethanol Crops Will Widen Gulf of Mexico Dead Zone." Environment News Service, Match 11, 2008. http://www.ens-newswire.com/ens/mar2008/2008-03-11-091.asp.

Corson, Walter H., ed. *The Global Ecology Handbook: What You Can Do about the Environmental Crisis.* Washington, DC: The Global Tomorrow Coalition, 1990.

Cortese, Amy. "Friend of Nature? Let's See Those Shoes." *New York Times*, March 7, 2007. http://www.nytimes.com/2007/03/06/business/businessspecial2/07label-sub.html.

Cotton, William R., and Roger A. Pielke Sr. *Human Impacts on Weather and Climate,* 2nd ed. Cambridge: Cambridge University Press, 2007.

Coulter, Ann. "Global Warming Theology." *Pittsburgh Tribune Review*, March 25, 2007, n.p.(LEXIS).

"Court Rebukes Administration in Global Warming Case." Associated Press in *New York Times*, April 2, 2007. http://www.nytimes.com/aponline/business/AP-Scotus-Greenhouse-Gase.html.

Couzin, Jennifer. "Landscape Changes Make Regional Climate Run Hot and Cold." *Science* 283 (January 15, 1999):317–318.

Couzin, Jennifer. "Opening Doors to Native Knowledge." *Science* 315 (March 16, 2007):1518–1519.

Cowell, Adam. "Britain Drafts Laws to Slash Carbon Emissions." *New York Times*, March 14, 2007. http://www.nytimes.com/2007/03/14/world/europe/14britain.html.

Cowell, Alan. "Sweden Offers Itself as a Model on Energy." *New York Times* in *International Herald-Tribune*, February 3, 2006. http://www.iht.com/articles/2006/02/03/news/sweden.php.

Cowen, Robert C. "Into the Cold? Slowing Ocean Circulation Could Presage Dramatic—and Chilly—Climate Change." *Christian Science Monitor*, September 26, 2002a, 14.

Cowen, Robert C. "One Large, Overlooked Factor in Global Warming: Tropical Forest Fires." *Christian Science Monitor*, November 7, 2002b, 14.

Cowie, Jonathan. *Climate Change: Biological and Human Aspects.* Cambridge: Cambridge University Press, 2007.

Cox, P. M., R.A. Betts, M. Collins, P. P. Harris, C. Huntingford, and C. D. Jones. "Amazonian Forest Dieback Under Climate-carbon Cycle Projections for the 21st Century." *Theoretical and Applied Climatology* 78 (2004):137–156.

Cox, P. M., R. A. Betts, C. D. Jones, S. A. Spall, and I. J. Totterdell. "Acceleration of Global Warming Due to Carbon-cycle Feedbacks in a Coupled Climate Model." *Nature* 408 (November 9, 2000):184–187.

Coy, Peter. "The Hydrogen Balm? Author Jeremy Rifkin Sees a Better, Post-Petroleum World." *Business Week*, September 30, 2002, 83.

Crable, Ad. "For Armadillos, It's Pa. [Pennsylvania] or Bust." *Lancaster New Era* (Pennsylvania), September 5, 2006, C-5.

Cramb, Auslan. "Highland River Salmon 'on Verge of Extinction.'" *London Daily Telegraph*, July 15, 2002, 7.

Crenson, Matt. "Louisiana Sinking: One State's Environmental Nightmare Could Become Common Problem." Associated Press, August 10, 2002 (LEXIS).

Crilly, Rob. "2050 to Be Good Year for Scottish Wine: Global Warming Will Bring Grapes North." *Glasgow Herald*, November 20, 2002, 11.

Crowley, Thomas J. "Causes of Climate Change over the Past 1,000 Years." *Science* 289 (July 14, 2000):270–277.

Crutzen, Paul J. "The Antarctic Ozone Hole, a Human-caused Chemical Instability in the Stratosphere: What Should We Learn from It? In *Geosphere-Biosphere Interactions and Climate*, ed. Lennart O. Bengtsson and Claus U. Hammer, 1–11. Cambridge: Cambridge University Press, 2001.

Curran, Mark A. J., Tas D. van Ommen, Vin I. Morgan, Katrina L. Phillips, and Anne S. Palmer. "Ice Core Evidence for Antarctic Sea Ice Decline since the 1950s." *Science* 302 (November 14, 2003):1203–1206.

Currie, Michael J., Jonathan K. Mapel, Timothy D. Heidel, Shalom Goffri, and Marc A. Baldo. "High-efficiency Organic Solar Concentrators for Photovoltaics." *Science* 321 (July 11, 2008):226–228.

Curry, Ruth, Bob Dickson, and Igor Yashayaev. "A Change in the Freshwater Balance of the Atlantic Ocean over the Past Four Decades." *Nature* 426 (December 18, 2003):826–829.

Curry, Ruth, and Cecilie Mauritzen. "Dilution of the Northern North Atlantic Ocean in Recent Decades." *Science* 308 (June 17, 2005):1772–1774.

Cushman, John H., Jr. "Industrial Group Plans to Battle Climate Treaty." *New York Times*, April 26, 1998, A-1, A-24.

Dai, Aiguo, Kevin E. Trenberth, and Taotao Qian. "A Global Dataset of Palmer Drought Severity Index for 1870–2002: Relationship with Soil Moisture and Effects of Surface Warming." *Journal of Hydrometeorology* 5, no. 6 (December 2004):1117–1130.

Daley, Suzanne. "Battered by Fierce Weekend Storm, Western Europe Begins an Enormous Cleanup Job." *New York Times*, Tuesday, December 28, 1999, A-10. [International Edition.]

"Dallas Toots Its Green Horn." Environment News Service, January 21, 2008. http://www.ens-newswire.com/ens/jan2008/2008-01-21-093.asp.

Dalton, Alastair. "Ice Pack Clue to Climate-change Effects." *The Scotsman*, October 18, 2001, 7.

Dasgupta, Partha. "A Challenge to Kyoto." Review, Lomborg, *Cool It! Nature* 449 (September 13, 2007):143–144.

Davey, Monica. "Balmy Weather May Bench a Baseball Staple." *New York Times*, July 11, 2007a. http://www.nytimes.com/2007/07/11/us/11ashbat.html.

Davey, Monica. "Death, Havoc, and Heat Mar Chicago Race." *New York Times*, October 8, 2007b, A-1, A-14.

David, Laurie. "Science a la Joe Camel." *Washington Post*, November 226, 2006, B-1. http://www.washingtonpost.com/wp-dyn/content/article/2006/11/24/AR2006112400789_pf.html.

Davidson, Eric A., and A. I. Hirsch. "Carbon Cycle: Fertile Forest Experiments." *Nature* 411 (May 24, 2001):431–433.

Davidson, Eric A., and Ivan A. Janssens. "Temperature Sensitivity of Soil Carbon Decomposition and Feedbacks to Climate Change." *Nature* 440 (March 9, 2006):165–173.

Davidson, Keay. "Media Goofed on Antarctic Data: Global Warming Interpretation Irks Scientists." *San Francisco Chronicle*, February 4, 2002, A-8.

Davidson, Keay. "Going to Depths for Evidence of Global Warming: Heating Trend in North Pacific Baffles Researchers." *San Francisco Chronicle*, March 1, 2004a, A-4.

Davidson, Keay. "Film's Tale of Icy Disaster Leaves the Experts Cold." *San Francisco Chronicle*, June 1, 2004b, E-1.

Davidson, Paul. "Coal Plant to Test Capturing Carbon Dioxide." *USA Today*, February 27, 2008, 3-A.

Davis, Curt H., Yonghong Li, Joseph R. McConnell, Markus M. Frey, and Edward Hanna. "Snowfall-Driven Growth in East Antarctic Ice Sheet Mitigates Recent Sea-Level Rise." *Science* 308 (June 24, 2005):1898–1901.

Davis, Neil. *Permafrost: A Guide to Frozen Ground in Transition.* Fairbanks: University of Alaska Press, 2001.

Davis, Robert E., Paul C. Knappenberger, Patrick J. Michaels, and Wendy M. Novicoff. "Changing Heat-related Mortality in the United States." *Environmental Health Perspectives*, July 23, 2003, doi: 10.1289/ehp.6336.

Daviss, Bennett. "Green Sky Thinking: Could Maverick Technologies Turn Aviation into an Eco-success Story? Yes, but Time Is Running Out." *New Scientist*, February 24, 2007, 32–38.

Daynes, Byron W., and Glen Sussman. "The 'Greenless' Response to Global Warming." *Current History*, December 2005, 438–443.

Dayton, Leigh. "'Scary' Science Finds Earth Heating up Twice as Fast as Thought." *The Australian* (Sydney), January 27, 2005, 3.

Dean, Cornelia. "Louisiana's Marshes Fight for Their Lives." *New York Times*, November 15, 2005. http://www.nytimes.com/2005/11/15/science/earth/15marsh.html.

Dean, Cornelia. "Next Victim of Warming: The Beaches." *New York Times* June 20, 2006. http://www.nytimes.com/2006/06/20/science/earth/20sea.html.

Dean, Cornelia. "Scientists Criticize White House Stance on Climate Change Findings." *New York Times*, January 31, 2007a. http://www.nytimes.com/2007/01/31/washington/31interfere.html.

Dean, Cornelia. "Even Before Its Release, World Climate Report Is Criticized as Too Optimistic." *New York Times*, February 2, 2007b. http://www.nytimes.com/2007/02/02/science/02oceans.html.

de Angelis, Hernán, and Pedro Skvarca. "Glacier Surge after Ice Shelf Collapse." *Science* 299 (March 7, 2003):1560–1562.

De'ath, Glenn, Janice M. Louygh, and Katharina E. Fabricius. "Declining Coral Calcification on the Great Barrier Reef." *Science* 323 (January 2, 2009):116–119.

Deckert, Rudolf, and Martin Dameris. "From Ocean to Stratosphere." *Science* 322 (October 3, 2008):53–55.

"Deforestation Could Push Amazon Rainforest to Its End." UniSci: Daily University Science News, June 28, 2001. http://www.Unisci.com.

Del Genio, Anthony D., Mao-Sung Yao, and Jeffrey Jonas. "Will Moist Convection Be Stronger in a Warmer Climate?" *Geophysical Research Letters* 34, no. 16 (August 17, 2007), L16703, doi: 10.1029/2007GL030525.

Del Giorgio, Paul A., and Carlos M. Duarte. "Respiration in the Open Ocean." *Nature* 420 (November 28, 2002):379–384.

Denby, David. "Review, *An Inconvenient Truth*." *The New Yorker*, June 19, 2006, 23.

Derbyshire, David. "'Heatwave' in the Antarctic Halves Penguin Colony." *London Daily Telegraph*, May 10, 2001a, 6.

Derbyshire, David. "Global Warming Fails to Boost Butterfly Visitors." *London Daily Telegraph*, November 1, 2001b, 13.

Derbyshire, David. "Baffled Bumble Bee Lured out Early by Changing Climate." *London Daily Telegraph*, March 12, 2004, 15.

Dessler, Andrew E., and Edward A. Parson. *The Science and Politics of Global Climate Change: A Guide to the Debate.* New York: Cambridge University Press, 2006.

Deutsch, Claudia. "Companies Giving Green an Office." *New York Times*, July 3, 2007a. http://www.nytimes.com/2007/07/03/business/03sustain.html.

Deutsch, Claudia. "A Threat So Big, Academics Try Collaboration." *New York Times*, December 25, 2007b. http://www.nytimes.com/2007/12/25/business/25sustain.html.

Dewan, Shaila. "Feeling Warmth, Subtropical Plants Move North." *New York Times*, May 3, 2007. http://www.nytimes.com/2007/05/03/science/03flowers.html.

Diaz, Henry F., and Raymond S. Bradley. "Temperature Variations during the Last Century at High-elevation Sites." *Climatic Change* 36 (1997):253–279.

Diaz, Henry F., and Richard J. Murnane, eds. *Climate Extremes and Society.* New York: Cambridge University Press, 2008.

Dickens, Gerald R. "A Methane Trigger for Rapid Warming?" *Science* 299 (February 14, 2003):1017. (Review of James P. Kennett, Kevin G. Cannariato, Ingrid L. Hendy, and

Richard J. Behl, *Methane Hydrates in Quaternary Climate Change: The Clathrate Gun Hypothesis.* Washington, DC: American Geophysical Union, 2002.)

Dickens, Gerald R. "Global Change: Hydrocarbon-driven Warming." *Nature* 429 (June 3, 2004):513–515.

Dickson, B., I. Yashayaev, J. Meincke, B. Turrell, S. Dye, and J. Holfort. "Rapid Freshening of the Deep North Atlantic Ocean over the Past Four Decades." *Nature* 416 (April 25, 2002):832–836.

Dicum, Gregory. "Plugging into the Sun." *New York Times*, January 4, 2007. http://www.nytimes.com/2007/01/04/garden/04solar.html.

DiMento, Joseph F. C., and Pamela Doughman, eds. *Climate Change: What It Means for Us, Our Children, and Our Grandchildren.* Cambridge, MA: MIT Press, 2007.

Dionne, E. J., Jr. "Gore's Energy Oomph." *Washington Post*, July 18, 2008. http://www.washingtonpost.com/wp-dyn/content/article/2008/07/17/AR2008071701840_pf.html.

"Dirtiest Power Plants in the USA Named." Environment News Service, July 26, 2007. http://www.ens-newswire.com/ens/jul2007/2007-07-26-05.asp.

Dixon, Chris. "As Beaches Erode, So Do Solutions." *New York Times*, November 2, 2007. http://www.nytimes.com/2007/11/02/travel/escapes/02sand.html.

Dlugokencky, E. J., K. A. Masrie, P. M. Lang, and P. P. Tans. "Continuing Decline in the Growth Rate of the Atmospheric Methane Burden." *Nature* 393 (June 4, 1999):447–450.

"Does Oil Have a Future?" Editorial, *Atlantic Monthly*, October 2005, 31–32.

Domack, Eugene, Diana Duran, Amy Leventer, Scott Ishman, Sarah Doane, Scott McCallum, David Amblas, Jim Ring, Robert Gilbert, and Michael Prentice. "Stability of the Larsen B Ice Shelf on the Antarctic Peninsula during the Holocene Epoch." *Nature* 436 (August 4, 2005):681–685.

Doney, Scott C. "Plankton in a Warmer World." *Nature* 444 (December 7, 2006):695–696.

Donn, Jeff. "New England's Brilliant Autumn Sugar Maples—and Their Syrup—Threatened by Warmth." Associated Press, September 23, 2002 (LEXIS).

Donnelly, Jeffrey P., and Jonathan D. Woodruff. "Intense Hurricane Activity over the Past 5,000 Years Controlled by El Niño and the West African Monsoon." *Nature* 447 (May 24, 2007):465–468.

Doran, Peter T. "Cold, Hard Facts." *New York Times*, July 27, 2006. http://www.nytimes.com/2006/07/27/opinion/27doran.html.

Doran, Peter T., John C. Priscu, W. Berry Lyons, John E. Walsh, Andrew G. Fountain, Diane M. McKnight, Daryl L. Moorhead, Ross A. Virginia, Diana H. Wall, Gary D. Clow, Christian H. Fritsen, Christopher P. McKay, and Andrew N. Parsons. "Antarctic Climate Cooling and Terrestrial Ecosystem Response." *Nature* 415 (January 30, 2002):517–520.

Dorell, Oren. "Report: Tropical Storm Activity at 30-Year Low." *USA Today*, November 12, 2008, 3-A.

Doucet, Clive. *Urban Meltdown: Cities, Climate Change, and Politics as Usual.* Gabriola Island, B.C.: New Society Publishers, 2007.

Dow, Kim, and Thomas E. Downing. *The Atlas of Climate Change: Mapping the World's Greatest Challenge.* Berkeley: University of California Press, 2006.

Downie, David Leonard and Terry Fenge, eds. *Northern Lights AGAINST POPs: Combatting Toxic Threats in the Arctic.* Montreal and Kingston: McGill-Queen's University Press, 2003.

"Drier, Warmer Springs in U.S. Southwest Stem from Human-caused Changes in Winds." NASA Earth Observatory, August 19, 2008. http://earthobservatory.nasa.gov/Newsroom/MediaAlerts/2008/2008081927359.html.

"Drought, Excessive Heat Ruining Harvests in Western Europe." Associated Press in Daytona Beach, Florida, *News-Journal*, August 5, 2003, 3-A.

"Drought Forces Barcelona to Ship in Drinking Water." *Wall Street Journal*, May 14, 2008, A-13.

"Drought Forecast to Intensify Across California, Southwest." Environment News Service, March 19, 2007. http://www.ens-newswire.com/ens/mar2007/2007-03-19-09.asp#anchor1.

"Drought in Iraq." NASA Earth Observatory, June 4, 2008. http://earthobservatory.nasa.gov/Newsroom/NewImages/images.php3?img_id=18046.

Dube, Francine. "North America's Growing Season 12 Days Longer: 'What Is Good for Plants Is Not Necessarily Good for the Planet.'" *National Post* (Canada), September 5, 2001.

Duff-Brown, Beth. "Arctic Inuit Argue U.S. Pollution Devastates Centuries-old Hunting Traditions." Associated Press Worldstream, March 1, 2007 (LEXIS).

Duffy, Andrew. "Global Warming Why Arctic Town Is Sinking: Permafrost Is Melting under Sachs: Inuit Leader." *Montreal Gazette*, April 18, 2000, A-13.

Dunne, Nancy. "Climate Change Research Sparks Hawaii Protests." *London Financial Times*, June 20, 2002, 2.

Duplessy, J. C., D. M. Roche, and M. Kageyama. "The Deep Ocean during the Last Interglacial Period." *Science* 316 (April 6, 2007):89–91.

Dyurgerow, Mark B., and Mark F. Meier. "Twentieth-century Climate Change: Evidence from Small Glaciers." *Proceedings of the National Academy of Sciences of the United States of America* 97, no. 4 (February 15, 2000):1351–1354.

"Earth Out of Balance." Editorial, *Indian Country Today*, Feb. 9, 2000, A-4.

Easterbrook, Gregg. "Global Warming: Who Loses—and Who Wins?" *The Atlantic*, April 2007, 52–64.

Easterling, David R., Briony Horton, Phillip D. Jones, Thomas C. Peterson, Thomas R. Karl, David E. Parker, M. James Salinger, Vyacheslav Razuvayev, Neil Plummer, Paul Jamason, and Christopher K. Folland. "Maximum and Minimum Temperature Trends for the Globe." *Science* 277 (1997):364–366.

Eco Bridge. "What Can We Do about Global Warming?" n.d. http://www.ecobridge.org/content/g_wdo.htm.

"Economists' Letter on Global Warming." June 23, 1997. http://uneco.org/Global_Warming.html.

Edgerton, Lynne T., and the Natural Resources Defense Council. *The Rising Tide: Global Warming and World Sea Levels*. Washington, DC: Island Press, 1991.

Editorial. *Vancouver Province*, September 8, 2003, A-16.

"Editors' Choice: A Summary of Glaciation." *Science* 295 (January 18, 2002):401.

Egan, Timothy. "Alaska, No Longer so Frigid, Starts to Crack, Burn, and Sag." *New York Times*, June 16, 2002a, A-1.

Egan, Timothy. "On Hot Trail of Tiny Killer in Alaska." *New York Times*, June 25, 2002b, F-1.

Egan, Timothy. "The Race to Alaska before It Melts." *New York Times*. Travel Section, June 26, 2005.

Egan, Timothy. "Heat Invades Cool Heights over Arizona Desert." *New York Times*, March 27, 2007. http://www.nytimes.com/2007/03/27/us/27warming.html.

Eggert, David. "Shareholders File Global Warming Resolutions at Ford, GM." Associated Press, December 11, 2002 (LEXIS).

Eilperin, Juliet. "Warming Tied to Extinction of Frog Species." *Washington Post*, January 12, 2006a, A-1. http://www.washingtonpost.com/wp-dyn/content/article/2006/01/11/AR2006011102121_pf.html.

Eilperin, Juliet. "Antarctic Ice Sheet Is Melting Rapidly: New Study Warns of Rising Sea Levels." *Washington Post*, March 3, 2006b, A-1.

Eilperin, Juliet. "Study Reconciles Data in Measuring Climate Change." *Washington Post*, May 3, 2006c, A-3.

Eilperin, Juliet. "Growing Acidity of Oceans May Kill Corals." *Washington Post*, July 5, 2006d, A-1. http://www.washingtonpost.com/wp-dyn/content/article/2006/07/04/AR2006070400772_pf.html.

Eilperin, Juliet. "Cities, States Aren't Waiting for U.S. Action on Climate." *Washington Post*, August 11, 2006e, A-1. http://www.washingtonpost.com/wp-dyn/content/article/2006/08/10/AR2006081001492_pf.html.

Eilperin, Juliet. "Greenland's Melting Ice Sheet May Speed Rise in Sea Level: Study Finds No Boost in Antarctic Snowfall to Mitigate Problem." *Washington Post*, August 11, 2006f, A-3. http://www.washingtonpost.com/wp-dyn/content/article/2006/08/10/AR2006081001557_pf.html.

Eilperin, Juliet. "Scientists Disagree on Link between Storms, Warming: Same Data, Different Conclusions." *Washington Post*, August 20, 2006g, A-3. http://www.washingtonpost.com/wp-dyn/content/article/2006/08/19/AR2006081900354_pf.html.

Eilperin, Juliet. "White House Outlines Global Warming Fight: Technology and Voluntary Cutbacks Urged." *Washington Post*, September 21, 2006h, A-12. http://www.washingtonpost.com/wp-dyn/content/article/2006/09/20/AR2006092001697_pf.html.

Eilperin, Juliet. "U.S. Wants Polar Bears Listed as Threatened." *Washington Post*, December 27, 2006i, A-1. http://www.washingtonpost.com/wp-dyn/content/article/2006/12/26/AR2006122601034_pf.html.

Eilperin, Juliet. "U.S., China Got Climate Warnings Toned Down." *Washington Post*, April 7, 2007a, A-5. http://www.washingtonpost.com/wp-dyn/content/article/2007/04/06/AR2007040600291_pf.html.

Eilperin, Juliet. "Clues to Rising Seas Are Hidden in Polar Ice." *Washington Post*, July 16, 2007b, A-6. http://www.washingtonpost.com/wp-dyn/content/article/2007/07/15/AR2007071500882_pf.html.

Eilperin, Juliet. "Warming May Be Hurting Gray Whales' Recovery." *Washington Post*, September 11, 2007c, A-12. http://www.washingtonpost.com/wp-dyn/content/article/2007/09/10/AR2007091002143_pf.html.

Eilperin, Juliet. "Climate Is a Risky Issue for Democrats: Candidates Back Costly Proposals." *Washington Post*, November 6, 2007d, A-1. http://www.washingtonpost.com/wp-dyn/content/article/2007/11/05/AR2007110502106_pf.html.

Eilperin, Juliet. "Global Warming Skeptics Insist Humans Not at Fault." *Washington Post*, March 4, 2008a, A-16. http://www.washingtonpost.com/wp-dyn/content/article/2008/03/03/AR2008030302781_pf.html.

Eilperin, Juliet. "Carbon Output Must Near Zero to Avert Danger, New Studies Say." *Washington Post*, March 10, 2008b, A-1. [http://www.washingtonpost.com/wp-dyn/content/article/2008/03/09/AR2008030901867_pf.html]

Eilperin, Juliet. "Carbon Is Building Up in Atmosphere Faster Than Predicted." *Washington Post*, September 26, 2008c, A-2. http://www.washingtonpost.com/wp-dyn/content/article/2008/09/25/AR2008092503989_pf.html.

Eilperin, Juliet. "Study Ties Tree Deaths to Change in Climate." *Washington Post*, January 23, 2009, A-8. http://www.washingtonpost.com/wp-dyn/content/article/2009/01/22/AR2009012202473_pf.html.

Einhorn, Arthur. Personal communication. March 23, 2006.

Einhorn, Arthur. "Sticker Shock." October 10, 2007.

"Ellesmere Island Ice Breaks Up." NASA Earth Observatory, September 9, 2008. http://earthobservatory.nasa.gov/Study/Ellesmere.

Elliott, Christopher. "Skiing Goes Downhill." *National Geographic*, December 2007, 12.

Elliott, Valerie. "Polar Bears Surviving on Thin Ice." *London Times*, October 30, 2003, n.p.(LEXIS).

Ellison, Christopher R. W., Mark R. Chapman, and Ian R. Hall. "Surface and Deep Ocean Interactions during the Cold Climate Event 8200 Years Ago." *Science* 312 (June 30, 2006):1929–1932.

El Nasser, Haya. "'Green' Efforts Embrace Poor." *USA Today*, November 24, 2008, 3-A.

Elsner, James B. "Evidence in Support of the Climate Change—Atlantic Hurricane Hypothesis." *Geophysical Research Letters* 33 (August 23, 2006), L16705.

Elsner, James B. "Tempests in Time." *Nature* 447 (June 7, 2007):647–648.

Elsner, James B., James P. Kossin, and Thomas H. Jagger. "The Increasing Intensity of the Strongest Tropical Cyclones." *Nature* 455 (September 4, 2008):92–95.

Emanuel, Kerry A. "An Air-sea Interaction Theory for Tropical Cyclones." Part I: Steady-state Maintenance." *Journal of the Atmospheric Sciences* 43 (1986):585–604.

Emanuel, Kerry A. "The Dependence of Hurricane Intensity on Climate." *Nature* 326, no. 2 (April 1987):483–485.

Emanuel, Kerry A. "The Maximum Intensity of Hurricanes." *Journal of the Atmospheric Sciences* 45 (1988a):1143–1156.

Emanuel, Kerry A. "Toward a General Theory of Hurricanes." *American Scientist* 76 (1988b):370–379.

Emanuel, Kerry A. "Comments on 'Global Climate Change and Tropical Cyclone': Part I." *Bulletin of the American Meteorological Society* 76 (1995):2241–2243.

Emanuel, Kerry A. "Thermodynamic Control of Hurricane Intensity." *Nature* 401 (October 14, 1999):665–669.

Emanuel, Kerry A. "Increasing Destructiveness of Tropical Storms over the Past 30 Years." *Nature* 436 (August 4, 2005):686–688.

Emanuel, Kerry A. *What We Know about Climate Change*. Cambridge, MA: MIT Press, 2007.

Emanuel, W. R., H. H. Shugart, and M. P. Stevenson. "Climatic Change and the Broad-scale Distribution of Terrestrial Eco-system Complexes." *Climatic Change* 7 (1985): 29–43.

"Eminent Scientists Warn of Disastrous, Permanent Global Warming." Environment News Service, February 19, 2007. http://www.ens-newswire.com/ens/feb2007/2007-02-19-03.asp.

"Energy and Equity: The Full Report." Illinois Environmental Protection Agency, n.d. http://www.cnt.org/ce/energy&equity.htm.

Engeler, Elaine. "U.N. Says 2006 Set Record for Greenhouse Gases in Atmosphere." Associated Press, November 23, 2007 (LEXIS).

Enger, Tim. "If This Is Spring, How Come I'm Still Shoveling Snow?" Letter to the Editor, *Edmonton Journal*, May 8, 2003, A-19.

English, Andrew. "Feeding the Dragon: How Western Car-makers Are Ignoring Ecological Dangers in Their Rush to Exploit a Wide-open Market." *London Daily Telegraph*, October 30, 2004, 1.

English, Philip. "Ross Island Penguins Struggling." *New Zealand Herald*, January 9, 2003, n.p.(LEXIS).

"EPA Petitioned to Limit Greenhouse Gases from Ships." Environment News Service, October 5, 2007. http://www.ens-newswire.com/ens/oct2007/2007-10-05-094.asp.

Epstein, Paul R. "Climate, Ecology, and Human Health." December 18, 1998. http://www.iitap.iastate.edu/gccourse/issues/health/health.html.

Epstein, Paul R. "Profound Consequences: Climate Disruption, Contagious Disease and Public Health." *Native Americas* 16, no. 3/4 (Fall/Winter 1999):64–67. http://nativeamericas.aip.cornell.edu/fall99/fall99epstein.html.

Epstein, Paul R., Henry F. Diaz, Scott Elias, Georg Grabherr, Nicohlas E. Graham, Willem J. M. Martens, Ellen Mosley-Thompson, and Joel Susskind. "Biological and Physical Signs of Climate Change: Focus on Mosquito-borne Diseases." *Bulletin of the American Meteorological Society* 79, Part 1 (1998):409–417.

Epstein, Paul R., Georg Grabbher, Tom Karl, Ellen Mosley-Thompson, Kevin Trenberth, and George M. Woodwell. *Current Effects: Global Climate Change. An Ozone Action Roundtable.* Washington, DC, June 24, 1996. http://www.ozone.org/curreff.html.

Erbacher, Jochen, Brian T. Huber, Richard D. Norris, and Molly Markey. "Increased Thermohaline Stratification as a Possible Cause for an Ocean Anoxic Event in the Cretaceous Period." *Nature* 409 (January 18, 2001):325–327.

Erickson, Jim. "Boulder Team Sees Obstacle to Saving Ozone Layer: 'Rocks' in Arctic Clouds Hold Harmful Chemicals." *Rocky Mountain News* (Denver), February 9, 2001, 37-A.

Erickson, Jim. "Glaciers Doff Their Ice Caps, and as Frozen Fields Melt, Anthropological Riches Are Revealed." *Rocky Mountain News*, August 22, 2002, 6-A.

Erickson, Jim. "Going, Going, Gone? Front-range Glaciers Declining: Researchers Point to a Warming World." *Rocky Mountain News*, October 26, 2004, 5-A.

Erlanger, Steven. "A New Fashion Catches on in Paris: Cheap Bicycle Rentals." *New York Times*, July 13, 2008. http://www.nytimes.com/2008/07/13/world/europe/13paris.html.

"Ethanol Production Threatens Plains States with Water Scarcity." Environment News Service, September 21, 2007. http://www.ens-newswire.com/ens/sep2007/2007-09-21-091.asp.

Etter, Lauren. "With Corn Prices Rising, Pigs Switch to Fatty Snacks." *Wall Street Journal*, May 21, 2007, A-1, A-14.

Eunjung Cha, Ariana. "China Embraces Nuclear Future: Optimism Mixes with Concern as Dozens of Plants Go Up." *Washington Post*, May 29, 2007.

"E.U. Plans to Become First Hydrogen Economy Superpower." *Industrial Environment* 12, no. 13 (December 2002):n.p.(LEXIS).

"European Marine Species Displaced by Warming Climate." Environment News Service, March 5, 2007. http://www.ens-newswire.com/ens/mar2007/2007-03-05-01.asp.

"European Scientists Stunned by Arctic Ice Melt." Environmental News Service, September 20, 2006.

"Europe to Cut Greenhouse Gases 20 Percent by 2020." Environment News Service, March 8, 2007. http://www.ens-newswire.com/ens/mar2007/2007-03-08-04.asp.

Evans-Pritchard, Ambrose. "Dutch Have Only Years before Rising Seas Reclaim Land: Dikes No Match against Global Warming Effects." *London Daily Telegraph* in *Ottawa Citizen*, September 8, 2004, A-6.

Everts, C. H. "Effects of Sea Level Rise and Net Sand Volume Change on Shoreline Position at Ocean City, Maryland." In *Potential Impacts of Sea Level Rise on the Beach at Ocean City, Maryland,* ed. J. G. Titus. Washington, DC: U.S. Environmental Protection Agency, 1985.

Ewen, Alexander. "Consensus Denied: Holy War over Global Warming." *Native Americas* 16, no, 3/4 (Fall/Winter 1999):26–33. http://nativeamericas.aip.cornell.edu.

"Expert Fears Warming Will Doom Bears." Canadian Press in *Victoria Times-Colonist,* January 5, 2003, C-8.

"Extreme Rainstorms a New Texas Trend." Environment News Service, December 3, 2007. http://www.ens-newswire.com/ens/dec2007/2007-12-04-095.asp.

Fackler, Martin. "A Global Comeback for Coal." *New York Times,* May 22, 2008, C-1, C-4.

Fagan, Brian. *The Great Warming: Climate Change and the Rise and Fall of Civilizations.* New York: Bloomsbury Press, 2008.

Fahey, D. W., R. S. Gao, K. S. Carslaw, J. Kettleborough, P. J. Popp, M. J. Northway, J. C. Holecek, S. C. Ciciora, R. J. McLaughlin, T. L. Thompson, R. H. Winkler, D. G. Baumgardner, B. Gandrud, P. O. Wennberg, S. Dhaniyala, K. McKinney, T. Peter, R. J. Salawitch, T. P. Bui, J. W. Elkins, C. R. Webster, E. L. Atlas, H. Jost, J. C. Wilson, R. L. Herman, A. Kleinböhl, and M. von König. "The Detection of Large HNO3-Containing Particles in the Winter Arctic Stratosphere." *Science* 291 (February 9, 2001):1026–1031.

Fahrenthold, David A. "Washington Warming to Southern Plants." *Washington Post,* December 20, 2006, A-1. http://www.washingtonpost.com/wp-dyn/content/article/2006/12/19/AR2006121901769_pf.html.

Fahrenthold, David A. "D.C. Area Outpaces Nations in Pollution: High Carbon Emission Blamed on Coal Plants." *Washington Post,* September 30, 2007, C-1. http://www.washingtonpost.com/wp-dyn/content/article/2007/09/29/AR2007092900959.html.

Fahrenthold, David A. "There's a Gold Mine in Environmental Guilt: Carbon-Offset Sales Brisk Despite Financial Crisis." *Washington Post,* October 6, 2008, A-1. http://www.washingtonpost.com/wp-dyn/content/article/2008/10/05/AR2008100502518_pf.html.

Fahrenthold, David A., and Steven Mufson. "Cost of Saving the Climate Meets Real-World Hurdles." *Washington Post,* August 16, 2007, A-1. http://www.washingtonpost.com/wp-dyn/content/article/2007/08/15/AR2007081502432_pf.html.

Fahy, Declan. "Nature Charts Its Own Change: Irish Researchers Are Finding Signs of Climate Change in Trees and Bird Species." *Irish Times,* September 13, 2001, n.p.(LEXIS).

"Failure to Manage Global Warming Would Cripple World Economy." Environment News Service, October 30, 2006. http://www.ens-newswire.com/ens/oct2006/2006-10-30-06.asp.

"Failure to Tackle Global Warming Could Cost Trillions." Environmental News Service, October 13, 2006. http://www.ens-newswire.com/ens/oct2006/2006-10-13-03.asp.

Faiola, Anthony, and Robin Shulman. "Cities Take Lead on Environment as Debate Drags at Federal Level." *Washington Post,* June 9, 2007, A-1. http://www.washingtonpost.com/wp-dyn/content/article/2007/06/08/AR2007060802779_pf.html.

Fairless, Daemon. "Renewable Energy: Energy Go-Round." *Nature* 447 (June 28, 2007a):1046–1048.

Fairless, Daemon. "Biofuel: The Little Shrub That Could—Maybe." *Nature* 449 (October 11, 2007b):652–655.

Falkowski, P., R. J. Scholes, E. Boyle, J. Canadell, D. Canfield, J. Elser, N. Gruber, K. Hibbard, P. Högberg, S. Linder, F. T. Mackenzie, B. Moore III, T. Pedersen, Y. Rosenthal, S. Seitzinger, V. Smetacek, and W. Steffen. "The Global Carbon Cycle: A Test of Our Knowledge of Earth as a System." *Science* 290 (October 13, 2000):291–296.

Fallows, James. "Turn Left at Cloud 109." *New York Times Sunday Magazine,* November 21, 1999, 84–89.

Fan, S., M. Gloor, J. Mahlman, S. Pacala, J. Sarmiento, T. Takahashi, P. Tans. "A Large Terrestrial Carbon Sink in North America Implied by Atmospheric and Oceanic Carbon Dioxide Data and Models." *Science* 282 (October 16, 1998):442–446.

Fang, Jingyun, Anping Chen, Changhui Peng, Shuqing Zhao, and Longjun Ci. "Changes in Forest Biomass Carbon Storage in China between 1949 and 1998." *Science* 292 (June 22, 2001):2320–2322.

Fankhauser, Samuel. *Valuing Climate Change: The Economics of the Greenhouse.* London: Earthscan Publications, Ltd., 1995.

Fargione, Joseph, Jason Hill, David Tilman, Stephen Polasky, and Peter Hawthorne. "Land Clearing and the Biofuel Carbon Debt." *Science* 319 (February 29, 2008):1235-1238.

Faris, Stephan. "The Real Roots of Darfur." *The Atlantic*, April 2007, 67–69.

Faris, Stephan. "Phenomenon: Ice Free." *New York Times Sunday Magazine*, July 27, 2008. http://www.nytimes.com/2008/07/27/magazine/27wwln-phenom-t.html.

Farman, J. C., B. G. Gardiner, and J. D. Shanklin. "Large Losses of Total Ozone Reveal Seasonal Cl0x/NOx Interaction." *Nature* 315 (1985):207–210.

Fauber, John, and Tom Vanden Brook. "Global Warming May Take Great Lakes Gulp: Plunge in Coming Century Would Have Significant Ripple Effect, Reports Say." *Milwaukee Journal-Sentinel*, June 14, 2000, 1-A.

Feder, Barnaby J. "The Showhouse That Sustainability Built." *New York Times*, March 26, 2008. http://www.nytimes.com/2008/03/26/business/businessspecial2/26boulder.html.

Fedorov, A. V., P. S. Dekens, M. McCarthy, A. C. Ravelo, P. B. deMenocal, M. Barreiro, R. C. Pacanowski, and S. G. Philander. The Pliocene Paradox (Mechanisms for a Permanent El Niño). *Science* (June 9, 2006):1485–1489.

Feely, Richard A., Christopher L. Sabine, Kitack Lee, Will Berelson, Joanie Kleypas, Victoria J. Fabry, and Frank J. Millero. "Impact of Anthropogenic CO_2 on the $CaCO_3$ System in the Oceans." *Science* 305 (July 16, 2004):362–366.

Feely, Richard A., Christopher L. Sabine, J. Martin Hernandez-Ayon, Debby Ianson, and Burke Hales. "Evidence for Upwelling of Corrosive 'Acidified' Water onto the Continental Shelf." *Science* 320 (June 13, 2008):1490–1492.

Feresin, Emiliano. "Europe Looks to Draw Power from Africa." *Nature* 450 (November 29, 2007):595.

Ferguson, H. L. "The Changing Atmosphere: Implications for Global Security." In *The Challenge of Global Warming*, ed. Dean Edwin Abrahamson, 48–62. Washington, DC: Island Press, 1989.

Fialka, John J. "U.S. Study on Global Warming may Overplay Dire Side." *Wall Street Journal*, May 26, 2000, A-24.

Fialka, John J. "Soot Storm: A Dirty Discovery over Indian Ocean Sets off a Fight." *Wall Street Journal*, May 6, 2003, A-1, A-6.

Fialka, John J. "Energy Independence: A Dry Hole?" *Wall Street Journal*, July 5, 2006a, A-4.

Fialka, John J. "Renewable Fuels May Provide 25 Percent of U.S. Energy by 2025." *Wall Street Journal*, November 13, 2006b, A-10.

Fialka, John J. "U.S. Plots New Climate Tactic." *Wall Street Journal*, September 7, 2007, A-8.

Fidelman, Charlie. "Longer, Stronger Blazes Forecast." *Montreal Gazette*, July 11, 2002, A-4.

Finch, Gavin. "Falklands Penguins Dying in Thousands.' *London Independent*, June 19, 2002, 12.

"Findings." *Washington Post*, April 30, 2004, A-30.

Finn, Peter. "In Balmy Europe, Feverish Choruses of 'Let It Snow.'" *Washington Post*, December 20, 2006, A01. http://www.washingtonpost.com/wp-dyn/content/article/2006/12/19/AR2006121901681_pf.html.

Finney, Paul Burnham. "U.S. Business Travelers Let the Train Take Way the Strain." *International Herald-Tribune*, April 24, 2007, 16.

"Fire Ecologists Adopt Climate Change Declaration." Environment News Service, November 17, 2006. http://www.ens-newswire.com/ens/nov2006/2006-11-17-09.asp#anchor6.

"First Half of Year Was Warmest on Record for U.S." Associated Press, July 14, 2006 (LEXIS).

"Fished to the Point of Ruin, North Sea Cod Stocks So Low as to Spell Disaster." *Glasgow Herald* (Scotland), November 7, 2000, 18.

Fitter, A. H., and R. S. R. Fitter. "Rapid Changes in Flowering Time in British Plants." *Science* 296 (May 31, 2002):1689–1691.

Fitz, Don. "What's Possible in the Military Sector?" Z-Net, April 30, 2007. http://www.zmag.org/content/showarticle.cfm?ItemID=12705.

Flam, Faye. "It's Hot Now, but Scientists Predict There's an Ice Age Coming." *Philadelphia Inquirer*, August 23, 2002, n.p.(LEXIS).

Flannery, Tim. *The Weather Markers: How Man Is Changing the Climate and What It Means for Life on Earth*. New York: Atlantic Monthly Press, 2005.

Fleck, John. "Jack Frost's Nip Arrives a Bit Later." *Albuquerque Journal*, October 26, 2002, A-1.

Fleck, John. "Dry Days, Warm Nights." *Albuquerque Journal*, December 28, 2003, B-1.

Flesher, John. "The Great Loss: Lakes' Water Drop Incites Debate on Cause, Concern about Impact." *Toledo Blade*, May 21, 2000, B-1, B-3.

Foley, Jonathan A. "Tipping Points in the Tundra." *Science* 310 (October 28, 2005):627–628.

Fong, Petti. "Greenhouse Gas Chokes Sky after Wildfires." CanWest News Service in *Calgary Herald*, September 23, 2003, A-8.

Forero, Juan. "As Andean Glaciers Shrink, Water Worries Grow." *New York Times*, November 24, 2002, A-3.

Fountain, Henry. "Observatory: Threat to Rice Crops." *New York Times*, December 12, 2000, F-5.

Fountain, Henry. "Observatory: Early Birds and Worms." *New York Times*, May 22, 2001, F-4.

Fountain, Henry. "Observatory: Rice and Warm Weather." *New York Times*, June 29, 2004, F-1.

Fountain, Henry. "Observatory: Lobsters Quarantining Lobsters." *New York Times*, May 30, 2006. http://www.nytimes.com/2006/05/30/science/30observ.html.

Fountain, Henry. "More Acidic Ocean Hurts Reef Algae as Well as Corals." *New York Times*, January 8, 2008, D-3.

Fowler, C., W. J. Emery, and J. Maslanik. "Satellite-derived Evolution of Arctic Sea Ice Age: October 1978 to March 2003." *Geoscience and Remote Sensing Letters* 1, no. 2 (2004): 71–74.

Francis, Justin. Responsibletravel.com. United Kingdom. Accessed April 30, 2006. http://www.responsibletravel.com/Copy/Copy101993.htm.

Frank, Peter. "Ten Great Endangered Places to See While You Still Can." *USA Today*, April 18, 2008, 3-D.

Frank, Robert. "How Thailand Became the 'Detroit of the East,'" *Wall Street Journal*, December 8, 1999, B-1, B-5.

Frank, Robert. "The Wealth Report: Living Large While Being Green." *Wall Street Journal*, August 24, 2007, W-2.

Frank, Thomas. "Planes Fly More, Emit Less Greenhouse Gas." *USA Today*, May 9, 2008, 1-B.

Frazier, M. R., Raymond B. Huey, and David Berrigan. "Thermodynamics Constrains the Evolution of Insect Population Growth Rates: 'Warmer Is Better.'" *The American Naturalist* 168, no. 4 (October 2006):512–520.

"Fred Thompson Embraces Bad Science on Global Warming." *Irregular Times*, June 24, 2007. http://irregulartimes.com/index.php/archives/2007/06/24/fred-thompson-solar-activity.

Freeman, C., N. Fenner, N. J. Ostle, H. Kang, D. J. Dowrick, B. Reynolds, M. A. Lock, D. Sleep, S. Hughes, and J. Hudson. "Export of Dissolved Carbon from Peatlands under Elevated Carbon Dioxide Levels." *Nature* 430 (July 8, 2004):195–198.

Freeman, James, and Eleanor Cowie. "Pollutants Threaten the Great Barrier Reef." *Glasgow Herald*, January 25, 2002, 7.

Freeman, Sholnn. "Pollution in Overdrive: New Report Cites U.S. Motorists for Production of Greenhouse Gases." *Washington Post*, June 28, 2006, D-1. http://www.washingtonpost.com/wp-dyn/content/article/2006/06/27/AR2006062701757_pf.html.

Freemantle, Tony. "Global Warming Likely to Hit Texas: Scientists Say Temperature Rise Will Change Rainfall, Gulf Coast Region." *Houston Chronicle*, October 24, 2001, 32.

Freer-Smith, Peter H., Mark S. J. Broadmeadow, and Jim M. Lynch, eds. *Forestry and Climate Change*. Wallingford, UK: CABI, 2007.

Freier, J. E., D. J. Rogers, M. J. Packer, N. Nicholls, and J. Almendares. "Vector-borne Diseases." *Lancet* 342 (December 4, 1993):1464–1469.

Frey, Darcy. "George Divoky's Planet." *New York Times Sunday Magazine*, January 6, 2002, 26–30.

Fried, Jeremy S., Margaret S. Torn, and Evan Mills. "The Impact of Climate Change on Wildfire Severity." *Climatic Change* 64 (May 2004):169–191.

Friedl, Randall R. "Perspectives: Atmospheric Chemistry" Unraveling Aircraft Impacts." *Science* 286 (October 1, 1999):57–58.

Friedman, Thomas L. "Doha and Dalian." *New York Times*, September 19, 2007, A-27.

Friedman, Thomas L. *Hot, Flat, and Crowded. Why We Need a Green Revolution—And How It Can Renew America.* New York: Farrar, Straus, and Giroux, 2008a.

Friedman, Thomas L. "Learning to Speak Climate." *New York Times*, August 6, 2008b. http://www.nytimes.com/2008/08/06/opinion/06friedman.html.

Frosch, Dan. "Citing Need for Assessments, U.S, Freezes Solar Energy Projects." *New York Times*, June 27, 2008, A-13.

"Fueling Jets with Animal Fat." Environment News Service, July 18, 2007. http://www.ens-newswire.com/ens/jul2007/2007-07-18-09.asp#anchor7.

Fukasawa, Masao, Howard Freeland, Ron Perkin, Tomowo Watanabe, Hiroshi Uchida, and Ayako Nishina. "Bottom Water Warming in the North Pacific Ocean." *Nature* 427 (February 26, 2004):825–827.

Funk, Chris, Michael D. Dettinger, Joel C. Michaelsen, James P. Verdin, Molly E. Brown, Mathew Barlow, and Andrew Hoell. "Warming of the Indian Ocean Threatens Eastern and Southern African Food Security but Could Be Mitigated by Agricultural Development." *Proceedings of the National Academy of Sciences* 105 (August 12, 2008): 11,081–11,086.

Funk, McKenzie. "Cold Rush: The Coming Fight for the Melting North." *Harper's Magazine*, September 2007, 56–55.

"FutureGen, World's Cleanest Coal Plant, Sited in Illinois." Environment News Service, December 18, 2007. http://www.ens-newswire.com/ens/dec2007/2007-12-18-01.asp.

Gaarder, Nancy. "State Enjoying 'Exceptional' Warmth." *Omaha World-Herald*, December 4, 2001, A-1, A-2.

Gaarder, Nancy. "Winds of Change Blowing in Iowa." *Omaha World-Herald*, May 13, 2007a, D-1, D-2.

Gaarder, Nancy. "Many Digging Deep for Cheaper Energy." *Omaha World-Herald*, May 29, 2007b, A-1, A-2.

Gaffen, Dian J. "Falling Satellites, Rising Temperature?" *Nature* 394 (August 13, 1998):615–616.

Galbraith, Kate. "Economy Shifts, and the Ethanol Industry Reels." *New York Times*, November 5, 2008, B-1, B-7.

Galloway, Elaine, and Chloe Rhodes. "Warm Spell Brings Early Start to Hay-fever Misery." *London Evening Standard*, April 14, 2003, 16.

Gange, A. C., E. G. Gange, T. H. Sparks, and L. Boddy. "Rapid and Recent Changes in Fungal Fruiting Patterns." *Science* 316 (April 6, 2007):71.

Ganopolski, Andrey, and Stefan Rahmstorf. "Rapid Changes of Glacial Climate Simulated in a Coupled Climate Model." *Nature* 409 (January 11, 2001):153–158.

Gardiner, Beth. "Report: Extreme Weather on the Rise, Likely to Get Worse." Associated Press Worldstream, International News, London. February 27, 2003 (LEXIS).

Gardner, Toby A., Isabelle M. Cote, Jennifer A. Gill, Alastair Grant, and Andrew R. Watkinson. "Long-Term Region-Wide Declines in Caribbean Corals." *Science* 301 (August 15, 2003):958–960. http://www.scienceexpress.org.

Garreau, Joel. "A Dream Blown Away: Climate Change Already Has a Chilling Effect on Where Americans Can Build Their Homes." *Washington Post*, December 2, 2006, C-1. http://www.washingtonpost.com/wp-dyn/content/article/2006/12/01/AR2006120101759_pf.html.

Gartner, John. "Climbers Bring Climate Change from Mountaintop to Laptops." Environment News Service, July 31, 2007a. http://www.ens-newswire.com/ens/jul2007/2007-07-31-04.asp.

Gartner, John. "Fossil Fuels' Free Ride Is Over." Environment News Service, December 3, 2007b. http://www.ens-newswire.com/ens/nov2007/2007-11-30-02.asp.

"Gators Spotted in Mississippi River Backwaters at Memphis." Associated Press in *Daytona Beach News-Journal*, May 13, 2006, n.p.

Gautier, Catherine. *Oil, Water, and Climate: An Introduction.* New York: Cambridge University Press, 2008.

Gelbspan, Ross. "The Heat Is On: The Warming of the World's Climate Sparks a Blaze of Denial." *Harper's Magazine*, December 1995. http://www.dieoff.com/page82.htm. Gelbspan, Ross. "A Global Warming." *The American Prospect* 31 (March/April 1997a). http://www.prospect.org/archives/31/31gelbfs.html.

Gelbspan, Ross. *The Heat Is On: The High Stakes Battle over Earth's Threatened Climate.* Reading, MA: Addison-Wesley Publishing Co., 1997b.

Gelbspan, Ross. "Beyond Kyoto." *The Amicus Journal* (Winter 1998). http://www.nrdc.org/nrdc/nrdc/nrdc/eamicus/98win/toc.html.

Gelbspan, Ross. "Boiling Point." *The Nation*, August 16, 2004a, 24–27.

Gelbspan, Ross. *Boiling Point: How Politicians, Big Oil and Coal, Journalists, and Activists Have Fueled the Climate Crisis—and What We Can Do to Avert Disaster.* New York: Basic Books (Perseus), 2004b.

Gelling, Peter. "Forest Loss in Sumatra Becomes a Global Issue." *New York Times,* December 6, 2007. http://www.nytimes.com/2007/12/06/world/asia/06indo.html.

Geoghegan, John. "Long Ocean Voyage Set for Vessel That Runs on Wave Power." *New York Times*, March 11, 2007, D-3.

George, Douglas M. Personal communication, October 13, 2006.

George, Jane. "Global Warming Threatens Nunavut's National Parks." *Nunatsiaq News*, May 19, 2000. http://www.nunatsiaq.com/archives/nunavut000531/nvt20519_18.html.

"Georgia Judge Yanks Coal Power Permit on Climate Concerns." Environment News Service, Jun 30, 2008. http://www.ens-newswire.com/ens/jun2008/2008-06-30-091.asp.

"Germans Blame Ethanol Boom for—Oh Mein Gott!—Rising Beer Prices." Associated Press in *Omaha World-Herald*, June 3, 2007, A-18.

Gerrard, Michael, ed. *Global Climate Change and U.S. Law.* Chicago: American Bar Association, 2007.

Gertner, Jon. "The Future Is Drying Up." *New York Times Sunday Magazine*, October 21, 2007. http://www.nytimes.com/2007/10/21/magazine/21water-t.html.

Ghazi, Polly, and Rachel Lewis. *The Low-Carbon Diet: Slim Down, Chill Out, and Save the World.* London: Short, 2007.

Giannini, A., R. Saravanan, and P. Chang. "Oceanic Forcing of Sahel Rainfall on Interannual to Interdecadal Time Scales." *Science* 302 (November 7, 2003):1027–1030.

"Giant Squid Film Team Makes Spectacular Catch." Agence France Presse, September 14, 2002.

Giardina, Christian P., and Michael G. Ryan. "Evidence That Decomposition Rates of Organic Carbon in Mineral Soil Do Not Vary with Temperature." *Nature* 404 (April 20, 2000):858–861.

Gibbard, Seran, Kenneth Caldeira, Govindasamy Bala, Thomas J. Phillips, and Michael Wickett. "Climate Effects of Global Land Cover Change." *Geophysical Research Letters* 23 (December 8, 2005), L23705, doi: 10.1029/2005GL024550.

Gibbs, Mark T., Karen L. Bice, Eric J. Barron, and Lee R. Kump. "Glaciation in the Early Paleozoic 'Greenhouse:' The Roles of Paleo-geography and Atmospheric CO_2." In *Warm Climates in Earth History*, ed. Brian T. Huber, Kenneth G. MacLeod, and Scott L. Wing, 386–422. Cambridge: Cambridge University Press, 2000.

Gibbs, Walter. "Scientists Back Off Theory of a Colder Europe in a Warming World." *New York Times*, May 15, 2007. http://www.nytimes.com/2007/05/15/science/earth/15cold.html.

Gibbs, Walter, and Sarah Lyall. "Gore Shares Peace Prize for Climate Change Work." *New York Times*, October 13, 2007, A-1, A-13.

Gifford, Roger M., Damian J. Barrett, Jason L. Lutze, and Ananda B. Samarakoon. "The CO_2 Fertilizing Effect: Relevance to the Global Carbon Cycle." In *The Carbon Cycle*, ed. T. M. L. Wigley and D. S. Schimel, 77–92. Cambridge: Cambridge University Press, 2000.

Giles, Jim. "The Outlook for Amazonia Is Dry." *Nature* 442 (August 17, 2006a):726–727.

Giles, Jim. "How Much Will It Cost to Save the World?" *Nature* 444 (November 2, 2006b):6–7.

Giles, Jim. "Methane Quashes Green Credentials of Hydropower." *Nature* 444 (November 30, 2006c):524–525.

Gill, R. A., H. W. Polley, H. B. Johnson, L. J. Anderson, H. Maherali, and R. B. Jackson. "Nonlinear Grassland Responses to Past and Future Atmospheric CO_2." *Nature* 417 (May 16, 2002):279–282.

Gill, Richardson Benedict. *The Great Maya Droughts: Water, Life, and Death.* Albuquerque: University of New Mexico Press, 2000.

Gille, Sarah T. "Warming of the Southern Ocean since the 1950s." *Science* 295 (February 15, 2002):1275–1277.

Gillett, Nathan P., and David W. J. Thompson. "Simulation of Recent Southern Hemisphere Climate Change." *Science* 302 (October 10, 2003):273–275.

Gillis, Justin. "A New Outlet for Venter's Energy: Genome Maverick to Take on Global Warming." *Washington Post*, April 30, 2002, E-1.

Gillon, Jim. "The Water Cooler." *Nature* 404 (April 6, 2000):555.

Gingrich, Newt, and Terry L. Maple. *A Contract with the Earth.* Baltimore: Johns Hopkins University Press, 2007.

"GISS 2007 Temperature Analysis." Goddard Institute for Space Studies. January 15, 2008. http://www.columbia.edu/~jeh1/mailings/20080114_GISTEMP.pdf.

Gjerdrum, Carina, Anne M. J. Vallee, Colleen Cassady St. Clair, Douglas F. Bertram, John L. Ryder, and Gwylim S. Blackburn. "Tufted Puffin Reproduction Reveals Ocean Climate Variability." *Proceedings of the National Academy of Sciences* 100, no. 16 (August 5, 2003):9377–9382.

"Glacial Retreat Seen Worldwide." Environment News Service, May 30, 2002. http://ens-news. com/ens/may2002/2002-05-30-09.asp#anchor2; see also http://www.gsfc.nasa.gov/ topstory/20020530glaciers.html.

"Glaciers and Ice Caps Quickly Melting Into the Seas." Environment News Service, July 20, 2007. http://www.ens-newswire.com/ens/jul2007/2007-07-20-03.asp.

Glantz, Michael H. *Climate Affairs: A Primer.* Washington, DC: Island Press, 2003.

Glick, Daniel. "The Heat Is On: Geosigns." *National Geographic,* September 2004, 12–33.

Glick, Patricia. *Global Warming: The High Costs of Inaction.* San Francisco: Sierra Club, 1998. http://www.sierraclub.org/global-warming/inaction.html.

"Global Air Traffic Rose, Easing Fuel Cost Blow." *Wall Street Journal,* July 2, 2006, A-7.

"Global Climate Shift Feeds Spreading Deserts." Environment News Service, June 17, 2002. http://ens-news.com/ens/jun2002/2002-06-17-03.asp.

"Global Warming." Editorial, *London Financial Times,* March 11, 1997. http://benetton.dkrz. de:3688/homepages/georg/kimo/0254.html.

"Global Warming Alarms Infectious Disease Experts." Environment News Service, May 23, 2007. http://www.ens-newswire.com/ens/may2007/2007-05-23-03.asp.

"Global Warming and Freak Winds Combine to Allow Explorers through Northeast Passage." *London Independent,* October 11, 2002, 14.

"Global Warming Blamed for Rising Sea Levels." Associated Press in *Omaha World-Herald,* November 25, 2001, 20-A.

"Global Warming Could Hamper Ocean Sequestration." Environment News Service, December 4, 2002. http://ens-news.com/ens/dec2002/2002-12-04-09.asp.

"Global Warming Could Persist for Centuries." Environment News Service, February 18, 2002. http://ens-news.com/ens/feb2002/2002L-02-18-09.html.

"Global Warming Could Spread Extreme Drought." Environment News Service, October 5, 2006. http://www.ens-newswire.com/ens/oct2006/2006-10-05-01.asp.

"Global Warming Goes to Court." Editorial, *New York Times,* November 28, 2006.

"Global Warming Is Changing Tropical Forests." Environment News Service, August 7, 2002. http://ens-news.com/ens/aug2002/2002-08-07-01.asp.

"Global Warming Linked to Increase in Kidney Stones." Environment News Service, May 19, 2008. http://www.ens-newswire.com/ens/may2008/2008-05-19-091.asp.

"Global Warming Makes China's Glaciers Shrink by Equivalent of Yellow River." Agence France Presse, August 23, 2004 (LEXIS).

"Global Warming Means More Snow for Great Lakes Region." Ascribe Newswire, November 4, 2003 (LEXIS).

"Global Warming Pests and Pestilence." *What's Hot: World Climate Report* 2, no. 2 (1995–2000). http://www.nhes.com/back_issues/Vol1and2/WH/hot.html.

"Global Warming Predicted to Create Novel Climates." Environment News Service, March 27, 2007. http://www.ens-newswire.com/ens/mar2007/2007-03-27-03.asp.

"Global Warming Signal from the Ocean." *Dallas Morning News* in *New Orleans Times-Picayune,* December 31, 2000, 4.

"Global Warming's Sooty Smokescreen Revealed." New Scientist.com, June 3, 2003.

"Global Warming Sticker Shock." Environment News Service, May 23, 2008. http://www.ens-newswire.com/ens/may2008/2008-05-23-01.asp.

"Global Warming Threatens California Water Supplies." Environment News Service, June 4, 2002. http://ens-news.com/ens/jun2002/2002-06-04-09.asp#anchor3.

"Global Warming Troubles Qinghai-Tibet Railway Construction." Xinhua (Chinese News Agency), April 30, 2003 (LEXIS).

"Global Warming Will Dramatically Alter U.S. Northeast." Environment News Service, October 4, 2006. http://www.ens-newswire.com/ens/oct2006/2006-10-04-03.asp.

"Global Warning on Climate." *Montreal Gazette,* February 27, 2001, B-2.

"Global Wind Power Generated Record Year in 2006." Environment News Service, February 12, 2007. http://www.ens-newswire.com/ens/feb2007/2007-02-12-04.asp.

Goddard Institute for Space Sciences. "Global Temperature Trends: 1998 Global Surface Temperature Smashes Record." 1999. http://www.giss.nasa.gov/research/observe/surftemp.

Goes, Joaquim I., Prasad G. Thoppil, Helga do R. Gomes, and John T. Fasullo. "Warming of the Eurasian Landmass Is Making the Arabian Sea More Productive." *Science* 308 (April 22, 2005):545–547.

"Going Green Is Hard Sell in Texas." Associated Press in *Omaha World-Herald*, January 20, 2008, 13-A.

Goldman, Erica. "Even in the High Arctic, Nothing Is Permanent." *Science* 297 (August 30, 2002):1493–1494.

Goodale, Christine L., and Eric A. Davidson. "Carbon Cycle: Uncertain Sinks in the Shrubs." *Nature* 418 (August 8, 2002):601.

Goodall, Chris. *How to Live a Low-Carbon Life: The Individual's Guide to Stopping Climate Change.* London: Earthscan Publications, Ltd., 2007.

Goodall, Chris. "Analysis of 27 Footnoted Sources in Lomborg's *Cool It!*" Carbon Commentary, October 15, 2007. http://www.carboncommentary.com/2007/10/15/29.

Goodell, Jeff. *Big Coal: The Dirty Secret behind America's Energy Future.* Boston: Houghton-Mifflin, 2006a.

Goodell, Jeff. "Our Black Future." *New York Times,* June 23, 2006b. http://www.nytimes.com/2006/06/23/opinion/23goodell.html.

Goodell, Jeff. "The Ethanol Scam." *Rolling Stone*, August 9, 2007, 48–53.

Goodstein, Eban. *Fighting for Love in the Century of Extinction: How Passion and Politics Can Stop Global Warming.* Burlington: University of Vermont Press, 2007.

Goodstein, Laurie. "Eighty-six Evangelical Leaders Join to Fight Global Warming." *New York Times*, February 8, 2006. http://www.nytimes.com/2006/02/08/national/08warm.htm.

Göran, Ekström, Meredith Nettles, and Victor C. Tsai. "Seasonality and Increasing Frequency of Greenland Glacial Earthquakes." *Science* 311 (March 24, 2006):1756–1758.

Gordon, Anita, and David Suzuki. *It's a Matter of Survival.* Cambridge, MA: Harvard University Press, 1991.

Gordon, Susan. "U.S. Pacific Northwest Gets Reduced Supply of Snow, Climate Study Says." *Tacoma News-Tribune*, February 7, 2003, n.p.(LEXIS).

Gore, Albert, Jr. *Earth in the Balance: Ecology and the Human Spirit.* Boston: Houghton Mifflin Co., 1992.

Gore, Albert, Jr. *An Inconvenient Truth: The Planetary Emergency of Global Warming and What We Can Do About It.* Emmaus, PA: Rodale Books, 2006.

Gore, Albert, Jr. "A Generational Challenge to Repower America." July 17, 2008. http://www.wecansolveit.org/content/pages/304.

Goreau, Thomas J., Raymond L. Hayes, Jenifer W. Clark, Daniel J. Basta, and Craig N. Robertson. "Elevated Sea-surface Temperatures Correlate with Caribbean Coral Reef Bleaching." In *A Global Warming Forum: Scientific, Economic, and Legal Overview*, ed. Richard A. Geyer, 225–262. Boca Raton, FL: CRC Press, 1993.

"Gore Warns of Sub-Prime Carbon Catastrophe." Environment News Service, September 27, 2008. http://www.ens-newswire.com/ens/sep2008/2008-09-27-01.asp.

Goswami, B. N.,V. Venugopal, D. Sengupta, M. S. Madhusoodanan, and Prince K. Xavier. "Increasing Trend of Extreme Rain Events over India in a Warming Environment." *Science* 314 (December 1, 2006):1442–1445.

Gough, Robert. "Stress on Stress: Global Warming and Aquatic Resource Depletion." *Native Americas* 16, no. 3/4 (Fall/Winter 1999):46–48. http://nativeamericas.aip.cornell.edu.

Government of Sweden. Ministry of Sustainable Development. 2006. http://www.sweden.gov.se/sb/d/2066.

Grabherr, G., N. Gottfried, and H. Pauli. "Climate Effects on Mountain Plants." *Nature* 369 (1994):448–451.

Grace, John, and Mark Rayment. "Respiration in the Balance." *Nature* 404 (April 20, 2000):819–820.

Grady, Denise. "Managing Planet Earth: On an Altered Planet, New Diseases Emerge as Old Ones Re-emerge." *New York Times*, August 20, 2002, F-2.

Graedel, Thomas E., and Jennifer A. Howard-Grenville. *Greening the Industrial Facility: Perspectives, Approaches, and Tools.* New York: Springer, 2005.

Graf, Hans-F. "The Complex Interaction of Aerosols and Clouds." *Science* 303 (February 27, 2004):1309–1311.

Grant, Christine. "Swelling Seas Eating Away at Country's Monuments." *The Scotsman*, December 24, 2001, 5.

Grant, Paul M. "Hydrogen Lifts Off – with a Heavy Load: The Dream of Clean, Usable Energy Needs to Reflect Practical Reality." *Nature* 424 (July 10, 2003):129–130.

Gray, William M. "Hurricanes and Hot Air." *Wall Street Journal*, July 26, 2007, A-12.

"Great Barrier Reef Is Springing Back to Life." *Western Daily Press* (Australia), December 18, 2002, 11.

Grebmeier, Jacqueline M., James E. Overland, Sue E. Moore, Ed V. Farley, Eddy C. Carmack, Lee W. Cooper, Karen E. Frey, John H. Helle, Fiona A. McLaughlin, and S. Lyn McNutt. "A Major Ecosystem Shift in the Northern Bering Sea." *Science* 311 (March 10, 2006):1461–1464.

Greenaway, Norma. "Disaster Toll from Weather up Tenfold: Droughts, Floods Need More Damage Control, Report Says." *Edmonton Journal*, February 28, 2003, A-5.

"Greening the Planet." Climate Change Debate, December 8, 1998. http://climatechangedebate.com/archive/12-08_12-15_1998.txt.

"Greenland's Ice Melt Accelerating." Environment News Service. September 21, 2006. http://www.ens-newswire.com/ens/sep2006/2006-09-21-01.asp.

"Greens Want Global Warming Examined in Bushfire Inquiry." Australian Associated Press, January 21, 2003 (LEXIS).

Gregg, Watson W., and Margarita E. Conkright. "Decadal Changes in Global Ocean Chlorophyll." *Geophysical Research Letters* 29, no. 15 (2002), doi: 10.1029/2002GL014689.

Gregory, Angela. "Fear of Rising Seas Drives More Tuvaluans to New Zealand." *New Zealand Herald*, February 19, 2003, n.p.(LEXIS).

Gregory, Jonathan M., Philippe Huybrechts, and Sarah C. B. Raper. "Threatened Loss of the Greenland Ice Sheet." *Nature* 428 (April 8, 2004):616.

Gribbin, John. *Hothouse Earth: The Greenhouse Effect and Gaia*. London: Bantam Press, 1990.

Gribbin, John. *The Origins of the Future: Ten Questions for the Next Ten Years*. New Haven, CT: Yale University Press, 2007.

Griscom-Little, Amanda. "Detroit Takes Charge." *Outside*, April 2007, 60.

"Growth in China's CO_2 Emissions Double Previous Estimates." Environment News Service, March 11, 2008. http://www.ens-newswire.com/ens/mar2008/2008-03-11-01.asp.

"Growth of Global Greenhouse Gas Emissions Accelerating." Environment News Service, November 29, 2006. http://www.ens-newswire.com/ens/nov2006/2006-11-29-02.asp.

Gugliotta, Guy. "In Antarctica, No Warming Trend: Scientists Find Temperatures Have Gotten Colder in Past Two Decades." *Washington Post*, January 14, 2002, A-2.

Gugliotta, Guy. "Warming May Threaten 37 Percent of Species by 2050." *Washington Post*, January 8, 2004, A-1. http://www.washingtonpost.com/wp-dyn/articles/A63153-2004Jan7.html.

"Guide Delivers Live Earth Global Warming Survival Skills." Environment News Service, June 19, 2007. http://www.ens-newswire.com/ens/jun2007/2007-06-19-02.asp.

Gunter, G. "An Example of Oyster Production Decline with a Change in the Salinity Characteristics of an Estuary—Delaware Bay, 1800–1973." *Proceedings of the National Shellfish Association* 65 (1974):3–13.

Haarsma, R. J., J. F. B. Mitchell, and C. A. Senior. "Tropical Disturbances in a G[lobal] C[limate] M[odel]." *Climate Dynamics* 8 (1993):247–257.

Haeberli, W. "Possible Effects of Climatic Change on the Evolution of Alpine Permafrost." In *Greenhouse-Impact on Cold-Climate Ecosystems and Landscapes*, ed. M. Boer and E. Koster, 23–35. *Selected Papers of a European Conference on Landscape Ecological Impact of Climatic Change*, Lunteren, the Netherlands, December 3–7, 1989. CARTENA Supplement 22. Cremlingen, Germany: Catena Verlag, 1992.

Haigh, Joanna D. "Climate Variability and the Influence of the Sun." *Science* 294 (December 7, 2001):2109–2111.

Haines, Andrew. "The Implications for Health." In *Global Warming: The Greenpeace Report*, ed. Jeremy Leggett, 149–162. New York: Oxford University Press, 1990.

Hakim, Danny. "Ford Stresses Business, but Disappoints Environmentalists." *New York Times*, August 20, 2002, C-4.

Hakim, Danny. "Several States Likely to Follow California on Car Emissions." *New York Times*, June 11, 2004a, C-4.

Hakim, Danny. "Ford Executives Adopt Ambitious Plan to Rein in Global Warming." *New York Times* in *International Herald-Tribune*, October 5, 2004b, 11.

Häkkinen, Sirpa, Andrey Proshutinsky, and Igor Ashik. "Sea Ice Drift in the Arctic since the 1950s." *Geophysical Research Letters* 35 (2008), L19704, doi: 10.1029/2008GL034791.

Häkkinen, Sirpa, and Peter B. Rhines. "Decline of Subpolar North Atlantic Circulation during the 1990s." *Science* 304 (April 23, 2004):555–559.

Hale, Ellen. "Climate Changes Could Devastate the Netherlands." *USA Today*, November 22, 2000, 27-A.

Hale, Ellen. "Seas Create Real Water Hazard: Changing Climate at Root of Erosion That's Putting Links Courses in Jeopardy." *USA Today*, July 18, 2001, 3-C.

"Half U.S. Climate Warming Due to Land Use Changes." Environment News Service, May 28, 2003. http://ens-news.com/ens/may2003/2003-05-28-01.asp.

Hall, Alan. "Postcards a Tell-tale for Icy Retreat." *The Scotsman*, August 3, 2002, 8.

Hall, Alex, and Ronald J. Stouffer. "An Abrupt Climate Event in a Coupled Ocean-Atmosphere Simulation without External Forcing." *Nature* 409 (January 11, 2001):171–174.

Hall, Carl T. "Spring Scorches the Record Books: It Was the Hottest in U.S. History." *San Francisco Chronicle*, June 17, 2000, A-1.

Hall, Carl T. "Ocean Tells the Story—Earth Is Heating Up: Human Activity, Not Variables in Nature, Cited as Culprit." *San Francisco Chronicle*, April 29, 2005, A-1.

Hallam, Anthony, and Paul Wignall. *Mass Extinctions and Their Aftermath*. Oxford: Oxford University Press, 1997.

Hall-Spencer, Jason M., Riccardo Rodolfo-Metalpa, Sophie Martin, Emma Ransome, Maoz Fine, Suzanne M. Turner, Sonia J. Rowley, Dario Tedesco, and Maria-Cristina Buia. "Volcanic Carbon Dioxide Vents Show Ecosystem Effects of Ocean Acidification." *Nature* 454 (July 3, 2008):96–99.

Halweil, Brian. "The Irony of Climate." *World Watch*, March/April 2005, 18–23.

Hamilton, Martha M. "British Petroleum Sets Goal of 10 Percent Cut in 'Greenhouse' Gases." *Washington Post*, September 18, 1998, A-6. http://climatechangedebate.com/archive/09-18_10-27_1998.txt.

Hann, Judith. "Spring Wakes Early, but Will Autumn Lie in Late Again? What Will Tomorrow's World Look Like?" United Kingdom Woodland Trust, 2002. http://www.woodland-trust.org.uk/news/subindex.asp?aid=328.

Hanna, Edward, Philippe Huybrechts, Konrad Steffen, John Cappelen, Russell Huff, Christopher Shuman, Tristram Irvine-Flynn, Stephen Wise, and Michael Griffiths. "Increased Runoff from Melt from the Greenland Ice Sheet: A Response to Global Warming." *Journal of Climate* 21, no. 2 (January 15, 2008):331–341.

"Hans Blix's Greatest Fear." *New York Times*, March 16, 2003, D-2.

Hansen, Bogi, William R. Turrell, and Svein Sterhus. "Decreasing Overflow from the Nordic Seas into the Atlantic Ocean through the Faroe Bank Channel since 1950." *Nature* 411 (June 21, 2001):927–930.

Hansen, James E. "The Greenhouse, the White House, and Our House." Typescript of a speech at the International Platform Association, Washington, DC, August 3, 1989.

Hansen, James E. "Defusing the Global Warming Time Bomb." *Scientific American* 290, no. 3 (March 2004a):68–77.

Hansen, James E. "Dangerous Anthropogenic Interference: A Discussion of Humanity's Faustian Climate Bargain and the Payments Coming Due." Presentation at the Distinguished Public Lecture Series at the Department of Physic and Astronomy, University of Iowa, Iowa City, October 26, 2004b.

Hansen, James E. "A Slippery Slope: How Much Global Warming Constitutes Dangerous Anthropogenic Interference?" *Climatic Change* 68 (2005a):269–279.

Hansen, James E. "Is There Still Time to Avoid 'Dangerous Anthropogenic Interference' with Global Climate? A Tribute to Charles David Keeling." A paper delivered to the American Geophysical Union, San Francisco, December 6, 2005b. http://www.columbia.edu/~jeh1/keeling_talk_and_slides.pdf.

Hansen, James E. "The Threat to the Planet." *New York Review of Books*, July 2006a, 12–16.

Hansen, James E. "Declaration of James E. Hansen." *Green Mountain Chrysler-Plymouth-Dodge-Jeep, et al., Plaintiffs v. Thomas W. Torti, Secretary of the Vermont Agency of Natural Resources, et al., Defendants*. Case Nos. 2:05-CV-302 and 2:05-CV-304, Consolidated. United States District Court for the District of Vermont. August 14, 2006b. http://www.giss.nasa.gov/~dcain/recent_papers_proofs/vermont_14aug20061_textwfigs.pdf.

Hansen, James E. "Special Interests Are the One Big Obstacle." *London Times*, March 12, 2007a. http://business.timesonline.co.uk/tol/business/columnists/article1499726.ece.

Hansen, James E. "Political Interference with Government Climate Change Science." Testimony of James E. Hansen, to Committee on Oversight and Government Reform. United States House of Representatives, March 19, 2007b.

Hansen, James E. "Coal Trains of Death." E-mail list, July 23, 2007c.

Hansen, James E. "Goodbye Miami: NASA's James Hansen on Catastrophic Sea-level Rise." *New Scientist,* July 28, 2007d, 30–34.

Hansen, James E. Personal communication, November 27, 2007e.

Hansen, James E. "Dear Chancellor [Merkel, of Germany], Perspective of a Younger Generation." E-mail list, January 23, 2008a.

Hansen, James E. "Cold Weather." Accessed March 3, 2008b. http://www.columbia.edu/~jeh1/mailings/20080303_ColdWeather.pdf

Hansen, James E. "Trip Report." August 5, 2008c. http://www.columbia.edu/~jeh1/mailings/20080804_TripReport.pdf.

Hansen, James E. "Written Testimony, Kingsnorth Case." Accessed September 11, 2008d. http://www.columbia.edu/~jeh1/mailings/20080910_Kingsnorth.pdf.

Hansen, James E. Personal communication, September 21, 2008e.

Hansen, James E., D. Johnson, A. Lacis, S, Lebendeff, D. Rind, and G. Russell. "Climate Impact of Increasing Atmospheric Carbon Dioxide." *Science* 213 (1981):957–966.

Hansen, James E., A. Lacis, D. Rind, G. Russell, P. Stone, I Fung, R. Ruedy, and J. Lerner. "Climate Sensitivity: Analysis of Feedback Mechanisms." *Geophysical Monograph* 29 (1984):130–163.

Hansen, James E., A. Lacis, R. Ruedy, and M. Sato. "Potential Climate Impact of the Mount Pinatubo Eruption." *Geophysical Research Letters* 19, no. 2 (January 24, 1992):215–218.

Hansen, James E., and Larissa Nazarenko. "Soot Climate Forcing via Snow and Ice Albedos." *Proceedings of the National Academy of Sciences* 101, no. 2 (January 13, 2004):423–428.

Hansen, James E., Larissa Nazarenko, Reto Ruedy, Makiko Sato, Josh Willis, Anthony Del Genio, Dorothy Koch, Andrew Lacis, Ken Lo, Surabi Menon, Tica Novakov, Judith Perlwitz, Gary Russell, Gavin A. Schmidt, and Nicholas Tausnev. "Earth's Energy Imbalance: Confirmation and Implications." *Science* 308 (June 3, 2005):1431–1435.

Hansen, James E., R. Ruedy, J. Glascoe, and M. Sato. "GISS Analysis of Surface Temperature Change." *Journal of Geophysical Research* 104 (December 27, 1999):30,997–31,022.

Hansen, James E., R. Ruedy, A. Lacis, D. Koch, I. Tegen, T. Hall, D. Shindell, B. Santer, P. Stone, T. Novakov, et al. "Climate Forcings in Goddard Institute for Space Studies SI2000 Simulations." *Journal of Geophysical Research* 107, no. D18 (2002),4347, doi: 10.1029/2001JD001143.

Hansen, James E., G. Russell, D. Rind, P. Stone, A. Lacis, S. Lebedeff, R. Ruedy, and L. Travis. "Efficient Three-dimensional Global Models for Climate Studies: Models I and II." *Monthly Weather Review* 111 (1983):609–662.

Hansen, James E., and M. Sato. "Trends of Measured Climate Forcing Agents." *Proceedings of the National Academy of Sciences* 98 (December 18, 2001):14,778–14,783.

Hansen, James E., M. Sato, P. Kharecha, D. Beerling, V. Masson-Delmotte, M. Pagani, M. Raymo, D. Royer, and J. C. Zachos. "Target Atmospheric CO_2: Where Should Humanity Aim?" *Open Atmospheric Science Journal* 2, no. 15 (2008):217–231. http://dx.doi.org/10.2174/1874282300802010217.

Hansen, James E., M. Sato, P. Kharecha, G. Russell, D. W. Lea, and M. Siddall." Climate Change and Trace Gases." *Philosophical Transactions of the Royal Society of London* A365 (2007):1925–1954.

Hansen, James E., M. Sato, and R. Ruedy. "The Missing Climate Forcing." *Philosophical Transactions of the Royal Society of London* B352 (1997):231–240.

Hansen James E., M. Sato, R. Ruedy, P. Kharecha, A. Lacis, R. Miller, L. Nazarenko, K. Lo, G. A. Schmidt, G. Russell, et al.. "Dangerous Human-made Interference with Climate: A GISS Model Study." In *Atmospheric Chemistry and Physics Discussions,* March 27, 2007. http://www.cosis.net/copernicus/EGU/acpd/6/S7350/acpd-6-S7350_p.pdf.

Hansen, James E., M. Sato, R. Ruedy, K. Lo, D. W. Lea, and M. Medina-Elizade. "Global Temperature Change." *Proceedings of the National Academy of Sciences* 103 (September 25, 2006):14288–14293, doi: 10.1073/pnas.0606291103. http://pubs.giss.nasa.gov/abstracts/2006/Hansen_etal_1.html.

Harden, Blaine. "Tree-Planting Drive Seeks to Bring a New Urban Cool: Lower Energy Costs Touted as Benefit." *Washington Post,* September 4, 2006, A-1. http://www.washingtonpost.com/wp-dyn/content/article/2006/09/03/AR2006090300926_pf.html.

Harden, Blaine. "Air, Water Powerful Partners in Northwest: Region's Hydro-Heavy Electric Grid Makes for Wind-Energy Synergy." *Washington Post*, March 21, 2007, A-3. http://www.washingtonpost.com/wp-dyn/content/article/2007/03/20/AR2007032001634_pf.html.

Harden, Blaine. "Filipinos Draw Power from Buried Heat." *Washington Post,* October 4, 2008, A-1. http://www.washingtonpost.com/wp-dyn/content/article/2008/10/03/AR2008100303843_pf.html.

Harden, Blaine, and Juliet Eilperin. "On the Move to Outrun Climate Change: Self-Preservation Forcing Wild Species, Businesses, Planning Officials to Act." *Washington Post*, November 25, 2006, A-3.

Hardin, Garrett, and John Baden, eds. *Managing the Commons.* San Francisco, W. H. Freeman, 1977.

Hariss, Robert C., Terry Bensel, and Denise Blaha. "Methane Emissions to the Global Atmosphere from Coal Mining." In *A Global Warming Forum: Scientific, Economic, and Legal Overview*, ed. Richard A. Geyer, 339–346. Boca Raton, FL: CRC Press, 1993.

Harper, Laura Lee. *Blame It on the Rain: How the Weather Has Changed History.* New York: HarperCollins, 2006.

"Harper's Index." *Harper's Magazine*, September 2006, 17.

Harris, Paul G. *Global Warming and East Asia: The Domestic and International Politics of Climate Change.* London: Routledge, 2003.

Hartmann, Dennis L., John M. Wallace, Varavut Limpasuvan, David W. J. Thompson, and James R. Holton. "Can Ozone Depletion and Global Warming Interact to Produce Rapid Climate Change?" *Proceedings of the National Academy of Sciences of the United States of America* 97, no. 4 (February 15, 2000):1412–1417.

Harvell, C. D., K. Kim, J. M. Buckholder, R. R. Colwell, P. R. Epstein, D. J. Grimes, E. E. Hofmann, E. K. Lipp, A. D. M. E. Osterhaus, R. M. Overstreet, J. W. Porter, G. W. Smith, and G. R. Vasta. "Emerging Marine Diseases—Climate Links and Anthropogenic Factors." *Science* 285, no. 3 (September 3, 1999):1505–1510.

Harvell, C. D., C. E. Mitchell, J. R. Ward, S. Altizer, A. P. Dobson, R. S. Ostfeld, and M. D. Samuel. "Climate Warming and Disease Risks for Terrestrial and Marine Biota." *Science* 296 (June 21, 2002):2158–2162.

Harvey, Fiona. "Arctic May Have No Ice in Summer by 2070, Warns Climate Change Report." *London Financial Times*, November 2, 2004, 1.

Hatsuhisa, Takashima. "Climate." *Journal of Japanese Trade and Industry*, September 1, 2002, n.p.(LEXIS).

Hawkes, Nigel. "Giant Cities Are Creating Their Own Weather." *The Times of London*, February 23, 2000, n.p.(LEXIS).

Hay, S. I., J. Cox, D. J. Rogers, S. E. Randolph, D. I. Stern, G. D. Shanks, M. F. Myers, and R. W. Snow. "Climate Change and the Resurgence of Malaria in the East African Highlands." *Nature* 425 (February 21, 2002):905–909.

Hayes, Constance L. "Coca-Cola Experiments with Vending Machines." *New York Times* in *Milwaukee Journal-Sentinel*, October 28, 1999, 3-A.

Hayhoe, Katherine, Daniel Cayan, Christopher B. Field, Peter C. Frumhoff, Edwin P. Maurer, Norman L. Miller, Susanne C. Moser, Stephen H. Schneider, Kimberly Nicholas Cahill, Elsa E. Cleland, Larry Dale, Ray Drapek, R. Michael Hanemann, Laurence S. Kalkstein, James Lenihan, Claire K. Lunch, Ronald P. Neilson, Scott C. Sheridan, and Julia H. Verville. "Emissions Pathways, Climate Change, and Impacts on California." *Proceedings of the National Academy of Sciences* 101, no. 34 (August 24, 2004):12422–12427.

Healy, Patrick. "Warming Waters: Lobstermen on Cape Cod Blame Light Hauls on Higher Ocean Temperatures." *Boston Globe*, August 30, 2002, B-1.

Heartland Institute. "Background." Accessed May 10, 2008. http://www.heartland.org/NewYork08/background.cfm.

"Heavy Rains Threaten Flood-Prone Venus." *The Straits Times* (Singapore), June 8, 2002, n.p.(LEXIS).

Hegeri, Gabriele. "Climate Change: the Past as Guide to the Future." *Nature* 392 (April 23, 1998):758–759.

Heilprin, John. "Study Sees Economic Benefits of Reducing Global Warming." Associated Press, July 13, 2001. http://www.worldwildlife.org/climate/climate.cfm.

Heilprin, John. "Study Says Black Carbon Emissions in China and India Have Climate Change Effects." Associated Press, September 26, 2002 (LEXIS).

Heilprin, John. "Earth Hottest It's Been in 2,000 Years." Associated Press, June 23, 2006 (LEXIS).

Heinberg, Richard. *Peak Everything: Waking Up to the Century of Declines.* Gabriola, BC: New Society Publishers, 2007.

Hendee, David. "Peril Is Seen to State's Water Table." *Omaha World-Herald*, October 22, 2006, A-1, A-2.

Henderson, Casper. "The Other CO_2 Problem." *New Scientist*, August 5, 2006, 28–33.

Henderson, Mark. "Positive Winds Keeping Arctic Winters at Bay." *London Times*, July 6, 2001a, n.p.(LEXIS).

Henderson, Mark. "World Has 15 Years to Stop Global Warming." *London Times*, July 21, 2001b, n.p.(LEXIS).

Henderson, Mark. "Past Ten Summers Were the Hottest in 500 Years." *London Times*, March 5, 2004a, 10.

Henderson, Mark. "Hot News from 740,000 Years Ago Tells Us to Get Ready for Catastrophic Climate Change." *London Times*, June 10, 2004b, 4.

Henderson, Mark. "Southern Krill Decline Threatens Whales, Seals." *London Times* in *Calgary Herald*, November 4, 2004c, A-11.

Hendriks, C. A., K. Blok, and W. C. Turkenburg. "The Recovery of Carbon Dioxide from Power Plants." In *Climate and Energy: The Feasibility of Controlling CO_2 Emissions,* ed. P. A. Okken, R. J. Swart, and S. Zwerver, 125–142. Dordrecht, Germany: Kluwer Academic Publishers, 1989.

Henry, Tom. "Global Warming Grips Greenland, Leaving Lasting Mark." *Toledo Blade,* October 12, 2008. http://www.toledoblade.com/apps/pbcs.dll/article?AID=2008810109858.

Henson, Robert. *The Rough Guide to Climate Change.* London: Rough Guides, 2006.

Herrick, Thaddeus. "The New Texas Wind Rush: Oil Patch Turns to Turbines, as Ranchers Sell Wind Rights." *Wall Street Journal*, September 23, 2002, B-1, B-3.

Hertsgaard, Mark. "Will We Run Out of Gas? No, We'll Have Plenty of Carbon-based Fuel to See Us through the Next Century. That's the Problem." *Time*, November 5, 1999, 110–111.

Hertsgaard, Mark. "It's Much Too Late to Sweat Global Warming." *San Francisco Chronicle*, February 13, 2005, n.p.

Hesselbo, Stephen P., Darren R. Grocke, Hugh C. Jenkyns, Christian J. Bjerrum, Paul Farrimond, Helen S. Morgans Bell, and Owen R. Green. "Massive Dissociation of Gas Hydrate during a Jurassic Oceanic Anoxic Event." *Nature* 406 (July 27, 2000):392–395.

Hickman, Martin. "The Prince of Emissions." *London Independent*, April 1, 2006, 1.

Higgins, Adrian. "Kept from Hibernation by a Lingering Warmth." *Washington Post*, January 5, 2006, B-1. http://www.washingtonpost.com/wp-dyn/content/article/2007/01/04/AR2007010401973_pf.html.

Higgins, Matt. "Warm Temperatures Chill the Ice Fishing Season." *New York Times*, January 10, 2007. http://www.nytimes.com/2007/01/10/sports/othersports/10outdoors.html.

Higgins, Michelle. "Machu Picchu, Without Roughing It." *New York Times*, Travel, August 12, 2006, 6.

Highfield, Roger. "Winter Floods 'Five Times More Likely.'" *London Daily Telegraph*, January 31, 2002, 8.

Hileman, Bette. "Climate Observations Substantiate Global Warming Models." *Chemical and Engineering News*, November 27, 1995. http://pubs.acs.org/hotartcl/cenear/951127/pg1.html.

Hillman, Mayer, Tina Fawcett, and Sudhir Chella Rajan. *The Suicidal Planet: How to Prevent Global Climate Catastrophe.* New York: St. Martin's/Dunne, 2007.

Hinrichs, Kai-Uwe, Laura R. Hmelo, and Sean P. Sylva. "Molecular Fossil Record of Elevated Methane Levels in Late Pleistocene Coastal Waters." *Science* 299 (February 21, 2003):1214–1217.

Hirsch, Jerry. "Damage to Coral Reefs Mounts, Study Says: Broad Survey Cites Human Causes Such as Over-fishing and Pollution." *Los Angeles Times*, August 26, 2002, 14.

Hodge, Amanda. "Patagonia's Big Melt 'Sign of Global Warming.'" *The Australian* (Sydney), February 12, 2004, 8.

Hodges, Glenn. "The New Cold War: Stalking Arctic Climate Change by Submarine." *National Geographic*, March 2000, 30–41.

Hoegh-Guldberg, O., R. J. Jones, S. Ward, and W. K. Loh. "Is Coral Bleaching Really Adaptive?" *Nature* 415 (February 7, 2001):601–602.

Hoegh-Guldberg, O., P. J. Mumby, A. J. Hooten, R. S. Steneck, P. Greenfield, E. Gomez, C. D. Harvell, P. F. Sale, A. J. Edwards, K. Caldeira, N. Knowlton, C. M. Eakin, R. Iglesias-Prieto, N. Muthiga, R. H. Bradbury, A. Dubi, and M. E. Hatziolos. "Coral Reefs under Rapid Climate Change and Ocean Acidification." *Science* 318 (December 14, 2007):1737–1742.

Hoerling, Martin, and Arun Kumar. "The Perfect Ocean for Drought." *Science* 299 (January 31, 2003):691–694.

Hoffert, Martin, Ken Caldeira, Gregory Benford, David R. Criswell, Christopher Green, Howard Herzog, Atul K. Jain, Haroon S. Kheshgi, Klaus S. Lackner, John S. Lewis, et al. "Advanced Technology Paths to Global Climate Stability: Energy for a Greenhouse Planet." *Science* 298 (November 1, 2002):981–987.

Hoffert, Martin, Ken Caldeira, Curt Covey, Philip P. Duffy, and Benjamin D. Santer. "Solar Variability and the Earth's Climate." *Nature* 401 (October 21, 1999):764–765.

Hoffheiser, Chuck. "Previous Hot Theory Was Global Cooling." Letter to the Editor, *Wall Street Journal*, June 19, 2001, A-23.

Hoffman, Ian. "Iron Curtain over Global Warming: Ocean Experiment Suggests Phytoplankton May Cool Climate." *Daily Review* (Hayward, CA), April 17, 2004, n.p.(LEXIS).

Hoffman, Ian. "Rising Temps—Declining Crop Yields." *Inside Bay Area* (California), March 16, 2007.

Hogan, Treacy. "Still Raining in Costa del Ireland." *Belfast Telegram*, November 26, 2002, n.p.(LEXIS).

Hogarth, Murray. "Sea-warming Threatens Coral Reefs." *Sydney Morning Herald* (Australia), November 26, 1998. http://www.smh.com.au/news/9811/26/text/national13.html.

Holland, Jennifer S. "The Acid Threat: As CO_2 Rises, Shelled Animals May Perish." *National Geographic*, November 2001, 110–111.

Hollingsworth, Jan. "Global Warming Studies Put Heat on State: Tampa Bay Area Labeled Extremely Vulnerable." *Tampa Tribune*, October 24, 2001, 1.

Holly, Chris. "World CO_2 Emissions Up 13 Percent from 1990–2000." *Energy Daily* 30, no. 206 (October 25, 2002), n.p.(LEXIS).

Holly, Chris. "Sea-level Rise Seen as Key Global Warming Threat." *The Energy Daily* 32, no. 36 (February 25, 2004), n.p.(LEXIS).

Horovitz, Bruce. "Can Eateries Go Green, Earn Green?" *USA Today*, May 16, 2008, 1-B, 2-B.

"Hot Polymer Catches Carbon Dioxide" Environment News Service, May 29, 2002. http://ens-news.com/ens/may2002/2002-05-29-05.asp.

"Hot Times in the City Getting Hotter." Environment News Service, September 27, 2002. http://ens-news.com/ens/sep2002/2002-09-27-09.asp#anchor8.

Hotz, Robert Lee. "Greenland's Ice Sheet Is Slip, Sliding Away." *Los Angeles Times*, June 24, 2006, n.p.

Hough, Tom P. *Recent Developments in Solar Energy*. Hauppauge, NY: Nova Science, 2007.

Houghton, J. T., Y. Ding, D. J. Griggs, M. Noguer, P. J. van der Linden, X. Dai, K. Maskell, and C. A. Johnson. *Climate Change 2001: The Scientific Basis. Contribution of Working Group I to the Third Assessment Report of the Intergovernmental Panel on Climate Change.* Cambridge: Cambridge University Press, 2001.

Houghton, John. *Global Warming: The Complete Briefing.* Cambridge: Cambridge University Press, 1997.

Houghton, R. A., D. L. Skole, Carlos A. Nobre, J. L. Hackler, K. T. Lawrence, and W. H. Chomentowski. "Annual Fluxes of Carbon from Deforestation and Regrowth in the Brazilian Amazon." *Nature* 403 (January 20, 2000):301–304.

Houlder, Vanessa. "Faster Global Warming Predicted: Met Office Research Has 'Mind-blowing' Implications.'" *London Financial Times*, November 9, 2000, 2.

Houlder, Vanessa. "Royal Society Calls for Carbon Levy or Permits." *London Financial Times*, November 18, 2002a, 4.

Houlder, Vanessa. "Rise Predicted in Aviation Carbon Dioxide Emissions." *London Financial Times*, December 16, 2002b, 2.

"How about That Weather? The Answer Is Blowing in the Wind, Rain, Snow …" *Washington Post*, January 26, 2005, C-16.

Howat, I. M., I. R. Joughin, and T. A. Scambos. "Rapid Changes in Ice Discharge from Greenland Outlet Glaciers." Science Express, February 8, 2007, 10.1126/science.1138478. www.scienceexpress.org.

Howat, I. M., I. R. Joughin, S. Tulaczyk, and S. Gogineni. "Rapid Retreat and Acceleration of Helheim Glacier, East Greenland." *Geophysical Research Letters* 32 (2005), L22502, doi: 10.1029/2005GL024737.

Howden, S. Mark, Jean Francois Soussana, Francesco Tubiello, Netra Chhetri, Michael Dunlop, and Holger Meinke. "Adapting Agriculture to Climate Change." *Proceedings of the National Academy of Sciences* 104 (December 11, 2007):19,691–19,696.

"How Insurers Can Help Reduce Warming Risks." *Wall Street Journal,* June 4, 2007, B-9.

Howl, Joanne. "Siberia 2008: Kotuykan River Expedition." NASA Earth Observatory, July 9, 2008. http://earthobservatory.nasa.gov/Study/SiberiaBlog2008.

"How to Combat Global Warming: In the End, the Only Real Solution May Be New Energy Technologies." *Business Week,* August 16, 2004, 108.

Hoyos, C. D., P. A. Agudelo, P. J. Webster, and J. A. Curry. "Deconvolution of the Factors Contributing to the Increase in Global Hurricane Intensity." *Science* 312 (April 7, 2006): 94–97.

Hu, Feng Sheng, Darrell Kaufman, Sumiko Yoneji, David Nelson, Aldo Shemesh, Yongsong Huang, Jian Tian, Gerard Bond, Benjamin Clegg, and Thomas Brown. "Cyclic Variation and Solar Forcing of Holocene Climate in the Alaskan Subarctic." *Science* 301 (September 26, 2003):1890–1893.

Huang, Shaopeng, Henry N. Pollack, and Po-Yu Shen. "Temperature Trends over the Past Five Centuries Reconstructed from Borehole Temperatures." *Nature* 403 (February 17, 2000):756–758.

Huber, Brian T., Kenneth G. MacLeod, and Scott L. Wing. *Warm Climates in Earth History.* Cambridge: Cambridge University Press, 2000.

Huber, Matthew. "A Hotter Greenhouse?" *Science* 321 (July 18, 2008):353–354.

Huber, Matthew, and Rodrigo Caballero. "Eocene El Niño: Evidence for Robust Tropical Dynamics in the 'Hothouse.'" *Science* 299 (February 7, 2003):877–881.

Huber, Uli M., Harald K. M. Bugmann, and Mel A. Reasoner, eds. *Global Change and Mountain Regions: An Overview of Current Knowledge.* Dordrecht, Netherlands: Springer, 2005.

Hudson, Kris. "Whose Beach Is This, Anyway?" *Wall Street Journal,* December 12, 2007, B-1, B-8.

Hughes, Kathleen. "To Fight Global Warming, Some Hang a Clothesline." *New York Times,* April 12, 2007. http://www.nytimes.com/2007/04/12/garden/12clothesline.html.

Hughes, T. *The Stability of the West Antarctic Ice Sheet: What Has Happened and What Will Happen.* Washington, DC: U.S. Department of Energy, 1983.

Hughes, T. J., J. L. Fastook, and G. H. Denton. *Climatic Warming and the Collapse of the West Antarctic Ice Sheet.* Orono: University of Maine Press, 1979.

Hughes, T. P., A. H. Baird, D. R. Bellwood, M. Card, S. R. Connolly, C. Folke, R. Grosberg, O. Hoegh-Guldberg, J. B. C. Jackson, J. Kleypas, et al. "Climate Change, Human Impacts, and the Resilience of Coral Reefs." *Science* 31 (August 15, 2003):929–933.

Human, Katy. "Disappearing Arctic Ice Chills Scientists: A University of Colorado Expert on Ice Worries That the Massive Melting Will Trigger Dramatic Changes in the World's Weather." *Denver Post,* October 5, 2004, B-2.

Human, Katy, and Kim McGuire. "Ozone Decline Stuns Scientists." *Denver Post,* March 2, 2005, A-8.

Hume, Stephen. "A Risk We Can't Afford: The Summer of Fire and the Winter of the Deluge Should Prove to the Nay-sayers That if We Wait too Long to React to Climate Change We'll Be in Grave Peril." *Vancouver Sun,* October 23, 2003, A-13.

"Hundreds Pose Naked on Melting Swiss Glacier." Environment News Service, August 22, 2007. http://www.ens-newswire.com/ens/aug2007/2007-08-22-04.asp.

Hungate, Bruce A., Peter D. Stiling, Paul Dijkstra, Dale W. Johnson, Michael E. Ketterer, Graham J. Hymus, C. Ross Hinkle, and Bert G. Drake. "CO_2 Elicits Long-Term Decline in Nitrogen Fixation." *Science* 304 (May 28, 2004):1291.

"Hunger, Water Scarcity Displaces Thousands of Afghans." Reuters in *New York Times,* June 4, 2008. http://www.nytimes.com/reuters/world/international-afghan-displacement.html.

Huntley, Brian. "Lessons from the Climates of the Past." In *Global Warming: The Greenpeace Report,* ed. Jeremy Leggett, 133–148. New York: Oxford University Press, 1990.

Huq, Saleemul. "Climate Change and Bangladesh." *Science* 294 (November 23, 2001):1617.

"Hydrogen Car Has Far to Go." *Houston Chronicle* in *Omaha World-Herald*, September 9, 2007, 2-D.

"Hydrogen Leakage Could Expand Ozone Depletion." Environment News Service, June 13, 2003. http://ens-news.com/ens/jun2003/2003-06-13-09.asp.

Hymon, Steve. "Early Snowmelt Ignites Global Warming Worries: Scientists Have Known Rising Temperatures Could Deplete Water Sources, but Data Show It May Already Be Happening." *Los Angeles Times*, June 28, 2004, B-1.

"Ice a Scarce Commodity on Arctic Rinks: Global Warming Blamed for Shortened Hockey Season." *Financial Post* (Canada), January 7, 2003, A-3.

"Ice Sheet of Greenland Melting Away at Faster Pace." Bloomberg News in *Omaha World-Herald*, January 20, 2008, 18-A.

Idso, Sherwood B. *Carbon Dioxide: Friend or Foe?* Tempe, AZ: IBR Press, 1982.

Idso, Sherwood B., R. C. Balling Jr., and R. S. Cerveny. "Carbon Dioxide and Hurricanes: Implications of Northern Hemispheric Warming for Atlantic/Caribbean Storms." *Meteorology and Atmospheric Physics* 42 (1990):259–263.

Iglesias-Rodriguez, M. Debora, Paul R. Halloran, Rosalind E. M. Rickaby, Ian R. Hall, Elena Colmenero-Hidalgo, John R. Gittins, Darryl R. H. Green, Toby Tyrrell, Samantha J. Gibbs, Peter von Dassow, Eric Rehm, E. Virginia Armbrust, and Karin P. Boessenkool. "Phytoplankton Calcification in a High-CO_2 World." *Science* 320 (April 18, 2008):336–340.

Indermuhle, A., T. F. Stocker, F. Joos, H. Fischer, H. J. Smith, M. Wahlen, B. Deck, D. Mastroianni, J. Tschumi, T. Blunier, R. Meyer, and B. Stauffer. "Holocane Carbon-cycle Dynamics Based on CO_2 Trapped in Ice at Taylor Dome, Alaska." *Nature* 398 (March 11, 1999):121–126.

"Industrial Countries' Greenhouse Gas Emissions Going Up." Environment News Service, October 31, 2006. http://www.ens-newswire.com/ens/oct2006/2006-10-31-02.asp.

"Indy 500 Race Cars to Run on 100 Percent Ethanol." Environment News Service, May 25, 2007. http://www.ens-newswire.com/ens/may2007/2007-05-25-09.asp#anchor6.

Ingham, John. "Fears for the Earth: All Down to Us—Hotter by the Minute." *London Express*, September 2, 2003, 19.

Ingham, John. "Stingers Thrive as the Country Gets Warmer: Invasion of the Scorpions." *London Express*, June 18, 2004, 40.

Ingham, John, and Barry Keevins. "U.K. Set to Be 6° C Warmer by 2099: London May Be Just Like Madrid." *London Express*, February 24, 2003, 17.

"Inhofe Calls Global Warming Warnings a Hoax." Associated Press, Oklahoma State and Local Wire, July 29, 2003 (LEXIS).

Inkley, D. B., M. G. Anderson, A. R. Blaustein, V. R. Burkett, B. Felzer, B. Griffith, J. Price, and T. L. Root. *Global Climate Change and Wildlife in North America.* Washington, DC: The Wildlife Society, 2004. http://www.nwf.org/news.

Inman, Mason. "Methane Bubbling Up from Undersea Permafrost?" National Geographic News, December 19, 2008. http://news.nationalgeographic.com/news/2008/12/081219-methane-siberia.html.

"Inter-American Commission Considers Global Warming-Human Rights Link." Environment News Service, February 6, 2007. http://www.ens-newswire.com/ens/feb2007/2007-02-06-09.asp#anchor1.

"Investors Sizing Up 'Carbon Footprints.'" Cox News Service in *Omaha World-Herald*, September 30, 2007, D-1.

IPCC (Intergovernmental Panel on Climate Change). *Climate Change 2007: Synthesis Report.* Geneva, Switzerland: IPCC, 2007.

"Iron Link to CO_2 Reductions Weakened." Environment News Service, April 10, 2003. http://ens-news.com/ens/apr2003/2003-04-10-09.asp#anchor8.

"Is It Already Too Late to Stop Global Warming?" *Insurance Day*, July 14, 2004, n.p.(LEXIS).

"Italy to Build World's First Hydrogen-Fired Power Plant." Environment News Service, December 18, 2006. http://www.ens-newswire.com/ens/dec2006/2006-12-18-05.asp.

Iven, Chris. "Heat Strangles Fish." *Syracuse Post-Standard*, July 9, 2002. http://www.syracuse.com/news/poststandard/index.ssf?/base/news-0/10262073067431.xml.

Ivins, Molly. "Environment Is Ticking, but Will We Hear It?" *Chicago Sun-Times*, July 13, 2000, 39.

Ivins, Molly. "Ignoring Problem Works—For a While." *Charleston Gazette* (West Virginia), June 28, 2003, 4-A.

Izrael, Yu. A. "Climate Change Impact Studies: The IPCC Working Group II Report." In *Climate Change: Science, Impacts, and Policy,* ed. J. Jager and H. L. Ferguson, 83–86. Proceedings of the Second World Climate Conference. Cambridge: Cambridge University Press, 1991.

Jablonski, L. M., X. Wang, and P. S. Curtis. "Plant Reproduction under Elevated CO_2 Conditions: A Meta-analysis of Reports on 79 Crop and Wild Species." *New Phytologist* 156 (2002):9–26.

Jaccard, Mark. *Sustainable Fossil Fuels: The Unusual Suspect in the Quest for Clean and Enduring Energy.* Cambridge: Cambridge University Press, 2006.

Jackson, Derrick Z. "Sweltering in a Winter Wonderland." *Boston Globe,* December 5, 2001, A-23.

Jackson, Robert B., Jay L. Banner, Esteban G. Jobbagy, William T. Pockman, and Diana H. Wall. "Ecosystem Carbon Loss with Woody Plant Invasion of Grasslands." *Nature* 418 (August 8, 2002):623–626.

Jacobs, Andrew. "Report Sees New Pollution Threat." *New York Times,* November 14, 2008. http://www.nytimes.com/2008/11/14/world/14cloud.html.

Jacobs, S. S., C. F. Giulivi, and P. A. Mele. "Freshening of the Ross Sea during the Late 20th Century." *Science* 297 (July 19, 2002):386–389.

Jacobson, Mark. "Strong Radiative Heating Due to the Mixing State of Black Carbon in Atmospheric Aerosols." *Nature* 409 (February 8, 2001):695–697.

Jager, J., and H. L. Ferguson. *Climate Change: Science, Impacts, and Policy. Proceedings of the Second World Climate Conference.* Cambridge: Cambridge University Press, 1991.

Jager, Jill. "Developing Policies for Responding to Climate Change." In *The Challenge of Global Warming,* ed. Dean Edwin Abrahamson, 96–109. Washington, DC: Island Press, 1989.

Jans, Nick. "Living with Oil: The Real Price." In *The Last Polar Bear: Facing the Truth of a Warming World,* ed. Steven Kazlowski, 147–161. Seattle: Braided River, 2008.

Jardine, Kevin. "The Carbon Bomb: Climate Change and the Fate of the Northern Boreal Forests." Ontario, Canada: Greenpeace International, 1994. http://dieoff.org/page129.htm.

Jayaraman, K. S. "Monsoon Rains Start to Ease India's Drought." *Nature* 423 (June 12, 2003):673.

Jeffries, M. O. "Ellesmere Island Ice Shelves and Ice Islands." *Satellite Image Atlas of Glaciers of the World: Glaciers of North America—Glaciers of Canada,* J147-J164. Washington, DC: U.S. Geological Survey, 2002.

"Jellyfish Attack Destroys Salmon." British Broadcasting Corporation, November 21, 2007. http://news.bbc.co.uk/2/hi/uk_news/northern_ireland/7106631.stm.

"Jellyfish Swarms Invade Ecosystems Out of Balance." Environment News Service, December 16, 2008. http://www.ens-newswire.com/ens/dec2008/2008-12-16-01.asp.

Jenkyns, Hugh C., Astrid Forster, Stefan Schouten, and Jaap S. Sinninghe Damste. "High Temperatures in the Late Cretaceous Arctic Ocean." *Nature* 432 (December 16, 2004):888–892.

Jenney, L. L. "Reducing Greenhouse-gas Emissions from the Transportation Sector." In *Limiting Greenhouse Effects: Controlling Carbon-dioxide Emissions,* ed. G. I. Pearman, 283–302. *Report of the Dahlem Workshop on Limiting the Greenhouse Effect,* Berlin, December 9–14, 1990. New York: John Wiley & Sons, 1991.

Jensen, Elizabeth. "Activists Take 'The Day After' for a Spin." *Los Angeles Times,* May 26, 2004 (LEXIS).

Johannessen, Ola M., Kirill Khvorostovsky, Martin W. Miles, and Leonid P. Bobylev. "Recent Ice-Sheet Growth in the Interior of Greenland." *Science* 310 (November 11, 2005):1013–1016.

Johansen, Bruce E. "Pristine No More: The Arctic, Where Mother's Milk Is Toxic." *The Progressive,* December 2000, 27–29.

Johansen, Bruce E. "Arctic Heat Wave." *The Progressive,* October 2001, 18–20.

Johansen, Bruce E. *The Global Warming Desk Reference.* Westport, CT: Greenwood, 2006a.

Johansen, Bruce E. "The Paul Revere of Global Warming." *The Progressive,* August 2006b, 26–28.

Johansen, Bruce E. "Media Literacy and 'Weather Wars:' Hard Science and Hardball Politics at NASA." *Studies in Media and Information Literacy Education* (SIMILE) 6, no. 3 (August 2006c). http://www.utpjournals.com/simile/issue23/Johansen6.html.

Johansen, Bruce E. *Global Warming in the 21st Century.* 3 vols. Westport, CT: Praeger, 2006.

Johansen, Bruce E. "Scandinavia Gets Serious on Global Warming." *The Progressive,* July 2007, 22–24.

Johansen, Bruce E. "Hansen: Cut CO_2 10 Percent." *Nebraska Report,* May/June 2008a, 5.

Johansen, Bruce E. "Hansen Battles Climatic Self-deception World-wide." *Nebraska Report,* October 2008b, 7.

John Paul II. *On the Hundredth Anniversary of Rerum Novarum (Centesimus Annus).* No. 32. Washington, DC: United States Catholic Conference, 1991.

Johnson, Andrew. "Climate to Bring New Gardening Revolution: Hot Summers and Wet Winters Could Kill Our Best-loved Plants." *London Independent,* May 12, 2002, 5.

Johnson, Chalmers. *Nemesis: The Last Days of the American Republic.* New York: Metropolitan Books/Henry Holt, 2006.

Johnson, Keith. "Renewable Power Might Yield Windfall." *Wall Street Journal,* March 22, 2007, A-8.

Johnson, Tim. "World out of Balance: In a Prescient Time Native Prophecy Meets Scientific Prediction." *Native Americas* 16, no. 3/4 (Fall/Winter 1999):8–25. http://nativeamericas. aip.cornell.edu.

Johnston, David Cay. "Some Need Hours to Start Another Day at the Office." *New York Times* in *Omaha World-Herald,* February 6, 2000, 1-G.

Joling, Dan. "Study: Polar Bears May Turn to Cannibalism." Associated Press, June 14, 2006 (LEXIS).

Joling, Dan. "Walruses Abandon Ice for Alaska Shore." Associated Press in *Washington Post,* October 4, 2007a. http://www.washingtonpost.com/wp-dyn/content/article/2007/10/04/ AR2007100402299_pf.html.

Joling, Dan. "Thousands of Pacific Walruses Die: Global Warming Blamed." Associated Press, December 14, 2007b (LEXIS).

Jones, C. D., P. M. Cox, R. L. H. Essery, D. L. Roberts, and M. J. Woodage. "Strong Carbon Cycle Feedbacks in a Climate Model with Interactive CO_2 and Sulphate Aerosols." *Geophysical Research Letters* 30, no. 9 (2003), 16,867, doi: 0.1029/2003GL0.

Jones, Charisse. "Transit Systems Travel 'Green' Track." *USA Today,* May 8, 2008, 3-A.

Jones, George, and Charles Clover. "Blair Warns of Climate Catastrophe: 'Shocked' Prime Minister Puts Pressure on U.S. and Russia over Emissions." *London Daily Telegraph,* September 15, 2004, 2.

Jordan, Steve. "Wary Insurers Suspect Climate Change." *Omaha World-Herald,* September 1, 2005, 7-A.

Jordan, Steve. "Becoming Greener: Environmental Concerns, Lower Costs Push Drive for Energy-efficient Buildings." *Omaha World-Herald,* July 22, 2007a, D-1, D-2.

Jordan, Steve. "Coal-emission Cleanup a Challenge for Utilities." *Omaha World-Herald,* October 8, 2007b, D-1, D-2.

Jordan, Steve. "Tenaska to Build Clean Power Plant." *Omaha World-Herald,* February 20, 2008, D-1, D-2.

Joughin, Ian. "Greenland Rumbles Louder as Glaciers Accelerate." *Science* 311 (March 24, 2006):1719–1720.

Joughin, Ian, W. Abdalati, and M. Fahnestock. 2004. "Large Fluctuations in Speed on Greenland's Jakobshavn Isbrae Glacier." *Nature* 432 (December 2, 2004):608–610.

Joughin, Ian, Sarah B. Das, Matt A. King, Ben E. Smith, Ian M. Howat, and Twila Moon. "Seasonal Speedup along the Western Flank of the Greenland Ice Sheet." *Science* 320 (May 9, 2008):781–783.

Joughin, Ian, and Slawek Tulaczyk. "Positive Mass Balance of the Ross Ice Streams, West Antarctica." *Science* 295 (January 18, 2002):476–480.

"Jumbo Squid Has a Message for Us: Changing Global Patterns Are Going to Bring Different Species into Our Waters." *Victoria Times-Colonist* (British Columbia), October 8, 2004, A-10.

Jurgensen, John. "The Weather End Game: The Climate-change Disaster at the Heart of 'Day After Tomorrow' May Be Overplayed, but the Global-warming Threat Is Real." *Hartford Courant,* May 27, 2004, D-1.

Kafarowski, Joanna. "Cloutier, Sheila-Watt." In *Encyclopedia of the Arctic.* London: Routledge, 2007. Sample Proofs. http://www.routledge-ny.com/ref/arctic/watt.html.

Kahn, Joseph, and Jim Yardley. "As China Roars, Pollution Reaches Deadly Extremes." *New York Times*, August 26, 2006. http://www.nytimes.com/2007/08/26/world/asia/26china.html.

Kaiser, Jocelyn. "Possibly Vast Greenhouse Gas Sponge Ignites Controversy." *Science* 282 (October 16, 1998):386–387.

Kaiser, Jocelyn. "Glaciology: Warmer Ocean Could Threaten Antarctic Ice Shelves." *Science* 302 (October 31, 2003):759.

Kaiser, Jocelyn. "Reproductive Failure Threatens Bird Colonies on North Sea Coast." *Science* 305 (August 20, 2004):1090.

Kakutani, Michiko. "Beware! Tree-Huggers Plot Evil to Save World." *New York Times*, December 13, 2004, E-1.

Kalkstein, Laurence S. "Direct Impacts in Cities." *Lancet* 342 (December 4, 1993):1397–1400.

Kalnay, Eugenia, and Ming Cai. "Impact of Urbanization and Land-use Change on Climate." *Nature* 423 (May 29, 2003):528–531.

Kambayashi, Takehiko. "World Weather Prompts New Look at Kyoto." *Washington Times*, September 5, 2003, A-17.

Kana, T. W., J. Michel, M. O. Hayes, and J. R. Jenson. 1984. "The Physical Impact of Sea Level Rise in the Area of Charleston, South Carolina." In *Greenhouse Effect and Sea Level Rise: A Challenge for This Generation,* ed. M. C. Barth and J. G. Titus, 105–150. New York: Van Nostrand Reinhold Company, 1984.

Kane, R. L., and D. W. South. "The Likely Roles of Fossil Fuels in the Next 15, 50, and 100 Years, with or without Active Controls on Greenhouse-gas Emissions." In *Limiting Greenhouse Effects: Controlling Carbon-dioxide Emissions,* ed. G. I. Pearman, 189–27. *Report of the Dahlem Workshop on Limiting the Greenhouse Effect. Berlin, December 9–14, 1990.* New York: John Wiley & Sons, 1991.

"Kansans Rallied to Resist Coal-Burning Power Plants." Environment News Service, March 12, 2008. http://www.ens-newswire.com/ens/mar2008/2008-03-12-092.asp.

"Kansas Gets First U.S. Cellulosic Ethanol Plant" Environment News Service, August 28, 2007. http://www.ens-newswire.com/ens/aug2007/2007-08-28-097.asp.

Kanter, James. "Across the Atlantic, Slowing Breezes." *New York Times*, March 7, 2007. http://www.nytimes.com/2007/03/07/business/businessspecial2/07europe.html.

Kanter, James, and Andrew C. Revkin. "World Scientists Near Consensus on Warming." *New York Times*, January 30, 2007a. http://www.nytimes.com/2007/01/30/world/30climate.htm.

Kanter, James, and Andrew C. Revkin. "Scientists Detail Climate Changes, Poles to Tropics." *New York Times*, April 7, 2007b. http://www.nytimes.com/2007/04/07/science/earth/07climate.html.

Kaplow, Larry. "Solar Water Heaters: Israel Sets Standard for Energy." *Atlanta Journal-Constitution*, August 5, 2001, 1-P.

Karen, Mattias. "Sweden Outdoes Bush with Goal to go Oil-free by 2020." Associated Press in *Seattle Times*, February 8, 2006. http://seattletimes.nwsource.com/html/nationworld/2002791274_sweden08.html.

Karl, Thomas R., P. D. Jones, R. W. Knight, G. Kukla, N. Plummer, V. Razuvayev, K. P. Gallo, J. Lindsay, R. J. Charlson, and T. C. Peterson. "A New Perspective on Recent Global Warming: Asymmetric Trends of Daily Maximum and Minimum Temperature." *Bulletin of the American Meteorological Society* 74 (1993):1007–1023.

Karl, Thomas R., Richard W. Knight, and Bruce Baker. "The Record-breaking Global Temperatures of 1997 and 1998: Evidence for an Increase in the Rate of Global Warming." *Geophysical Research Letters* 27 (March 1, 2000):719–722.

Karl, Thomas R., Neville Nicholls, and Jonathan Gregory. "The Coming Climate: Meteorological Records and Computer Models Permit Insights Into Some of the Broad Weather Patterns of a Warmer World." *Scientific American* 276 (1997):79–83. http://www.scientificamerican.com/0597issue/0597karl.htm.

Karl, Thomas R., and Kevin E. Trenberth. "Modern Global Climate Change." *Science* 302 (December 5, 2003):1719–1723.

Karling, Horace M. *Global Climate Change Revisited.* Hauppauge, NY: Nova Science, 2007.

Karoly, David J. "Ozone and Climate Change." *Science* 302 (October 10, 2003):236–237.

Kaser, Georg, Douglas R. Hardy, Thomas M. A. Org, Raymond S. Bradley, and Tharsis M. Hyera. "Modern Glacier Retreat on Kilimanjaro as Evidence of Climate Change: Observations and Facts." *International Journal of Climatology* 24 (2004):329–339.

Katz, Miriam E., Dorothy K. Pak, Gerald R. Dickens, and Kenneth G. Miller. "The Source and Fate of Massive Carbon Input during the Latest Paleocene Thermal Maximum." *Science* 286 (November 19, 1999):1531–1533.

Kaufman, Marc. "Warming Arctic Is Taking a Toll: Peril to Walrus Young Seen as Result of Melting Ice Shelf." *Washington Post,* April 15, 2006, A-7. http://www.washingtonpost.com/wp-dyn/content/article/2006/04/14/AR2006041401368_pf.html.

Kaufman, Marc. "Climate Experts Worry as 2006 Is Hottest Year on Record in U.S." *Washington Post,* January 10, 2007a. http://www.washingtonpost.com/wp-dyn/content/article/2007/01/09/AR2007010901949_pf.html.

Kaufman, Marc. "Antarctic Glaciers' Sloughing of Ice Has Scientists at a Loss." *Washington Post,* March 16, 2007b, A-2. http://www.washingtonpost.com/wp-dyn/content/article/2007/03/15/AR2007031501063_pf.html.

Kaufman, Marc. "Southwest May Get Even Hotter, Drier: Report on Warming Warns of Droughts." *Washington Post,* April 6, 2007c, A-3. http://www.washingtonpost.com/wp-dyn/content/article/2007/04/05/AR2007040501180_pf.html.

Kaufman, Marc. "Report Warns of a Much Warmer Northeast Effects Could Be Disastrous, Says Two-year Study." *Washington Post,* July 12, 2007d, A-4. http://www.washingtonpost.com/wp-dyn/content/article/2007/07/11/AR2007071100942_pf.html.

Kaufman, Marc. "Escalating Ice Loss Found in Antarctica: Sheets Melting in an Area Once Thought to Be Unaffected by Global Warming." *Washington Post,* January 14, 2008a. http://www.washingtonpost.com/wp-dyn/content/article/2008/01/13/AR2008011302753_pf.html.

Kaufman, Marc. "Decline in Snowpack Is Blamed on Warming: Water Supplies in West Affected." *Washington Post,* February 1, 2008b, A-1. http://www.washingtonpost.com/wp-dyn/content/article/2008/01/31/AR2008013101868_pf.html.

Kaufman, Marc. "Perennial Arctic Ice Cover Diminishing, Officials Say." *Washington Post,* March 19, 2008c, A-3. http://www.washingtonpost.com/wp-dyn/content/article/2008/03/18/AR2008031802903_pf.html.

Kazlowski, Steven, with Theodore Roosevelt IV, Charles Wohlforth, Daniel Glick, Richard Nelson, Nick Jans, and Frances Beinecke. *The Last Polar Bear: Facing the Truth of a Warming World. A Photographic Journey.* Seattle: Braided River Books, 2008.

Keates, Nancy. "Building a Better Bike Lane." *Wall Street Journal,* May 4, 2007, W-1, W-10.

Keeling, Charles D., and Timothy P. Whorf. "The 1,800-year Oceanic Tidal Cycle: A Possible Cause of Rapid Climate Change." *Proceedings of the National Academy of Sciences of the United States of America* 97, no. 8 (April 11, 2000):3814–3819.

Keen, Judy. "Neighbors at Odds Over Noise from Wind Turbines." *USA Today,* November 4, 2008, A-3.

Keenlyside, N. S., M. Latif, J. Jungclaus, L. Kornblueh, and E. Roeckner. "Advancing Decadal-scale Climate Prediction in the North Atlantic Sector." *Nature* 453 (May 1, 2008):84–88.

Kelleher, Lynne. "Look Who's Here: Tropical Fish Warming to Waters around Ireland." *London Sunday Mirror,* October 20, 2002, 15.

Kelley, Kate. "City Approves 'Carbon Tax' in Effort to Reduce Gas Emissions." *New York Times,* November 18, 2006. http://www.nytimes.com/2006/11/18/us/18carbon.html.

Kellogg, William W. "Theory of Climate Transition from Academic Challenge to Global Imperative." In *Greenhouse Glasnost: The Crisis of Global Warming,* ed. Terrell J. Minger, 99. New York: Ecco Press, 1990.

Kelly, Anne E., and Michael L. Goulden. "Rapid Shifts in Plant Distribution with Recent Climate Change." *Proceedings of the National Academy of Sciences* 105, no. 33 (August 19, 2008):11823–11826.

Kelly, Mick. "Halting Global Warming." In *Global Warming: The Greenpeace Report,* ed. Jeremy Leggett, 83–112. New York: Oxford University Press, 1990.

"Kelp Points to Worrying Sea Change." *Canberra Times,* August 30, 2004, A-8.

Kennedy, Donald. "The Biofuels Conundrum." *Science* 316 (April 27, 2007):515.

Kennett, James P., Kevin G. Cannariato, Ingrid L. Hendy, and Richard J. Behl. "Carbon Isotopic Evidence for Methane Hydrate Instability during Quaternary Interstadials." *Science* 288 (April 7, 2000):128–133.

Kennett, James P., Kevin G. Cannariato, Ingrid L. Hendy, and Richard J. Behl. *Methane Hydrates in Quaternary Climate Change: The Clathrate Gun Hypothesis.* Washington, DC: American Geophysical Union, 2003.

Kerr, Richard A. "West Antarctica's Weak Underbelly Giving Way?" *Science* 281 (July 24, 1998a):499–500.

Kerr, Richard A. "Among Global Thermometers, Warming Still Wins Out." *Science* 281 (September 25, 1998b):1948–1949.

Kerr, Richard A. "Deep Chill Triggers Record Ozone Hole." *Science* 282 (October 16, 1998c):391.

Kerr, Richard A. "Big El Niños Ride the Back of Slower Climate Change." *Science* 283 (February 19, 1999a):1108–1109.

Kerr, Richard A. "In North American Climate, a More Local Control." *Science* 283 (February 19, 1999b):1109.

Kerr, Richard A. "Oceanography: Has a Great River in the Sea Slowed Down?" *Science* 286 (November 5, 1999c):1061–1062.

Kerr, Richard A. "A Smoking Gun for an Ancient Methane Discharge." *Science* 286 (November 19, 1999d):1465.

Kerr, Richard A. "Globe's 'Missing Warming' Found in the Ocean." *Science* 287 (March 24, 2000a):2126–2127.

Kerr, Richard A. "Global Warming: Draft Report Affirms Human Influence." *Science* 288 (April 28, 2000b):589–590.

Kerr, Richard A. "Dueling Models: Future U.S. Climate Uncertain." *Science* 288 (June 23, 2000c):2113.

Kerr, Richard A. "A Variable Sun Paces Millennial Climate." *Science* 294 (November 16, 2001):1431–1432.

Kerr, Richard A. "A Single Climate Mover for Antarctica." *Science* 296 (May 3, 2002a):825–826.

Kerr, Richard A. "A Warmer Arctic Means Change for All." *Science* 297 (August 30, 2002b):1490–1492.

Kerr, Richard A. "European Climate: Mild Winters Mostly Hot Air, Not Gulf Stream." *Science* 297 (September 27, 2002c):2202.

Kerr, Richard A. "A Perfect Ocean for Four Years of Globe-Girdling Drought." *Science* 299 (January 31, 2003a):636.

Kerr, Richard A. "Warming Indian Ocean Wringing Moisture from the Sahel." *Science* 302 (October 10, 2003b):210–211.

Kerr, Richard A. "Climate Change: Sea Change in the Atlantic." *Science* 303 (January 2, 2004):35.

Kerr, Richard A. "Looking Back for the World's Climatic Future." *Science* 312 (June 9, 2005a):1456–1457.

Kerr, Richard A. "El Niño or La Niña? The Past Hints at the Future." *Science* 309 (July 29, 2005b):687.

Kerr, Richard A. "Is Katrina a Harbinger of Still More Powerful Hurricanes?" *Science* 309 (September 16, 2005c):1807.

Kerr, Richard A. "Latest Forecast: Stand By for a Warmer, but Not Scorching, World." *Science* 312 (April 21, 2006a):351.

Kerr, Richard A. "Yes, It's Been Getting Warmer in Here since the CO_2 Began to Rise." *Science* 312 (June 30, 2006b):1854.

Kerr, Richard A. "Atlantic Mud Shows How Melting Ice Triggered an Ancient Chill." *Science* 312 (June 30, 2006c):1860.

Kerr, Richard A. "Pollute the Planet for Climate's Sake? *Science* 314 (October 20, 2006d):401–403.

Kerr, Richard A. "False Alarm: Atlantic Conveyor Belt Hasn't Slowed Down After All." *Science* 314 (November 17, 2006d):1064.

Kerr, Richard A. "Could Mother Nature Give the Warming Arctic a Reprieve? *Science* 315 (January 5, 2007a):36.

Kerr, Richard A. "Is a Thinning Haze Unveiling the Real Global Warming?" *Science* 315 (March 16, 2007b):1480.

Kerr, Richard A. "Pushing the Scary Side of Global Warming." *Science* 316 (June 8, 2007c):1412–1415.

Kerr, Richard A. "Is Battered Arctic Sea Ice Down for the Count?" *Science* 318 (October 5, 2007d):33–34.

Kerr, Richard A. "How Urgent Is Climate Change?" *Science* 318 (November 23, 2007e):1230–1231.

Kerr, Richard A. "More Climate Wackiness in the Cretaceous Supergreenhouse?" *Science* 319 (January 11, 2008a):145.

Kerr, Richard A. "Climate Tipping Points Come In from the Cold." *Science* 319 (January 11, 2008b):153.

Kerr, Richard A. "Greenland Ice Slipping Away but Not All That Quickly." *Science* 320 (April 18, 2008c):301.

Kerr, Richard A. "Climate Change Hot Spots Mapped Across the United States." Science 321 (August 15, 2008d):909.

Kerr, Richard A. "Winds, Not Just Global Warming, Eating Away at the Ice Sheets." *Science* 322 (October 3, 2008e):33.

Kever, Jeannie. "They're Makin' Tracks: The Armadillo, That Official Small Mammal of Texas, Is Waddling Its Way North." *Houston Chronicle*, September 4, 2006, 1.

Keys, David. "Methane Threatens to Repeat Ice-age Meltdown: 55 Million Years Ago, a Massive Blast of Gas Drove up Earth's Temperature 7° C, and Another Explosion Is in the Cards, Say the Experts." *London Independent*, June 16, 2001, n.p.(LEXIS).

Kher, Unmesh. "Pay for Your Carbon Sins." *TIME*, April 9, 2007. http://www.time.com/time/printout/0,8816,1603737,00.html.

Kiesecker, Joseph M., Andrew R. Blaustein, and Lisa K. Belden. "Complex Causes of Amphibian Population Declines." *Nature* 410 (April 5, 2001):681–684.

King, David A. "Climate Change Science: Adapt, Mitigate, or Ignore?" *Science* 303 (January 9, 2004):176–177.

Kintisch, Eli. "Floods? Droughts? More Storms? Predictions Vary, but Scientists Agree Our Climate Will Change: Midwesterners Could End Up Wet, or Dry." *St. Louis Post-Dispatch*, May 30, 2004, B-1.

Kintisch, Eli. "Hot Times for the Cretaceous Oceans." *Science* 311 (February 24, 2006a):1095.

Kintisch, Eli. "As the Seas Warm." *Science* 313 (August 11, 2006b):776–779.

Kintisch, Eli. "Climate Change Goes to Court." *ScienceNOW Daily News*, August 31, 2006c. http://sciencenow.sciencemag.org/cgi/content/full/2006/831/1.

Kintisch, Eli. "Report Backs More Projects to Sequester CO_2 from Coal." *Science* 315 (March 16, 2007a):1481.

Kintisch, Eli. "Making Dirty Coal Plants Cleaner." *Science* 317 (July 13, 2007b):184–186.

Kintisch, Eli. "Light-Splitting Trick Squeezes More Electricity Out of Sun's Rays." *Science* 317 (August 3, 2007c):583–584.

Kintisch, Eli. "Rules for Ocean Fertilization Could Repel Companies." *Science* 322 (November 7, 2008):835.

Kintisch, Eli, and Erik Stokstad. "Ocean CO_2 Studies Look Beyond Coral." *Science* 319 (February 22, 2008):1029.

Kirby, Alex. "Climate Change: It's the Sun and Us." British Broadcasting Corp. News, November 26, 1998. http://news.bbc.co.uk/hi/english/sci/tech/newsid_222000/222437.stm.

Kirby, Alex. "Costing the Earth." October 26, 2000. British Broadcasting Corporation, Radio Four. http://news.bbc.co.uk/hi/english/sci/tech/newsid_990000/990391.stm.

Kirchner, Stephanie. "Be Aggressive about Passive [Solar Power]." *TIME*, April 9, 2007. http://www.time.com/time/printout/0,8816,1603747,00.html.

Kirk-Davidoff, Daniel, Daniel P. Schrag, and James G. Anderson. "On the Feedback of Stratospheric Clouds on Polar Climate." *Geophysical Research Letters* 29, no. 11 (2002):14,659–14,663.

Kitney, Geoff. "Global Warming Back on the Agenda." *Australian Financial Review*, May 29, 2004, 30.

Kizzia, Tom. "Seal Hunters Await Late Ice." *Anchorage Daily News*, November 28, 1998. http://www.adn.com/stories/T98112872.html.

Kleiner, Kurt. "The Shipping Forecast." *Nature* 449 (September 20, 2007):272–273.

Kleiven, Helga (Kikki) Flesche, Catherine Kissel, Carlo Laj, Ulysses S. Ninnemann, Thomas O. Richter, and Elsa Cortijo. "Reduced North Atlantic Deep Water Coeval with the Glacial Lake Agassiz Freshwater Outburst." *Science* 3129 (January 4, 2008):60-64.

Kling, G. W., G. W. Kipphut, and M. C. Miller. "The Flux of Carbon Dioxide and Methane from Lakes and Rivers in Arctic Alaska." *Hydrobiologia* 240 (1992):23–36.

Klug, Edward C. "Global Warming: Melting Down the Facts about This Overheated Myth." CFACT Briefing Paper No. 105, November 1997. http://www.cfact.org/IssueArchive/greenhouse.bp.n97.txt.

Knoblauch, Jessica A. "Have It Your (the Sustainable) Way." *Environmental Journalism* (Spring 2007):28–30, 46.

Knorr, Gregory, and Gerrit Lohmann. "Southern Ocean Origin for the Resumption of Atlantic Thermohaline Circulation during Deglaciation." *Nature* 424 (July 31, 2003):532–536.

Knorr, W., I. C. Prentice, J. I. House, and E. A. Holland. "Long-term Sensitivity of Soil Carbon Turnover to Warming." *Nature* 433 (January 20, 2005):298–301.

Knutson, Thomas R., and Robert E. Tuleya. "Impact of CO_2-Induced Warming on Simulated Hurricane Intensity and Precipitation: Sensitivity to the Choice of Climate Model and Convective Parameterization." *Journal of Climate* 17, no. 18 (September 15, 2004):3477–3495.

Knutson, Thomas R., Robert E. Tuleya, and Yoshio Kurihara. "Simulated Increase of Hurricane Intensities in a CO_2-warmed Climate." *Science* 297 (February 13, 1998):1018–1020.

Knutson, Thomas R., Robert E. Tuleya, Weixing Shen, and Isaac Ginis. "Impact of CO_2-induced Warming on Hurricane Intensities as Simulated in a Hurricane Model with Ocean Coupling." *Journal of Climate* 14 (2001):2458–2469.

Knutson, Thomas R., Joseph J. Sirutis, Stephen T. Garner, Gabriel A. Vecchi, and Isaac M. Held. "Simulated Reduction in Atlantic Hurricane Frequency under Twenty-first-century Warming Conditions." *Nature Geoscience* (May 18, 2008), doi: 10.1038/ngeo202. http://www.nature.com/ngeo/journal/vaop/ncurrent/abs/ngeo202.html.

Knutti, R., J. Fluckiger, T. F. Stocker, and A. Timmermann. "Strong Hemispheric Coupling of Glacial Climate through Freshwater Discharge and Ocean Circulation." *Nature* 430 (August 19, 2004):851–856.

Kolbert, Elizabeth. *Field Notes from a Catastrophe: Man, Nature, and Climate Change.* New York: Bloomsbury, 2006a.

Kolbert, Elizabeth. "The Darkening Sea: What Carbon Emissions Are Doing to the Oceans." *The New Yorker*, November 20, 2006b, 66–75.

Kolbert, Elizabeth. "Don't Drive, He Said." (Talk of the Town) *The New Yorker*, May 7, 2007a, 23–24.

Kolbert, Elizabeth. "Running on Fumes: Does the Car of the Future have a Future?" *The New Yorker*, November 5, 2007b, 87–90.

Kolbert, Elizabeth. "Unconventional Crude: Canada's Synthetic-fuels Boom." *The New Yorker*, November 12, 2007c, 46–51.

Kolbert, Elizabeth. "Testing the Climate." (Talk of the Town) *The New Yorker*, December 24 and 31, 2007d, 43–44.

Kolbert, Elizabeth. "The Island in the Wind: A Danish Community's Victory over Carbon Emissions." *The New Yorker*, July 7, 2008. http://www.newyorker.com/reporting/2008/07/07/080707fa_fact_kolbert.

Kondratyev, Kirill, Vladimir F. Krapivin, and Costas A. Varotsos. *Global Carbon Cycle and Climate Change.* Berlin, Germany: Springer/Praxis, 2004.

Konviser, Bruce I. "Glacier Lake Puts Global Warming on the Map." *Boston Globe*, July 16, 2002, C-1.

Korner, C. "Slow In, Rapid Out—Carbon Flux Studies and Kyoto Targets. *Science* 300 (May 23, 2003):1242–1243.

Koslov, Mikhail, and Natalia G. Berlina. "Decline in Length of the Summer Season on the Kola Peninsula, Russia." *Climatic Change* 54 (September 2002):387–398.

Koster, E. A., and M. E. Nieuwenhuyzen. "Permafrost Response to Climatic Change." In *Greenhouse-Impact on Cold-Climate Ecosystems and Landscapes,* ed. M. Boer and E. Koster, 23–35. *Selected Papers of a European Conference on Landscape Ecological Impact of Climatic Change,* Lunteren, the Netherlands, December 3–7, 1989. CARTENA Supplement 22. Cremlingen, Germany: Catena Verlag, 1992.

Krabill, W., W. Abdalati, E. Frederick, S. Manizade, C. Martin, J. Sonntag, R. Swift, R. Thomas, W. Wright, and J. Yungel. "Greenland Ice Sheet: High-elevation Balance and Peripheral Thinning." *Science* 289 (July 21, 2000):428–430.

Krajick, Kevin. "Arctic Life, on Thin Ice." *Science* 291 (January 19, 2001a):424–425.

Krajick, Kevin. "Tracing Icebergs for Clues to Climate Change." *Science* 292 (June 22, 2001b):2244–2245.

Krajick, Kevin. "Thriving Arctic Bottom Dwellers Could Get Strangled by Warming." *Science* 315 (March 16, 2007):1527.

Krauss, Clifford. "Bear Hunting Caught in Global Warming Debate." *New York Times*, May 27, 2006. http://www.nytimes.com/2006/05/27/world/americas/27bears.html.

Krauss, Clifford. "Venture Capitalists Want to Put Some Algae in Your Tank." *New York Times*, March 7, 2007a. http://www.nytimes.com/2007/03/07/business/07algae.html.

Krauss, Clifford. "As Ethanol Takes Its First Steps, Congress Proposes a Giant Leap." *New York Times*, December 18, 2007b. http://www.nytimes.com/2007/12/18/washington/18ethanol.html.

Krauss, Clifford. "Move Over, Oil, There's Money in Texas Wind." *New York Times*, February 23, 2008. http://www.nytimes.com/2008/02/23/business/23wind.html.

Kraus, Clifford, Steven Lee Myers, Andrew C. Revkin, and Simon Romero. "As Polar Ice Turns to Water, Dreams of Treasure Abound." *New York Times*, October 10, 2005. http://www.nytimes.com/2005/10/10/science/10arctic.html.

Kristal-Schroder, Carrie. "British Adventurer's Polar Trek Foiled by Balmy Arctic." Ottawa *Citizen*, May 17, 2004, A-1.

Kristof, Nicholas. "For Pacific Islanders, Global Warming Is No Idle Threat." *New York Times*, March 2, 1997. http://sierraactivist.org/library/990629/islanders.html.

Kristof, Nicholas. "Baked Alaska on the Menu?" *New York Times* in *Alameda Times-Star* (California), September 14, 2003, n.p.(LEXIS).

Kristof, Nicholas. "The Storm Next Time." *New York Times*, September 1, 2005. http://www.nytimes.com/2005/09/11/opinion/11kristof.html.

Kristof, Nicholas. "Extended Forecast: Bloodshed." *New York Times*, April 13, 2008. http://www.nytimes.com/2008/04/13/opinion/13kristof.html.

Krugman, Paul. "A Test of Our Character." *New York Times*, May 26, 2006, n.p.

Krugman, Paul. "The Sum of All Ears." *New York Times*, January 29, 2007, A-23.

Krupp, Fred, and Miriam Horn. *Earth, the Sequel: The Race to Reinvent Energy and Stop Global Warming*. New York: W.W. Norton, 2008.

Kump, Lee R. "What Drives Climate?" *Nature* 408 (December 7, 2000):651–652.

Kump, Lee R. "Chill Taken out of the Tropics." *Nature* 413 (October 4, 2001):470–471.

Kunzig, Robert. "Drying of the West." *National Geographic*, February 2008, 90–113.

Kunzig, Robert, and Wallace Broecker. *Fixing Climate*. New York: Hill and Wang, 2008.

Kurz, W. A., M. J. Apps, B. J. Stocks, and J. A. Volney. "Global Climate Change: Disturbance Regimes and Biospheric Feedbacks of Temperate and Boreal Forests." In *Biotic Feedbacks in the Global Climate System: Will the Warming Feed the Warming?* ed. George M. Woodwell and Fred T. MacKenzie, 119–133. New York: Oxford University Press, 1995.

Kurz, W. A., C. C. Dymond, G. Stinson, G. J. Rampley, E. T. Neilson, A. L. Carroll, T. Ebata, and L. Safranyik. "Mountain Pine Beetle and Forest Carbon Feedback to Climate Change." *Nature* 452 (April 24, 2008):987–991.

Kyper, T., and R. Sorensen. "Potential Impacts of Selected Sea-level Rise Scenarios on the Beach and Coastal Works at Sea Bright, New Jersey." In *Coastal Zone '85*, ed. O. T. Magoon, H. Converse, D. Miner, D. Clark, and L. T. Tobin. New York: American Society of Civil Engineers, 1985.

Labatt, Sonia, and Rodney R. White. *Carbon Finance: The Financial Implications of Climate Change*. Hoboken, NJ: Wiley, 2007.

Lachenbruch, A. H., and B. V. Marshall. "Changing Climate: Geothermal Evidence from Permafrost in the Alaskan Arctic." *Science* 234 (1986):689–696.

Lackner, Kalus S. "A Guide to CO_2 Sequestration." *Science* 300 (June 13, 2003):1677–1678.

LaDeau, Shannon L., and James S. Clark. "Rising CO_2 Levels and the Fecundity of Forest Trees." *Science* 292 (April 6, 2001):95–98.

Lai, R. "Soil Carbon Sequestration Impacts on Global Climate Change and Food Security." *Science* 304 (June 11, 2004):1623–1627.

Landauer, Robert. "Big Changes in Our China Suburb." *Sunday Oregonian*, October 20, 2002, F-4.

Landler, Mark. "Global Warming Poses Threat to Ski Resorts in the Alps." *New York Times*, December 16, 2006. http://www.nytimes.com/2006/12/16/world/europe/16austria.html.

Landler, Mark. "Call for Speed Limit Has German Blood at 178 m.p.h. Boil." *New York Times*, March 16, 2007. http://www.nytimes.com/2007/03/16/world/europe/16autobahn.html.

Landler, Mark. "Solar Valley Rises in an Overcast Land." *New York Times*, May 16, 2008. http://www.ens-newswire.com/ens/may2008/2008-05-14-01.asp.

Landsea, C. W. "A Climatology of Intense (or Major) Atlantic Hurricanes." *Monthly Weather Review* 121 (1993):1703–1713.

Landsea, C. W. "NOAA: Report on Intensity of Tropical Cyclones." Miami, Florida, August 12, 1999. http://www.aoml.noaa.gov/hrd/tcfaq/tcfaqG.html#G3.

Landsea, C. W., Bruce A. Harper, Karl Hoarau, and John A. Knaff. "Can We Detect Trends in Extreme Tropical Cyclones?" *Science* 313 (July 26, 2006):452–454.

Landsea, C. W., N. Nicholls, W. M. Gray, and L. A. Avila. "Downward Trends Atlantic Hurricanes during the Past Five Decades." *Geophysical Research Letters* 23 (1996):1697–1700.

Lane, Anthony. "Cold Comfort: 'The Day After Tomorrow.'" *The New Yorker*, June 7, 2004, 102–103.

Lane, Nick. "Climate Change: What's in the Rising Tide?" *Nature* 449 (October 18, 2007):778–780.

"Late Snow Smoothes the Way for Iditarod Sled-dog Race." Reuters in *Ottawa Citizen*, March 2, 2001, B-8.

Latif, M., C. Boning, J. Willebrand, A. Biastoch, J. Dengg, N. Keenlyside, U. Schweckendiek, and G. Madec. "Is the Thermohaline Circulation Changing?" *Journal of Climate* 19, no. 18 (September 15, 2006):4631–4637.

Laukaitis, Algis J. "Talmage Man: No Kiddo … It's an Armadillo!" *Lincoln Journal Star*, August 18, 2005, A-1.

Laurance, Jeremy. "Climate Change to Kill Thousands, Ministers Warned." *London Independent*, February 9, 2001, 2.

Laurance, William F. "Switch to Corn Promotes Amazon Deforestation." Letter to the Editor, *Science* 318 (14 December 14, 2007):1721.

Laurance, William F., Alexandre A. Oliveira, Susan G. Laurance, Richard Condit, Henrique E. M. Nascimento, Ana G. Sanchez-Torin, Thomas E. Lovejoy, Ana Andrade, Sammya D'Angelo, Jose E. Ribeiro, and Christopher W. Dick. "Pervasive Alteration of Tree Communities in Undisturbed Amazonian Forests." *Nature* 428 (March 11, 2004):171–175.

Lavers, Chris. *Why Elephants Have Big Ears*. New York: St. Martin's Press, 2000.

Law, Kathy S., and Andreas Stohl. "Arctic Air Pollution: Origins and Impacts." *Science* 315 (March 16, 2007):1537–1540.

Lawrence, David M., and Andrew G. Slater. "A Projection of Severe Near-surface Permafrost Degradation during the 21st Century." *Geophysical Research Letters* 32 (December 17, 2005), L24401, doi: 10.1029/2005GL025080.

Lawrence, Mark G., and Paul J. Crutzen. "Influence of Nitrous Oxide Emissions from Ships on Tropospheric Photochemistry and Climate." *Nature* 402 (November 11, 1999):167–168.

Lawson, Nigel. *An Appeal to Reason: A Cool Look at Global Warming*. London: Gerald Duckworth, 2008.

Lawton, R. O., U. S. Nair, R. A. Pielke Sr., and R. M. Welch. "Climatic Impact of Tropical Lowland Deforestation on Nearby Montane Cloud Forests." *Science* 294 (October 19, 2001):584–587.

Lazaroff, Cat. "Melting Arctic Permafrost May Accelerate Global Warming." Environment News Service, February 7, 2001a. http://ens-news.com/ens/feb2001/2001L-02-07-06.html.

Lazaroff, Cat. "Aerosol Pollution Could Drain Earth's Water Cycle." Environment News Service, December 7, 2001b. http://ens-news.com/ens/dec2001/2001L-12-07-06.html.

Lazaroff, Cat. "Climate Change Threatens Global Biodiversity." Environment News Service, February 7, 2002a. http://ens-news.com/ens/feb2002/2002L-02-07-06.html.

Lazaroff, Cat. "Replacing Grass with Trees May Release Carbon." Environment News Service, August 8, 2002b. http://ens-news.com/ens/aug2002/2002-08-08-07.asp.

Lazaroff, Cat. "Land Use Rivals Greenhouse Gases in Changing Climate." Environment News Service, October 2, 2002c. http://ens-news.com/ens/oct2002/2002-10-02-06.asp.

Lazaroff, Cat. "Loggerhead Turtle Sex Ratio Raises Concerns." Environment News Service, December 18, 2002d. http://ens-news.com/ens/dec2002/2002-12-18-06.asp.

Lazo, Alejandro. "A Shorter Link between the Farm and Dinner Plate: Some Restaurants, Grocers Prefer Food Grown Locally." *Washington Post*, July 29, 2007, A-1. http://www.washingtonpost.com/wp-dyn/content/article/2007/07/28/AR2007072801255.html?wpisrc=newsletter.

Lea, David W. "The 100,000-year Cycle in Tropical SST [Sea-surface Temperature], Greenhouse Forcing, and Climate Sensitivity." *Journal of Climate* 17, no. 11 (June 1, 2004):2170–2179.

Leake, Jonathan. "Fiery Venus Used to Be Our Green Twin." *London Sunday Times*, December 15, 2002, 11.

Lean, Geoffrey. "Quarter of World's Corals Destroyed." *London Independent*, January 7, 2001a, 7.

Lean, Geoffrey. "Experts Prove How Warming Changes World." *London Independent*, February 18, 2001b, 12.

Lean, Geoffrey. "We Regret to Inform You That the Flight to Malaga Is Destroying the Planet: Air Travel Is Fast Becoming One of the Biggest Causes of Global Warming." *London Independent*, August 26, 2001c, 23.

Lean, Geoffrey. "Antarctic Becomes Too Hot for the Penguins: Decline of 'Dinner Jacket' Species Is a Warning to the World." *London Independent*, February 3, 2002a, 9.

Lean, Geoffrey. "U.K. Homes Face Huge New Threat from Floods." *London Independent*, September 15, 2002b, 1.

Lean. Geoffrey. "Hot Summer Sparks Global Food Crisis." *London Independent*, August 31, 2003, 4.

Lean, Geoffrey. "Worst U.S. Drought in 500 Years Fuels Raging California Wildfires." *London Independent*, July 25, 2004a, 20.

Lean, Geoffrey. "Global Warming Will Redraw Map of the World." *London Independent*, November 7, 2004b, 8.

Lean, Geoffrey. "The Big Thaw: Global Disaster Will Follow if the Ice Cap on Greenland Melts." *London Independent*, November 20, 2005 (in Common Dreams News Center). http://www.commondreams.org/headlines05/1120-03.htm.

Lean, J., J. Beer, and R. Bradley. "Reconstruction of Solar Irradiance since 1610: Implications for Climate Change." *Geophysical Research Letters* 22 (1995):3195–3198.

Leatherman, Stephen P. "Coastal Land Loss in the Chesapeake Bay Region: An Historical Analog Approach to Global Change Analysis." In *The Regions and Global Warming: Impacts and Response Strategies,* ed. Jurgan Schmandt and Judith Clarkson, 17–27. New York: Oxford University Press, 1992.

Lederer, Edith M. "U.N. Report Says Planet in Peril." Associated Press, August 13, 2002.

Leggett, Jeremy, ed. *Global Warming: The Greenpeace Report.* New York: Oxford University Press, 1990.

Leggett, Jeremy. *The Carbon War: Global Warming and the End of the Oil Era.* New York: Routledge, 2001.

Leggett, Karby. "In Rural China, General Motors Sees a Frugal but Huge Market: It Bets Tractor Substitute Will Look Pretty Good to Cold, Wet Farmers." *Wall Street Journal*, January 16, 2001, A-19.

Leidig, Michael, and Roya Nikkhah. "The Truth about Global Warming—It's the Sun That's to Blame: The Output of Solar Energy Is Higher Now Than for 1,000 years, Scientists Calculate." *London Sunday Telegraph*, July 18, 2004, 5.

Lenoir, J., J. C. Gègout, P. A. Marquet, P. de Ruffray, and H. Brisse. "A Significant Upward Shift in Plant Species Optimum Elevation during the 20th Century." *Science* 320 (June 27, 2008):1768–1771.

Leroux, Marcel, and Jacques Comby. *Global Warming—Myth or Reality? The Erring Ways of Climatology.* Berlin: Springer-Praxis Books, 2005.

Levitus, Sydney, John I. Antonov, Timothy P. Boyer, and Cathy Stephens. "Warming of the World Ocean." *Science* 287 (2000):2225–2229.

Levitus, Sydney, John I. Antonov, Julian Wang, Thomas L. Delworth, Keith W. Dixon, and Anthony J. Broccoli. "Anthropogenic Warming of Earth's Climate System." *Science* 292 (April 13, 2001):267–270.

Lieberman, D., M. Jonas, M., Z. Nahorski, S. Nilsson, eds. *Accounting for Climate Change: Uncertainty in Greenhouse Gas Inventories—Verification, Compliance, and Trading.* Dordrecht, Netherlands: Springer, 2008.

Liesman, Steve. "Texaco Appears to Moderate Stance on Global Warming: New Hires and Investments Move the Oil Giant in a Greener Direction." *Wall Street Journal*, May 15, 2000, B-4.

Liesman, Steve, and Jacob M. Schlesinger. "The Price of Oil Has Doubled This Year: So, Where's the Recession?" *Wall Street Journal*, December 15, 1999, A-1, A-10.

Lighthill, J., G. Holland, W. Gray, C. Landsea, G. Craig, J. Evans, Y. Kurihara, and C. Guard. "Global Climate Change and Tropical Cyclones." *Bulletin of the American Meteorological Society* 75 (1994):2147–2157.

"Limiting Methane, Soot Could Quickly Curb Global Warming." Environment News Service, January 16, 2002. http://ens-news.com/ens/jan2002/2002L-01-16-01.html.

Linden, Eugene. "The Big Meltdown: As the Temperature Rises in the Arctic, It Sends a Chill around the Planet." *Time* 156, no. 10 (September 4, 2000). http://www.time.com/time/magazine/articles/0,3266,53418,00.html.

Linden, Eugene. *The Winds of Change: Climate, Weather, and the Destruction of Civilizations.* New York: Simon & Schuster, 2006.

Lindroth, A., A. Grelle, and A. S. Moren. "Long-term Measurements of Boreal Forest Carbon Balance Reveal Large Temperature Sensitivity." *Global Change Biology* 4 (April 1998):443–450.

Lippert, John. "General Motors Chief Weighs Future of Fuel Cells." *Toronto Star*, September 27, 2002, E-3.

"Liquid CO_2 Dump in Norwegian Sea Called Illegal." Environment News Service, July 11, 2002. http://ens-news.com/ens/jul2002/2002-07-11-02.asp.

Liska, Adam J., Haishun S. Yang, Virgil R. Bremer, Terry J. Klopfenstein, Daniel T. Walters, Galen E. Erickson, and Kenneth G. Cassman. "Improvements in Life-Cycle Energy Efficiency and Greenhouse Gas Emissions of Corn-Ethanol." *Journal of Industrial Ecology* (January 2009), doi: 10.1111/j.1530-9290.2008.00105.x.

"Live Earth to Kick Off in Australia." Associated Press in *New York Times*, July 6, 2007. http://www.nytimes.com/aponline/world/AP-Music-Live-Earth.html.

Loeb, V. "Effects of Sea-ice Extent and Krill or Salp Dominance on the Antarctic Food Web." *Nature* 387 (1997):897–900.

"Loggerhead Turtle Nesting Down by Half since 1998." Environment News Service, November 8, 2007. http://www.ens-newswire.com/ens/nov2007/2007-11-08-094.asp.

Lohr, Steve. "The Energy Challenge: "The Cost of an Overheated Planet." *New York Times*, December 12, 2006. http://www.nytimes.com/2006/12/12/business/worldbusiness/12warm.html.

Lomborg, Bjorn. "Entertaining Discredited Ideas of a Climatic Catastrophe." *The Australian*, May 27, 2004, n.p.(LEXIS).

Lomborg, Bjorn. *Cool It: The Skeptical Environmentalist's Guide to Global Warming.* New York: Knopf, 2007.

Lomborg, Bjorn. "A Better Way Than Cap and Trade." *Washington Post*, June 26, 2008, A19. http://www.washingtonpost.com/wp-dyn/content/article/2008/06/25/AR2008062501946_pf.html.

Long, Stephen P., Elizabeth A. Ainsworth, Andrew D. B. Leakey, Josef Nösberger, and Donald R. Ort. "Food for Thought: Lower-Than-Expected Crop Yield Stimulation with Rising CO_2 Concentrations." *Science* 312 (June 30, 2006):1918–1921.

Lorey, David E. *Global Environmental Challenges of the Twenty-First Century: Resources, Consumption, and Sustainable Solutions.* Woodbridge, CT: Scholarly Resource, Inc., 2002.

"Lost Manatee Rescued in Mass. Dies on Trip to Fla." Associated Press, October 2, 2008 (LEXIS).

Lovejoy, Thomas E., and Lee Hannah, eds. *Climate Change and Biodiversity.* New Haven, CT: Yale University Press, 2006.

Lovelock, James. *The Revenge of Gaia: Why the Earth Is Fighting Back—And How We Can Still Save Humanity.* London: Allen Lane, 2006.

Lovett, Richard A. "Global Warming: Rain Might Be Leading Carbon Sink Factor." *Science* 296 (June 7, 2002):1787.

Lovins, A. B. and L. H. Lovins. "Least-cost Climatic Stabilization." In *Limiting Greenhouse Effects: Controlling Carbon-dioxide Emissions*, ed. G. I. Pearman, 351–442. *Report of the Dahlem Workshop on Limiting the Greenhouse Effect*, Berlin, September 9–14, 1990. New York: John Wiley & Sons, 1991.

Lovins, Amory. "More Profit with Less Carbon." *Scientific American*, September 2005, 74, 76–83.

Low, Pak Sum. *Climate Change and Africa.* Cambridge: Cambridge University Press, 2005.

Lowell, Thomas V. "As Climate Changes, So Do Glaciers." *Proceedings of the National Academy of Sciences of the United States of America* 97, no. 4 (February 15, 2000):1351–1354.

Lowy, Joan. "Effects of Climate Warming Are Here and Now." Scripps-Howard News Service, May 5, 2004 (LEXIS).

Loya, Wendy M., and Paul Grogan. "Carbon Conundrum on the Tundra." *Nature* 431 (September 23, 2004):406–407.

Lublin, Joann S. "Environmentalism Sprouts Up on Corporate Boards." *Wall Street Journal*, August 11, 2006, B-6.

Lucht, Wolfgang, I. Colin Prentice, Ranga B. Myneni, Stephen Sitch, Pierre Friedlingstein, Wolfgang Cramer, Philippe Bousquet, Wolfgang Buermann, and Benjamin Smith. "Climatic Control of the High-latitude Vegetation Greening Trend and Pinatubo Effect." *Science* 296 (May 31, 2002):1687–1689.

Luhnow, David, and Geraldo Samor. "As Brazil Fills Up on Ethanol, It Weans off Energy Imports." *Wall Street Journal*, January 9, 2006, A-1, A-8.

Luterbacher, Jurg, Daniel Dietrich, Elena Xoplaki, Martin Grosjean, and Heinz Wanner. "European Seasonal and Annual Temperature Variability, Trends, and Extremes since 1500." *Science* 303 (March 5, 2004):1499–1503.

Luthcke, S. B., H. J. Zwally, W. Abdalati, D. D. Rowlands, R. D. Ray, R. S. Nerem, F. G. Lemoine, J. J. McCarthy, and D. S. Chinn. "Recent Greenland Ice Mass Loss by Drainage System from Satellite Gravity Observations." *Science* 314 (November 24, 2006):1286–1289.

Luyssaert, Sebastiaan, E. Detlef Schulze, Annett Börner, Alexander Knohl, Dominik Hessenmöller, Beverly E. Law, Philippe Ciais, and John Grace. "Old-growth Forests as Global Carbon Sinks." *Nature* 455 (September 11, 2008):213–215.

Lyall, Sarah. "At Risk from Floods, but Looking Ahead with Floating Houses." *New York Times*, April 3, 2007a. http://www.nytimes.com/2007/04/03/science/earth/03clim.html.

Lyall, Sarah. "Warming Revives Flora and Fauna in Greenland." *New York Times*, October 28, 2007b. http://www.nytimes.com/2007/10/28/world/europe/28greenland.html.

Lydersen, Kari. "Great Lakes' Lower Water Levels Propel a Cascade of Hardships." *Washington Post*, January 27, 2008, A-4. http://www.washingtonpost.com/wp-dyn/content/article/2008/01/26/AR2008012601748_pf.html.

Lyman, Rick. "Rising Ocean Temperatures Threaten Florida's Coral Reef." *New York Times*, May 22, 2006. http://www.nytimes.com/2006/05/22/us/22coral.html.

Lynas, Mark. *High Tide: The Truth about Our Climate Crisis*. New York: Picador/St. Martins, 2004a.

Lynas, Mark. "Meltdown: Alaska Is a Huge Oil Producer and Has Become Rich on the Proceeds—but It Has Suffered the Consequences." *London Guardian* (Weekend Magazine), February 14, 2004b, 22.

Lynas, Mark. "Vanishing Worlds: A Family Snap[shot] of a Peruvian Glacier Sent Mark Lynas on a Journey of Discovery—with the Ravages of Global Warming, Would It Still Exist 20 Years Later?" *London Guardian*, March 31, 2004c, 12.

Lynas, Mark. "Fly and Be Damned." *New Statesman* (London), April 3, 2006, 12–15. www.newstatesman.com/200604030006.

Lynas, Mark. *Six Degrees: Our Future on a Hotter Planet*. London: Fourth Estate, 2007.

Lynch-Stieglitz, Jean, Jess F. Adkins, William B. Curry, Trond Dokken, Ian R. Hall, Juan Carlos Herguera, Joîl J.-M. Hirschi, Elena V. Ivanova, Catherine Kissel, Olivier Marchal, Thomas M. Marchitto, I. Nicholas McCave, Jerry F. McManus, Stefan Mulitza, Ulysses Ninnemann, Frank Peeters, Ein-Fen Yu, and Rainer Zahn. "Atlantic Meridional Overturning Circulation during the Last Glacial Maximum." *Science* 316 (April 6, 2007): 66–69.

Macalister, Terry. "Confused Esso Tries to Silence Green Critics." *London Guardian*, June 25, 2002, 21.

Macalister, Terry. "Shell Chief Delivers Global Warming Warning to Bush in His Own Back Yard." *London Guardian*, March 12, 2003, 19.

Macdougall, Doug. *Frozen Earth: The Once and Future Story of Ice Ages*. Berkeley: University of California Press, 2004.

Macey, Richard. "Climate Change Link to Clearing." *Sydney Morning Herald*, June 29, 2004, 2.

Machalaba, Daniel. "Crowds Heeds Amtrak's 'All Aboard.'" *Wall Street Journal*, August 23, 2007, B-1, B-22.

Machlis, Gary E., and Thor Hanson. "War Ecology." *BioScience* 58, no. 8 (September 2008):729–736.

Mack, Michelle C., Edward A. G. Schuur, M. Syndonia Bret-Harte, Gaius R. Shaver, and F. Stuart Chapin III. "Ecosystem Carbon Storage in Arctic Tundra Reduced by Long-term Nutrient Fertilization." *Nature* 432 (September 23, 2004):440–443.

Macken, Julie. "The Big Dry: Bushfires Re-ignite Heated Debate on Global Warming." *Australian Financial Review*, February 17, 2003, 68.

Macken, Julie. "The Double-Whammy Drought." *Australian Financial Review*, May 4, 2004, 61.

MacKenzie, James J., and Michael P. Walsh. *Driving Forces: Motor Vehicle Trends and Their Implications for Global Warming, Energy Strategies, and Transportation Planning.* Washington, DC: World Resources Institute, 1990.

MacLeod, Calum. "China Envisions Environmentally Friendly Eco-city." *USA Today*, February 16, 2007, 9-A.

Magnuson, John J., Dale M. Robertson, Barbara J. Benson, Randolph H. Wynne, David M. Livingstone, Tadashi Arai, Raymond A. Assel, Roger G. Barry, Virginia Card, Esko Kuusisto, et al. "Historical Trends in Lake and River Ice Cover in the Northern Hemisphere." *Science* 289 (September 8, 2000):1743–1746.

Mahtab, Fasih Uddin. "The Delta Regions and Global Warming: Impact and Response Strategies for Bangladesh." In *The Regions and Global Warming: Impacts and Response Strategies,* ed. Jurgan Schmandt, Alan Jones, and Judith Clarkson, 28–43. New York: Oxford University Press, 1992.

"Maine Sets Global Warming Reduction Goals." Environment News Service, June 26, 2003. http://ens-news.com/ens/jun2003/2003-06-26-09.asp#anchor4.

Majeed, A. "The Impact of Militarism on the Environment: An Overview of Direct and Indirect Effects." Ottawa, Ontario, Canada: Physicians for Global Survival, 2004.

"Major Temperature Rise Recorded in Arctic This Year: German Scientists." Agence France Presse, August 27, 2004 (LEXIS).

Malhi, Yadvinder, J. Timmons Roberts, Richard A. Betts, Timothy J. Killeen, Wenhong Li, and Carlos A. Nobre. "Climate Change, Deforestation, and the Fate of the Amazon." *Science* 319 (January 11, 2008):169–172.

Mallaby, Sebastian. "A Dated Carbon Approach." *Washington Post,* July 10, 2006, A-17. http://www.washingtonpost.com/wp-dyn/content/article/2006/07/09/AR2006070900537_pf.html.

Mallaby, Sebastian. "Carbon Policy That Works: Avoiding the Pitfalls of Kyoto Cap-and-Trade." *Washington Post,* July 23, 2007, A-17. http://www.washingtonpost.com/wp dyn/content/article/2007/07/22/AR2007072200884_pf.html.

Malmer, Nils. "Peat Accumulation and the Global Carbon Cycle." In *Greenhouse-Impact on Cold-Climate Ecosystems and Landscapes,* ed. M. Boer and E. Koster, 97–110. *Selected Papers of a European Conference on Landscape Ecological Impact of Climatic Change,* Lunteren, the Netherlands, December 3–7, 1989. CARTENA Supplement 22. Cremlingen, Germany: Catena Verlag, 1992.

Malone, Thomas F., Edward D. Goldberg, and Walter H. Munk. "Roger Randall Dougan Revelle, 1909–1991." Biographical Memoirs of the National Academy of Sciences. Cited in Deborah Day. "Roger Randall Dougan Revelle Biography." Accessed January 28, 2009. http://www.repositories.cdlib.org/cgi/viewcontent.cgi?article=1084&context=sio/arch.

Maloney, Peter. "Solar Projects Draw New Opposition." *New York Times* (Business of Green), September 24, 2008, H-2.

Manabe, S., and R. J. Stouffer. "Century-scale Effects of Increased Atmospheric CO_2 on the Ocean-atmosphere System." *Nature* 364 (1993):215–218.

"Mandatory U.S. Greenhouse Gas Cap Wins New Corporate Supporters." Environment News Service, May 8, 2007. http://www.ens-newswire.com/ens/may2007/2007-05-08-01.asp.

Mankiw, N. Gregory. "One Answer to Global Warming: A New Tax." *New York Times*, September 16, 2007. http://www.nytimes.com/2007/09/16/business/16view.html.

Mann, Barbara. Personal communication, August 3, 1999.

Mann, Michael E., Raymond S. Bradley, and Michael K. Hughes. "Global-scale Temperature Patterns and Climate Forcing over the Past Six Centuries." *Nature* 392 (April 23, 1998):779–787.

Mann, Michael E., and Philip D. Jones. "Global Surface Temperatures over the Past Two Millennia." *Geophysical Research Letters* 30, no. 15 (August 2003), doi: 10.1029/2003GL017814. http://www.ngdc.noaa.gov/paleo/pubs/mann2003b/mann2003b.html.

Mann, Michael E., and Lee Kump. *Dire Predictions. Understanding Global Warming: The Illustrated Guide to the Findings of the IPCC.* New York: DK, 2008.

"Many U.S. Industry Giants Ignoring Global Warming." Environment News Service, July 9, 2003. http://ens-news.com/ens/jul2003/2003-07-09-11.asp.

Marris, Emma. "Sugar Cane and Ethanol: Drink the Best and Drive the Rest." *Nature* 444 (December 7, 2006):670–672.

Marris, Emma, and Daemon Fairless. "Wind Farms' Deadly Reputation Hard to Shift." *Nature* 447 (May 10, 2007):126.

Marsh, Peter. "Progress Is a Long and Windy Road: A Once-bankrupt Danish Crane Parts Maker Is Now the World's Biggest Maker of Wind Turbines." *London Financial Times*, May 30, 2002, 30.

Marsh, Virginia. "Australia Expected to Become Hotter." *London Financial Times*, May 9, 2001, 18.

Marshall, George. "Sleepwalking into Disaster—Are We in a State of Denial about Climate Change?" Climate Outreach and Information Network, Oxford, 2005. http://coinet.org.uk/discussion/perspectives/marshall.

Marshall, John, and R. Alan Plumb. *Atmosphere, Ocean, and Climate Dynamics: An Introductory Text*. Burlington, MA: Elsevier Academic, 2007.

Martens, Pim. "How Will Climate Change Affect Human Health?" *American Scientist* 87, no.6 (November/December 1999):534–541.

Martens, Willem J. M., Theo H. Jetten, and Dana A. Focks. "Sensitivity of Malaria, Schistosomiasis, and Dengue to Global Warming." *Climatic Change* 35 (1997):145–156.

Martin, Andrew. "In Eco-friendly Factory, Low-Guilt Potato Chips." *New York Times*, November 15, 2007, A-1. A-22.

Martin, Andrew. "Fuel Choices, Food Crises and Finger-Pointing." *New York Times*, April 15, 2008a. http://www.nytimes.com/2008/04/15/business/worldbusiness/15food.html.

Martin, Andrew. "Food Report Criticizes Biofuel Policies." *New York Times*, May 30, 2008b. http://www.nytimes.com/2008/05/30/business/worldbusiness/30food.html.

Maskell, Kathy, and Irving M. Mintzer. "Basic Science of Climate Change." *Lancet* 342 (1993):1027–1032.

Maslin, Mark, and Stephen J. Burns. "Reconstruction of the Amazon Basin Effective Moisture Availability over the Past 14,000 Years." *Science* 290 (December 22, 2000):2285–2287.

Mason, John, Jack A. Bailey, and Ardea London. "Doomsday for Butterflies as Britain Warms Up: Dozens of Native Species at Risk of Extinction as Habitats Come under Threat." *London Independent*, September 29, 2002, 12.

Masson, Gordon. "Eco-Friendly Movement Growing in Music Biz." *Billboard*, March 15, 2003, 1.

Mathews-Amos, Amy, and Ewann A. Berntson. "Turning up the Heat: How Global Warming Threatens Life in the Sea." World Wildlife Fund and Marine Conservation Biology Institute, 1999. http://www.worldwildlife.org/news/pubs/wwf_ocean.htm.

Matthews, H. Damon, and Ken Caldeira. "Stabilizing Climate Requires Near-zero Emissions." *Geophysical Research Letters* 35 (February 27, 2008), L04705, doi: 10.1029/2007GL032388.

Maxwell, Barrie. "Arctic Climate: Potential for Change Under Global Warming." *Arctic Ecosystems in a Changing Climate: An Ecophysiological Perspective*, Chapin, F. Stuart III, Robert L. Jefferies, James F. Reynolds, Gaius R. Shaver, and Josef Svoboda, 11–34. San Diego: Academic Press, 1992.

Maxwell, Fordyce. "Climate Warning for Scotland's Wildlife." *The Scotsman*, November 14, 2001, 7.

May, Wilhelm, Reinhard Voss, and Erich Roeckner. "Changes in the Mean and Extremes of the Hydrological Cycle in Europe under Enhanced Greenhouse Gas Conditions in a Global Time-slice Experiment." In *Climatic Change: Implications for the Hydrological Cycle and for Water Management*, Martin Beniston, 1–30. Dordrecht, Germany: Kluwer Academic Publishers, 2002.

Maynard, Roger. "Climate Change Bringing More Floods to Australia." *The Straits Times* (Singapore), March 14, 2001, 17.

McCarthy, Michael. "Ford Predicts End of Car Pollution: Boss Predicts the End of Petrol." *London Independent*, October 6, 2000a, 10.

McCarthy, Michael. "Climate Change Will Bankrupt the World." *London Independent*, November 24, 2000b, 6.

McCarthy, Michael. "Warm Spell Sees Nature Defying the Seasons: As 150 Countries Meet in Morocco to Discuss Climate Change, Britain's Natural World Responds to Record Temperatures." *London Independent*, October 30, 2001, 12.

McCarthy, Michael. "Climate Change Provides Exotic Sea Life with a Warm Welcome to Britain." *London Independent*, January 24, 2002, 13.

McCarthy, Michael. "'Rainforests of the Sea' Ravaged: Over-fishing and Pollution Kill 80 Percent of Coral on Caribbean Reefs." *London Independent*, July 18, 2003a, 3.

McCarthy, Michael. "The Four Degrees: How Europe's Hottest Summer Shows Global Warming Is Transforming Our World." *London Independent*, December 8, 2003b, 3.

McCarthy, Michael. "Countdown to Global Catastrophe." *London Independent*, January 24, 2005, 1.

McCord, Joel. "Marshes in Decay Haunt the Bay." *Baltimore Sun*, December 6, 2000, 1-B.

McCrea, Steve. "Air Travel: Eco-tourism's Hidden Pollution." *San Diego Earth Times*, August 1996. http://www.sdearthtimes.com/et0896/et0896s13.html.

McEwen, Bill. "The West's Dying Forests." Letter to the Editor, *New York Times*, August 2, 2004, A-16.

McFadden, Robert D. "New Orleans Begins a Search for Its Dead: Violence Persists." *New York Times*, September 5, 2005. http://www.nytimes.com/2005/09/05/national/nationalspecial/05storm.html.

McFarling, Usha Lee. "Scientists Warn of Losses in Ozone Layer over Arctic." *Los Angeles Times*, May 27, 2000, A-20.

McFarling, Usha Lee. "Studies Point to Human Role in Global Warming." *Los Angeles Times*, April 13, 2001a, A-1.

McFarling, Usha Lee. "Warmer World Will Starve Many, Report Says." *Los Angeles Times*, July 11, 2001b, A-3.

McFarling, Usha Lee. "Fear Growing over a Sharp Climate Shift." *Los Angeles Times*, July 13, 2001c, A-1.

McFarling, Usha Lee. "Scientists Now Fear 'Abrupt' Global Warming Changes: Severe and 'Unwelcome' Shifts Could Come in Decades, not Centuries, a National Academy Says in an Alert." *Los Angeles Times*, December 12, 2001d, A-30.

McFarling, Usha Lee. "Study Links Warming to Epidemics: The Survey Lists Species Hit by Outbreaks and Suggests That Humans Are Also in Peril." *Los Angeles Times*, June 21, 2002a, A-7.

McFarling, Usha Lee. "Glacial Melting Takes Human Toll: Avalanche in Russia and Other Disasters Show That Global Warming Is Beginning to Affect Areas Much Closer to Home." *Los Angeles Times*, September 25, 2002b, A-4.

McFarling, Usha Lee. "Shrinking Ice Cap Worries Scientists." *Los Angeles Times* in *Edmonton Journal*, December 8, 2002c. http://www.canada.com/regina/story.asp?id={54910725-535A-4B0E-9A7E-FD7176D9C392}.

McFarling, Usha Lee. "NASA Finds 2002 Second Warmest Year on Record." *Los Angeles Times* in Calgary *Herald*, December 12, 2002d, A-5.

McFarling, Usha Lee. "A Tiny 'Early Warning' of Global Warming's Effect: The Population of Pikas, Rabbit-like Mountain Dwellers, Is Falling, a Study Finds." *Los Angeles Times*, February 26, 2003, A-17.

McFarling, Usha Lee, and Kenneth R. Weiss. "A Whale of a Food Shortage." *Los Angeles Times*, June 24, 2002, 1.

McGuire, Bill. *Surviving Armageddon: Solutions for a Threatened Planet*. New York: New York: Oxford University Press, 2005.

McGuire, Bill. *What Everyone Should Know about the Future of Our Planet: And What We Can Do about It*. London: Weidenfeld & Nicolson Ltd., 2007.

McGuire, Bill. *Seven Years to Save the Planet*. London: Weidenfeld & Nicolson, 2008.

McKay, Paul. "Ford Leads Big Three in Green Makeover." *Ottawa Citizen*, May 28, 2001, A-1.

McKibben, Bill. *The End of Nature*. New York: Random House, 1989.

McKibben, Bill. "Bush in the Greenhouse." Review of *IPCC Third Assessment*. *New York Review of Books*, July 5, 2001, 35–38.

McKibben, Bill. "Worried? Us?" *Granta* 83 (Fall 2003):7–12.

McKibben, Bill. "Welcome to the Climate Crisis: How to Tell Whether a Candidate Is Serious about Combating Global Warming." *Washington Post*, May 27, 2006, A-25. http://www.washingtonpost.com/wp-dyn/content/article/2006/05/26/AR2006052601549_pf.html.

McKibben, Bill. "The Race against Warming." *Washington Post*, September 29, 2007, A-19. http://www.washingtonpost.com/wp-dyn/content/article/2007/09/28/AR2007092801400_pf.html.

McKie, Robin. "Dying Seas Threaten Several Species: Global Warming Could Be Tearing Apart the Delicate Marine Food Chain, Spelling Doom for Everything from Zooplankton to Dolphins." *London Observer*, December 2, 2001, 14.

McKie, Robin. "Decades of Devastation Ahead as Global Warming Melts the Alps: A Mountain of Trouble as Matterhorn Is Rocked by Avalanches." *London Observer*, July 20, 2003, 18.

McMichael, A. J. *Planetary Overload: Global Environmental Change and the Health of the Human Species.* Cambridge: Cambridge University Press, 1993.

McNeill, J. R. *Something New Under the Sun: An Environmental History of the Twentieth-century World.* New York: W.W. Norton, 2000.

McNulty, Sheila. "U.S. Power Generation Answer Is Blowing in the Wind." *London Financial Times*, April 24, 2007, 7.

McPhaden, Michael J., Stephen E. Zebiak, and Michael H. Glantz. "ENSO as an Integrating Concept in Earth Science." *Science* 314 (December 15, 2006):1740–1745.

McPhaden, Michael J., and D. Zhang. "Slowdown of the Meridional Overturning Circulation in the Upper Pacific Ocean." *Nature* 415 (February 6, 2002):603–607.

McWilliams, Brendan. "Study of Plants Confirms Global Warming." *Irish Times*, November 1, 2001, 26.

Meacher, Michael. "This Is the World's Chance to Tackle Global Warming." The *London Times*, September 3, 2000, n.p.(LEXIS).

Mead, Margaret, and William Kellogg, eds. *The Atmosphere: Endangered and Endangering.* Tunbridge Wells, England: Castle House, 1980.

Meagher, John. "Look What the Changing Climate Dragged in ..." *Irish Independent*, July 9, 2004, n.p.(LEXIS).

Meehl, Gerald A., and Claudia Tebaldi. "More Intense, More Frequent, and Longer Lasting Heat Waves in the 21st Century." *Science* 305 (August 13, 2004):994–997.

Meek, James. "Wildflowers Study Gives Clear Evidence of Global Warming." *London Guardian*, May 31, 2002a, 6.

Meek, James. "Global Warming Gives Pests Taste for Life in London." *London Guardian*, October 8, 2002b, 6.

Meier, Mark F., and Mark B. Dyurgerov. "Sea-level Changes: How Alaska Affects the World." *Science* 297 (July 19, 2002):350–351.

Meier, Mark F., Mark B. Dyurgerov, Ursula K. Rick, Shad O'Neel, W. Tad Pfeffer, Robert S. Anderson, Suzanne P. Anderson, and Andrey F. Glazovsky. "Glaciers Dominate Eustatic Sea-Level Rise in the 21st Century." *Science* 317 (August 24, 2007):1064–1067.

Meier, W., J. Stroeve, F. Fetterer, T. Arbetter, W.N. Meier, J. Maslanik, and K. Knowles. "Reductions in Arctic Sea Ice Cover No Longer Limited to Summer." *EOS: Transactions of the American Geophysical Union* 86 (2005):326.

Melillo, J. M. "Warm, Warm on the Range." *Science* 283 (January 8, 1999):183.

Melillo, J. M., P. A. Steudler, J. D. Aber, K. Newkirk, H. Lux, F. P. Bowles, C. Catricala, A. Magill, T. Ahrens, and S. Morrisseau. "Soil Warming and Carbon-cycle Feedbacks to the Climate System." *Science* 298 (December 13, 2002):2173–2176.

"Melting Andean Glaciers Could Leave 30 Million High and Dry." Environment News Service, April 28, 2008. http://www.ens-newswire.com.

"Melting Antarctic Glacier Could Raise Sea Level." Reuters. July 24, 1998. http://benetton.dkrz. de:3688/homepages/georg/kimo/0254.html.

"Melting Faster." Editors' Choice. *Science* 316, no. 1 (May 18, 2007):955.

"Melting Ice May Release Frozen Influenza Viruses." Environment News Service, November 27, 2006. http://www.ens-newswire.com/ens/nov2006/2006-11-27-09.asp#anchor3.

"Melting Planet: Species Are Dying Out Faster Than We Have Dared Recognize, Scientists Will Warn This Week." *London Independent*, October 2, 2005. http://news.independent.co.uk/world/environment/article316604.ece

Melvin, Don. "There'll Always Be an England? Study of Global Warming Says Sea Is Winning." *Atlanta Journal-Constitution*, June 5, 2004a, 3-A.

Melvin, Don. "Storm over Wind Energy: Britain's Renewable Power Push Stirs Turbulent Debate." *Atlanta Journal-Constitution*, July 5, 2004b, 8-A.

Mendelson, Joseph R. III, Karen R. Lips, Ronald W. Gagliardo, George B. Rabb, James P. Collins, James E. Diffendorfer, Peter Daszak, Roberto Ibáñez D., Kevin C. Zippel, Dwight P. Lawson, Kevin M. Wright, Simon N. Stuart, Claude Gascon, Hélio R. da Silva, Patricia A. Burrowes, Rafael L. Joglar, Enrique La Marca, Stefan Lötters, Louis H. du Preez, Ché Weldon, Alex Hyatt, José Vicente Rodriguez-Mahecha, Susan Hunt, Helen Robertson, Brad Lock, Christopher J. Raxworthy, Darrel R. Frost, Robert C. Lacy, Ross A. Alford, Jonathan A. Campbell, Gabriela Parra-Olea, Federico Bolaños, José Joaquin Calvo Domingo, Tim

Halliday, James B. Murphy, Marvalee H. Wake, Luis A. Coloma, Sergius L. Kuzmin, Mark Stanley Price, Kim M. Howell, Michael Lau, Rohan Pethiyagoda, Michelle Boone, Michael J. Lannoo, Andrew R. Blaustein, Andy Dobson, Richard A. Griffiths, Martha L. Crump, David B. Wake, and Edmund D. Brodie, Jr. "Confronting Amphibian Declines and Extinctions." *Science* 313 (July 7, 2006):48.

Menon, Surabi, James Hansen, Larissa Nazarenko, and Yunfeng Luo. "Climate Effects of Black Carbon Aerosols in China and India." *Science* 297 (September 27, 2002):2250–2253.

Menzel, A., and P. Fabian. "Growing Season Extended in Europe." *Nature* 397 (1999):659–662.

Merzer, Martin. "Study: Global Warming Likely Making Hurricanes Stronger." *Miami Herald*, August 1, 2005, n.p.(LEXIS).

"Methane Didn't Act Alone." *Nature* 453 (May 15, 2008):260.

"Methane Emissions Increasing and Could Hasten Global Warming." Environment News Service, September 28, 2006. http://www.ens-newswire.com/ens/sep2006/2006-09-28-01.asp.

"Methane from Dams: Greenhouse Gas to Power Source." Environment News Service, May 9. 2007. www.ens-newswire.com/ens/archives/2007/may2007archive.asp.

Metz, Bert, ed. *Climate Change 2007: Mitigation of Climate Change.* New York: Cambridge University Press, 2008.

Meuvret, Odile. "Global Warming Could Turn Siberia into Disaster Zone: Expert." Agence France Presse, October 2, 2003.

Michael, Daniel, and Susan Carey. "Airlines Feel Pressure as Pollution Fight Takes Off." *Wall Street Journal*, December 12, 2006, A-6.

Michaels, Patrick J. "Solar Energy." In "What's Hot." *World Climate Report Archives: 1995–2000* 1, no. 15 (1996). http://www.nhes.com/back_issues/Vol1and2/WH/hot.html.

Michaels, Patrick J. "Kyoto Protocol: A Useless Appendage to an Irrelevant Treaty." Testimony of Patrick J. Michaels, Professor of Environmental Sciences, University of Virginia, and Senior Fellow in Environmental Studies at Cato Institute before the Committee on Small Business, United States House of Representatives, Washington, DC, July 29, 1998. http://climatechangedebate.com/archive/09-18_10-27_1998.txt.

Michaels, Patrick J. "'Day after Tomorrow': A Lot of Hot Air." *USA Today*, May 25, 2004a, 21-A.

Michaels, Patrick J. *Meltdown: The Predictable Distortion of Global Warming by Scientists, Politicians, and the Media.* Washington, DC: Cato Institute, 2004b.

Michaels, Patrick J. *Shattered Consensus: The True State of Global Warming.* Lanham, MD: Rowman & Littlefield, 2005.

Michaels, Patrick J., and Robert C. Balling, Jr. *The Satanic Gases: Clearing the Air about Global Warming.* Cato Institute, 2000.

"Mideast Snow Disrupts Life, Prayers." *Omaha World-Herald*, January 29, 2000, 4.

Mieszkowski, Katharine. "Bush: Global Warming Is Just Hot Air." Salon.com, September 10, 2004 (LEXIS).

Mieszkowski, Katherine. "Did Al Gore Get the Science Right?" Salon.com, June 10, 2006 (LEXIS).

"Migratory Species and Climate Change." United Nations Environmental Programme (UNEP), 2006. http://www.cms.int/news/PRESS/nwPR2006/november/cms_ccReport.htm.

Miles, Lera, and Valerie Kapos. "Reducing Greenhouse Gas Emissions from Deforestation and Forest Degradation: Global Land-Use Implications." *Science* 320 (June 13, 2008):1454–1455.

Miles, Paul. "Fiji's Coral Reefs Are Being Ruined by Bleaching." The *London Daily Telegraph*, June 2, 2001, 4.

Miller, T., J. C. Walker, G. T. Kingsley, and W. A. Hyman. "Impact of Global Climate Change on Urban Infrastructure." In *Potential Effects of Global Climate Change on the United States: Appendix H, Infrastructure*, ed. J. B. Smith and D. A. Tirpak. Washington, DC: U.S. Environmental Protection Agency, 1989.

Milly, P. C. D., R. T. Wetherald, K. A. Dunne, and T. L. Delworth. "Increasing Risk of Great Floods in a Changing Climate." *Nature* 415 (January 30, 2002):514–517.

Milmo, Cahal, and Elizabeth Nash. "Fish Farms Push Atlantic Salmon towards Extinction." *London Independent*, June 1, 2001, 11.

Minnis, Patrick, J. Kirk Ayres, Rabindra Palikonda, and Dung Phan. "Contrails, Cirrus Trends, and Climate." *Journal of Climate* (April 5, 2004): 1671–1685.

Mitchell, John G. "Down the Drain: The Incredible Shrinking Great Lakes." *National Geographic*, September 2002, 34–51.

Mittelstaedt, Martin. "World Faces Perpetual Food Crisis: Study." *Globe and Mail* (Toronto), January 8, 2009. http://www.theglobeandmail.com/servlet/story/RTGAM.20090108.wcli-mate0108/BNStory/International/home.

Moberg, Anders, Dmitry M. Sonechkin, Karin Holmgren, Nina M. Datsenko, and Wibjörn Karlén. "Highly Variable Northern Hemisphere Temperatures Reconstructed from Low- and High-resolution Proxy Data." *Nature* 433 (February 10, 2005):613–617.

Modie, Neil. "Mount Kilimanjaro: On Africa's Roof, Still Crowned with Snow." *New York Times*, January 20, 2008. http://travel.nytimes.com/2008/01/20/travel/20Explorer.html.

"Mogul Pledges Billions against Global Warming." Associated Press in *New York Times*, September 21, 2006. http://www.nytimes.com/aponline/us/AP-Clinton-Global-Initiative.html.

Møller, Anders P., Wolfgang Fielder, and Peter Berthold, ed. *Birds and Climate Change.* Burlington, MA: Academic (Elsevier), 2006.

Monaghan, Andrew J., David H. Bromwich, Ryan L. Fogt, Sheng-Hung Wang, Paul A. Mayewski, Daniel A. Dixon, Alexey Ekaykin, Massimo Frezzotti, Ian Goodwin, Elisabeth Isaksson, Susan D. Kaspari, Vin I. Morgan, Hans Oerter, Tas D. Van Ommen, Cornelius J. Van der Veen, and Jiahong Wen. "Insignificant Change in Antarctic Snowfall since the International Geophysical Year." *Science* 313 (August 11, 2006):827–831.

Monastersky, Richard. "The Long Goodbye: Alaska's Glaciers Appear to Be Disappearing before Our Eyes. Are They a Sign of Things to Come?" *New Scientist*, April 14, 2001, 30–32.

Monbiot, George. *Heat: How to Stop the Planet from Burning.* Toronto: Doubleday Canada, 2006a.

Monbiot, George. "We Are All Killers: Until We Stop Flying." *London Guardian*, February 28, 2006b. http://www.monbiot.com/archives/2006/02/28/we-are-all-killers.

"Monster Iceberg Heads into Antarctic Waters." Agence France Presse, October 22, 2002 (LEXIS).

Montague, Peter, ed. "Rachel's Environment and Health Weekly No. 300. Global Warming Part I: How Global Warming Is Sneaking Up on Us." Environmental Research Foundation, Annapolis, MD, August 26, 1992. http://www.monitor.net/rachel/r300.html.

Montaigne, Fen. "The Heat Is On: Ecosigns." *National Geographic*, September 2004, 34–55.

Montanez, Isabel P., Neil J. Tabor, Deb Niemeier, William A. DiMichele, Tracy D. Frank, Christopher R. Fielding, John L. Isbell, Lauren P. Birgenheier, and Michael C. Rygel. "CO_2-forced Climate and Vegetation Instability during Late Paleozoic Deglaciation." *Science* 315 (January 5, 2007):87–91.

Mooney, C. "Blinded by Science: How 'Balanced' Coverage Lets the Scientific Fringe Hijack Reality." *Columbia Journalism Review*, November/December 2004. http://www.cjr.org/issues/2004/6/mooney-science.asp.

Mooney, Chris. *Storm World: Hurricanes, Politics, and the Battle over Global Warming.* New York: Harcourt, 2007.

Moore, Molly. "In Northern France, Warming Presses Fall Grape Harvest into Summertime." *Washington Post*, September 2, 2007, A-1. http://www.washingtonpost.com/wp-dyn/content/article/2007/09/01/AR2007090101360_pf.html.

Moore, Thomas Gale. "Why Global Warming Would Be Good for You." *Public Interest*, January 1, 1995, 83. http://www.cycad.com/cgi-bin/Upstream/Issues/science/WARMIN.html.

Moore, Thomas Gale. "Health and Amenity Effects of Global Warming." Working Papers in Economics, E-96-1. Stanford, CA: Hoover Institution, 1996.

Moore, Thomas Gale. "Happiness Is a Warm Planet." *Wall Street Journal*, October 7, 1997. http://www.freerepublic.com/forum/a182.htm.

"More Carbon Dioxide Could Reduce Crop Value." Environment News Service, October 3, 2002. http://ens-news.com/ens/oct2002/2002-10-03-09.asp#anchor2.

"More CEOs Call for Climate Action." *Solutions* 38, no. 5 (November 2007):9. Environmental Defense. http://www.environmentaldefense.org.

Moreno, Fidel. "In the Arctic, Ice Is Life, and It's Disappearing." *Native Americas* 16, no. 3/4 (Fall/Winter 1999):42–45. http://nativeamericas.aip.cornell.edu.

"More Than 80 Percent of Spain's Pyrenean Glaciers Melted Last Century." Agence France Presse. September 29, 2004 (LEXIS).

Morrison, John, and Alex Sink. "The Climate Change Peril That Insurers See." *Washington Post*, September 27, 2007, A-25. http://www.washingtonpost.com/wp-dyn/content/article/2007/09/26/AR2007092602070_pf.html.

Morton, Oliver. "The Tarps of Kilimanjaro." *New York Times*, November 17, 2003. http://www.nytimes.com/2003/11/17/opinion/17MORT.html.

Moser, Susanne C., and Lisa Dilling, eds. *Creating a Climate for Change: Communicating Climate Change and Facilitating Social Change*. Cambridge: Cambridge University Press, 2007.

Moss, Stephen. "Casualties." *London Guardian*, April 26, 2001, 18.

"Most Serious Greenhouse Gas Is Increasing, International Study Finds." *Science Daily*, April 27, 2001. http://www.sciencedaily.com/releases/2001/04/010427071254.htm.

Motavalli, Jim, ed. *Feeling the Heat: Dispatches from the Front Lines of Climate Change*. London: Routledge, 2004.

Mote, Philip, and Georg Kaser. "The Shrinking Glaciers of Kilimanjaro: Can Global Warming Be Blamed?" *American Scientist*, July-August 2007. http://www.americanscientist.org/template/AssetDetail/assetid/55553.

Mouawad, Jad. "Industries Allied to Cap Carbon Differ on the Details." *New York Times*, June 2, 2008. http://www.nytimes.com/2008/06/02/business/02trade.html.

Mudelsee, Mandred, Michael Borngen, Gerd Tetzlaff, and Uwe Grunewald. "No Upward Trends in the Occurrence of Extreme Floods in Central Europe." *Nature* 425 (September 11, 2003):166–169.

Mufson, Steven. "A Sunnier Forecast for Solar Energy: Still Small, Industry Adds Capacity and Jobs to Compete with Utilities." *Washington Post*, November 20, 2006, D-1 http://www.washingtonpost.com/wp-dyn/content/article/2006/11/19/AR2006111900688_pf.html.

Mufson, Steven. "Europe's Problems Color U.S. Plans to Curb Carbon Gases." *Washington Post*, April 9, 2007a, A-1. http://www.washingtonpost.com/wp-dyn/content/article/2007/04/08/AR2007040800758_pf.html.

Mufson, Steven. "In Battle for U.S. Carbon Caps, Eyes and Efforts Focus on China." *Washington Post*, June 6, 2007b, D-1. http://www.washingtonpost.com/wp-dyn/content/article/2007/06/05/AR2007060502546_pf.html.

Mufson, Steven. "On Capitol Hill, a Warmer Climate for Biofuels." *Washington Post*, June 15, 2007c, D-1. http://www.washingtonpost.com/wp-dyn/content/article/2007/06/14/AR2007061402089_pf.html.

Mufson, Steven. "Florida's Governor to Limit Emissions." *Washington Post*, July 12, 2007d, D-1. http://www.washingtonpost.com/wp-dyn/content/article/2007/07/11/AR2007071102139_pf.html.

Mufson, Steven. "Climate Change Debate Hinges on Economics: Lawmakers Doubt Voters Would Fund Big Carbon Cuts." *Washington Post*, July 15, 2007e, A-1. http://www.washingtonpost.com/wp-dyn/content/article/2007/07/14/AR2007071401246_pf.html.

Mufson, Steven. "Coal Rush Reverses, Power Firms Follow Plans for New Plants Stalled by Growing Opposition." *Washington Post*, September 4, 2007f, D-1. http://www.washingtonpost.com/wp-dyn/content/article/2007/09/03/AR2007090301119_pf.html.

Mufson, Steven. "Siphoning Off Corn to Fuel Our Cars." *Washington Post*, April 30, 2008a, A-1. http://www.washingtonpost.com/wp-dyn/content/article/2008/04/29/AR2008042903092_pf.html.

Mufson, Steven. "Power-sector Emissions of China to Top U.S." *Washington Post*, August 27, 2008b, D-1. http://www.washingtonpost.com/wp-dyn/content/article/2008/08/26/AR2008082603096_pf.html.

Mufson, Steven. "The Car of the Future—but at What Cost? Hybrid Vehicles Are Popular, but Making Them Profitable Is a Challenge." *Washington Post*, November 25, 2008c, A-1. http://www.washingtonpost.com/wp-dyn/content/article/2008/11/24/AR2008112403211_pf.html.

Mufson, Steven, and Juliet Eilperin. "Energy Firms Come to Terms with Climate Change." *Washington Post*, November 25, 2006, A-1. http://www.washingtonpost.com/wpdyn/content/article/2006/11/24/ AR2006112401361_pf.html.

Mufson, Steven, and Blaine Harden. "Coal Can't Fill World's Burning Appetite with Supplies Short, Price Rise Surpasses Oil and U.S. Exporters Profit." *Washington Post*, March 20, 2008, A-1. http://www.washingtonpost.com/wpdyn/content/article/2008/03/19/AR2008031903859_pf.html.

Mufson, Steven, and Dan Morgan. "Switching to Biofuels Could Cost Lots of Green." *Washington Post*, June 8, 2007, D-1. http://www.washingtonpost.com/wp-dyn/content/article/2007/06/07/AR2007060702176_pf.html.

Munro, Margaret. "Global Warming Affecting Squirrels' Genes, Study Finds: Research in Yukon: 'Phenomenal Change' Seen as Rodents Breeding Earlier." *National Post* (Canada), February 12, 2003a, A-2.

Munro, Margaret. "Puffin Colony Threatened by Warming: A Few Degrees Can Be Devastating. Thousands of Triangle Island Chicks Die When Heat Drives off Their Favoured Fish." Montreal *Gazette*, July 15, 2003b, A-12.

Munro, Margaret. "Earth's 'Big Burp' Triggered Warming: Prehistoric Release of Methane a Cautionary Tale for Today." *Edmonton Journal*, June 3, 2004, A-10.

Murgia, Joe. "NASA: Greenland's Glaciers Are Shrinking: A New Study Suggests That Rapid Thinning and Excess Run-off from Greenland's Southeastern Glaciers May Be Partly Caused by Climate Changes." GS Report, March 10, 1999. http://www.gsreport.com/articles/art000078.html.

Murphy, Dean. "Study Finds Climate Shift Threatens California." *New York Times*, August 17, 2004. http://www.nytimes.com/2004/08/17/national/17heat.html.

Murphy, Kim. "Front-row Exposure to Global Warming: Climate: Engineers Say Alaskan Village Could Be Lost as Sea Encroaches." *Los Angeles Times*, July 8, 2001, A-1.

Murray, Alan. "The American Century: Is It Going or Coming?" *Wall Street Journal*, December 27, 1999, 1.

Murray, Tavi. "Greenland's Ice on the Scales." *Nature* 443 (September 21, 2006):277–278.

Mydans, Seth. "Reports from Four Fronts in the War on Warming." *New York Times*, April 3, 2007. http://www.nytimes.com/2007/04/03/science/earth/03clim.html.

Nader, Ralph, and Toby Heaps. "We Need a Global Carbon Tax." *Wall Street Journal*, December 3, 2008, A-17.

Naftz, David L., David D. Susong, Paul F. Schuster, L. DeWayne Cecil, Michael D. Dettinger, Robert L. Michel, and Carol Kendall. "Ice Core Evidence of Rapid Air Temperature Increases since 1960 in the Alpine Areas of the Wind River Range, Wyoming, United States." *Journal of Geophysical Research* 107 (2002):4171. http://www.agu.org/pubs/crossref/2002/2001JD000621.shtml.

Nagourney, Adam, Margorie Connelly, and Dalia Sussman. "Polls Find Voters in Early States Weighing Issues vs. Electability." *New York Times*, November 14, 2007, A-1, A-19.

Naik, Gautam. "Global Warming May Be Spurring Allergy, Asthma." *Wall Street Journal*, May 3, 2007a, A-1, A-13.

Naik, Gautam. "Arctic Becomes Tourism Hot Spot, but Is It Cool?" *Wall Street Journal*, September 24, 2007b, A-1, A-12. Naik, Gautam, and Geraldo Samor. "Drought Spotlights Extent of Damage in Amazon Basin." *Wall Street Journal*, October 21, 2005, A-12.

Naish, T. R., K. J. Woolfe, P. J. Barnett, G. S. Wilson, C. Atkins, S. M. Bohaty, C. J. Backer, M. Claps, F. J. Davey, G. B. Dunbar, A. G. Dunn., C. R. Fielding, F. Florindo, M. J. Hannah, D. M. Harwood, S. A. Henrys, L. A. Krissek, M. Lavelle, J. van der Meer, W. C. McIntosh, F. Niessen, S. Passchier, R. D. Powell, A. P. Roberts, L. Sagnotti, R. P. Scherer, C. P. Strong, F. Talarico, K. L. Verosub, G. Villa, D. K. Watkins, P.-N. Webb, and T. Wonik. "Orbitally Induced Oscillations in the East Antarctic Ice Sheet at the Oligocene/Miocene Boundary." *Nature* 413 (October 18, 2001):719–723.

Nance, John J. *What Goes Up: the Global Assault on Our Atmosphere.* New York: William Morrow and Co., 1991.

Naqvi, S. W. A., D. A. Jayakumar, P. V. Narvekar, H. Naik, V. V. S. S. Sarma, W. D'Souza, S. Joseph, and M. D. George. "Increased Marine Production of N2O Due to Intensifying Anoxia on the Indian Continental Shelf." *Nature* 408 (November 16, 2000):346–349.

"NASA: Vast Areas of West Antarctica Melted in 2005." Environment News Service, May 28, 2007. http://www.ens-newswire.com/ens/may2007/2007-05-28-09.asp#anchor1.

"NASA Data Show Some African Drought Linked to Warmer Indian Ocean." NASA Earth Observatory, August 5, 2008. http://earthobservatory.nasa.gov/Newsroom/NasaNews/2008/2008080527314.html

"NASA Researchers Find Snowmelt in Antarctica Creeping Inland." NASA Press Release, September 20, 2007. http://earthobservatory.nasa.gov/Newsroom/NasaNews/2007/2007092025613.html.

"NASA Study Finds Rising Arctic Storm Activity Sways Sea Ice, Climate." NASA Earth Observatory, October 6, 2008. http://earthobservatory.nasa.gov/Newsroom/NasaNews/2008/2008100627645.html.

"NASA Study Finds World Warmth Edging Ancient Levels." NASA, September 25, 2006. http://www.giss.nasa.gov/research/news/20060925.

"NASA Study Predicts More Severe Storms with Global Warming." NASA Earth Observatory, August 30, 2007. http://www.nasa.gov/centers/goddard/news/topstory/2007/moist_convection.html.

National Academy of Sciences. *Policy Implications of Greenhouse Warming*. Washington, DC: National Academy Press, 1991.

"National Ban on New Power Plants without CO_2 Controls Proposed." Environment News Service, March 12, 2008. http://www.ens-newswire.com/ens/mar2008/2008-03-12-091.asp.

Navarro, Mireya. "Environment Blamed in Western Tree Deaths." *New York Times*, January 23, 2009. http://www.nytimes.com/2009/01/23/us/23trees.html.

Naveira Garabato, Alberto C., David P. Stevens, Andrew J. Watson, and Wolfgang Roether. "Short-circuiting of the Overturning Circulation in the Antarctic Circumpolar Current." *Nature* 447 (May 10, 2007):194–197

Nemani, Ramakrishna R., Charles D. Keeling, Hirofumi Hashimoto, William M. Jolly, Stephen C. Piper, Compton J. Tucker, Ranga B. Myneni, and Steven W. Running. "Climate-Driven Increases in Global Terrestrial Net Primary Production from 1982 to 1999." *Science* 300 (June 6, 2003):1560–1563.

Nepstad, Daniel C. "Report from the Amazon." Woods Hole Research Center, May 1998. http://terra.whrc.org/science/tropfor/fire/report2.htm.

Nesmith, Jeff. "Sewage off Keys Cripples Coral: Bacteria Causes Deadly Disease." Atlanta *Journal-Constitution*, June 18, 2002, 3-A.

Nesmith, Jeff. "Is the Earth too Hot? A New Study Says No, but Then It Was Funded by Big Oil Companies." *Atlanta Journal-Constitution* in *Hamilton Spectator* (Ontario, Canada), May 30, 2003a, n.p.(LEXIS).

Nesmith, Jeff. "Dirty Snow Spurs Global Warming: Study Says Soot Blocks Reflection, Hurries Melting." *Atlanta Journal-Constitution*, December 23, 2003b, 3-A.

Nesmith, Jeff. "Antarctic Glacier Melt Increases Dramatically." *Atlanta Journal-Constitution*, September 22, 2004, 9-A.

"New Belgium: Our Story." May 3, 2007. http://www.newbelgium.com/sustainability.php.

"New Cracks in the Wilkins Ice Shelf." NASA Earth Observatory, December 9, 2008. http://earthobservatory.nasa.gov/IOTD/view.php?id=36060.

"New Look at Satellite Data Supports Global Warming Trend." AScribe Newswire, April 30, 2003 (LEXIS).

Newman, Cathy. "Prescott Warns U.S. over Climate." *London Financial Times*, April 27, 2000, 7.

Newman, Paul A. "Preserving Earth's Stratosphere." *Mechanical Engineering*, October 1998. http://www.memagazine.org/backissues/october98/features/stratos/stratos.html.

"New NASA Satellite Sensor and Field Experiment Shows Aerosols Cool the Surface but Warm the Atmosphere." National Aeronautics and Space Administration Public Information Release, August 15, 2001. http://earthobservatory.nasa.gov/Newsroom/MediaResources/Indian_Ocean_Experiment/indoex_release.html.

"New Research Links Global Warming to Wildfires across the West." *Los Angeles Times* in *Omaha World-Herald*, November 5, 2004, 11-A.

"New Research Reveals 50-year Sustained Antarctic Ice Decline." Agence France Presse, November 14, 2003 (LEXIS).

"New Study Shows Global Warming Trend Greater without El Niño and Volcanic Influences." Environmental Journalists' Bulletin Board, December 13, 2000. environmental journalists@egroups.com.

"New Wave of Bleaching Hits Coral Reefs Worldwide." Environment News Service, October 29, 2002. http://ens-news.com/ens/oct2002/2002-10-29-19.asp#anchor1.

"New York Lakes Fail to Freeze." Environment News Service, March 21, 2002. http://ens-news.com/ens/mar2002/2002L-03-21-09.html#anchor1.

"New York's Mayor Plans Hybrid Taxi Fleet." Associated Press in *New York Times*, May 22, 2007. http://www.nytimes.com/aponline/us/AP-Green-Taxis.html.

Nicholls, Neville. "The Changing Nature of Australian Droughts." *Climatic Change* 63 (2004):323–336.

Nickerson, Colin. "An Early Melting Hurts Seals, Hunters in Canada." *Boston Globe*, April 1, 2002, A-1.

Nisbet, E. G. "The End of the Ice Age." *Canadian Journal of Earth Sciences* 27 (1990):148–157.

Nisbet, E. G., and B. Ingham. "Methane Output from Natural and Quasinatural Sources: A Review of the Potential for Change and for Biotic and Abiotic Feedbacks." In *Biotic Feedbacks in the Global Climate System: Will the Warming Feed the Warming?* ed. George M. Woodwell and Fred T. MacKenzie, 188–218. New York: Oxford University Press, 1995.

"No Climate Benefit Gained by Planting Temperate Forests." Environmental News Service, December 12, 2006. http://www.ens-newswire.com.

Nordhaus, William D. "Economic Approaches to Greenhouse Warming." In *Global Warming: Economic Policy Reponses,* ed. Rudiger Dornbusch and James M. Poterba, 33–66. Cambridge, MA: MIT Press, 1991.

Normile, Dennis. "Some Coral Bouncing Back from El Niño." *Science* 288 (May 12, 2000):941–942.

Norris, Richard D., and Ursula Röhl. "Carbon Cycling and Chronology of Climate Warming during the Palaeocene/Eocene Transition." *Nature* 401 (October 21, 1999):775–778.

North, Gerald R., Jurgen Schmandt, and Judith Clarkson. *The Impact of Global Warming on Texas.* Austin: University of Texas Press, 1995.

"North Pole Had Subtropical Seas because of Global Warming." Agence France Presse, September 7, 2004 (LEXIS).

"North Pole Mussels Point to Warming, Scientists Say." Reuters in *Canada National Post,* September 18, 2004, A-16.

"Northwest Passage Nearly Open." NASA Earth Observatory, August 27, 2007. http://earthobservatory.nasa.gov/Newsroom/NewImages/images.php3?img_id=17752.

"Norway Says No to Controversial Plan to Store CO_2 on Ocean Floor." Agence France Presse, August 22, 2002.

Nosengo, Niccola. "Venice Floods: Save Our City!" *Nature* 424 (August 7, 2003):608–609.

"No Sugar-beet Answer." Editorial, *Omaha World-Herald,* April 7, 2007, 6-B.

Nowak, Rachel. "The Continent That Ran Dry." *The New Scientist,* June 16, 2007, 8–11.

Nunes, Flavia, and Richard D. Norris. "Abrupt Reversal in Ocean Overturning during the Palaeocene/Eocene Warm Period." *Nature* 439 (January 5, 2006):60–63.

Nussbaum, Alex. "The Coming Tide: Rise in Sea Level Likely to Increase N.J. Floods." *Bergen County Record* (New Jersey), September 4, 2002, A-1.

Nuttall, Nick. "Global Warming 'Will Turn Rainforests into Deserts.'" *London Times,* November 3, 1998. http://bonanza.lter.uaf.edu/~davev/nrm304/glbxnews.htm.

Nuttall, Nick. "Climate Change Lures Butterflies Here Early." *London Times,* May 24, 2000a, n.p.

Nuttall, Nick. "Experts Are Poles Apart over Ice Cap." *London Times,* August 21, 2000b, n.p.(LEXIS).

Nuttall, Nick. "Coral Reefs 'on the Edge of Disaster.'" *London Times,* October 25, 2000c, n.p.(LEXIS).

Nuttall, Nick. "Global Warming Boosts El Niño." *London Times,* October 26, 2000d, n.p.(LEXIS).

Nyberg, Johan, Bjorn A. Malmgren, Amos Winter, Mark R. Jury, K. Halimeda Kilbourne, and Terrence M. Quinn. "Low Atlantic Hurricane Activity in the 1970s and 1980s Compared to the Past 270 Years." *Nature* 447 (June 7, 2007):698–701.

O'Brien, S. T., B. P. Hayden, and H. H. Shugart. "Global Climatic Change, Hurricanes, and a Tropical Forest." *Climatic Change* 22 (1992):175–190.

O'Carroll, Cynthia. "NASA Blames Greenhouse Gases for Wintertime Warming." *UniSci,* April 24, 2001. http://unisci.com/stories/20012/0424011.htm.

"Ocean Temperatures Reach Record Highs." Associated Press, September 9, 2002 (LEXIS).

O'Connell, Sanjida. "Power to the People." *London Times,* May 20, 2002, n.p.(LEXIS).

Odling-Smee, Lucy. "Biofuels Bandwagon Hits a Rut." *Nature* 446 (March 29, 2007):483.

O'Driscoll, Patrick. "2005 Is Warmest Year on Record for Northern Hemisphere, Scientists Say." *USA Today,* December 16, 2005, 2-A.

Ogle, Andy. "Squirrels Get Squirrelier Earlier: Climate Change to Blame—Breeding Season in Yukon Advances 18 Days in Decade." *Edmonton Journal* (Canwest News Service) in *Montreal Gazette,* February 12, 2003, A-12.

O'Harra, Doug. "Marine Parasite Infects Yukon River King Salmon: Fish Are Left Inedible, Scientists Study Overall Impacts." *Anchorage Daily News,* January 28, 2004. A-1.

"Oilman Pickens Promotes the Wind." Environment News Service, July 8, 2008. http://www.ens-newswire.com/ens/jul2008/2008-07-08-091.asp.

Olsen, Jan M. "Europe Is Warned of Changing Climate." Associated Press, August 19, 2004 (LEXIS).

Olson, Jeremy. "Flash Flooding Closes I-80." *Omaha World-Herald*, July 7, 2002, A-1.

O'Malley, Brendan. "Global Warming Puts Rainforest at Risk." *Cairns Courier-Mail* (Australia), July 24, 2003, 14.

"100 Still Missing in Russian Avalanche." *Los Angeles Times*, September 24, 2002, A-5.

O'Neill, Graeme. "The Heat Is On." *Sunday Herald-Sun* (Sydney, Australia), March 31, 2002.

"One in Five ExxonMobil Shareholders Want Climate Action." Environment News Service. May 28, 2003. http://ens-news.com/ens/may2003/2003-05-28-09.asp#anchor3.

Oppenheimer, Michael, and Robert H. Boyle: *Dead Heat: The Race against the Greenhouse Effect.* New York: Basic Books, 1990.

O'Reilly, Catherine M., Simone R. Alin, Pierre-Denis Plisnier, Andrew S. Cohen, and Brent A. McKee. "Climate Change Decreases Aquatic Ecosystem Productivity of Lake Tanganyika, Africa." *Nature* 424 (August 14, 2003):766–768.

Oren, R., D. S. Ellsworth, K. H. Johnsen, N. Phillips, B. E. Ewers, C. Maier, K. V. Schafer, H. McCarthy, G. Hendrey, S. G. McNulty, and G. G. Katul. "Soil Fertility Limits Carbon Sequestration by Forest Ecosystems in a CO_2-enriched Atmosphere." *Nature* 411 (May 24, 2001):469–472.

Orr, James C., Victoria J. Fabry, Olivier Aumont, Laurent Bopp, Scott C. Doney, Richard A, Feely, Anand Gnanadesikan, Nicolas Gruber, Akio Ishida, Fortunat Joos, Robert M. Key, Keith Lindsay, Ernst Maier-Reimer, Richard Matear, Patrick Monfray, Anne Mouchet, Raymond G. Najjar, Gian-Kasper Plattner, Keith B. Rodgers, Christopher L. Sabine, Jorge L. Sarmiento, Reiner Schlitzer, Richard D. Slater, Ian J. Totterdell, Marie-France Weirig, Yasuhiro Yamanaka, and Andrew Yooi. "Anthropogenic Ocean Acidification over the Twenty-first Century and Its Impact on Calcifying Organisms." *Nature* 437 (September 29, 2005):681–686.

Otto-Bliesner, Bette L., Shawn J. Marshall, Jonathan T. Overpeck, Gifford H. Miller, Aixue Hu, and CAPE Last Interglacial Project Members. "Simulating Arctic Climate Warmth and Icefield Retreat in the Last Interglaciation." *Science* 311 (March 24, 2006):1751–1753.

Ouroussoff, Nicolai. "Why Are They Greener Than We Are?" *New York Times Sunday Magazine*, May 20, 2007. http://www.nytimes.com/2007/05/20/magazine/20europe-t.html.

"Over 80 Percent of Indonesia's Coral Reefs under Threat." *Jakarta Post*, September 13, 2001, n.p.(LEXIS).

Overpeck, Jonathan T., and Julia E. Cole. "Lessons from a Distant Monsoon." *Nature* 445 (January 12, 2007):270–271.

Overpeck, Jonathan T., Bette L. Otto-Bliesner, Gifford H. Miller, Daniel R. Muhs, Richard B. Alley, and Jeffrey T. Kiehl. "Paleoclimatic Evidence for Future Ice-Sheet Instability and Rapid Sea-Level Rise." *Science* 311 (March 24, 2006):1747–1750.

Overpeck, J. T., M. Strum, J. A. Francis, D. K. Perovich, M. C. Serreze, R. Benner, E. C. Carmack. F. S. Chapin III, S. C. Gerlach, L. C. Hamilton, et al. "Arctic System on Trajectory to New, Seasonally Ice-Free State." *EOS: Transactions of the American Geophysical Union* 86, no. 34 (August 23, 2005):309, 312.

Owen, David. "The Dark Side: Making War on Light Pollution." *The New Yorker*, August 20, 2007, 28–33.

"Ozone Hole of 2008." NASA Earth Observatory, October 28, 2008. http://earthobservatory.nasa.gov/Newsroom/NewImages/images.php3?img_id=18192.

"Ozone Loss Reaches New Record." Environment News service, October 2, 2006. http://www.ens-newswire.com/ens/oct2006/2006-10-02-01.asp.

Pacala, S., G. C. Hurtt, D. Baker, P. Peylin, R. A. Houghton, R. A. Birdsey, L. Heath, E. T. Sundquist, R. F. Stallard, P. Ciais, et al. "Consistent Land- and Atmosphere-Based U.S. Carbon Sink Estimates." *Science* 292 (June 22, 2001):2316–2320.

Pacala, S., and R. Socolow. "Stabilization Wedges: Solving the Climate Problem for the Next 50 Years with Current Technologies." *Science* 305 (August 13, 2004):968–972.

"Pacific Coast Turning More Acidic." Earth Observatory. Media Alerts Stories Archive. May 22, 2008. http://earthobservatory.nasa.gov/Newsroom/MediaAlerts/2008/2008052226903.html.

"Pacific Too Hot for Corals of World's Largest Reef." Environment News Service, May 23, 2002. http://ens-news.com/ens/may2002/2002-05-23-01.asp.

Packard, Kimberly O'Neill, and Forest Reinhardt. "What Every Executive Needs to Know About Global Warming." *Harvard Business Review* 78, no. 4 (July/August 2000):128–135.

Padden, Brian. "Native Alaskans Feel the Heat of Global Warming." Voice of America. August 11, 2006.

Paerl, Hans W., and Jef Huisman. "Blooms Like It Hot." *Science* 320 (April 4, 2008): 57–58.

Pagani, M., M. A. Arthur, and K. H. Freeman. "Miocene Evolution of Atmospheric Carbon Dioxide." *Paleoceanography* (1999):273–292.

Page, Susan E., Florian Siegert, John O. Rieley, Hans-Dieter V. Boehm, Adi Jaya, and Suwido Limin. "The Amount of Carbon Released from Peat and Forest Fires in Indonesia during 1997." *Nature* 420 (November 7, 2002):61–65.

Paillard, Didler. "Glacial Hiccups." *Nature* 409 (January 11, 2001):147–148.

Palmer, T. N., and J. Ralsanen. "Quantifying the Risk of Extreme Seasonal Precipitation Events in a Changing Climate." *Nature* 415 (January 30, 2002):512–514.

"Panel Says Humans 'Very Likely' Cause of Global Warming." Associated Press in *New York Times*, February 3, 2007. http://www.nytimes.com/aponline/science/AP-France-Climate.

Paraskevas, Joe. "Glaciers in the Canadian Rockies Shrinking to Their Lowest Level in 10,000 Years." *National Post* (Canada), December 4, 2003, A-8.

Parker, Cindy L., and Steven M. Shapiro. *Climate Chaos: Your Health at Risk: What You Can Do to Protect Yourself and Your Family.* Westport, CT: Praeger, 2008.

Parker-Pope, Tara. "Climate Changes Are Making Poison Ivy More Potent." *Wall Street Journal* June 26, 2007, D-1.

Parmesan, Camille, Nils Ryrholm, Constanti Stefanescu, Jane K. Hill, Chris D. Thomas, Henri Descimon, Brian Huntley, Lauri Kaila, Jaakko Kulberg, Toomas Tammaru, et al. "Poleward Shifts in Geographical Ranges of Butterfly Species Associated with Regional Warming." *Nature* 399 (June 10, 1999):579–583.

Parmesan, Camille, and Gary Yohe. "A Globally Coherent Fingerprint of Climate Change Impacts across Natural Systems." *Nature* 421 (January 2, 2003):37–42.

Parry, Martin. *Climate Change and World Agriculture.* London: Earthscan, 1990.

Parry, Martin, and Zhang Jiachen. "The Potential Effect of Climate Changes on Agriculture." In *Climate Change: Science, Impacts, and Policy,* ed. J. Jager and H. L. Ferguson, 279–289. Proceedings of the Second World Climate Conference. Cambridge: Cambridge University Press, 1991.

Parsons, Michael L. *Global Warming: The Truth behind the Myth.* New York: Plenum Press/ Insight, 1995.

Paterson, W. S. B., and N. Reeh. "Thinning of the Ice Sheet in Northwest Greenland over the Past Forty Years." *Nature* 414 (November 1, 2001):60–62.

Patterson, Kathryn L., James W. Porter, Kim B. Ritchie, Shawn W. Polson, Erich Mueller, Esther C. Peters, Deborah L. Santavy, and Garriet W. Smith. "The Etiology of White Pox, a Lethal Disease of the Caribbean Elkhorn Coral, *Acropora Palmata.*" *Proceedings of the National Academy of Sciences* 99, no. 13 (June 25, 2002):8725–8730.

Patzek, Tad W. "The Real Biofuel Cycles." Letter to the Editor, *Science* 312 (June 23, 2006):1747.

Pauli, H., M. Gottfried, and M. Grabherr. "Effects of Climate Change on Mountain Ecosystems—Upward Shifting of Alpine Plants." *World Resources Review* 8 (1996):382–390.

Payne, A. J., A. Vieli, A. P. Shepherd, D. J. Wingham, and E. Rignot. "Recent Dramatic Thinning of Largest West Antarctic Ice Stream Triggered by Oceans." *Geophysical Research Letters* 31, no. 23 (December 9, 2004), L23401, doi: 10.1029/2004GL021284.

Pearce, Fred. "Nature Plants Doomsday Devices." *The Guardian* (England), November 25, 1998. http://go2.guardian.co.uk/science/912000568-disast.html.

Pearce, Fred. "Massive Peat Burn Is Speeding Climate Change." NewScientist.com, November 3, 2004. http://www.newscientist.com/news/news.jsp?id=ns99996613.

Pearce, Fred. "Failing Ocean Current Raises Fears of Mini Ice Age." NewScientist.com, November 30, 2005. http://www.newscientist.com/article.ns?id=dn8398.

Pearce, Fred. "But Here's What They Didn't Tell Us." *New Scientist,* February 10–16, 2007a, 6–9.

Pearce, Fred. *With Speed and Violence: Why Scientists Fear Tipping Points in Climate Change.* Boston: Beacon Press, 2007b.

"Pearl Jam Offsets Climate Footprint of 2006 World Tour." Environment News Service, July 10, 2006.

Pearson, Paul N., Peter W. Ditchfield, Joyce Singano, Katherine G. Harcourt-Brown, Christopher J. Nicholas, Richard K. Olsson, Nicholas J. Shackleton, and Mike A. Hall. "Warm Tropical Sea-surface Temperatures in the Late Cretaceous and Eocene Epochs." *Nature* 413 (October 4, 2001):481–487.

Pearson, Paul N., and Martin R. Palmer. "Atmospheric Carbon Dioxide Concentrations over the Past 60 Million Years." *Nature* 406 (August 17, 2000):695–699.

Pegg. J. R. "Plants Prospering from Climate Change." Environment News Service, June 5, 2003. http://ens-news.com/ens/jun2003/2003-06-06-10.asp.

Pegg, J. R. "The Earth Is Melting, Arctic Native Leader Warns." Environment News Service, September 16, 2004. www.ens-newswire.com/ens/sep2004/2004-09-16-10.asp.

Pegg, J. R. "Climate Change Increases Food Security Concerns." Environment News Service, December 5, 2006. http://www.ens-newswire.com/ens/dec2006/2006-12-05-01.asp.

Pegg, J. R. "U.S. Congress Warming to Climate Debate." Environment News Service, January 30, 2007a. http://www.ens-newswire.com/ens/jan2007/2007-01-30-10.asp.

Pegg, J. R. "Effects of Bush Climate Science Censorship Linger." Environment News Service, February 7, 2007b. http://www.ens-newswire.com/ens/feb2007/2007-02-07-10.asp.

Pegg, J. R. "U.S. Lawmakers Hear Stern Warnings on Climate Change." Environment News Service, February 14, 2007c. http://www.ens-newswire.com/ens/feb2007/2007-02-14-10.asp.

Pegg, J. R. "Gore Urges Immediate U.S. Freeze on Warming Emissions." Environment News Service, March 21, 2007d. http://www.ens-newswire.com/ens/mar2007/2007-03-21-11.asp.

Pegg, J. R. "Warming Climate Adds to U.S. Flood Fears." Environment News Service, July 2, 2008. http://www.ens-newswire.com/ens/jul2008/2008-07-02-10.asp.

Pelton, Tom. "New Maps Highlight Vanishing Eastern Shore: Technology Provides a Stark Forecast of the Combined Effect of Rising Sea Level and Sinking Land along the Bay." *Baltimore Sun*, July 30, 2004, 1-A.

Peng, Shaobing, Jianliang Huang, John E. Sheehy, Rebecca C. Laza, Romeo M. Visperas, Xuhua Zhong, Grace S. Centeno, Gurdev S. Khush, and Kenneth G. Cassman. "Rice Yields Decline with Higher Night Temperature from Global Warming." *Proceedings of the National Academy of Sciences* 101, no. 27 (July 6, 2004):9971–9975.

Pennell, William, and Tim Barnett, eds. "Special Issue: The Effects of Climate Change on Water Resources in the West." *Climatic Change* 62, no. 1–3 (January and February 2004).

Pennisi, Elizabeth. "U.S. Weighs Protection for Polar Bears." *Science* 315 (January 5, 2007):25.

Pennisi, Elizabeth. "Calcification Rates Drop in Australian Reefs." *Science* 323 (January 2, 2009a):27.

Pennisi, Elizabeth. "Western U.S. Forests Suffer Death by Degrees." *Science* 323 (January 23, 2009b):447.

Pennisi, Elizabeth, Jesse Smith, and Richard Stone. "Momentous Changes at the Poles." *Science* 315 (March 16, 2007):1513.

Peretz, Lawrence N., ed. *Climate Change Research Progress.* Hauppauge, NY: Nova Science, 2008.

Perlman, David. "Decline in Oceans' Phytoplankton Alarms Scientists: Experts Pondering Whether Reduction of Marine Plant Life Is Linked to Warming of the Seas." *San Francisco Chronicle*, October 6, 2003, A-6.

Perlman, David. "Shrinking Glaciers Evidence of Global Warming: Differences Seen by Looking at Photos from 100 Years Ago." *San Francisco Chronicle*, December 17, 2004, A-18.

Perry, Michael. "Global Warming Devastates World's Coral Reefs." Reuters, November 26, 1998. http://www.gsreport.com/articles/art000023.html.

Peterka, Amanda E. "Sustainable Hospitality." *EJ* [Environmental Journalism], Michigan State University Knight Center for Environmental Journalism 7, no. 1 (Fall 2007):20–21, 38.

Peters, Robert L. "Effects of Global Warming on Biological Diversity." In *The Challenge of Global Warming*, ed. Edwin Abrahamson, 82–95. Washington, DC: Island Press, 1989.

Peterson, Bruce J., Robert M. Holmes, James W. McClelland, Charles J. Vorosmarty, Richard B. Lammers, Alexander I. Shiklomanov, Igor A. Shiklomanov, and Stefan Rahmstorf. "Increasing River Discharge to the Arctic Ocean." *Science* 298 (December 13, 2002):2171–2173.

Peterson, Matthew. *Global Warming and Global Politics.* London: Routledge, 1996.

Petit, Charles W. "Polar Meltdown: Is the Heat Wave on the Antarctic Peninsula a Harbinger of Global Climate Change?" *U.S. News and World Report*, February 28, 2000, 64–74.

Petit, Charles W. "Arctic Thaw." *U.S. News and World Report,* November 8, 2004, 66–69.
Petit, Charles W. "In the Rockies, Pines Die and Bears Feel It." *New York Times,* January 30, 2007. http://www.nytimes.com/2007/01/30/science/30bear.html.
Petit, J. R., J. Jouzel, D. Raynaud, N.I. Barkov, J.-M. Barnola, I. Basile, M. Benders, I. Chappellaz, M. Davis, G. Delaygue, M. Delmotte, V. M. Kotlyakov, M. Legrand, V. Y. Lipenkov, C. Lorius, L. Pepin, C. Ritz, E. Saltzman, and M. Stievenard. "Climate and Atmospheric History of the Past 420,000 Years from the Vostok Ice Core, Antarctica." *Nature* 399 (June 3, 1999):429–436.
Petition to the Inter-American Commission on Human Rights Seeking Relief from Violations Resulting from Global Warming Caused by Acts and Omissions of the United States. Submitted by Sheila Watt-Cloutier with the Support of the Inuit Circumpolar Conference, and Behalf of All Inuit in the Arctic. Iqaluit, Nunavut, December 7, 2005. http://inuitcircumpolar.com/index.php?ID=316&Lang=En.
Petrillo, Lisa. "Turning the Tide in Venice." Copley News Service, April 28, 2003 (LEXIS).
Pew Center on Global Climate Change. "Experts Say Global Warming More Than Predicted: A New Study Released by the Pew Center on Global Climate Change Foresees Greater Global Warming Than Previously Predicted, Along with Greater Extremes of Weather and Faster Sea-level Rise." July 10, 1999. http://www.gsreport.com/articles/art000175.html.
Pfeffer W. T., J. T. Harper, and S. O'Neel. "Kinematic Constraints on Glacier Contributions to 21st-Century Sea-Level Rise." *Science* 321 (September 5, 2008):1340–1343.
Phelan, Amanda. "Turning up the Heat." *Sydney Sunday Telegraph,* August 18, 2002, 47.
"Phew, What a Scorcher—and It's Going to Get Worse." Agence France Presse, December 1, 2004 (LEXIS).
Philander, S. George. *Is the Temperature Rising? The Uncertain Science of Global Warming.* Princeton, NJ: Princeton University Press, 1998.
Philander, S. George. *Our Affair with El Niño: How We Transformed an Enchanting Peruvian Current into a Global Climate Hazard.* Princeton, NJ: Princeton University Press, 2006.
Philander, S. George, ed. *Encyclopedia of Global Warming and Climate Change.* 4 vols. Los Angeles: Sage, 2008.
Phillips, John. "Tropical Fish Bask in Med's Hot Spots." *London Times,* July 15, 2000, n.p.
Phillips, Oliver L., Yadvinder Malhi, Niro Higuchi, William F. Laurance, Percy V. Nuñez, Rodolfo M. Vasquez, Susan G. Laurance, Leandro V. Ferreira, Margaret Stern, Sandra Brown, and John Grace. "Changes in the Carbon Balance of Tropical Forests: Evidence from Long-term Plots." *Science* 282 (October 16, 1998):439–442.
Pianin, Eric. "A Baltimore without Orioles? Study Says Global Warming May Rob Maryland, Other States of Their Official Birds." *Washington Post,* March 4, 2002a, A-3.
Pianin, Eric. "Study Fuels Worry over Glacial Melting: Research Shows Alaskan Ice Mass Vanishing at Twice Rate Previously Estimated." *Washington Post,* July 19, 2002b, A-14.
Pianin, Eric. "On Global Warming, States Act Locally: At Odds with Bush's Rejection of Mandatory Cuts, Governors and Legislatures Enact Curbs on Greenhouse Gases." *Washington Post,* November 11, 2002c, A-3.
Pielke, Roger. "Land Use Changes and Climate Change." *Philosophical Transactions of the Royal Society of London: Mathematical, Physical, and Engineering Sciences,* August 2002.
Pilewskie, Peter. "Aerosols Heat Up." *Nature* 448 (August 3, 2007):541–542.
Pilkey, Orrin H., and Andrew G. Cooper. "Society and Sea Level Rise." *Science* 303 (March 19, 2004):1781–1782.
Pincock, Steve. "Showdown in a Sunburnt Country." *Nature* 450 (November 15, 2007):336–338.
"Pine Beetles Changing Rocky Mountain Air Quality, Weather." Environment News Service, October 1, 2008. http://www.ens-newswire.com/ens/oct2008/2008-10-01-091.asp.
Pittman, Craig. "Global Warming Report Warns: Seas Will Rise." *St. Petersburg Times,* October 24, 2001, 3-B.
"Planting Northern Forests Would Increase Global Warming." NewScientist.com, July 11, 2001. http://www.newscientist.com/news/news.jsp?id=ns99991003.
Pleven, Liam, Ian McDonald, and Karen Richardson. "As Hurricane Season Starts, Disaster Insurance Runs Short." *Wall Street Journal,* July 10, 2006, A-1, A-8.
Poggioli, Sylvia. "Venice Struggling with Increased Flooding." National Public Radio Morning Edition, November 29, 2002 (LEXIS).
Pohl, Otto. "New Jellyfish Problem Means Jellyfish Are Not the Only Problem." *New York Times,* May 21, 2002, F-3.

Polakovic, Gary. "States Taking the Initiative to Fight Global Warming: Unhappy with Bush's Policies, Local Officials Work to Slow Climate Change." *Los Angeles Times*, October 7, 2001a, A-1.

Polakovic, Gary. "Deforestation Far Away Hurts Rain Forests, Study Says: Downing Trees on Costa Rica's Coastal Plains Inhibits Cloud Formation in Distant Peaks." *Los Angeles Times*, October 19, 2001b, A-1.

Polakovic, Gary. "Airborne Soot Is Significant Factor in Global Warming, Study Says." *Los Angeles Times*, May 15, 2003, A-30.

"Polar Bears Shift Dens." *Townsville Bulletin* (Australia), January 25, 2007, 12.

Polgreen, Lydia. "New Depths: A Godsend for Darfur, or a Curse?" *New York Times*, July 22, 2007. http://www.nytimes.com/2007/07/22/weekinreview/22polgreen.html.

"Poll: Global Warming as Big a Threat as Terrorism." Environment News Service, March 29, 2007. http://www.ens-newswire.com/ens/mar2007/2007-03-29-09.asp#anchor2.

Pomerance, Rafe. "The Dangers from Climate Warming: A Public Awakening." In *The Challenge of Global Warming*, ed. Edwin Abrahamson, 259–269. Washington, DC: Island Press, 1989.

Portner, Hans O., and Rainer Knust. "Climate Change Affects Marine Fishes through the Oxygen Limitation of Thermal Tolerance." *Science* 315 (January 5, 2007):95–97.

"Portugal Celebrates Massive Solar Plant" Associated Press in *New York Times*, March 28, 2007. http://www.nytimes.com/aponline/technology/AP-Portugal-Solar-Power-Plant.html.

Poterba, James. "Tax Policy to Combat Global Warming." In *Global Warming: Economic Policy Reponses*, Rutiger Dornbusch and James M. Poterba, 71–98. Cambridge, MA: MIT Press, 1991.

Potter, Thomas D., and Bradley R. Colman, eds. *Handbook of Weather, Climate, and Water: Dynamics, Climate, Physical Meteorology, Weather Systems, and Measurements*. Hoboken, NJ: Wiley-Interscience, 2003.

Pounds, J. Alan. "Climate and Amphibian Decline." *Nature* 410 (April 5, 2001):639–640.

Pounds, J. Alan, Martin R. Bustamante, Luis A. Coloma, Jamie A. Consuegra, Michael P. L. Fogden, Pru N. Foster, Enrique La Marca, Karen L. Masters, Andres Merino-Viteri, Robert Puschendorf, et al. "Widespread Amphibian Extinctions from Epidemic Disease Driven by Global Warming." *Nature* 439 (January 12, 2006):161–167.

Pounds, J. Alan, and Robert Puschendorf. "Clouded Futures." *Nature* 427 (January 8, 2004):107–108.

Powell, James Lawrence. *Dead Pool: Lake Powell, Global Warming, and the Future of Water in the West*. Berkeley: University of California Press, 2009.

Powell, Michael. "Northeast Seen Getting Balmier: Studies Forecast Altered Scenery, Coast." *Washington Post*, December 17, 2001, A-3.

Powell, Michael. "The End of Eden: James Lovelock Says This Time We've Pushed the Earth Too Far." *Washington Post*, September 2, 2006, C-1. http://www.washingtonpost.com/wp-dyn/content/article/2006/09/01/AR2006090101800_pf.html.

"Power Games: Britain Was a World Leader in Wind Farm Technology. So Why Are All Our Windmills Made in Denmark?" *London Guardian*, July 16, 2001, 2.

Powlson, David. "Will Soil Amplify Climate Change?" *Nature* 433 (January 20, 2005):204–205.

Prasad, Monica. "On Carbon, Tax and Don't Spend." *New York Times*, March 25, 2008. http://www.nytimes.com/2008/03/25/opinion/25prasad.html.

Prather, Michael J. "An Environmental Experiment with H2?" *Science* 302 (October 24, 2003):581–582.

"Prehistoric Extinction Linked to Methane." Associated Press in *Omaha World-Herald*, July 27, 2000, 9.

Prigg, Mark. "Despite All the Heavy Rain, That was the Hottest June for 28 Years." *London Evening Standard*, July 1, 2004, A-9.

Pringle, Laurence P. *Global Warming: The Threat of Earth's Changing Climate*. New York: SeaStar Books, 2001.

Prins, Gwyn, and Steve Rayner. "Time to Ditch Kyoto." *Nature* 449 (October 25, 2007):973–975.

Putkonen, J. K., and G. Roe. "Rain-on-snow Events Impact Soil Temperatures and Affect Ungulate Survival." *Geophysical Research Letters* 30, no. 4 (2003):1188–1192.

Quadfasel, Detlef. "Oceanography: The Atlantic Heat Conveyor Slows." *Nature* 438 (December 1, 2005):565–566.

Quammen, David. "Planet of Weeds: Tallying the Losses of Earth's Animals and Plants." *Harpers*, October 1998, 57–69.

Quayle, Wendy C., Lloyd S. Peck, Helen Peat, J. C. Ellis-Evans, and P. Richard Harrigan. "Extreme Responses to Climate Change in Antarctic Lakes." *Science* 295 (January 25, 2002):645.

"Quebec's Smoky Warning." Editorial, *Baltimore Sun*, July 9, 2002, 10-A

"Rachel's No. 466: Warming and Infectious Diseases." November 2, 1995. Environmental Research Foundation, Annapolis, Maryland. http://www.igc.apc.org/awea/wew/othersources/rachel466.html.

Radford, Tim. "Greenhouse Buildup Worst for 20m[Million] Years." *London Guardian*, August 17, 2000a, 9.

Radford, Tim. "World May Be Warming up Even Faster: Climate Scientists Warn New Forests Would Make Effects Worse." *London Guardian*, November 9, 2000b, 10.

Radford, Tim. "Antarctic Ice Cap Is Getting Thinner: Scientists' Worries That the South Polar Ice Sheet Is Melting May Be Confirmed by the Dramatic Retreat of the Region's Biggest Glacier." *London Guardian*, February 2, 2001a, 9.

Radford, Tim. "As the World Gets Hotter, Will Britain Get Colder? Plunging Temperatures Feared after Scientists Find Gulf Stream Changes." *London Guardian*, June 21, 2001b, 3.

Radford, Tim. "Coral Reefs Face Total Destruction within 50 Years." *London Guardian*, September 6, 2001c, 9.

Radford, Tim. "Ten Key Coral Reefs Shelter Much of Sea Life: American Association Scientists Identify Vulnerable Marine 'Hot Spots' with the Richest Biodiversity on Earth." *London Guardian*, February 15, 2002a, 12.

Radford, Tim. "World Sickens as Heat Rises: Infections in Wildlife Spread as Pests Thrive in Climate Change." *London Guardian*, June 21, 2002b, 7.

Radford, Tim. "85 Percent of Alaskan Glaciers Melting at 'Incredible Rate.'" *London Guardian*, July 19, 2002c, 9.

Radford, Tim. "Scientists Discover the Harbinger of Drought: Subtle Temperature Changes in Tropical Seas May Trigger Northern Hemisphere's Long, Dry Spells." *London Guardian*, January 31, 2003, 18.

Radford, Tim. "2020: The Drowned World." *London Guardian*, September 11, 2004, 10.

Radowitz, John von. "Calmer Sun Could Counteract Global Warming." Press Association, October 5, 2003 (LEXIS).

Radowitz, John von. "Global Warming 'Smoking Gun' Found in the Oceans." Press Associated Ltd., February 18, 2005 (LEXIS).

Rahmstorf, Stefan. "Shifting Seas in the Greenhouse?" *Nature* 399 (June 10, 1999):523–524.

Rahmstorf, Stefan. "Thermohaline Circulation: The Current Climate." *Nature* 421 (February 13, 2003):699.

Rahmstorf, Stefan. "A Semi-Empirical Approach to Projecting Future Sea-Level Rise." *Science* 215 (January 19, 2007):368–370.

Rahmstorf, Stefan, Anny Cazenave, John A. Church, James E. Hansen, Ralph F. Keeling, David E. Parker, and Richard C. J. Somerville. "Recent Climate Observations Compared to Projections." *Science* 216 (May 4, 2007):709.

Rajeev, K., and V. Ramanathan. "Direct Observations of Clear-Sky Aerosol Radiative Forcing from Space during the Indian Ocean Experiment." *Journal of Geophysical Research/Atmospheres* 106, no. D15 (August 16, 2001):17,221.

Ralston, Greg. "Study Admits Arctic Danger." *Yukon News*, November 15, 1996. http://yukonweb.com/community/yukon-news/1996/nov15.htmld/#study.

Ramanathan, V. "Observed Increases in Greenhouse Gases and Predicted Climatic Changes." In *The Challenge of Global Warming*, ed. Edwin Abrahamson, 239–247. Washington, DC: Island Press, 1989.

Ramanathan, V. "Trace Gas Trends and Change." Testimony before the Senate Subcommittee on Environmental Protection, January 23, 1987.In *The Rising Tide: Global Warming and World Sea Levels*, Lynne T. Edgerton and the Natural Resources Defense Council, 11,108. Washington, DC: Island Press, 1991.

Ramanathan, V., P. J. Crutzen, J. T. Kiehl, and D. Rosenfeld. "Aerosols, Climate, and the Hydrological Cycle." *Science* 294 (December 7, 2001):2119–2124.

Ramanathan, V., and Y. Feng. "On Avoiding Dangerous Anthropogenic Interference with the Climate System: Formidable Challenges Ahead." *Proceedings of the National Academy of Sciences* 105, no. 38 (September 23, 2008):14,245–14,250.

Ramanathan, V., M. V. Ramana, G. Roberts, D. Kim, C. Corrigan, C. Chung, and D. Winker. "Warming Trends in Asia Amplified by Brown Cloud Solar Absorption." *Nature* 448 (August 2, 2007):575–578.

Ramaswamy, V., M. D. Schwarzkopf, W. J. Randel, B. D. Santer, B. J. Soden, and G. L. Stenchikov. "Anthropogenic and Natural Influences in the Evolution of Lower Stratospheric Cooling." *Science* 311 (February 24, 2006):1138–1141.

Ramseur, Jonathan L., Larry Parker, and Brent D. Yacobucci, eds. *Greenhouse Gases: Management, Reduction, and Impact.* Huntington, NY: Novinka (Nova Science), 2008.

Randall, C. E., V. L. Harvey, G. L. Manney, Y. Orsolini, M. Codrescu, C. Sioris, S. Brohede, C. S. Haley, L. L. Gordley, J. M. Zawodny, and J. M. Russell III. "Stratospheric Effects of Energetic Particle Precipitation in 2003–2004." *Geophysical Research Letters* 32 (March 2, 2005), L05802, doi: 10.1029/2004GL022003.

Randerson, J. T., H. Liu, M. G. Flanner, S. D. Chambers, Y. Jin, P. G. Hess, G. Pfister, M. C. Mack, K. K. Treseder, L. R. Welp, F. S. Chapin, J. W. Harden, M. L. Goulden, E. Lyons, J. C. Neff, E. A. G. Schuur, and C. S. Zender. "The Impact of Boreal Forest Fire on Climate Warming." *Science* 314 (November 17, 2006):1,130–1.132.

Range, Stacey. "Climatologists Say Midlands in Dust Bowl-like Drought." *Omaha World-Herald*, January 12, 2000, 18.

Rappaport, Ann, and Sarah Hammond Creighton. *Degrees That Matter: Climate Change and the University.* Cambridge, MA: Massachusetts Institute of Technology Press, 2007.

"Rare Sighting of Wasp North of Arctic Circle Puzzles Residents." Canadian Broadcasting Corporation, September 9, 2004. http://www.cbc.ca/story/science/national/2004/09/09/wasp040909.html.

Raupach, Michael R., Gregg Marland, Philippe Ciais, Corinne Le Quere, Josep G. Canadell, Gernot Klepper, and Christopher B. Field. "Global and Regional Drivers of Accelerating CO_2 Emissions." *Proceedings of the National Academy of Sciences* 104, no. 24 (June 12, 2007):10,288–10,293.

Raver, Anne. "It Eats CO_2 for Breakfast." *New York Times*, July 17, 2008. http://www.nytimes.com/2008/07/17/garden/17garden.html.

Ray, G. Carlston. "Dispatches from the Bering Sea: The Icy Seascape." *University of Virginia Today*, May 11, 2007. http://www.virginia.edu/uvatoday/beringsea/bering-sea4.html.

Ray, G. Carleton, Gary L. Hufford, Igor I. Krupnik, James E. Overland, Benjamin S. Halpern, Carrie V. Kappel, Fiorenza Micheli, Kimberly A. Selkoe, Caterina d'Agrosa, John Bruno, et al. "Diminishing Sea Ice." *Science* 321 (September 12, 2008):1443–1445.

Raymond, Peter A., and Jonathan J. Cole. "Increase in the Export of Alkalinity from North America's Largest River." *Science* 301 (July 4, 2003):88–91.

Recar, Paul. "Study: Elements Can Stunt Plant Growth." Associated Press Online, December 5, 2002 (LEXIS).

"Recent Warming of Arctic May Affect World-wide Climate." National Aeronautics and Space Administration Press Release, October 23, 2003. http://www.gsfc.nasa.gov/topstory/2003/1023esuice.html.

"Record Rainfall Floods India." *New York Times*, July 28, 2005, A-12.

Reed, Leslie. "Research: Ethanol Isn't So Wasteful." *Omaha World-Herald*, January 27, 2009, 1-A, 2-A.

Reed, Nicholas. "Mild Winter Stirs Wildlife to Early Thoughts of Love." *Vancouver Sun*, February 12, 2003, B-1.

Regalado, Antonio. "Skeptics on Warming Are Criticized." *Wall Street Journal*, July 31, 2003, A-3, A-4.

Regalado, Antonio. "The Ukukus Wonder Why a Sacred Glacier Melts in Peru's Andes." *Wall Street Journal*, June 17, 2005, A-1, A-10.

Reid, K., and J. P. Croxall. "Environmental Response of Upper Trophic-level Predators Reveals a System Change in an Antarctic Marine Ecosystem." *Proceedings of the Royal Society of London* B268 (2001):377–384.

Reid, T. R. "As White Cliffs Are Crumbling, British Want Someone to Blame." *Washington Post*, May 13, 2001, A-23.

Remington, Robert. "Goodbye to Glaciers: Thanks to Global Warming, Mountains—the World's Water Towers—Are Losing Their Ice. As It Disappears, so Does an Irreplaceable Source of Water." *Financial Post* (Canada), September 6, 2002, A-19. *Renewing the Earth: An Invitation to Reflection and Action on Environment in Light of Catholic Social Teaching.* Washington, DC: United States Catholic Conference, n.d.

"A Renewed Focus on Passenger Trains." Associated Press in *New York Times*, November 3, 2008, SA-15.

"Report: Last Ice Age Had Quick End." Associated Press in *Omaha World-Herald*, October 29, 1999, 15.

"Report Says Oceans Hit by Carbon Dioxide Use." *Boston Globe* in *Omaha World-Herald*, July 17, 2004, 5-A.

"Research Casts Doubt on China's Pollution Claims." *Washington Post*, August 15, 2001, A-16. http://www.washingtonpost.com/wp-dyn/articles/A10645-2001Aug14.html.

"Researchers Warm Up to Melt's Role in Greenland Ice Loss." NASA Earth Observatory, April 17, 2008. http://earthobservatory.nasa.gov/Newsroom/NasaNews/2008/2008041726522.html.

"Research Finds That Earth's Climate Is Approaching 'Dangerous' Point." Press Release, NASA Goddard Institute for Space Studies, New York City, May 29, 2007.

Revelle, R., and H. E. Suess. "Carbon Dioxide Exchange between Atmosphere and Ocean and the Question of an Increase of Atmospheric CO_2 during the Past Decades." *Tellus* 9 (1957):18–27.

Revkin, Andrew C. "Study Faults Humans for Large Share of Global Warming." *New York Times*, July 14, 2000a, A-12.

Revkin, Andrew C. "Planting New Forests Can't Match Saving Old Ones in Cutting Greenhouse Gases, Study Finds." *New York Times*, September 22, 2000b, A-23.

Revkin, Andrew C. "Antarctic Test Raises Hope on a Global-Warming Gas." *New York Times*, October 12, 2000c, A-18.

Revkin, Andrew C. "A Message in Eroding Glacial Ice: Humans Are Turning Up the Heat." *New York Times*, February 19, 2001a, A-1.

Revkin, Andrew C. "Two New Studies Tie Rise in Ocean Heat to Greenhouse Gases." *New York Times*, April 13, 2001b, A-15.

Revkin, Andrew C. "Both Sides Now: New Way That Clouds May Cool." *New York Times*, June 19, 2001c, F-4.

Revkin, Andrew C. "A Chilling Effect on the Great Global Melt." *New York Times*, January 18, 2002a, A-17.

Revkin, Andrew C. "Forecast for a Warmer World: Deluge and Drought." *New York Times*, August 28, 2002b, A-10.

Revkin, Andrew C. "Scientists Say a Quest for Clean Energy Must Begin Now." *New York Times*, November 1, 2002c, A-6.

Revkin, Andrew C. "Climate Talks Shift Focus to How to Deal with Changes." *New York Times*, November 3, 2002d, A-8.

Revkin, Andrew C. "Study of Antarctic Points to Rising Sea Levels." *New York Times*, March 7, 2003, A-8.

Revkin, Andrew C. "Climate Debate Gets Its Icon: Mt. Kilimanjaro." *New York Times*, March 23, 2004a. http://www.nytimes.com/2004/03/23/science/earth/23CLIM.html.

Revkin, Andrew C. "An Icy Riddle as Big as Greenland." *New York Times*, June 8, 2004b. http://www.nytimes.com/2004/06/08/science/earth/08gree.html.

Revkin, Andrew C. "Antarctic Glaciers Quicken Pace to Sea: Warming Is Cited." *New York Times*, September 24, 2004c, A-24.

Revkin, Andrew C. "Global Warming Is Expected to Raise Hurricane Intensity." *New York Times*, September 30, 2004d, A-20.

Revkin, Andrew C. "Bush vs. the Laureates: How Science Becomes a Partisan." *New York Times*, October 19, 2004e, F-1.

Revkin, Andrew C. "2004 Was Fourth-Warmest Year Ever Recorded." *New York Times*, February 10, 2005a. http://www.nytimes.com/2005/02/10/science/10warm.html.

Revkin, Andrew C. New Climate Model Highlights Arctic's Vulnerability." *New York Times*, October 31, 2005b. http://www.nytimes.com/2005/10/31/science/earth/01warm_web.html

Revkin, Andrew C. *The North Pole Was Here: Puzzles and Perils at the Top of the World.* New York: Kingfisher, 2006a.

Revkin, Andrew C. "Studies Portray Tropical Arctic in Distant Past." *New York Times*, June 1, 2006b. http://www.nytimes.com/2006/06/01/science/earth/01climate.html.

Revkin, Andrew C. "*An Inconvenient Truth*: Al Gore's Fight against Global Warming. *New York Times*, May 22, 2006c. http://www.nytimes.com/2006/05/22/movies/22gore.html.

Revkin, Andrew C. "NASA's Goals Delete Mention of Home Planet." *New York Times*, July 22, 2006d, A-1, A-10.

Revkin, Andrew C. "Climate Experts Warn of More Coastal Building." *New York Times*, July 25, 2006e, D-2.

Revkin, Andrew C. "Team Looks at Seafloor as Gas Trap." *New York Times*, August 8, 2006f. http://www.nytimes.com/2006/08/08/science/earth/08carbon.html.

Revkin, Andrew C. "Last 7 Months Were Warmest Stretch on Record." *New York Times*, August 8, 2006g. http://www.nytimes.com/2006/08/08/science/earth/08brfs-006.html.

Revkin, Andrew C. "Study Links Tropical Ocean Warming to Greenhouse Gases." *New York Times*, September 12, 2006h. http://www.nytimes.com/2006/09/12/science/12ocean.html.

Revkin, Andrew C. "NASA Scientists See New Signs of Global Warming." *New York Times*, September 14, 2006i. http://www.nytimes.com/2006/09/14/science/earth/14climate.html.

Revkin, Andrew C. "Gore Calls for Immediate Freeze on Heat-Trapping Gas Emissions." *New York Times*, September 19, 2006j. http://www.nytimes.com/2006/09/19/washington/19gore.html.

Revkin, Andrew C. "Budgets Falling in Race to Fight Global Warming." *New York Times*, October 30, 2006k. http://www.nytimes.com/2006/10/30/business/worldbusiness/30energy.html.

Revkin, Andrew C. "By 2040, Greenhouse Gases Could Lead to an Open Arctic Sea in Summers." *New York Times*, December 12, 2006l. http://www.nytimes.com/2006/12/12/science/earth/12arcti.html.

Revkin, Andrew C. "Arctic Ice Shelf Broke Off Canadian Island." *New York Times*, December 30, 2006m. http://www.nytimes.com/2006/12/30/science/earth/30ice.html.

Revkin, Andrew C. "Reports from Four Fronts in the War on Warming." *New York Times*, April 3, 2007a. http://www.nytimes.com/2007/04/03/science/earth/03clim.html.

Revkin, Andrew C. "Carbon-neutral Is Hip, but Is It Green? *New York Times*, April 29, 2007b. http://www.nytimes.com/2007/04/29/weekinreview/29revkin.html.

Revkin, Andrew C. "Climate Panel Reaches Consensus on the Need to Reduce Harmful Emissions." *New York Times*, May 4, 2007c. http://www.nytimes.com/2007/05/04/science/04climate.html.

Revkin, Andrew C. "Analysis Finds Large Antarctic Area Has Melted." *New York Times*, May 16, 2007d. http://www.nytimes.com/2007/05/16/science/earth/16melt.html.

Revkin, Andrew C. "Study Finds Hurricanes Frequent in Some Cooler Periods." *New York Times*, May 24, 2007e. http://www.nytimes.com/2007/05/24/science/earth/24storm.html.

Revkin, Andrew C. "Cyclone Nears Iran and Oman." *New York Times*, June 6, 2007f. http://www.nytimes.com/2007/06/06/world/middleeast/06storm.html.

Revkin, Andrew C. "Many Arctic Plants Have Adjusted to Big Climate Changes, Study Finds." *New York Times*, June 15, 2007g. http://www.nytimes.com/2007/06/15/science/15arctic.html.

Revkin, Andrew C. "Arctic Melt Unnerves the Experts." *New York Times*, October 2, 2007h. http://www.nytimes.com/2007/10/02/science/earth/02arct.html.

Revkin, Andrew C. "Grim Outlook for Polar Bears." *New York Times*, October 2, 2007i, D-4.

Revkin, Andrew C. "Challenges to Both Left and Right on Global Warming." *New York Times*, November 13, 2007j. http://www.nytimes.com/2007/11/13/science/earth/13book.html.

Revkin, Andrew C. "Arctic Update: Resilient Bears, Shrinking Ice." *Dot.Earth, New York Times*, December 12, 2007k. http://dotearth.blogs.nytimes.com/2007/12/12/arctic-update-resilient-bears-vanishing-ice/index.html.

Revkin, Andrew C. "In Greenland, Ice and Instability." *New York Times*, January 8, 2008a. http://www.nytimes.com/2008/01/08/science/earth/08gree.html.

Revkin, Andrew C. "Skeptics on Human Climate Impact Seize on Cold Spell." *New York Times*, March 2, 2008b. http://www.nytimes.com/2008/03/02/science/02cold.html.

Revkin, Andrew C. "Link to Global Warming in Frogs' Disappearance Is Challenged." *New York Times*, March 25, 2008c, D-3.

Revkin, Andrew C. "In a New Climate Model, Short-Term Cooling in a Warmer World." *New York Times*, May 1, 2008d. http://www.nytimes.com/2008/05/01/science/earth/01climate.html.

Revkin, Andrew C. "5 Countries Agree to Talk, Not Compete, Over the Arctic." *New York Times*, May 29, 2008e. http://www.nytimes.com/2008/05/29/science/earth/29arctic.html.

Revkin, Andrew C. "Reports: Energy Thirst Still Topping Climate Risks." *Dot.Earth*, *New York Times*, June 25, 2008f. http://dotearth.blogs.nytimes.com/2008/06/25/reports-energy-thirst-still-topping-climate-risks/index.html.

Revkin, Andrew C. "After Applause Dies Down, Global Warming Talks Leave Few Concrete Goals." *New York Times*, July 10, 2008g. http://www.nytimes.com/2008/07/10/science/earth/10assess.html

Revkin, Andrew C. "Tropical Warming Tied to Flooding Rains." *New York Times*, August 8, 2008h. http://www.nytimes.com/2008/08/08/science/earth/08rain.html.

Revkin, Andrew C. "Arctic Ocean Ice Retreats Less Than Last Year." *New York Times*, September 17, 2008i. http://www.nytimes.com/2008/09/17/science/earth/17ice.html.

Revkin, Andrew C. "Will the Next Ice Age Be Permanent?." *Dot.Earth*, *New York Times*, November 12, 2008j. http://dotearth.blogs.nytimes.com/2008/11/12/will-next-ice-age-be-permanent.

Revkin, Andrew C., and Felicity Barringer. "Bills on Climate Move to Spotlight in New Congress." *New York Times*, January 18, 2007. http://www.nytimes.com/2007/01/18/washington/18climate.html.

Revkin, Andrew C., and Patrick Healy. "Coalition to Make Buildings Energy-Efficient." *New York Times*, May 17, 2007. http://www.nytimes.com/2007/05/17/us/17climate.html.

Revkin, Andrew C., and Matthew L. Wald. "Material Shows Weakening of Climate Reports." *New York Times*, March 20, 2007a. http://www.nytimes.com/2007/03/20/washington/20climate.html.

Revkin, Andrew C., and Matthew L. Wald. "Solar Power Captures Imagination, Not Money." *New York Times*, July 16, 2007b. http://www.nytimes.com/2007/07/16/business/16solar.html.

Revkin, Andrew C., and Timothy Williams. "Global Warming Called Security Threat." *New York Times*, April 15, 2007. http://www.nytimes.com/2007/04/15/us/15warm.html.

Rex, Markus, P. von der Gathen, Alfred Wegener, R. J. Salawitch, N. R. P. Harris, M. P. Chipperfield, and B. Naujokat. "Arctic Ozone Loss and Climate Change." *Geophysical Research Letters* 31 (March 10, 2004). http://www.eurekalert.org/pub_releases/2004-03/agu-ajh031004.php.

Reynolds, James. "Earth Is Heading for Mass Extinction in Just a Century." *The Scotsman*, June 18, 2003, 6.

"Ribbon Seal of the Bering Sea Losing Icy Habitat." Environment News Service, December 26, 2007. http://www.ens-newswire.com/ens/dec2007/2007-12-26-094.asp.

Richards, Bill. "A Good Combination: Biofuels, Smart Tilling." *Omaha World-Herald*, March 11, 2008, 7-B.

Richardson, Franci. "Sharks Take the Bait: Sightings in Maine an 'Unusual Circumstance.'" *Boston Herald*, August 11, 2002, 3.

Richardson, Michael. "Indonesian Peat Fires Stoke Rise of Pollution." *International Herald-Tribune*, December 13, 2002, 5.

"Rich Countries' Greenhouse Gas Emissions Ballooning." Environment News Service, June 9, 2003. http://ens-news.com/ens/jun2003/2003-06-09-02.asp.

Richtel, Matt, and John Markoff. "A Green Energy Industry Takes Root under the California Sun." *New York Times*, February 1, 2008. http://www.nytimes.com/2008/02/01/technology/01solar.html.

Rickaby, R. E. M., and P. Halloran. "Cool La Niña during the Warmth of the Pliocene?" *Science* 307 (March 25, 2005):1948–1952.

Ridenour, David. "Hypocrisy in Buenos Aires: Millions of Gallons of Fuel to Be Burned By Those Seeking Curbs on Fuel Use." National Policy Analysis: A Publication of the National Center for Public Policy Research, No. 217, October 1998. http://nationalcenter.org/NPA217.html.

Riebesell, U., K. G. Schulz, R. G. J. Bellerby, M. Botros, P. Fritsche, M. Meyerhöfer, C. Neill, G. Nondal, A. Oschlies, J. Wohlers, and E. Zöllner. "Enhanced Biological Carbon Consumption in a High CO_2 Ocean." *Nature* 450 (November 22, 2007):545–548.

Rifkin, Jeremy. *The Hydrogen Economy: The Creation of the Worldwide Energy Web and the Redistribution of Power on Earth.* New York: Jeremy P. Tarcher/Putnam, 2002.

Rignot, E. "Fast Recession of a West Antarctic Glacier." *Science* 281 (July 24, 1998):549–551.

Rignot, E., Jonathan L. Bamber, Michiel R. van den Broeke, Curt Davis, Yonghong Li, Willem Jan van de Berg, and Erik van Meijgaard. "Recent Antarctic Ice Mass Loss from Radar Interferometry and Regional Climate Modelling." *Nature Geoscience* 1 (2008):106–110. http://www.nature.com/ngeo/journal/vaop/ncurrent/abs/ngeo102.html.

Rignot, E., G. Casassa, P. Gogineni, W. Krabill, A. Rivera, and R. Thomas. "Accelerated Ice Discharge from the Antarctic Peninsula Following the Collapse of Larsen B Ice Shelf." *Geophysical Research Letters* 31, no. 18 (September 22, 2004), doi: 10.1029/2004GL020697.

Rignot, E., Andrès Rivera, and Gino Casassa. "Contribution of the Patagonia Icefields of South America to Sea Level Rise." *Science* 302 (October 17, 2003):434–437.

Rignot, E., and R. H. Thomas. "Mass Balance of Polar Ice Sheets." *Science* 297 (August 30, 2002):1502–1506.

Rimer, Sara. "How Green Is the College? Time the Showers." *New York Times*, May 26, 2008. http://www.nytimes.com/2008/05/26/education/26green.html.

Rind, D., and J. Overpeck. "Hypothesized Causes of Decade-to-Century-Scale Climate Variability: Climate Model Results." *Quaternary Science Reviews* 12 (1993):357–374.

"Rising Tide: Who Needs Essex Anyway." *London Guardian*, June 12, 2003, 4.

Rittenour, Tammy M., Julie Brigham-Grette, and Michael E. Mann. "El Niño-like Climate Teleconnections in New England during the Late Pleistocene." *Science* 288 (May 12, 2000):1039–1042.

Ritter, Malcolm. "This Is Winter? Much of Nation Basked in Warm January." Associated Press, February 3, 2006 (LEXIS).

Rivers, Patrick N., ed. *Leading Edge Research in Solar Energy.* Hauppauge, NY: Nova Science, 2007.

Robbins, Jim. "Snow in July Is a Mixed Blessing for the Northern Rockies." *New York Times*, July 2, 2008a, A-9, A-10.

Robbins, Jim. "Bark Beetles Kill Millions of Acres of Trees in West." *New York Times*, November 18, 2008b. http://www.nytimes.com/2008/11/18/science/18trees.html.

Roberts, Callum M., Colin J. McClean, John E. N. Veron, Julie P. Hawkins, Gerald R. Allen, Don E. McAllister, Cristina G. Mittermeier, Frederick W. Schueler, Mark Spalding, Fred Wells, Carly Vynne, and Timothy B. Werner. "Marine Biodiversity Hotspots and Conservation Priorities for Tropical Reefs." *Science* 295 (February 15, 2002):1280–1284.

Roberts, Greg. "Great Barrier Grief as Warm-water Bleaching Lingers." *Sydney Morning Herald*, January 20, 2003, 4.

Roberts, Paul. *The End of Oil: The Edge of a Perilous New World.* Boston: Houghton-Mifflin, 2004.

Robertson, G. Philip, Virginia H. Dale, Otto C. Doering, Steven P. Hamburg, Jerry M. Melillo, Michele M. Wander, William J. Parton, Paul R. Adler, Jacob N. Barney, Richard M. Cruse, Clifford S. Duke, Philip M. Fearnside, Ronald F. Follett, Holly K. Gibbs, Jose Goldemberg, David J. Mladenoff, Dennis Ojima, Michael W. Palmer, Andrew Sharpley, Linda Wallace, Kathleen C. Weathers, John A. Wiens, and Wallace W. Wilhelm. "Sustainable Biofuels Redux." *Science* 322 (October 3, 2008):49–50.

Robock, Alan. "Whither Geoengineering?" *Science* 320 (May 30, 2008):1166–1167.

Robson, Seth. "Glaciers Melting." *Christchurch Press* (New Zealand), February 26, 2003, 13.

"Rock Measurements Suggest Warming Is Global." Environment News Service, April 16, 2002. http://ens-news.com/ens/apr2002/2002L-04-16-09.html.

Roe, Nicholas. "Show Me a Home Where the Reindeer Roam." *London Times*, November 10, 2001, n.p.(LEXIS).

Rogers, Paul. "Solar Energy Heats Up." *Omaha World-Herald*, October 15, 2006, 1-RE, 2-RE.

Rohling, E. J., K. Grant, Ch. Hemleben, M. Siddall, B. A. A. Hoogakker, M. Bolshaw, and M. Kucera. "High Rates of Sea-level Rise during the Last Interglacial Period." *Nature Geoscience* (December 16, 2007), doi:10.1038/ngeo.2007.28. http://www.nature.com.

Rohter, Larry. "Punta Arenas Journal: In an Upside-Down World, Sunshine Is Shunned." *New York Times*, December 27, 2002, A-4.

Rohter, Larry. "Deep in the Amazon Forest, Vast Questions about Global Climate Change." *New York Times*, November 4, 2003. http://nytimes.com/2003/11/04/science/earth/04AMAZ.html.

Rohter, Larry. "Antarctica, Warming, Looks Ever More Vulnerable." *New York Times*, January 25, 2005a. http://www.nytimes.com/2005/01/25/science/earth/25ice.html.

Rohter, Larry. "A Record Amazon Drought, and Fear of Wider Ills." *New York Times*, December 11, 2005b. http://www.nytimes.com/2005/12/11/international/americas/11amazon.html.

Rohter, Larry. "With Big Boost from Sugar Cane, Brazil Is Satisfying Its Fuel Needs." *New York Times*, April 10, 2006. http://www.nytimes.com/2006/04/10/world/americas/10brazil.html.

Rohter, Larry. "Brazil, Alarmed, Reconsiders Policy on Climate Change." *New York Times*, July 31, 2007. http://www.nytimes.com/2007/07/31/world/americas/31amazon.html.

Rolfe, Christopher. "Comments on the British Columbia Greenhouse Gas Action Plan." West Coast Environmental Law Association. A Presentation to the Air and Water Management Association, April 17, 1996. http://www.wcel.org/wcelpub/11026.html.

Romero, Simon. "Bolivia's Only Ski Resort Is Facing a Snowless Future." *New York Times*, February 2, 2007. http://www.nytimes.com/2007/02/02/world/americas/02bolivia.html.

Romm, Joseph J. *Hell and High Water—the Solution and the Politics—and What We Should Do.* New York: William Morrow, 2007.

Root, Terry L., Jeff T. Price, Kimberly L, Hall, Stephen H. Schneider, Cynthia Rosenzweig, and J. Alan Pounds. "Fingerprints of Global Warming on Wild Animals and Plants." *Nature* 421 (January 2, 2003):57–60.

Rosen, Yereth. "Alaska's Not-so-Permanent Frost." *Christian Science Monitor,* October 7, 2003, 1.

Rosenbloom, Stephanie. "Giant Retailers Look to Sun for Energy Savings." *New York Times*, August 11, 2008. http://www.nytimes.com/2008/08/11/business/11solar.html.

Rosenthal, Elisabeth. "As the Climate Changes, Bits of England's Coast Crumble." *New York Times,* May 4, 2007a. http://www.nytimes.com/2007/05/04/world/europe/04erode.html.

Rosenthal, Elisabeth. "Likely Spread of Deserts to Fertile Land Requires Quick Response, U.N. Report Says." *New York Times*, June 28, 2007b. http://www.nytimes.com/2007/06/28/world/28deserts.html.

Rosenthal, Elisabeth. "Vatican Penance: Forgive Us Our Carbon Output." *New York Times*, September 17, 2007c. http://www.nytimes.com/2007/09/17/world/europe/17carbon.html.

Rosenthal, Elisabeth. "U.N. Chief Seeks More Climate Change Leadership." *New York Times*, November 18, 2007d. http://www.nytimes.com/2007/11/18/science/earth/18climatenew.html.

Rosenthal, Elisabeth. "As Earth Warms Up, Tropical Virus Moves to Italy." *New York Times*, December 23, 2007e. http://www.nytimes.com/2007/12/23/world/europe/23virus.html.

Rosenthal, Elisabeth. "Studies Deem Biofuels a Greenhouse Threat." *New York Times*, February 8, 2008a. http://www.nytimes.com/2008/02/08/science/earth/08wbiofuels.html.

Rosenthal, Elisabeth. "Europe Turns to Coal Again, Raising Alarms on Climate." *New York Times*, April 12, 2008b. http://www.nytimes.com/2008/04/23/world/europe/23coal.html.

Rosenthal, Elisabeth. "New Trend in Biofuels Has New Risks." *New York Times*, May 21, 2008c. http://www.nytimes.com/2008/05/21/science/earth/21biofuels.html.

Rosenthal, Elisabeth. "Water Is New Battleground in Drying Spain." *New York Times*, June 3, 2008d, A-1, A-12.

Rosenthal, Elisabeth. "China Increases Lead as Biggest Carbon Dioxide Emitter." *New York Times*, June 14, 2008e. http://www.nytimes.com/2008/06/14/world/asia/14china.html.

Rosenthal, Elisabeth. "Air Travel and Carbon on Increase in Europe." *New York Times*, June 22, 2008f. http://www.nytimes.com/2008/06/22/world/europe/22fly.html.

Rosenthal, Elisabeth. "Stinging Tentacles Offer Hint of Oceans' Decline." *New York Times*, August 3, 2008g. http://www.nytimes.com/2008/08/03/science/earth/03jellyfish.html.

Rosenthal, Elisabeth. "European Support for Bicycles Promotes Sharing of the Wheels." *New York Times*, November 10, 2008h, A-10.

Rosenthal, Elisabeth. "As More Eat Meat, a Bid to Cut Emissions." *New York Times,* December 4, 2008i. http://www.nytimes.com/2008/12/04/science/earth/04meat.html.

Rosenthal, Elisabeth. "No Furnaces but Heat Aplenty in 'Passive Houses.'" *New York Times*, December 27, 2008j. http://www.nytimes.com/2008/12/27/world/europe/27house.html.

Rosenthal, Elisabeth, and Andrew C. Revkin. "Climate Panel Issues Urgent Warning to Curb Gases." *New York Times*, February 2, 2007. http://www.nytimes.com/2007/02/02/science/earth/02cnd-climate.htm.

Rosenzweig, Cynthia, and Daniel Hillel. *Climate Variability and the Global Harvest Impacts of El Niño and Other Oscillations on Agro-Ecosystems.* New York: Oxford University Press, 2008.

Rosenzweig, Cynthia, David Karoly, Marta Vicarelli, Peter Neofotis, Qigang Wu, Gino Casassa, Annette Menzel, Terry L. Root, Nicole Estrella, Bernard Seguin, et al. "Attributing Physical and Biological Impacts to Anthropogenic Climate Change." *Nature* 453 (May 15, 2008):353–357.

Roston, Eric. *The Carbon Age: How Life's Core Element Has Become Civilization's Greatest Threat.* London: Walker & Co., 2008.

Rowan, Rob. "Thermal Adaptation in Reef Coral Symbionts." *Nature* 430 (August 12, 2004):742.

Rowe, Mark. "When the Music's Over ... a Forest Will Rise." *London Independent,* June 25, 2000, 5.

Rowland, Sherwood, and Mario Molina. "Stratospheric Sink for Chlorofluoromethanes: Chlorine Atom-Catalyzed Destruction of Ozone." *Nature* 249 (June 28, 1974):810–812.

Rowlands, Ian H. *The Politics of Global Atmospheric Change.* Manchester, UK: Manchester University Press, 1995.

Royse, David. "Scientists: Bush Global Warming Stance Invites Stronger Storms." Associated Press, October 25, 2004 (LEXIS).

Rubin, Daniel. "Venice Sinks as Adriatic Rises." Knight-Ridder News Service, July 1, 2003, n.p.(LEXIS).

Rubin, Josh. "Toronto's Blooming Warm: Gardens, Golfers Spring to Life as Record High Nears." *Toronto Star,* December 5, 2001, B-2.

Ruddiman, William F. *Plows, Plagues, and Petroleum: How Humans Took Control of Climate.* Princeton, NJ: Princeton University Press, 2005.

Rudmin, Floyd. Personal communication from Tromso, Norway, January 17, 2002.

Rudolf, John Collins. "The Warming of Greenland." *New York Times,* January 16, 2007. http://www.nytimes.com/2007/01/16/science/earth/16gree.html.

Ruhlemann, Carsten, Stefan Mulitza, Peter J. Muller, Gerold Wefer, and Rainer Zahn. "Warming of the Tropical Atlantic Ocean and Slowdown of Thermohaline Circulation during the Last Deglaciation." *Nature* 402 (December 2, 1999):511–514.

Running, Steven W. "Is Global Warming Causing More, Larger Wildfires?" *Science* 313 (August 18, 2006):927–928.

Russell, Cristine. "Climate Change: Now What? A Big Beat Grows More Challenging and Complex." *Columbia Journalism Review,* July/August 2008. http://www.cjr.org/feature/climate_change_now_what.php.

Russell, Sabin. "Glaciers on Thin Ice: Expert Says Melting to Be Faster Than Expected." *San Francisco Chronicle,* February 17, 2002, A-4.

Ruttimann, Jacqueline. "Oceanography: Sick Seas." *Nature* 442 (August 31, 2006):978–980.

Ryall, Julian. "Tokyo Plans City Coolers to Beat Heat." *London Times,* August 11, 2002, 20.

Ryan, Siobhain. "National Icons Feel the Heat." *Courier Mail* (Australia), February 4, 2002, 1.

Ryskin, G. "Methane-driven Oceanic Eruptions and Mass Extinctions." *Geology* 31 (2003):737–740.

Sabadini, Roberto. "Ice Sheet Collapse and Sea Level Change." *Science* 295 (March 29, 2002):2376–2377.

Sabine, Christopher L., Richard A. Feely, Nicolas Gruber, Robert M. Key, Kitack Lee, John L. Bullister, Rik Wanninkhof, C. S. Wong, Douglas W. R. Wallace, Bronte Tilbrook, Frank J. Millero, Tsung-Hung Peng, Alexander Kozyr, Tsueno Ono, and Aida F. Rios. "The Oceanic Sink for Anthropogenic CO_2." *Science* 305 (July 16, 2004):367–371.

Sadler, Richard, and Geoffrey Lean. "North Sea Faces Collapse of Its Ecosystem." *London Independent,* October 19, 2003, 12.

Sahlin, Monica. "Sweden First to Break Dependence on Oil! New Programme Presented." Swedish Ministry for Sustainable Development, October 1, 2005. http://www.sweden.gov.se/sb/d/3212/a/51058.

Salkin, Allen. "Before It Disappears." *New York Times,* December 16, 2007. http://www.nytimes.com/2007/12/16/fashion/16disappear.html.

Sample, Ian. "Warming Hits 'Tipping Point': Climate Change Alarm as Siberian Permafrost Melts for First Time since Ice Age." *Manchester Guardian Weekly,* August 18, 2005, 1. http://www.guardian.co.uk/guardianweekly/story/0,12674,1550685,00.html.

Sanders, Robert. "Standby Appliances Suck Up Energy." *Cal* [California] *Neighbors*, Spring 2001. http://communityrelations.berkeley.edu/CalNeighbors/Spring2001/appliances.html.

Sanderson, Katherine. "U.S. Biofuels: A Field in Ferment." *Nature* 444 (December 7, 2006):673–675.

Sanderson, Katherine. "Flights of Green Fancy." *Nature* 453 (May 15, 2008):264–265.

Santer, B. D., T. M. L. Wigley, P. J. Gleckler, C. Bonfils, M. F. Wehner, K. AchutaRao, T. P. Barnett, J. S. Boyle, W. Bruggemann, M. Fiorino, et al. "Forced and Unforced Ocean Temperature Changes in Atlantic and Pacific Tropical Cyclogenesis Regions." *Proceedings of the National Academy of Sciences* 103, no. 38 (September 19, 2006):13,905–13,910.

Santora, Marc. "Global Warming Starts to Divide G.O.P. Contenders." *New York Times*, October 17, 2007. http://www.nytimes.com/2007/10/17/us/politics/17climate.html.

"Sarahan Dust Has Chilling Effect on North Atlantic." NASA Earth Observatory Press Release, December 14, 2007. http://earthobservatory.nasa.gov/Newsroom/NasaNews/2007/2007121425986.html.

"Satellite Images Show Chunk of Broken Antarctic Ice Shelf." University of Colorado News and Events for the Media, April 16, 1997. http://cires.colorado.edu/news/press/1997/97-04-16.html.

"Satellite Images Show Continued Breakup of Greenland's Largest Glaciers, Predict Disintegration in Near Future." NASA Earth Observatory, August 20, 2008. http://earthobservatory.nasa.gov/Newsroom/MediaAlerts/2008/2008082027361.html.

Sato, Makiko, James Hansen, Dorothy Koch, Andrew Lacis, Reto Ruedy, Oleg Dubovik, Brent Holben, Mian Chin, and Tica Novakov. "Global Atmospheric Black Carbon Inferred from AERONET." *Proceedings of the National Academy of Sciences* 100, no. 11 (May 27, 2003):6319–6324.

Saulny, Susan. "As Oil Prices Soar, Restaurant Grease Thefts Rise." *New York Times*, May 30, 2008. http://www.nytimes.com/2008/05/30/us/30grease.html.

Saunders, Mark A., and Adam S. Lea. "Large Contribution of Sea Surface Warming to Recent Increase in Atlantic Hurricane Activity." *Nature* 451 (January 31, 2008):557–560.

"Saving Earth's Plant Diversity from Global Warming." Environment News Service, May 22, 2007. http://www.ens-newswire.com/ens/may2007/2007-05-22-02.asp.

Scambos, T. A., J. A. Bohlander, C. A. Shuman, and P. Skvarca. "Glacier Acceleration and Thinning after Ice Shelf Collapse in the Larsen B Embayment, Antarctica." *Geophysical Research Letters* 31, no. 18 (September 22, 2004), doi: 10.1029/2004GL020670.

Schar, Christoph, and Gerd Jendritzky. "Hot News from Summer 2003." *Nature* 432 (December 2, 2004):559–561.

Schellnhuber, Hans Joachim, ed. *Avoiding Dangerous Climate Change*. New York: Cambridge University Press, 2006.

Schellnhuber, Hans Joachim. "Global Warming: Stop Worrying, Start Panicking?" *Proceedings of the National Academy of Sciences* 105, no. 38 (September 23, 2008):14,239–14,240.

Schiermeier, Quirin. "The Oresmen." *Nature* 421 (January 9, 2003a):109–110.

Schiermeier, Quirin. "Gas Leak: Global Warming Isn't a New Phenomenon—Sea-bed Emissions of Methane Caused Temperatures to Soar in Our Geological Past, but No One Is Sure What Triggered the Release." *Nature* 423 (June 12, 2003b):681–682.

Schiermeier, Quirin. "A Rising Tide: The Ice Covering Greenland Holds Enough Water to Raise the Oceans Six Metres—and It's Starting to Melt." *Nature* 428 (March 11, 2004a):114–115.

Schiermeier, Quirin. "Researchers Seek to Turn the Tide on Problem of Acid Seas." *Nature* 430 (August 19, 2004b):820.

Schiermeier, Quirin. "Poles Lose Out as Ozone Levels Begin to Recover." *Nature* 437 (September 8, 2005):179.

Schiermeier, Quirin. "On Thin Ice: The Arctic Is the Bellwether of Climate Change." *Nature* 441 (May 10, 2007a):146–147.

Schiermeier, Quirin. "Ocean Circulation Noisy, Not Stalling." *Nature* 448 (August 23, 2007b):844–845.

Schiermeier, Quirin. "Europe to Capture Carbon." *Nature* 451 (January 17, 2008a):232.

Schiermeier, Quirin. "The Long Summer Begins." *Nature* 454 (July 17, 2008b):266–269.

Schiermeier, Quirin. "Global Warming Blamed for Growth in Storm Intensity." *Nature* 455 (September 3, 2008c). http://www.nature.com/news/2008/080903/full/news.2008.1079.html.

Schiermeier, Quirin. "Fears Surface over Methane Leaks." *Nature* 455 (October 2, 2008d):572.

Schimel, D. S., J. I. House, K. A. Hibbard, P. Bousquet, P. Cials, P. Peylin, B. H. Braswell, M. J. Apps, D. Baker, A. Bondeau, et al. "Recent Patterns and Mechanisms of Carbon Exchange by Terrestrial Ecosystems." *Nature* 414 (November 8, 2001):169–172.

Schimel, David. "Climate Change and Crop Yields: Beyond Cassandra." *Science* 312 (June 30, 2006):1889–1890.

Schimel, David, and David Baker. "Carbon Cycle: The Wildlife Factor." *Nature* 420 (November 7, 2002):29–30.

Schimel, David, Jerry Melillo, Hanqin Tian, A. David McGuire, David Kicklighter, Timothy Kittel, Nan Rosenbloom, Steven Running, Peter Thornton, Dennis Ojima, et al. "Contribution of Increasing CO_2 and Climate to Carbon Storage by Ecosystems in the United States." *Science* 287 (March 17, 2000):2004–2006.

Schleck, Dave. "High Fliers May Be Creating Clouds: Global Warming May Be Worsened by Contrails from Aircraft." *Montreal Gazette*, July 18, 2004, D-6.

Schleifstein, Mark. "The Gulf Will Rise, Report Predicts: Maybe 44 Inches, Scientists Say." *New Orleans Times-Picayune*, October 24, 2001, 1.

Schlesinger, James. "The Theology of Global Warming." *Wall Street Journal*, August 8, 2005, A-10.

Schlesinger, W. H., and J. Lichter. "Limited Carbon Storage in Soil and Litter of Experimental Forest Plots under Increased Atmospheric CO_2." *Nature* 411 (May 24, 2001):466–469.

Schmid, Randolph E. "Warming Climate Reduces Yield for Rice, One of World's Most Important Crops." Associated Press, June 28, 2004 (LEXIS).

Schmid, Randolph E. "Panel Sees Growing Threat in Melting Arctic." Associated Press, August 23, 2005 (LEXIS).

Schmid, Randolph E. "Climate Change Could Devastate U.S. Wineries." Associated Press, July 10, 2006a (LEXIS).

Schmid, Randolph E. "State of the Arctic Warming, with Widespread Melting." Associated Press, November 16, 2006b (LEXIS).

Schmidhuber, Josef, and Francesco Tubiello. "Global Food Security under Climate Change." *Proceedings of the National Academy of Sciences* 104 (December 11, 2007):19,703–19,708.

Schmit, Julie. "Small Electric Car Sales Could Get a Jolt." *USA Today*, December 11, 2007, 3-B.

Schmittner, Andreas. "Decline of the Marine Ecosystem Caused by a Reduction in the Atlantic Overturning Circulation." *Nature* 434 (March 31, 2005):628–633.

Schneider, Stephen H. "Talk Abstract: Surprises and Scaling Connections between Climatology and Ecology." Institute for Mathematics and Its Applications, n.d. http://www.ima.umn.edu/biology/wkshp_abstracts/schneider1.html.

Schneider, Stephen H. *Global Warming: Are We Entering the Greenhouse Century?* San Francisco: Sierra Club Books, 1989.

Schneider, Stephen H. "Detecting Climatic Change Signals: Are There Any Fingerprints?" *Science* 263 (January 21, 1994):341–347.

Schneider, Stephen H. "Modeling Climate Change Impacts and Their Related Uncertainties." Paper for Conference on Social Science and the Future Somerville College, Oxford, July 7–8, 1999.

Schneider, Stephen H. "No Therapy for the Earth: When Personal Denial Goes Global." In *Nature, Environment, and Me: Explorations of Self In a Deteriorating World*, ed. Michael Aleksiuk and Thomas Nelson. Montreal: McGill-Queens University Press, 2000a.

Schneider, Stephen H. "Kyoto Protocol: The Unfinished Agenda: An Editorial Essay." Unpublished ms., March 18, 2000b.

Schneider, Stephen H. Personal communication, March 18, 2000c.

Schneider, Stephen H., and R. S. Chen. "Carbon Dioxide Warming and Coastline Flooding: Physical Factors and Climatic Impact." *American Review of Energy* 5 (1980):107–140.

Schneider, Stephen H., and Terry Root, eds. *Wildlife Responses to Climate Change: North American Case Studies*. Washington, DC: Island Press, 2001.

Schrope, Mark. "Successes in Fight to Save Ozone Layer Could Close Holes by 2050." *Nature* 408 (December 7, 2000):627.

Schrope, Mark. "Global Warming: A Change of Climate for Big Oil." *Nature* 411 (May 31, 2001):516–518.

Schultz, Martin G., Thomas Diehl, Guy P. Brasseur, and Werner Zittel. "Air Pollution and Climate-Forcing Impacts of a Global Hydrogen Economy." *Science* 302 (October 24, 2003):624–627.

Schulze, Ernst-Detlef, Christian Wirth, and Martin Heimann. "Managing Forests after Kyoto." *Science* 289 (September 22, 2000):2058–2059.

Schwander, Dominique Raynaud, Valèrie Masson-Delmotte, and Jean Jouzel. "Atmospheric Methane and Nitrous Oxide of the Late Pleistocene from Antarctic Ice Cores." *Science* 310 (November 25, 2005):1317–1321.

Schwartzman, Stephen. "Reigniting the Rainforest: Fires, Development and Deforestation." *Native Americas* 16, no. 3/4 (Fall/Winter 1999):60–63.

"Scientific Consensus: Villach (Austria) Conference." In *The Challenge of Global Warming*, Edwin Abrahamson, 63–67. Washington, DC: Island Press, 1989.

"Scientist: Warming Threatens Koalas." Associated Press in *Omaha World-Herald*, May 8, 2008, 7-A.

"Scientists Predict Future of Weather Extremes." Environment News Service, October 20, 2006.

"Scientists Report Large Ozone Loss." *USA Today*, April 6, 2000, 3-A.

"Scientists' Statement on Global Climatic Disruption." Statements on Climate Change by Foreign Leaders at Earth Summit, June 23, 1997. http://uneco.org/Global_Warming.html.

Seabrook, Charles. "Amphibian Populations Drop." *Atlanta Journal-Constitution*, October 15, 2004, 1-C.

Seager, R., D. S. Battisti, J. Yin, N. Gordon, N. Naik, A. C. Clement, and M. A. Kane. "Is the Gulf Stream Responsible for Europe's Mild Winters?" *Quarterly Journal of the Royal Meteorological Society* 128 (2002):2563–2586.

Searchinger, Timothy, Ralph Heimlich, R. A. Houghton, Fengxia Dong, Amani Elobeid, Jacinto Fabiosa, Simla Tokgoz, Dermot Hayes, and Tun-Hsiang Yu. "Use of U.S. Croplands for Biofuels Increases Greenhouse Gases through Emissions from Land-Use Change." *Science* 319 (February 29, 2008):1238–1240.

Seelye, Katharine. "Environmental Groups Gain as Companies Vote on Issues." *New York Times*, May 29, 2003, C-1.

Seibel, Brad A., and Patrick J. Walsh. "Potential Impacts of C02 Injection on Deep-sea Biota." *Science* 294 (October 12, 2001):319–320.

Seidel, Dian, Qiang Fu, William J. Randel, and Thomas J. Reichler. "Widening of the Tropical Belt in a Changing Climate." *Nature Geoscience* (December 2, 2007), doi: 10.1038/ngeo.2007.38. http://www.nature.com.leo.lib.unomaha.edu/ngeo.

Semenov, Vladimir A., and Lennart Bengtsson. *Modes of Wintertime Arctic Temperature Variability*. Hamburg, Germany: Max Planck Institut fur Meteorologie, 2003.

Semmens, Grady. "Ecologists See Disaster in Dwindling Water Supply." *Calgary Herald*, November 27, 2003, A-14.

Semple, Robert B. "A Film That Could Warm Up the Debate on Global Warming." Editorial Observer, *New York Times*, May 27, 2004. http://nytimes.com/2004/05/27/opinion/27THU3.html.

Sengupta, Somini. "Sea's Rise in India Buries Islands and a Way of Life." *New York Times*, April 11, 2007a. http://www.nytimes.com/2007/04/11/world/asia/11india.html.

Sengupta, Somini. "Glaciers in Retreat." *New York Times*, July 17, 2007b. http://www.nytimes.com/2007/07/17/science/earth/17glacier.html.

Serreze, Mark C., and Roger G. Barry. *The Arctic Climate System*. Cambridge: Cambridge University Press, 2006.

Serreze, Mark C., Marika M. Holland, and Julienne Stroeve. "Perspectives on the Arctic's Shrinking Sea-Ice Cover." *Science* 315 (March 16, 2007):1533–1536.

Serreze, Mark C., J. A. Maslanik, T. A. Scambos, F. Fetterer, J. Stroeve, K. Knowles, C. Fowler, S. Drobot, R. G. Barry, and T. M. Haran. "A Record Minimum Arctic Sea Ice Extent and Area in 2002." *Geophysical Research Letters* 30 (2003), doi: 10.1029/2002GL016406.

Service, Robert F. "Solar Power: Can the Upstarts Top Silicon?" *Science* 319 (February 8, 2008a):718–720.

Service, Robert F. "Study Fingers Soot as Major Player in Global Warming." *Science* 319 (March 28, 2008b):1745.

Service, Robert F. "New Catalyst Marks Major Step in the March Toward Hydrogen Fuel." *Science* 321 (August 1, 2008c):620.

Sevunts, Levon. "Prepare for More Freak Weather, Experts Say." *Montreal Gazette*, May 10, 2001, A-4.

Sewall, J. O., and L. C. Sloan. "Disappearing Arctic Sea Reduces Available Water in the American West." *Geophysical Research Letters* 31 (2004), L06209, doi: 10.1029/2003Gl019133, 2004.

"Shanghai Mulls Building Dam to Ward off Rising Sea Levels." Agence France Presse, February 9, 2004 (LEXIS).

"Sharks in Alaskan Waters Could Herald Global Warming." Environment News Service, February 19, 2002. http://ens-news.com/ens/feb2002/2002L-02-19-09.html.

Sharp, David. "Study: New England's Winters Not What They Used to Be." Associated Press State and Regional News Feed, July 23, 2003 (LEXIS).

Shaw, Daniel. "Global Warming Pushes Ski Industry Downhill." *Grist Magazine*, June 19, 2006. http://www.alternet.org/story/31991.

Shaw, M. Rebecca, Erika S. Zavaleta, Nona R. Chiariello, Elsa E. Cleland, Harold A. Mooney, and Christopher B. Field. "Grassland Responses to Global Environmental Changes Suppressed by Elevated CO_2." *Science* 298 (December 6, 2002):1987–1990.

Shearman, David, and Joseph Wayne Smith. *The Climate Change Challenge and the Failure of Democracy*. Westport, CT: Praeger, 2007.

Sheehan, James J. *Where Have All the Soldiers Gone? The Transformation of Modern Europe*. Boston: Houghton-Mifflin, 2008.

Shelby-Biggs, Brooke. "Hard to Swallow." Mother Jones.com. March 15, 2001. http://www.motherjones.com/commentary/columns/2001/03/newshole2.html.

Shellenberger, Michael, and Ted Nordhaus. *Break Through: From the Death of Environmentalism to the Politics of Possibility*. New York: Houghton-Mifflin, 2007.

Shepherd, Andrew, and Duncan Wingham. "Recent Sea-Level Contributions of the Antarctic and Greenland Ice Sheets." *Science* 315 (March 16, 2007):1529–1532.

Shepherd, Andrew, Duncan Wingham, Justin A. D. Mansley, and Hugh F. J. Corr. "Inland Thinning of Pine Island Glacier, West Antarctica." *Science* 291 (February 2, 2001):862–864.

Shepherd, Andrew, Duncan Wingham, Tony Payne, and Pedro Skvarca. "Larsen Ice Shelf Has Progressively Thinned." *Science* 302 (October 31, 2003):856–859.

Sheppard, Charles R. C. "Predicted Recurrences of Mass Coral Mortality in the Indian Ocean." *Nature* 425 (September 18, 2003):294–297.

Shindell, D. T. "Local and Remote Contributions to Arctic Warming." *Geophysical Research Letters* 34, no. 14 (July 20, 2007), L14704, doi: 10.1029/2007GL030221.

Shindell, D. T., David Rind, and Patrick Lonergan. "Increased Polar Stratospheric Ozone Losses and Delayed Eventual Recovery Owing to Increasing Greenhouse-gas Concentrations." *Nature* 392 (April 9, 1998):589–592.

Shindell, D. T., and G. A. Schmidt. "Southern Hemisphere Climate Response to Ozone Changes and Greenhouse Gas Increases." *Geophysical Research Letters* 31 (2004), L18209, doi: 10.1029/2004GL020724.

Shindell, D. T., G. A. Schmidt, R. L. Miller, and D. Rind. "Northern Hemisphere Winter Climate Response to Greenhouse Gas, Ozone, Solar, and Volcanic Forcing." *Journal of Geophysical Research* 106 (2001):7193–7210.

Shine, Keith P., and William T. Sturges. "CO_2 Is Not the Only Gas." *Science* 315 (March 30, 2007):1804–1805.

"Shrinking Glaciers Thawed Faster in 2005." Environment News Service, January 29, 2007. http://www.ens-newswire.com/ens/jan2007/2007-01-29-05.asp.

Shulman, Robin. "N.Y. Hopes to Ensure Smooth Pedaling for Bike Commuters." *Washington Post*, May 25, 2008, A-2. http://www.washingtonpost.com/wp-dyn/content/article/2008/05/24/AR2008052401457_pf.html.

Shulman, Seth. *Undermining Science: Suppression and Distortion in the Bush Administration*. Berkeley: University of California Press, 2006.

Shwartz, Mark. "New Study Reveals a Major Cause of Global Warming—Ordinary Soot." Stanford University Departmental News. February 7, 2001. http://www.stanford.edu/dept/news.

Siegenthaler, Urs., and H. Oeschger. "Predicting Future Atmospheric Carbon Dioxide Levels." *Science* 199 (1978):388–395.

Siegenthaler, Urs, Thomas F. Stocker, Eric Monnin, Jakob Schwander, Bernhard Stauffer, Dominique Raynaud, Jean-Marc Barnola, Hubertus Fischer, Valèrie Masson-Delmotte, and

Jean Jouzel. "Stable Carbon Cycle Climate Relationship during the Late Pleistocene." *Science* 310 (November 25, 2005):1313–1317.

Siegert, F., G. Ruecker, A. Hinrichs, and A. A. Hoffmann. "Increased Damage from Fires in Logged Forests during Droughts Caused by El Niño." *Nature* 414 (November 22, 2001):437–440.

Sierra Club. "Global Warming: The High Costs of Inaction." 1999. http://www.sierraclub.org/global-warming/resources/innactio.htm.

Siggins, Lorna. "Warm-water Anchovies Landed by Trawlers in Donegal Bay." *Irish Times*, December 12, 2001, 1.

"Silenced—1,400 Times." Editorial, *Wall Street Journal*, March 27, 2007, A-18.

Silver, Cheryl Simon, and Ruth S. DeFries. *One Earth, One Future: Our Changing Global Environment.* Washington, DC: National Academy Press, 1990.

Simon, Stephanie. "Global Warming, Local Initiatives: Unhappy with Federal Resistance to World Standards, Communities Are Curbing Their Energy Use and Emissions." *Las Angeles Times*, December 10, 2005, A-1.

Simons, Paul. "Weatherwatch." *London Guardian.* November 26, 2001, 14.

Singer, S. Fred. "Global Warming Whining." *Washington Times*, April 16, 1999. http://www.cop5.org/apr99/singer.htm.

Singer, S. Fred. *Hot Talk, Cold Science: Global Warming's Unfinished Debate.* Oakland, CA: Independent Institute, 1999.

Singer, S. Fred, and Dennis T. Avery. *Unstoppable Global Warming Every 1,500 Years.* Lanham, MD: Rowman and Littlefield, 2006.

Sisario, Ben. "Songs for an Overheated Planet." *New York Times*, July 6, 2007. http://www.nytimes.com/2007/07/06/arts/music/06eart.html.

Slater, Dashka. "Public Corporations Shall Take Us Seriously." *New York Times Sunday Magazine*, August 12, 2007, 22–27.

"Slowing Ocean Currents Could Freeze Europe." Environment News Service, February 21, 2002. http://ens-news.com/ens/feb2002/2002L-02-21-09.html.

Sluijs, Appy, Stefan Schouten, Mark Pagani, Martijn Woltering, Henk Brinkhuis, Jaap S. Sinninghe Damst, Gerald R. Dickens, Matthew Huber, Gert-Jan Reichart, Ruediger Stein, Jens Matthiessen, Lucas J. Lourens, Nikolai Pedentchouk, Jan Backman8, Kathryn Moran, and the Expedition 302 Scientists. "Subtropical Arctic Ocean Temperatures during the Palaeocene/Eocene Thermal Maximum." *Nature* 441 (June 1, 2006):610–613.

Smith, Craig S. "One Hundred and Fifty Nations Start Groundwork for Global Warming Policies." *New York Times*, January 18, 2001, 7.

Smith, L. C., Y. Sheng, G. M. MacDonald, and L. D. Hinzman. "Disappearing Arctic Lakes." *Science* 308 (June 3, 2005):1429.

Smith, Lewis. "Falling Numbers Silence Cuckoo's Call of Spring." *London Times*, March 6, 2002, n.p.(LEXIS).

Smith, O. Glenn. "Harvest the Sun—From Space." *New York Times*, July 23, 2008. http://www.nytimes.com/2008/07/23/opinion/23smith.html.

Smith, Rebecca. "Utilities Amp Up Push to Slash Energy Use." *Wall Street Journal*, January 9, 2007a, A-1, A-12.

Smith, Rebecca. "Wind, Solar Power Gain Users." *Wall Street Journal*, January 18, 2008b, A-6.

Smith, Rebecca. "New Plants Fueled by Coal Are Put on Hold." *Wall Street Journal*, July 25, 2007c, A-1, A-10.

Smith, Rebecca. "Wind, Solar Power Gain Users." *Wall Street Journal*, January 18, 2008, A-6.

Smith, Stephen. "Comin' Ah-choo: Tepid Temperatures Speeding Allergy Season." *Boston Globe*, April 10, 2002, A-1.

Smucker, Philip. "Global Warming Sends Troops of Baboons on the Run: Rising Temperatures and Humans Encroaching on Grasslands Are Endangering the Ethiopian Primates." *Christian Science Monitor*, June 15, 2001, 7.

Socci, Anthony D. "The Climate Continuum: An Overview of Global Warming and Cooling Throughout the History of Planet Earth." In *A Global Warming Forum: Scientific, Economic, and legal Overview*, ed. Richard A. Geyer, 161–207. Boca Raton, FL: CRC Press, 1993.

Soden, Brian J., and Isaac M. Held. "An Assessment of Climate Feedbacks in Coupled Ocean-Atmosphere Models." *Journal of Climate* 19 (2006):3354–3360.

Soederlindh, Lisa Monique. "Sweden: Charging Tolls to Drive Downtown—Good Idea or Bad?" InterPress Service, August 23, 2006 (LEXIS).

Solanki, S. K., I. G. Usoskin, B. Kromer, M. Schussler, and J. Beer. "Unusual Activity of the Sun during Recent Decades Compared to the Previous 11,000 Years." *Nature* 431 (October 28, 2004):1084–1086.

Son, S.-W., L. M. Polvani, D. W. Waugh, H. Akiyoshi,4 R. Garcia, D. Kinnison, S. Pawson, E. Rozanov, T. G. Shepherd, K. Shibata. "The Impact of Stratospheric Ozone Recovery on the Southern Hemisphere Westerly Jet." *Science* 320 (June 13, 2008):1486–1489.

Sorensen, Eric. "The Letter We Can't See: Atmospheric Carbon Is One of the Worst Culprits in Global Warming." *Seattle Times*, April 22, 2001, A-1.

Sorokhtin, O. G., G. V., Chilingar, and L. F. Khilyuk. *Global Warming and Global Cooling: Evolution of Climate on Earth*. Amsterdam: Elsevier, 2007.

Souder, William. "Global Warming and a Toad Species' Decline." *Washington Post*, April 9, 2001, n.p.(LEXIS).

Sovacool, Benjamin K. *The Dirty Energy Dilemma: What's Blocking Clean Power in the United States*. Westport, CT: Praeger, 2008.

Spalding, Mark. "Coral Grief: Rising Temperatures, Pollution, Tourism and Fishing Have All Helped to Kill Vast Stretches of Reef in the Indian Ocean. Yet, with Simple Management, Says Mark Spalding, the Marine Life Can Recover." *London Guardian*, September 12, 2001a, 8.

Spalding, Mark. *World Atlas of Coral Reefs*. Berkeley: University of California Press, 2001b.

Spears, Tom. "Antarctica Rides Global 'Heatwave:' Continent's Warm Coast Causes Concern." *Ottawa Citizen*, August 8, 2001, A-1.

Spears, Tom. "Cold Spring Bucks the Trend." *Ottawa Citizen*, June 7, 2002, A-8.

Spector, Mike. "Can U.S. Adopt Europe's Fuel-Efficient Cars?" *Wall Street Journal*, June 26, 2007, B-1.

Spence, Chris. *Global Warming: Personal Solutions for a Healthy Planet*. New York: Palgrave/MacMillan, 2005.

Spencer, Jane. "Big Firms to Press Suppliers on Climate." *Wall Street Journal*, October 9, 2007, A-7.

Spencer, Roy. *Climate Confusion: How Global-warming Hysteria Leads to Bad Science, Pandering Politicians, and Misguided Policies That Hurt the Poor*. Lanham, MD: Encounter Books, 2008.

Speth, James Gustave. *Red Sky at Morning: America and the Crisis of the Global Environment*. New Haven, CT: Yale University Press, 2004.

Spotts, Peter N. "As Arctic Warms, Scientists Rethink Culprits." *Christian Science Monitor*, August 22, 2000, 4.

Spotts, Peter N. "Trees No Savior for Global Warming." *Christian Science Monitor*, May 25, 2001, n.p.(LEXIS).

"Spray Cans Warming the Planet, One Dust-busting Puff at a Time." *International Herald-Tribune* in *Herald Asahi* (Tokyo), May 27, 2004, n.p.(LEXIS).

"Spring 2000 Is Warmest on Record." Associated Press in *Omaha World-Herald*, June 17, 2000, A-1.

Sridhar, Venkataramana, David B. Loope, James B. Swinehart, Joseph A. Mason, Robert J. Oglesby, Clinton M. Rowe. "Large Wind Shift on the Great Plains during the Medieval Warm Period." *Science* 313 (July 21, 2006):345–347.

Staba, David. "Snowstorm Blankets Buffalo, Killing at Least 3." *New York Times*, October 14, 2006. http://www.nytimes.com/2006/10/14/nyregion/14storm.html.

Staba, David. "An Old Steel Mill Retools to Produce Clean Energy." *New York Times*, May 22, 2007. http://www.nytimes.com/2007/05/22/nyregion/22wind.html.

Stainforth, D. A., T. Aina, G. Christensen, M. Collins, N. Faull, D. J. Frame, J. A. Kettleborough, S. Knight, A. Martin, J. M. Murphy, et al. "Uncertainty in Predictions of the Climate Response to Rising Levels of Greenhouse Gases." *Nature* 433 (January 27, 2005):403–406.

Stark, Mike. "Assault by Bark Beetles Transforming Forests: Vast Swaths of West Are Red, Gray, and Dying—Drought, Fire Suppression, and Global, Warming Are Blamed." *Billings Gazette* in *Los Angeles Times*, October 6, 2002, B-1.

Steffen, Konrad. "Greenland Melt Accelerating, According to Colorado University-Boulder Study." Press Release, University of Colorado, December 11, 2007. NASA Earth Observatory, December 25, 2007. http://www.news.uiuc.edu/news/07/1210nitrogen.html.

Steffensen, Jørgen Peder, Katrine K. Andersen, Matthias Bigler, Henrik B. Clausen, Dorthe Dahl-Jensen,1 Hubertus Fischer, Kumiko Goto-Azuma, Margareta Hansson, Sigfús J. Johnsen, Jean Jouzel, Valérie Masson-Delmotte, Trevor Popp, Sune O. Rasmussen, Regine Röthlisberger, Urs Ruth, Bernhard Stauffer, Marie-Louise Siggaard-Andersen, Árny E. Sveinbjörnsdóttir, Anders Svensson, and James W. C. White. "High-Resolution Greenland Ice Core Data Show Abrupt Climate Change Happens in Few Years." *Science* 321 (August 1, 2008):680–684.

Steig, Eric J., David P. Schneider, Scott D. Rutherford, Michael E. Mann, Josefino C. Comiso, and Drew T. Shindell. "Warming of the Antarctic Ice-Sheet Surface since the 1957 International Geophysical Year." *Nature* 457 (January 22, 2009):459–462.

Stein, Rob, and Shankar Vedantam. Science: Notebook." *Washington Post*, November 12, 2001, A-9.

Steinhauer, Jennifer. "In California, Heat Is Blamed for 100 Deaths." *New York Times*, July 28, 2006. http://www.nytimes.com/2006/07/28/us/28heat.html.

Steinman, David. *Safe Trip to Eden: 10 Steps to Save Planet Earth from Global Warming Meltdown.* New York: Thunder's Mouth Press, 2007.

Steinman, David, and Wendy Gordon Rockefeller. *Safe Trip to Eden: 10 Steps to Save Planet Earth from Global Warming Meltdown.* New York: Thunder's Mouth Press, 2007.

Steitz, David E. "NASA Researchers Document Shrinking of Greenland's Glaciers." NASA Press Release 99–33, March 4, 1999. http://www.earthobservatory.nasa.gov/Newsroom/Nasa-News/19990304207.html.

Stephens, Bret. "Global Warming as Mass Neurosis." *Wall Street Journal*, July 1, 2008, A-15.

Sterman, John D. "Risk Communication on Climate: Mental Models and Mass Balance." *Science* 322 (October 24, 2008):532–533.

Stern, Nicholas. *The Economics of Climate Change: The Stern Report.* Cambridge: Cambridge University Press, 2007.

Stevens, William K. "Computers Model World's Climate, but How Well?" *New York Times*, April 11, 1997. http://benetton.dkrz.de:3688/homepages/georg/kimo/0254.html.

Stevens, William K. *The Change in the Weather: People, Weather, and the Science of Climate.* New York: Delacourte Press, 1999a.

Stevens, William K. "1998 and 1999 Warmest Years Ever Recorded." *New York Times*, December 19, 1999b, 1, 38.

Stevens, William K. "The Oceans Absorb Much of Global Warming, Study Confirms." *New York Times*, March 24, 2000, A-16.

Stiffler, Linda, and Robert McClure. "Effects Could Be Profound." *Seattle Post-Intelligencer*, November 13, 2003, A-8.

Stirling, I., and A. E. Derocher. "Possible Impacts of Climate Warming on Polar Bears." *Arctic* 46, no. 3 (1993):240–245.

Stirling, I., and N. J. Lunn. "Environmental Fluctuation in Arctic Marine Ecosystems as Reflected by Variability in Reproduction of Polar Bears and Ringed Seals." In *Ecology of Arctic Environments*, ed. S. J. Woodlin and M. Marquiss, 167–181. Oxford: Blackwell Science Ltd., 1997.

Stocker, Thomas F. "Global Change: South Dials North." *Nature* 424 (July 31, 2003):496–497.

Stocker, Thomas F., Reto Knutti, and Gian-Kasper Plattner. "The Future of the Thermohaline Circulation—A Perspective." In *The Oceans and Rapid Climate Change: Past, Present, and Future,* ed. Dan Seidov, Bernd J. Haupt, and Mark Maslin, 277–293. Washington, DC: American Geophysical Union, 2001.

Stocker, Thomas F., and A. Schmittner. "Influence of CO_2 Emission Rates on the Stability of the Thermohaline Circulation." *Nature* 388 (1997):862–864.

Stoddard, Ed. "Global Warming Threatens 'Living Fossil' Fish: Coelacanths Have Existed 400 Million Years." Reuters in *Ottawa Citizen*, July 14, 2001, B-4.

Stokstad, Erik. "Defrosting the Carbon Freezer of the North." *Science* 304 (June 11, 2004a):1618–1620.

Stokstad, Erik. "Global Survey Documents Puzzling Decline of Amphibians." *Science* 306 (October 15, 2004b):391.

Stokstad, Erik. "California Sets Goals for Cutting Greenhouse Gases." *Science* 308 (June 10, 2005):1530.

Stokstad, Erik. "Boom and Bust in a Polar Hot Zone." *Science* 315 (March 16, 2007):1522–1523.

Stoll, John D. "Visions of the Future: What Will the Car of Tomorrow Look Like? Perhaps Nothing Like the Car of Today." *Wall Street Journal*, April 17, 2006, R-8.

Stone, Richard. "Have Desert Researchers Discovered a Hidden Loop in the Carbon Cycle?" *Science* 320 (June 13, 2008):1409–1410.

Stott, Peter A., D. A. Stone, and M. R. Allen. "Human Contribution to the European Heatwave of 2003." *Nature* 432 (December 2, 2004):610–613.

Stott, Peter A., S. F. B. Tett, G. S. Jones, M. R. Allen, J. F. B. Mitchell, and G. J. Jenkins. "External Control of 20th Century Temperature by Natural and Anthropogenic Forcings." *Science* 290 (December 15, 2000):2133–2137.

Strassel, Kimberly. "Ethanol's Bitter Taste." *Wall Street Journal*, May 18, 2007, A-16.

Streets, David G., Kejun Jiang, Xiulian Hu, Jonathan E. Sinton, Xiao-Quan Zhang, Deying Xu, Mark Z. Jacobson, and James E. Hansen. "Recent Reductions in China's Greenhouse Gas Emissions." *Science* 294 (November 30, 2001):1835–1837.

Stroeve, J. C., M. C. Serreze, F. Fetterer, T. Arbetter, W. Meier, J. Maslanik, and K. Knowles. "Tracking the Arctic's Shrinking Ice Cover: Another Extreme September Minimum in 2004." *Geophysical Research Letters* 32, no. 4 (February 25, 2005), L04501, http://dx.doi.org/10.1029/2004GL021810.

Struck, Doug. "'Rapid Warming' Spreads Havoc in Canada's Forests: Tiny Beetles Destroying Pines." *Washington Post*, March 1, 2006a, A-1. http://www.washingtonpost.com/wp-dyn/content/article/2006/02/28/AR2006022801772_pf.html.

Struck, Doug. "Inuit See Signs in Arctic Thaw: String of Warm Winters Alarms 'Sentries for the Rest of the World.'" *Washington Post*, March 22, 2006b, A-1. http://www.washingtonpost.com/wp-dyn/content/article/2006/03/21/AR2006032101722_pf.html.

Struck, Doug. "Climate Change Drives Disease to New Territory: Viruses Moving North to Areas Unprepared for Them, Experts Say." *Washington Post*, May 5, 2006c, A-16.

Struck, Doug. "Earth's Climate Warming Abruptly, Scientist Says. Tropical-Zone Glaciers May Be at Risk of Melting." *Washington Post*, June 27, 2006d, A-3. http://www.washingtonpost.com/wp-dyn/content/article/2006/06/26/AR2006062601237_pf.html.

Struck, Doug. "On the Roof of Peru, Omens in the Ice Retreat of Once-Mighty Glacier Signals Water Crisis, Mirroring Worldwide Trend." *Washington Post*, July 29, 2006e, A-1. http://www.washingtonpost.com/wp-dyn/content/article/2006/07/28/AR2006072801994.html.

Struck, Doug. "Canada in Quandary over Gas Emissions: Push for New Policy Puts Oil-Rich West at Odds with Big Automakers in East." *Washington Post*, October 5, 2006f, A-28. http://www.washingtonpost.com/wp-dyn/content/article/2006/10/04/AR2006100401724_pf.html.

Struck, Doug. "In Far North, Peril and Promise: Great Forests Hold Fateful Role in Climate Change." *Washington Post*, February 22, 2007a, A-1. http://www.washingtonpost.com/wp-dyn/content/article/2007/02/21/AR2007022102095_pf.html.

Struck, Doug. "Icy Island Warms to Climate Change: Greenlanders Exploit 'Gifts from Nature' While Facing New Hardships." *Washington Post*, June 7, 2007b, A-1. http://www.washingtonpost.com/wp-dyn/content/article/2007/06/06/AR2007060602783.html?referrer=email.

Struck, Doug. "Warming Thins Herd for Canada's Seal Hunt: Pups Drown in Melting Ice—Government Reduces Quotas." *Washington Post*, April 4, 2007c, A-8. http://www.washingtonpost.com/wp-dyn/content/article/2007/04/03/AR2007040301754_pf.html.

Strum, Matthew, Josh Schimel, Gary Michaelson, Jeffery M. Welker, Steven F. Oberbauer, Glen E. Liston, Jace Fahnestock, and Vladimir E. Romanovsky. "Winter Biological Processes Could Help Convert Arctic Tundra to Shrubland." *BioScience* 55, no. 1 (January 2005):17–26.

Struzik, Ed. "Fiery Future in Store for Forests if Climate Warms: Western Landscape Vulnerable: Study." *Edmonton Journal*, March 16, 2003, A-1.

Stuart, Simon N., Janice S. Cranson, Neil A. Cox, Bruce E. Young, Ana S. L. Rodrigues, Debra L. Fischman, and Robert W. Walker. "Status and Trends of Amphibian Declines and Extinctions Worldwide." *Science* 306 (December 3, 2004):1783–1786.

Stuber, Nicola, Piers Forster, Gaby Radel, and Keith Shine. "The Importance of the Diurnal and Annual Cycle of Air Traffic for Contrail Radiative Forcing." *Nature* 441 (June 15, 2006):864–867.

"Study Confirms Antarctica Warming." Environment News Service, September 6, 2006. www.ens-newswire.com/ens/sep2006/2006-09-06-02.asp.

"Study of Ancient Air Bubbles Raises Concern about Today's Greenhouse Gases." Associated Press in *Omaha World-Herald*, November 25, 2005, 4-A.

"Study Reports Large-scale Salinity Changes in Oceans: Saltier Tropical Oceans, Fresher Ocean Waters near Poles Are Further Signs of Global Warming's Impacts on Planet." *Ascribe Newsletter*, December 17, 2003 (LEXIS).

"Study Reveals Increased River Discharge to Arctic Ocean: Finding Could Mean Big Changes to Global Climate." *Ascribe Newsletter*, December 12, 2002 (LEXIS).

"Study Says Glaciers Formed during a Very Warm Period." Reuters in *New York Times*, January 11, 2008. http://www.nytimes.com/2008/01/11/world/europe/11glacier.html.

"Study Shows Soil Warming May Stimulate Carbon Storage in Some Forests Effect Would Slow Rate of Climate Change." Ascribe Newsletter, December 12, 2002 (LEXIS).

Stukin, Stacie. "The Lean, Green Kitchen." *Vegetarian Times*, September 2007, 51–54.

Sturm, M., C. Racine, and K. Tape. "Climate Change: Increasing Shrub Abundance in the Arctic." *Nature* 411 (May 31, 2001):546–548.

Sudetic, Chuck. "As the World Burns." *Rolling Stone*, September 2, 1999, 97–106, 129.

Suffling, R. "Climate Change and Boreal Forest Fires in Fennoscandia and Central Canada." In *Greenhouse Impact on Cold-climate Ecosystems and Landscapes*, ed. M. Boer, and E. Koster, 111–132. *Selected Papers of a European Conference on Landscape Ecological Impact of Climatic Change*, Lunteren, the Netherlands, December 3–7, 1989. CARTENA Supplement 22. Cremlingen, Germany: Catena Verlag, 1992.

"Suffocation Suspected for Greatest Mass Extinction." NewScientist.com, September 9, 2003. http://www.newscientist.com/news/news.jsp?id=ns99994138.

Suhr, Jim. "Armadillos Making Northern March." Associated Press Illinois State Wire, September 16, 2006 (LEXIS).

Sullivan, Kevin. "In Britain, All Parties Want to Color the Flag Green." *Washington Post*, October 31, 2006, A-18.

"Summer Heat Wave in Europe Killed 35,000." United Press International, October 10, 2003 (LEXIS).

Suplee, Curt. "For 500 Million, a Sleeper on Greenland's Ice Sheet." *Washington Post*, July 10, 2000a, A-9.

Suplee, Curt. "Historical Records Provide a Growing Sense of Global Warmth." *Washington Post*, September 8, 2000b, A-2.

"Supreme Court Will Decide EPA's Authority Over Climate Gases." Environment News Service, November 27, 2006. http://www.ens-newswire.com/ens/nov2006/2006-11-27-02.asp.

Surendran, Aparna. "Fossil Fuel Cuts Would Reduce Early Deaths, Illness, Study Says." *Los Angeles Times*, August 17, 2001, A-20.

Svenning, Jens-Christian, and Richard Condit. "Biodiversity in a Warmer World." *Science* 322 (October 10, 2008):206–207.

Svensen, Henrik, Sverre Planke, Anders Malthe-Sorenssen, Bjorn Jamtveit, Reidun Myklebust, Torfinn Rasmussen Eidem, and Sebastin S. Rey. "Release of Methane from a Volcanic Basin as a Mechanism for Initial Eocene Global Warming." *Nature* 429 (June 3, 2004):542–545.

Svetsky, Benjamin. "How Al Gore Tamed Hollywood." *Entertainment Weekly*, July 21, 2006, 26–32.

"Sweden to Abandon Oil and Nuclear Power." *The Dominion: News from the Grassroots*. February 14, 2006. http://dominionpaper.ca/international_news/2006/02/14/sweden_to_.html.

"Swedish Bogs Flooding Atmosphere with Methane: Thawing Sub-arctic Permafrost Increases Greenhouse Gas Emissions." American Geophysical Union, February 10, 2004. http://www.scienceblog.com/community/article2366.html.

Sweet, William. *Kicking the Carbon Habit: Global Warming and the Case for Renewable and Nuclear Energy*. New York: Columbia University Press, 2006.

Takahashi, Taro. "The Fate of Industrial Carbon Dioxide." *Science* 305 (July 16, 2004):352–353.

Tangley, Laura. "Greenhouse Effects: High CO_2 Levels May Give Fast-Growing Trees an Edge." *Science* 292 (April 6, 2001):36–37.

Tanner, Lawrence H., John F. Hubert, Brian P. Coffey, and Dennis P. McInerney. "Stability of Atmospheric CO_2 Levels across the Triassic/Jurassic Boundary." *Nature* 411 (June 7, 2001):675–677.

Taubes, Gary. "Apocalypse Not." Junk Science: All the Junk That's Fit to Debunk, 1997. http://www.junkscience.com/news/taubes2.html.

Taylor, James M. "Hollywood's Fake Take on Global Warming." *Boston Globe*, June 1, 2004, A-11.

"Temperate Forests Could Worsen Global Warming." Press Release, Carnegie Institution, December 6, 2005.

"Thailand Weather: Hot Nights Ahead for Thailand, Bangkok." Deutsche Presse-Agentur." October 29, 2007 (LEXIS).

Thiessen, Mark. "Researchers Fighting 'Bad' Breath in Cattle." Associated Press, June 8, 2003 (LEXIS).

"Thin Polar Bears Called Sign of Global Warming." Environmental News Service, May 16, 2002. http://ens-news.com/ens/may2002/2002L-05-16-07.html.

"30 Years of Global Warming Has Altered the Planet." Environment News Service, May 16, 2008. http://www.ens-newswire.com/ens/may2008/2008-05-14-01.asp.

Thomas, Chris D., Alison Cameron, Rhys E. Green, Michael Bakkenes, Linda J. Beaumont, Yvonne C. Collingham, Barend F. N. Erasmus, Marinez Ferreira de Siqueira, Alan Grainger, Lee Hannah, et al. "Extinction Risk from Climate Change." *Nature* 427 (January 8, 2004):145–148.

Thomas, Chris D., and Jack J. Lennon. "Birds Extend Their Ranges Northwards." *Nature* 399 (May 20, 1999):213.

Thomas, R. "Future Sea-level Rise and Its Early Detection by Satellite Remote Sensing." In *Effects of Changes in Atmospheric Ozone and Global Climate*, Vol. 4. New York: United Nations Environment Programme/United States Environmental Protection Agency, 1986.

Thomas, R., E. Rignot, G. Casassa, P. Kanagaratnam, C. Acuna, T. Akins, H. Brecher, E. Frederick, P. Gogineni, W. Krabill, S. Manizade, H. Ramamoorthy, A. Rivera, R. Russell, J. Sonntag, R. Swift, J. Yungel, and J. Zwally. "Accelerated Sea-Level Rise from West Antarctica." *Science* 306 (October 8, 2004):255–258.

Thomas, William. "Salmon Dying in Hot Waters." Environment News Service, September 22, 1998. http://www.econet.apc.org/igc/en/hl/9809244985/hl11.html.

Thompson, Dan. "Experts Say California Wildfires Could Worsen with Global Warming." Associated Press, November 12, 2003 (LEXIS).

Thompson, David W. J., and Susan Solomon. "Interpretation of Recent Southern Hemisphere Climate Change." *Science* 296 (May 3, 2002):895–899.

Thompson, David W. J., and John M. Wallace. "Regional Climate Impacts of the Northern Hemisphere Annular Mode." *Science* 293 (July 6, 2001):85–89.

Thompson, Fred. "Plutonic Warming." *National Review* Online, March 22, 2007. http://article.nationalreview.com/?q=NTQzYWY1MGM5NTkyZTM2YWVlMDMzMDlhMzQwNThhNDU=1.

Thompson, Jake. "Nelson: Don't Waste the Waste." *Omaha World-Herald*, May 9, 2007, A-1, A-2.

Thompson, L. G., E. Mosley-Thompson, H. Brecher, M. Davis, B. León, D. Les, P.-N. Lin, T. Mashiotta, and K. Mountain. "Abrupt Tropical Climate Change: Past and Present." *Proceedings of the National Academy of Sciences* 103, no. 28 (July 11, 2006):10,536–10,543.

Thompson, L. G., T. Yao, E. Mosley-Thompson, M. E. Davis, K. A. Henderson, and P.-N. Lin. "A High-Resolution Millennial Record of the South Asian Monsoon from Himalayan Ice Cores." *Science* 289 (September 15, 2000):1916–1919.

Thorpe, R. B., J. M. Gregory, T. C. Johns, R. A. Wood, and J. F. B. Mitchell. "Mechanisms Determining the Atlantic Thermohaline Circulation Response to Greenhouse Gas Forcing in a Non-Flux-Adjusted Coupled Climate Model." *Journal of Climate* 14 (July 15, 2001):3102–3116.

Thurman, Judith. "In Fashion: Broad Stripes and Bright Stars: The Spring-summer Men's Fashion Shows in Milan and Paris." *The New Yorker*, July 28, 2003, 78–82.

"Tibetan Glaciers Melting at Stunning Rate." Discovery.com, November 24, 2008. http://dsc.discovery.com/news/2008/11/24/tibet-glaciers-warming-02.html.

Tidwell, Mike. *The Ravaging Tide: Strange Weather, Future Katrinas, and the Coming Death of America's Coastal Cities*. New York: Free Press, 2006.

Tilman, David, Jason Hill, and Clarence Lehman. "Carbon-negative Biofuels from Low-Input High-Diversity Grassland Biomass." *Science* 314 (December 8, 2006):1598–1600.

Tilmes, Simone, Rolf Müller, and Ross Salawitch. "The Sensitivity of Polar Ozone Depletion to Proposed Geoengineering Schemes." *Science* 320 (May 30, 2008):1201–1204.

Timmermann, A., J. Oberhuber, A. Bacher, M. Esch, M. Latif, and E. Roeckner. "Increased El Niño Frequency in a Climate Model Forced by Future Greenhouse Warming." *Nature* 398 (April 22, 1999):694–696.

Timmons, Heather. "British Science Group Says Exxon Misrepresents Climate Issues." *New York Times,* September 21, 2006. http://www.nytimes.com/2006/09/21/business/21green.html.

Titus, James G., and Vijay Narayanan. "The Risk of Sea Level Rise: A Delphic Monte Carlo Analysis in Which Twenty Researchers Specify Subjective Probability Distributions for Model Coefficients within their Respective Areas of Expertise." Washington, DC, U.S. Environmental Protection Agency, n.d. http://users.erols.com/jtitus/Risk/CC.html.

Tolbert, Margaret A., and Owen B. Toon. "Solving the P[olar] S[tratospheric] C[loud] Mystery." *Science* 292 (April 6, 2001):61–63.

Tollefson, Jeff. "UN Decision Puts Brakes on Ocean Fertilization." *Nature* 453 (June 5, 2008):704.

Tomsho, Robert. "Wind Shift in Energy Debate." *Wall Street Journal,* June 19, 2008, A-11.

Toner, Mike. "Huge Ice Chunk Breaks off Antarctica." *Atlanta Journal-Constitution,* March 20, 2002a, A-1.

Toner, Mike. "Meltdown in Montana: Scientists Fear Park's Glaciers May Disappear within 30 Years." *Atlanta Journal-Constitution,* June 30, 2002b, 4-A.

Toner, Mike. "Temperatures Indicate More Global Warming." *Atlanta Journal-Constitution,* July 11, 2002c, 12-A.

Toner, Mike. "Microscopic Ocean Life in Global Decline: Temperature Shifts a Cause or an Effect?" *Atlanta Journal-Constitution,* August 9, 2002d, 3-A.

Toner, Mike. "Warming Rearranges Life in Wild." *Atlanta Journal-Constitution,* January 2, 2003a, 1-A.

Toner, Mike. "Drought May Signal World Warming Trend." *Atlanta Journal-Constitution,* January 31, 2003b, 4-A.

Toner, Mike. "Oceans' Acidity Worries Experts: Carbon Dioxide on Rise, Marine Life at Risk." *Atlanta Journal-Constitution,* September 25, 2003c, n.p.(LEXIS).

Toner, Mike. "Arctic Ice Thins Dramatically, NASA Satellite Images Show." *Atlanta Journal-Constitution,* October 24, 2003d, 1-A.

Toniazzo, T., J. M. Gregory, and P. Huybrechts. "Climatic Impact of a Greenland Deglaciation and Its Possible Irreversibility." *Journal of Climate* 17, no. 1 (January 1, 2004):21–33.

"Top Scientists Warn of Water Shortages and Disease Linked to Global Warming." Associated Press in *New York Times,* March 12, 2007. http://www.nytimes.com/2007/03/12/science/earth/12climate.html.

"Top U.S. Newspapers' Focus on Balance Led to Skewed Coverage of Global Warming, Analysis Reveals." Ascribe Newsletter, August 25, 2004 (LEXIS).

"Tornado Kills at Least Nine in Northern Japan." Reuters, November 7, 2006. http://www.climatepatrol.com/forum/7/2093/pg1/index.php.

Townsend, Mark. "Monsoon Britain: As Storms Bombard Europe, Experts Say That What We Still Call 'Freak' Weather Could Soon Be the Norm." *London Observer,* August 11, 2002, 15.

"Transatlantic Plan to Cut Aircraft Emissions Lifts Off." Environment News Service, June 18, 2007. http://www.ens-newswire.com/ens/jun2007/2007-06-18-04.asp.

Trapp, Robert J., Noah S. Diffenbaugh, Harold E. Brooks, Michael E. Baldwin, Eric D. Robinson, and Jeremy S. Pal. "Changes in Severe Thunderstorm Environment Frequency During the 21st Century Caused by Anthropogenically Enhanced Global Radiative Forcing." *Proceedings of the National Academy of Sciences* 104 (December 11, 2007):19,719–19,723.

Travis, Davis J., Andrew M. Carleton, and Ryan G. Lauritsen. "Climatology: Contrails Reduce Daily Temperature Range." *Nature* 418 (August 8, 2002):593–594.

Treaster, Jospeh B. "Gulf Coast Insurance Expected to Soar." *New York Times,* September 24, 2005. http://www.nytimes.com/2005/09/24/business/24insure.html.

Trenberth, Kevin E., Aiguo Dai, Roy M. Rassmussen, and David B. Parsons. "The Changing Character of Precipitation." *Bulletin of the American Meteorological Society* 84, no. 9 (September 2003):1205–1217.

Trenberth, Kevin E., and Timothy J. Hoar. "The 1990–1995 El Niño-Southern Oscillation Event: Longest on Record." *Geophysical Research Letters* 23, no. 1 (January 1, 1996):57–60.

Trenberth, Kevin E., and Timothy J. Hoar. "El Niño and Climate Change." *Geophysical Research Letters* 24 (1997):3057–3060.

Treydte, Kerstin S., Gerhard H. Schleser, Gerhard Helle, David C. Frank, Matthias Winiger, Gerald H. Haug, and Jan Esper. "The Twentieth Century was the Wettest Period in Northern Pakistan over the Past Millennium." *Nature* 440 (April 27, 2006):1179–1182.

Tromp, Tracey K., Run-Lie Shia, Mark Allen, John M. Eiler, and Y. L. Yung. "Potential Environmental Impact of a Hydrogen Economy on the Stratosphere." *Science* 300 (June 13, 2003):1740–1742.

Truffer, Martin, and Mark Fahnestock. "Rethinking Ice Sheet Time Scales." *Science* 315 (March 16, 2007):1508–1510.

Trumbore, Susan E., and Claudia I. Czimczik. "An Uncertain Future for Soil Carbon." *Science* 321 (September 12, 2008):1455–1456.

Tsuda, Atsushi, Shigenobu Takeda, Hiroaki Saito, Jun Nishioka, Yukihiro Nojiri, Isao Kudo, Hiroshi Kiyosawa, Akihiro Shiomoto, Keiri Imai, Tsuneo Ono, Akifumi Shimamoto, Daisuke Tsumune, Takeshi Yoshimura, Tatsuo Aono, Akira Hinuma, Masatoshi Kinugasa, Koji Suzuki, Yoshiki Sohrin, Yoshifumi Noiri, Heihachiro Tani, Yuji Deguchi, Nobuo Tsurushima, Hiroshi Ogawa, Kimio Fukami, Kenshi Kuma, and Toshiro Saino. "A Mesoscale Iron Enrichment in the Western Subarctic Pacific Induces a Large Centric Diatom Bloom." *Science* 300 (May 9, 2003):958–961.

Tubiello, Francesco, Jean Francois Soussana, and S. Mark Howden. "Crop and Pasture Response to Climate Change." *Proceedings of the National Academy of Sciences* 104 (December 11, 2007):19,686–19,690.

"2007 Arctic Sea Ice Nearly Matches Record Low." Environment News Service, April 4, 2007. http://www.ens-newswire.com/ens/apr2007/2007-04-04-09.asp#anchor2.

"2006 Third Warmest Year on Record Across USA." Environment News Service. December 14, 2006. http://www.ens-newswire.com/ens/dec2006/2006-12-14-09.asp#anchor1.

Tysver, Robynn. "Populist Huckabee Gains Ground with Religious Approach." *Omaha World-Herald*, December 2, 2007, A-1, A-2.

Ulrich, Lawrence. "They're Electric, but Can They Be Fantastic?" *New York Times*, September 23, 2007. http://www.nytimes.com/2007/09/23/automobiles/23AUTO.html.

"UN: Industrial Countries' Greenhouse Gases Rose 2000–2006." Environment News Service, November 18. 2008. http://www.ens-newswire.com/ens/nov2008/2008-11-18-01.asp.

"UN Climate Change Impact Report: Poor Will Suffer Most." Environment News Service, April 6, 2007. http://www.ens-newswire.com/ens/apr2007/2007-04-06-01.asp.

Understanding the Science of Global Climate Change. Washington, DC: Environmental Media Services, 1997.

Unger, Mike. "Global Warming Hits Big Screen." *Annapolis Capital* (Maryland), May 28, 22004, A-1.

"University of Georgia Creates New Biofuel from Trees." Environment News Service, May 18, 2007. http://www.ens-newswire.com/ens/may2007/2007-05-18-09.asp#anchor5.

Unwin, Brian. "Tropical Birds and Exotic Sea Creatures Warm to Britain's Welcoming Waters." *London Independent*, August 20, 2001, 7.

Urban, Frank E., Julia E. Cole, and Jonathan T. Overpeck. "Influence on Mean Climate Variability from a 155-year Tropical Pacific Coral Record." *Nature* 407 (October 26, 2000):989–993.

Urquhart, Frank, and Jim Gilchrist. "Air Travel to Blame as Well." *The Scotsman*, October 8, 2002, n.p.(LEXIS).

"U.S. Electricity Sector Makes Twice as Much Greenhouse Gas as Europe: Report." Agence France Presse, October 21, 2002 (LEXIS).

U.S. Environmental Protection Agency. *Ecological Impacts from Climate Change: An Economic Analysis of Freshwater Recreational Fishing.* Washington, DC: EPA, April 1995.

U.S. Environmental Protection Agency. *The Cost of Holding Back the Sea.* Washington, DC: EPA, 1997. http://users.erols.com/jtitus/Holding/NRJ.html#causes.

U.S. Green Building Council Web site. http://www.usgbc.org/DisplayPage.aspx?CategoryID=19.

"U.S. Greenhouse Gas Emissions Lower in 2006." Environment News Service, April 15, 2008. http://www.ens-newswire.com/ens/apr2008/2008-04-15-01.asp.

"U.S. Mayors Seek Federal Help to Protect Climate." Environment News Service, November 5, 2007. http://www.ens-newswire.com/ens/nov2007/2007-11-05-01.asp.

"U.S. Mayors Take the Lead in Fighting Climate Change." Environment News Service, June 25, 2007. http://www.ens-newswire.com/ens/jun2007/2007-06-25-04.asp.

"U.S. NSF: Scientists Find Climate Change Is Major Factor in Drought's Growing Reach." M2 Presswire, January 12, 2005 (LEXIS).

Utton, Tim. "The Rat Rampage." *London Daily Mail*, January 22, 2003, n.p.(LEXIS).

"UV Radiation Linked to Deformed Amphibians." Environment News Service." June 21, 2002. http://www.ens-newswire.com/ens/jun2002/2002-06-21-09.asp#anchor4.

Valentini, R., G. Matteucci, A. J. Dolman, E.-D. Schulze, C. Rebmann, E. J. Moors, A. Granier, P. Gross, N. O. Jensen, K. Pilegaard, A. Lindroth, A. Grelle, C. Bernhofer, T. Grunwald, M. Aubinet, R. Ceulmans, A. S. Kowalski, T. Vesala, U. Rannik, P. Berbigler, D. Loustau, J. Guomundsson, H. Thorgiersson, A. Ibrom, K. Morgenstern, R. Clement, J. Moncrieff, L. Montagnani, S. Minerbi, and P. G. Jarvis. "Respiration as the Main Determinant of Carbon Balance in European Forests." *Nature* 404 (April 20, 2000):861–865.

Vanderkam, Laura. "Want to Save the Planet? Stay Home." *USA Today*, May 20, 2008, 9-A.

Van Mantgem, Phillip J., Nathan L. Stephenson, John C. Byrne, Lori D. Daniels, Jerry F. Franklin, Peter Z. Fulé, Mark E. Harmon, Andrew J. Larson, Jeremy M. Smith, Alan H. Taylor, and Thomas T. Veblen. "Widespread Increase of Tree Mortality Rates in the Western United States." *Science* 323 (January 23, 2009):521–524.

Vaughan, D. G., and R. Arthern. "Why Is It Hard to Predict the Future of Ice Sheets?" *Science* 315 (March 16, 2007):1503–1504.

Vaughan, D. G., G. J. Marshall, W. M. Connolley, C. L. Parkinson, R. Mulvaney, D. A. Hodgson, J. C. King, C. J. Pudsey, and J. Turner. "Recent Rapid Regional Climate Warming on the Antarctic Peninsula." *Climatic Change* 60, no. 3 (October 2003):243–274.

Vecchi, Gabriel A., Brian J. Soden, Andrew T. Wittenberg, Isaac M. Held, Ants Leetmaa, and Matthew J. Harrison. "Weakening of Tropical Pacific Atmospheric Circulation Due to Anthropogenic Forcing." *Nature* 441 (May 4, 2006):73–76.

Vecchi, Gabriel A., Kyle L. Swanson, and Brian J. Soden. "Whither Hurricane Activity?" *Science* 322 (October 31, 2008):687–689.

Vedantam, Shankar. "Glacier Melt Could Signal Faster Rise in Ocean Levels." *Washington Post*, February 17, 2006, A-1. http://www.washingtonpost.com/wp-dyn/content/article/2006/02/16/AR2006021601292_pf.html.

Veizer, Jan, Yves Godderis, and Louis M. Francois. "Evidence for Decoupling of Atmospheric CO_2 and Global Climate during the Phanerozoic Eon." *Nature* 408 (December 7, 2000):698–701.

Velicogna, Isabella, and John Wahr. "Measurements of Time-Variable Gravity Show Mass Loss in Antarctica." *Science* online March 2, 2006a, doi: 10.1126/science.1123785.

Velicogna, Isabella, and John Wahr. "Acceleration of Greenland Ice Mass Loss in Spring 2004." *Nature* 443 (September 21, 2006b):329–331.

Verburg, Piet, Robert E. Hecky, and Hedy Kling. "Ecological Consequences of a Century of Warming in Lake Tanganyika." *Science* 301 (July 25, 2003):505–507.

Vergano, Dan. "Global Warming May Leave West in the Dust By 2050, Water Supplies Could Plummet 30 Percent, Climate Scientists Warn." *USA Today*, November 21, 2002, 9-D.

Vergano, Dan. "Kidney Stone Cases Could Heat Up." *USA Today*, July 15, 2008, A-1.

Vermeij, Geerat J., and Peter D. Roopnarine. "The Coming Arctic Invasion." *Science* 321 (August 8, 2008):780–781.

Vernal, Anne de, and Claude Hillaire-Marcel. "Natural Variability of Greenland Climate, Vegetation, and Ice Volume during the Past Million Years." *Science* 320 (June 20, 2008):1622–1625.

Verschuren, Dirk. "Global Change: The Heat on Lake Tanganyika." *Nature* 424 (August 14, 2003):731–732.

Vick, Karl, and Sonya Geis. "California Fires Continue to Rage Evacuation May Be Largest, Officials Say." *Washington Post*, October 24, 2007, A-1. http://www.washingtonpost.com/wp-dyn/content/article/2007/10/23/AR2007102300347_pf.html.

Victor, David G. *Climate Change: Debating America's Policy Options.* New York: Council on Foreign Relations, 2004.

Vidal, John. "The Darling Buds of February: Daffodils Flower and Frogs Spawn as Spring Gets Earlier and Earlier." *London Guardian*, February 23, 2002a, 3.

Vidal, John. "Antarctica Sends Warning of the Effects of Global Warming: Scientists Stunned as Ice Shelf Falls Apart in a Month." *London Guardian*, March 20, 2002b, 3.

Vidal, John. "You Thought It Was Wet? Wait until the Asian Brown Cloud Hits Town: Extreme Weather Set to Worsen through Pollution and El Niño: Cloud with No Silver Lining." *London Guardian*, August 12, 2002c, 3.

Vidal, John. "Mountain Cultures in Grave Danger Says U.N.: Agriculture, Climate and Warfare Pose Dire Threat to Highland Regions around the World." *London Guardian,* October 24, 2002d, 7.

Vidal, John. "Better Get Used to It, Say Climate Experts." *London Guardian,* October 28, 2002e, 3.

Vidal, John. "Sweden Plans to Be World's First Oil-free Economy." *London Guardian,* February 8, 2006. http://www.peopleandplanet.net/doc.php?id=2662.

Vidal, John, and Terry Macalister. "Kyoto Protests Disrupt Oil Trading." *London Guardian,* February 17, 2005, 4.

Vincent, John, and Paul Brown. "Swoop to Conquer: Global Warming Brings Butterflies to Britain Earlier." *London Guardian,* May 24, 2000, 9.

Vinciguerra, Thomas. "At 90, an Environmentalist from the '70s Still Has Hope, *New York Times* June 19, 2007. http://www.nytimes.com/2007/06/19/science/earth/19conv.html.

Vinnikov, Konstantin Y., and Norman C. Grody. "Global Warming Trend of Mean Tropospheric Temperature Observed by Satellites." *Science* 302 (October 10, 2003):269–272.

Vinton, Nathaniel. "Changing Climate Is Forcing World Cup Organizers to Adapt." *New York Times,* November 27, 2006. http://www.nytimes.com/2006/11/27/sports/othersports/27ski.html.

Vitello, Paul. "Home Insurers Canceling in East." *New York Times,* October 16, 2007. http://www.nytimes.com/2007/10/16/nyregion/16insurance.html.

Vogel, Nancy. "Less Snowfall Could Spell Big Problems for State." *Los Angeles Times,* June 11, 2001, A-1.

Volk, Tyler. "Real Concerns, False Gods: Invoking a Wrathful Biosphere Won't Help Us Deal with the Problems of Climate Change." Review, James Lovelock, *The Revenge of Gaia,* 2006. *Nature* 440 (April 13, 2006):869–870.

Volk, Tyler. *CO_2 Rising: The World's Greatest Environmental Challenge.* Cambridge, MA: MIT Press, 2008.

Von Radowitz, John. "Antarctic Wildlife 'at Risk from Global Warming.'" Press Association News, September 9, 2002 (LEXIS).

"Vulnerable Communities Worldwide Adapt to Climate Change." Environment News Service, December 5, 2007. http://www.ens-newswire.com/ens/dec2007/2007-12-05-01.asp.

Waddell, Lynn. "Rising Insurance Rates Push Florida Homeowners to Brink." *New York Times,* June 29, 2006. http://www.nytimes.com/2006/06/29/us/29florida.html.

Wadman, Meredith. "Car Maker Joins Exodus from Anti-Kyoto Coalition." *Nature* 404 (2000):322.

Wagner, Angie. "Debate over Causes Aside, Warm Climate's Effects Striking in the West." Associated Press, April 27, 2004 (LEXIS).

Wald, Matthew L. "It's Free, Plentiful and Fickle." *New York Times,* December 28, 2006. http://www.nytimes.com/2006/12/28/business/28wind.html.

Wald, Matthew L. "What's So Bad about Big?" *New York Times,* March 7, 2007a. http://www.nytimes.com/2007/03/07/business/businessspecial2/07big.html.

Wald, Matthew L. "In a Test of Capturing Carbon Dioxide, Perhaps a Way to Temper Global Warming." *New York Times,* March 15, 2007b. http://www.nytimes.com/2007/03/15/business/15carbon.html.

Wald, Matthew L. "Turning Glare into Watts." *New York Times,* March 6, 2008a. http://www.nytimes.com/2008/03/06/business/06solar.html.

Wald, Matthew L. "Mounting Costs Slow the Push for Clean Coal." *New York Times,* May 30, 2008b. http://www.nytimes.com/2008/05/30/business/30coal.html.

Wald, Matthew L. "Georgia Judge Cites Carbon Dioxide in Denying Coal-plant Permit. *New York Times,* July 1, 2008c, C-4.

Wald, Matthew L. "Two Large Solar Plants Planned in California." *New York Times,* August 15, 2008d. http://www.nytimes.com/2008/08/15/business/15solar.html.

Wald, Matthew L. "Wind Energy Bumps into Power Grid's Limits." *New York Times,* August 27, 2008e. http://www.nytimes.com/2008/08/27/business/27grid.html.

Walker, Gabrielle. "The Tipping Point of the Iceberg." *Nature* 441 (June 15, 2006):802–805.

Walker, Gabrielle, and Sir David King. *The Hot Topic: What We Can Do about Global Warming.* Vancouver, B.C. and Toronto: Greystone Books/ Douglas & McIntyre Publishing Group, 2008.

"Walnuts and Vineyards." *London Times,* October 29, 2001, n.p.(LEXIS).

Walter, K. M., S. A. Zimov, J. P. Chanton, D. Verbyla, and F. S. Chapin, III. "Methane Bubbling from Siberian Thaw Lakes as a Positive Feedback to Climate Warming." *Nature* 414 (September 7, 2006):71–75.

Walther, Gian-Reto. "Weakening of Climatic Constraints with Global Warming and Its Consequences for Evergreen Broad-leaved Species." *Folia Geobotanica* 37 (2002):129–139.

Walther, Gian-Reto. "Plants in a Warmer World." *Perspectives in Plant Ecology, Evolution, and Systematics* 6, no. 3 (2003):169–185.

Walther, Gian-Reto, Emmanuel Gritti, Silje Berger, Thomas Hickler, Zhiyao Tang, and Martin T. Sykes. "Palms Tracking Climate Change." *Global Ecology and Biogeography* (2007), doi: 10.1111/j.1466-8238.2007.00328.x.

Walther, Gian-Reto, Eric Post, Peter Convey, Annette Menzel, Camille Parmesan, Trevor J. C. Beebee, Jean-Marc Fromentin, Ove Hoegh-Guldberg, and Franz Bairlein. "Ecological Responses to Recent Climate Change." *Nature* 416 (March 28, 2002):389–395.

Wang, Tobias, and Johannes Overgaard. "The Heartbreak of Adapting to Global Warming." *Science* 315 (January 25, 2007):49–50.

Wara, Michael W., Ana Christina Ravelo, and Margaret L. Delaney. "Permanent El Niño-Like Conditions during the Pliocene Warm Period." *Science* 309 (July 29, 2005):758–761.

Ward, Peter D. *Under a Green Sky: Global Warming, the Mass Extinctions of the Past, and What They Mean for Our Future.* Washington, DC: Smithsonian Institution Press, 2007.

"Warmer Climate Could Disrupt Water Supplies." Environment News Service, December 20, 2001. http://ens-news.com/ens/dec2001/2001L-12-20-09.html.

"Warmer Climate Hurting Birds, New IUCN Red List Shows." Environment News Service, May 19, 2008. http://www.ens-newswire.com/ens/may2008/2008-05-20-03.asp.

"Warming Climate Undermines World Food Supply." Environment News Service, December 3, 2007. http://www.ens-newswire.com/ens/dec2007/2007-12-03-05.asp.

"Warming Could Submerge Three of India's Largest Cities: Scientist." Agence France Presse, December 6, 2003 (LEXIS).

"Warming Doom for Great Barrier Reef." Australian Associated Press in *The Mercury* (Hobart, Australia), February 16, 2002, n.p.(LEXIS).

"Warming Oceans Put More Stress on Whales." Environment News Service, May 21, 2007. http://www.ens-newswire.com/ens/may2007/2007-05-21-04.asp.

"Warming Streams Could Wipe Out Salmon, Trout." Environment News Service, May 22, 2002. http://ens-news.com/ens/may2002/2002L-05-22-06.html.

"Warming Tropics Show Reduced Cloud Cover." Environment News Service, February 1, 2002. http://ens-news.com/ens/feb2002/2002L-02-01-09.html.

"Warm Water Surging into Arctic Ocean." Environment News Service, September 27, 2006. http://www.ens-newswire.com/ens/sep2006/2006-09-27-01.asp.

Warne, Kennedy. "Forests of the Tide." *National Geographic*, February 2007, 132–151.

Warren, M. S., J. K. Hill, J. A. Thomas, J. Asher, R. Fox, B. Huntley, D. B. Roy, M. G. Telfer, S. Jeffcoate, P. Harding, G. Jeffcoate, S. G. Willis, J. N. Greatorex-Davies, D. Moss, and C. D. Thomas. "Rapid Responses of British Butterflies to Opposing Forces of Climate and Habitat Change." *Nature* 414 (November 1, 2001):65–69.

Warrick, Jody. "Earth at Its Warmest in Past 12 Centuries: Scientist Says Data Suggest Human Causes." *Washington Post*, December 8, 1998. http://www.asoc.org/currentpress/1208post.htm.

Watson, A. J., D. C. E. Bakker, A. J. Ridgwell, P. W. Boyd, and C. S. Law. "Effect of Iron Supply on Southern Ocean CO_2 Uptake and Implications for Glacial Atmospheric CO_2." *Nature* 407 (October 12, 2000):730–733.

Watson, Jeremy. "Plan to Hold Back Tides of Venice Runs into Flood of Opposition from Greens." *Scotland on Sunday*, December 30, 2001, 18.

Watson, Traci. "China Leaves U.S. in Dust as the No. 1 CO_2 Offender." *USA Today*, May 1, 2008, A-5.

Watt, Nicholas. "Planet Is Running out of Time, says Meacher: U.S. Rejection of Kyoto Climate Plan 'Risks Uninhabitable Earth.'" *London Guardian*, May 16, 2002, 11.

Watt-Cloutier, Sheila. "Honouring Our Past, Creating Our Future: Education in Northern and Remote Communities." In *Aboriginal Education: Fulfilling the Promise*, ed. Lynne Davis, Louise Lahache, and Marlene Castellano. Vancouver: University of British Columbia Press, 2000.

Watt-Cloutier, Sheila. Personal communication, March 28, 2001.

Watt-Cloutier, Sheila. Speech to Conference of Parties to the United Nations Framework Convention on Climate Change, Milan, Italy, December 10, 2003a.

Watt-Cloutier, Sheila. "Speech Notes for Sheila Watt-Cloutier, Chair, Inuit Circumpolar Conference." Conference of Parties to the United Nations Framework Convention on Climate Change, Milan, Italy, December 10, 2003b. http://www.inuitcircumpolar.com/index.php?ID=242&Lang=En.

Watt-Cloutier, Sheila. Personal communication, from Iqaluit, Nunavut, January 4, 2004a.

Watt-Cloutier, Sheila. "Canada and Inuit: Addressing Global Environmental Challenges, Remarks by Sheila Watt-Cloutier, Chair, Inuit Circumpolar Conference at the Inaugural Environmental Protection Service Inuit Speaker Series." Ottawa, Ontario, January 16, 2004b. http://www.inuitcircumpolar.com/index.php?ID=250&Lang=En.

Watt-Cloutier, Sheila. "The Climate Change Petition by the Inuit Circumpolar Conference to the Inter-American Commission on Human Rights." Presentation at the Eleventh Conference of Parties to the UN Framework Convention on Climate Change, Montreal, December 7, 2005. http://inuitcircumpolar.com/index.php?ID=318&Lang=En.

Watt-Cloutier, Sheila. Personal communication, February 27, 2006a.

Watt-Cloutier, Sheila. Personal communication, March 1, 2006b.

Watt-Cloutier, Sheila. "The Arctic and the Global Environment: Making a Difference on Climate Change." Solutions for Communities Climate Summit, Hosted by Global Green USA. Beverly Hills, California, April 1, 2006c. http://inuitcircumpolar.com/index.php?ID=329&Lang=En.

Watt-Cloutier, Sheila, and Terry Fenge. "Poisoned by Progress: Will Next Week's Negotiations in Bonn Succeed in Banning the Chemicals That Inuit are Eating?" Press Release, Inuit Circumpolar Conference, March 14, 2000. http://www.inuitcircumpolar.com/index.php?ID=250&Lang=En.

Weart, Spencer R. *The Discovery of Global Warming*. Cambridge, MA: Harvard University Press, 2003.

Webb, Jason. "Mosquito Invasion as Argentina Warms." Reuters, 1998a. http://bonanza.lter.uaf.edu/~davev/nrm304/glbxnews.htm.

Webb, Jason. "Small Islands Say Global Warming Hurting Them Now." Reuters, 1998b. http://bonanza.lter.uaf.edu/~davev/nrm304/glbxnews.htm.

Webb, Jason. "World Forests Said Vulnerable to Global Warming." Reuters. November 4, 1998c. http://bonanza.lter.uaf.edu/~davev/nrm304/glbxnews.htm.

Webb, Jason. "World Temperatures Could Jump Suddenly." Reuters, November 4, 1998d. http://bonanza.lter.uaf.edu/~davev/nrm304/glbxnews.htm.

Weber, Bob. "Arctic Sea Ice Isn't Melting, Just Drifting Away: New Study." *Montreal Gazette*, April 25, 2001, A-8.

Webster, Ben. "Boeing Admits Its New Aircraft Will Guzzle Fuel." *London Times*, June 19, 2001, n.p.(LEXIS).

Webster, P. J., G. J. Holland, J. A. Curry, and H.-R. Chang. "Changes in Tropical Cyclone Number, Duration, and Intensity in a Warming Environment." *Science* 309 (September 16, 2005):1844–1846.

Weinburg, Bill. "Hurricane Mitch, Indigenous Peoples and Mesoamerica's Climate Disaster." *Native Americas* 16, no. 3/4 (Fall/Winter 1999):50–59. http://nativeamericas.aip.cornell.edu/fall99/fall99weinberg.html.

Weiner, Jonathan. *The Next One Hundred Years: Shaping the Fate of Our Living Earth*. New York: Bantam Books, 1990.

Weise, Elizabeth. "Warming Climate Makes Gardeners' Map Out-of-Date." *USA Today*, April 24, 2008, A-1.

Weiss, Rick. "Firms Seek Patents on 'Climate Ready' Altered Crops." *Washington Post*, May 13, 2008. http://www.washingtonpost.com/wp-dyn/content/article/2008/05/12/AR2008051202919_pf.html.

Welch, Craig. "Global Warming Hitting Northwest Hard, Researchers Warn." *Seattle Times*, February 14, 2004. http://seattletimes.nwsource.com/html/localnews/2001857961_warming14m.html.

Welsh, Jonathan. "Drive Buys: Honda Insight. A Car with an Extra Charge." *Wall Street Journal*, March 10, 2000, W-15C.

Wentz, Frank J., Lucrezia Ricciardulli, Kyle Hilburn, and Carl Mears. "How Much More Rain Will Global Warming Bring?" *Science* 317 (July 2007):233–235.

Wentz, Frank J., and M. Schabel. "Effects of Orbital Decay on Satellite-Derived Lower Tro-posheric Temperature Trends." *Nature* 394 (August 13, 1998):661–664.

Westerling, A. L., H. G. Hidalgo, D. R. Cayan, and T. W. Swetnam. "Warming and Earlier Spring Increase Western U.S. Forest Wildfire Activity." *Science* 313 (August 18, 2006):940–943.

"Western Wildfires Linked to Atlantic Ocean Temperatures." Environmental News Service, January 3, 2007. http://www.ens-newswire.com/ens/jan2007/2007-01-03-09.asp#anchor2.

Whalley, John, and Randall Wigle. "The International Incidence of Carbon Taxes." Paper Presented at a Conference on Economic Policy Responses to Global Warming. Rome, Italy, September 1990.

"What About Us?" Editorial, *New York Times*, July 28, 2006. http://www.nytimes.com/2006/07/28/opinion/28fri2.html.

Whipple, Dan. "Climate: The Arctic Goes Bush." United Press International, January 10, 2005 (LEXIS).

White, James C. "Do I Hear a Million?" *Science* 304 (June 11, 2004):1609–1610.

White, Joseph R. "An Ecotopian View of Fuel Economy." *Wall Street Journal*, June 26, 2006, D-4.

White, M. A., N. S. Diffenbaugh, G. V. Jones, J. S. Pal, and F. Gioirgi. "Extreme Heat Reduces and Shifts United States Premium Wine Production in the 21st Century." *Proceedings of the National Academy of Sciences* 103, no. 30 (July 25, 2006):11,217–11,222.

White, Martha C. "Enjoy Your Green Stay." *New York Times*, June 26, 2007. http://www.nytimes.com/2007/06/26/business/26green.html.

Whitlock, Craig. "Cloudy Germany a Powerhouse in Solar Energy." *Washington Post*, May 5, 2007, A-1. http://www.washingtonpost.com/wp-dyn/content/article/2007/05/04/AR2007050402466_pf.html.

Whoriskey, Peter, and Joby Warrick. "Report Revises Katrina's Force: Hurricane Center Downgrades Storm to Category 3 Strength." *Washington Post*, December 22, 2005, A-3. http://www.washingtonpost.com/wp-dyn/content/article/2005/12/21/AR2005122101960.html.

"Why We're All Being Caught on the Hop by Global Warming." *Irish Independent*, July 17, 2004, n.p.(LEXIS).

Wielicki, Bruce A., Takmeng Wong, Richard P. Allan, Anthony Slingo, Jeffrey T. Kiehl, Brian J. Soden, C. T. Gordon, Alvin J. Miller, Shi-Keng Yang, David A. Randall, Franklin Robertson, Joel Susskind, and Herbert Jacobowitz. "Evidence for Large Decadal Variability in the Tropical Mean Radiative Energy Budget." *Science* 295 (February 1, 2002):841–844.

Wigley, T. M. L. "ENSO, Volcanoes, and Record Breaking Temperatures." *Geophysical Research Letters* 27 (2000):4101–4104.

Wigley, T. M. L. "A Combined Mitigation/Geoengineering Approach to Climate Stabilization." *Science* 314 (October 20, 2006):452–454.

Wignall, Paul B., John M. McArthur, Crispin T. S. Little, and Anthony Hallam. "Methane Release in the Early Jurassic Period." *Nature* 441 (June 1, 2006):E5. [Arising from D. B. Kemp, A. L. Coe, A. S. Cohen, and L. Schwark. *Nature* 437 (2005):396–399.]

Wilber, Del Quentin. "U.S. Airlines under Pressure to Fly Greener: Carriers Already Trying to Save Fuel as Europe Proposes Plan." *Washington Post*, July 28, 2007, D-1. http://www.washingtonpost.com/wp-dyn/content/article/2007/07/27/AR2007072702256_pf.html.

"Wildfires Add Carbon to the Atmosphere." Environment News Service, December 9, 2002. http://ens-news.com/ens/dec2002/2002-12-09-09.asp.

Wiles, David. "The Road to Sweden's Oil-Free Future." *Sweden Today*, March 31, 2006. Sweden.Se, "The Official Gateway to Sweden." October 6, 2006. http://www.sweden.se/templates/cs/Article_14363.aspx.

Wilfred, John Noble. "Ages-old Polar Icecap Is Melting, Scientists Find." *New York Times*, August 19, 2000, 1.

Wilhite, Donald A. "Drought in the U.S. Great Plains." In *Handbook of Weather, Climate, and Water: Atmospheric Chemistry, Hydrology, and Societal Impacts*, ed. Thomas D. Potter and Bradley R. Colman, 743–758. Hoboken, NJ: Wiley Interscience, 2003.

Wilkin, Dwayne. "A Team of Glacial Ice Experts Say Mother Nature's Thermostat Has Kept the Eastern Arctic at about the Same Temperature since 1960." *Nunatsiaq News*, May 30, 1997. http://www.nunanet.com/~nunat/week/70530.html#7.

Wilkinson, Clive, Olof Linden, Herman Cesar, Gregor Hodgson, Jason Rubens, and Alan E. Strong. "Ecological and Socioeconomic Impacts of 1998 Coral Mortality in the Indian

Ocean: An ENSO Impact and a Warning of Future Change?" *Ambio* 28, no. 2 (March 1999):188–196.

Will, George F. "Fuzzy Climate Math." *Washington Post*, April 12, 2007, A-27. http://www. washingtonpost.com/wp-dyn/content/article/2007/04/11/AR2007041102109_pf.html.

Willett, Katharine M., Nathan P. Gillett, Philip D. Jones, and Peter W. Thorne. "Attribution of Observed Surface Humidity Changes to Human Influence." *Nature* 449 (October 11, 2007):710–712.

Willey, Zach, and Bill Chameides, eds. *Harnessing Farms and Forests in the Low-Carbon Economy: How to Create, Measure, and Verify Greenhouse Gas Offsets.* Durham, NC: Duke University Press, 2007.

Williams, Alex. "Buying Into the Green Movement." *New York Times*, July 1, 2007. http://www. nytimes.com/2007/07/01/fashion/01green.html.

Williams, Brian. "Reef Down to Half Its Former Self." *Courier Mail* (Queensland, Australia), June 24, 2004, 11.

Williams, Carol J. "Danes See a Breezy Solution: Denmark Has Become a Leader in Turning Offshore Windmills into Clean, Profitable Sources of Energy as Europe Races to Meet Emissions Goals." *Los Angeles Times*, June 25, 2001, A-1.

Williams, Frances. "Everest Hit by Effects of Global Warming." *London Financial Times*, June 6, 2002, 2.

Williams, Gisela. "Resorts Prepare for a Future Without Skis." *New York Times*, December 2, 2007. http://travel.nytimes.com/2007/12/02/travel/02skiglobal.htm.

Williams, Jack. "Rising Ocean Temperatures Aren't Breaking the Ice." *USA Today*, April 25, 2000, 10-D.

Wilmsen, Steven. "Critters Enjoy a Baby Boom: Mild Winter's Downside Is Proliferation of Vermin." *Boston Globe*, March 30, 2002, B-1.

Wilson, Paul A., and Richard D. Norris. "Warm Tropical Ocean Surface and Global Anoxia during the Mid-Cretaceous Period." *Nature* 412 (July 26, 2001):425–429.

Wilson, Scott. "Warming Shrinks Peruvian Glaciers: Retreat of Andean Snow Caps Threatens Future for Valleys." *Washington Post*, July 9, 2001, A-1.

Wines, Michael. "Rising Star Lost in Russia's Latest Disaster." *New York Times*, September 24, 2002, A-11.

Wines, Michael. "Dinner Disappears, and African Penguins Pay the Price." *New York Times*, June 4, 2007. http://www.nytimes.com/2007/06/04/world/africa/04robben.html.

Winestock, Geoff. "How to Cut Emissions? E.U. Can't Decide." *Wall Street Journal*, July 13, 2001, A-9.

"Winter and Summer, Arctic Sea Ice Is Shrinking." Environment News Service, November 29, 2006. http://www.ens-newswire.com/ens/nov2006/2006-11-29-09.asp.

"Wintertime Disintegration of Wilkins Ice Shelf." NASA Earth Observatory. July 22, 2008. http://earthobservatory.nasa.gov/Newsroom/NewImages/images.php3?img_id=18095.

Wiseman, Paul. "Australia Pushes New Climate Plan." *USA Today*, September 6, 2007, 10-A.

Witt, Howard. "Blasting A/C in the Arctic." Chicago *Tribune*, September 29, 2006. http://www. chicagotribune.com/news/nationworld/chi-0609290169sep29,1,1322189.story?ctrack=1& cset=true.

Witze, Alexandra. "Evidence Supports Warming Theory." *Dallas Morning News* in *Omaha World-Herald* (Metro Extra), January 12, 2000, 1.

Witze, Alexandra. "Climate Change: Losing Greenland." *Nature* 452 (April 17, 2008):798–802.

Wofsy, Steven C. "Where Has All the Carbon Gone?" *Science* 292 (June 22, 2001):2261–2263.

Wohlforth, Charles. *The Whale and the Supercomputer: On the Northern Front of Climate Change.* New York: North Point Press/Farrar, Strauss and Giroux, 2004.

Wolf, Martin. "Hot Air about Global Warming." *London Financial Times*, November 29, 2000, 27.

Wolff, Eric W. "Whither Antarctic Sea Ice?" *Science* 302 (November 14, 2003):1,164.

Wolff, Eric W., Laurent Augustin, Carlo Barbante, Piers R. F. Barnes, Jean Marc Barnola, Matthias Bigler, Emiliano Castellano, Olivier Cattani, Jerome Chappellaz, Dorthe Dahl-Jensen, et al. "Eight Glacial Cycles from an Antarctic Ice Core." *Nature* 429 (June 10, 2004):623–627.

Wood, Richard A., Anne B. Keen, John F.B. Mitchell, and Jonathan M. Gregory. "Changing Spatial Structure of the Thermohaline Circulation in Response to Atmospheric CO_2 Forcing in a Climate Model." *Nature* 399 (June 10, 1999):572–575.

Woodard, Colin. "Slowly, but Surely, Iceland Is Losing Its Ice: Global Warming Is Prime Suspect in Meltdown." *San Francisco Chronicle,* August 21, 2000, A-1.

Woodard, Colin. "Wind Turbines Sprout from Europe to U.S." *Christian Science Monitor,* March 14, 2001, 7.

Woodard, Colin. "Curbing Climate Change. Is the World Doing Enough?" *Congressional Quarterly Global Researcher* 1, no. 2 (February 2007):27–50. http://www.lib.iup.edu/depts/libsci/cfr/fieldnotes_files/curbing_climate_change.pdf. or http://www.globalresearcher.com.

Woodcock, John. "Coral Reefs at Risk from Man-made Cocktail of Poisons." *The Scotsman,* September 11, 2001, 4.

Woods, Audrey. "English Gardens Disappearing in Global Warmth: Will They Be Replaced by Palm Trees?" Associated Press in *Financial Post* (Canada), November 20, 2002, S-10.

Woodwell, George M. "The Effects of Global Warming." In *Global Warming: The Greenpeace Report,* ed. Jeremy Leggett, 116–132. New York: Oxford University Press, 1990.

Woodwell, George M. "The Global Warming Issue." 1999. http://www.gibbons.free online.co.uk/Articles/The_Global_warming_issue.htm.

Woodwell, George M., and Fred T. MacKenzie, eds. *Biotic Feedbacks in the Global Climate System: Will the Warming Feed the Warming?* New York: Oxford University Press, 1995.

Woodwell, George M., Fred T. MacKenzie, R. A. Houghton, Michael J. Apps, Eville Gorham, and Eric A. Davidson. "Will the Warming Speed the Warming?" In *Biotic Feedbacks in the Global Climate System: Will the Warming Feed the Warming?* ed. George M. Woodwell and Fred T. MacKenzie, 393–411. New York: Oxford University Press, 1995.

Woodwell, George M., Fred T. MacKenzie, R. A. Houghton, M Michael J. Apps, Eville Gorham, and Eric A. Davidson. "Biotic Feedbacks in the Warming of the Earth." *Climatic Change* 40 (1998):495–518.

World Bank. *International Trade and Climate Change: Economic, Legal, and Institutional Perspectives.* Washington, DC: 2007.

"World Breaks Temperature Records." *Guardian Unlimited,* March 16, 2007 (LEXIS).

"World Wind-power Capacity Marks Record Growth for 2002." Japan Economic Newswire, March 3, 2003, n.p.(LEXIS).

Wuethrich, Bernice. "Lack of Icebergs Another Sign of Global Warming?" *Science* 285 (July 2, 1999):37.

Wuethrich, Bernice. "How Climate Change Alters Rhythms of the Wild." *Science* 287 (February 4, 2000):793–795.

"Yale University: Vast Majority of Americans Believe Global Warming Is 'Serious Problem.'" M2 Presswire, May 29, 2004 (LEXIS).

Yang, Jing. "China's 4 Percent Fall in Electricity Output May Portend Worse Economic Slump." *Wall Street Journal,* November 14, 2008, A-11.

Yardley, William. "China Says Rich Countries Should Take Lead on Global Warming." *New York Times,* February 7, 2007a. http://www.nytimes.com/2007/02/07/world/asia/07china.html.

Yardley, William. "Engulfed by Climate Change, Town Seeks Lifeline." *New York Times,* May 27, 2007b. http://www.nytimes.com/2007/05/27/us/27newtok.html.

Yardley, William. "China Releases Climate Change Plan." *New York Times,* June 4, 2007c. http://www.nytimes.com/2007/06/04/world/asia/04cnd-china.html.

Yardley, William. "Mayors, Looking to Cities' Future, Are Told It Must Be Colored Green." *New York Times,* November 3, 2007d. http://www.nytimes.com/2007/11/03/us/03mayors.html.

Yoon, Carol Kaesuk. "Penguins in Trouble Worldwide." *New York Times,* June 26, 2001a, F-1.

Yoon, Carol Kaesuk. "Something Missing in Fragile Cloud Forest: The Clouds." *New York Times,* November 20, 2001b, F-5.

Young, Samantha. "Boxer Pledges Shift on Global Warming." Associated Press in *Daytona Beach News-Journal,* November 10, 2006. http://hosted.ap.org/dynamic/stories/B/BOXER_ENVIRONMENT.

Younge, Gary. "Bush U-turn on Climate Change Wins Few Friends." *London Guardian,* August 27, 2004, 18.

"Your Carbon Ration Card." Editorial, *Wall Street Journal,* July 7, 2008, A-11.

Yozwiak, Steve. "'Island' Sizzle: Valley an Increasingly Hot Spot." *Arizona Republic* (Phoenix), September 25, 1998. http://www.sepp.org/reality/arizrepub.html.

Zachos, James C., Ursula Röhl, Stephen A. Schellenberg, Appy Sluijs, David A. Hodell, Daniel C. Kelly, Ellen Thomas, Micah Nicolo, Isabella Raffi, Lucas J. Lourens, Heather McCarren, and Dick Kroon. "Rapid Acidification of the Ocean during the Paleocene-Eocene Thermal Maximum." *Science* 308 (June 10, 2005):1611–1615.

Zavaleta, Erika S., M. Rebecca Shaw, Nona R. Chiariello, Harold A. Mooney, and Christopher B. Field. "Additive Effects of Simulated Climate Changes, Elevated CO_2, and Nitrogen Deposition on Grassland Diversity." *Proceedings of the National Academy of Sciences* 100, no. 13 (June 24, 2003):7650–7654.

Zeller, Tom, Jr. "America's Breadbasket Moves to Canada." *New York Times*, December 5, 2006. http://thelede.blogs.nytimes.com/2006/12/05/americas-breadbasket-moves-to-canada.

Zeng, Ning. "Drought in the Sahel." *Science* 302 (November 7, 2003):999–1000.

Zezima, Katie. "Heat Blankets U.S., Causing Discomfort." *New York Times*, August 3, 2006. http://www.nytimes.com/2006/08/03/us/03swelter.html.

Zezima, Katie. "In New Hampshire, Towns Put Climate on the Agenda." *New York Times*, March 19, 2007. http://www.nytimes.com/2007/03/19/us/19climate.htm.

Zezima, Kate. "With Free Bikes, Challenging Car Culture on Campus." *New York Times*, October 20, 2008. http://www.nytimes.com/2008/10/20/education/20bikes.html.

Zhang, D. D., P. Brecke, H. F. Lee, Y.-Q. He, and J. Zhang. "Global Climate Change, War, and Population Decline in Recent Human History." *Proceedings of the Academy of the National Academy of Sciences* 104 (2007):19,214–19,219.

Zhang, Keqi, Bruce C. Douglas, and Stephen P. Leatherman. "Global Warming and Coastal Erosion." *Climatic Change* 64, nos. 1 and 2 (May 2004):41–58.

Zhang, Renyi, Xuexi Tie, and Donald W. Bond. "Impacts of Anthropogenic and Natural NOx Sources over the U.S. on Tropospheric Chemistry." *Proceedings of the National Academy of Sciences* 104 (February 18, 2007):1505–1509.

Zimmer, Carl. "A Radical Step to Preserve a Species: Assisted Migration." *New York Times*, January 23, 2007. http://www.nytimes.com/2007/01/23/science/23migrate.html.

Zimmermann, M., and W. Haeberli. "Climatic Change and Debris-flow Activity in High Mountain Areas: A Case Study in the Swiss Alps." In *Greenhouse-Impact on Cold-Climate Ecosystems and Landscapes,* ed. M. Boer and E. Koster, 59–72. *Selected Papers of a European Conference on Landscape Ecological Impact of Climatic Change,* Lunteren, the Netherlands, December 3–7, 1989. CARTENA Supplement 22. Cremlingen, Germany: Catena Verlag, 1992.

Zimov, Sergey A., Edward A. G. Schuur, and F. Stuart Chapin III. "Permafrost and the Global Carbon Budget." *Science* 312 (June 16, 2006):1612–1613.

Zremski, Jeremy. "A Chilling Forecast on Global Warming." *Buffalo News*, August 8, 2002, A-1.

Zwally, H. Jay, Waleed Abdalati, Tom Herring, Kristine Larson, Jack Saba, and Konrad Steffen. "Surface Melt-induced Acceleration of Greenland Ice-Sheet Flow." *Science* 297 (July 12, 2002):218–222.

Zwiers, Francis W., and Andrew J. Weaver. "The Causes of 20th Century Warming." *Science* 290 (December 15, 2000):2081–2083.

Index

About the Author

BRUCE E. JOHANSEN is Frederick W. Kayser Research Professor of Communication and Native American Studies at the University of Nebraska at Omaha. He is the author of twenty-six books, including *The Global Warming Combat Manual: Solutions for a Sustainable World* (Praeger, 2008), *Global Warming 101* (Greenwood, 2008), *Global Warming in the 21st Century,* 3 vols. (Praeger, 2006), *The Dirty Dozen: Toxic Chemicals and the Earth's Future* (Praeger, 2003), *Indigenous Peoples and Environmental Issues* (Greenwood, 2003), and *The Global Warming Desk Reference* (Greenwood, 2001). Johansen regularly contributes articles on environmental issues to such national periodicals as *The Nation,* the *Progressive,* the *Wall Street Journal,* and the *Atlantic Monthly.*